Recent Advances in Chaotic Systems and Synchronization

Recent Advances in Chaotic Systems and Synchronization

From Theory to Real World Applications

Edited by

Olfa Boubaker

Sajad Jafari

Academic Press is an imprint of Elsevier
125 London Wall, London EC2Y 5AS, United Kingdom
525 B Street, Suite 1650, San Diego, CA 92101, United States
50 Hampshire Street, 5th Floor, Cambridge, MA 02139, United States
The Boulevard, Langford Lane, Kidlington, Oxford OX5 1GB, United Kingdom

Library of Congress Cataloging-in-Publication Data

A catalog record for this book is available from the Library of Congress

British Library Cataloguing-in-Publication Data

A catalogue record for this book is available from the British Library

ISBN 978-0-12-815838-8

For information on all Academic Press publications visit our website at https://www.elsevier.com/books-and-journals

Publisher: Mara Conner
Acquisition Editor: Sonnini R. Yura
Editorial Project Manager: Gabriela D. Capille
Production Project Manager: Surya Narayanan Jayachandran
Cover Designer: Victoria Pearson

Typeset by SPi Global, India

Contents

Contributors xiii
Preface xvii

Part I
New Chaotic Systems: Design and Analysis

1. Experimental Observations and Circuit Realization of a Jerk Chaotic System With Piecewise Nonlinear Function 3

Abir Lassoued, Olfa Boubaker, Rachid Dhifaoui and Sajad Jafari

1 Introduction 3
2 Mathematical Model and Basic Properties 4
3 Dynamic Analysis 5
3.1 Equilibrium and Stability 5
3.2 Routes to Chaos 10
3.3 Lyapunov Exponents Analysis 11
3.4 Comparative Analysis 12
4 Circuit Design 13
5 Experimental Results 17
6 Conclusion 18
References 20

2. Analytical, Numerical and Experimental Analysis of an RC Autonomous Circuit With Diodes in Antiparallel 23

Sifeu Takougang Kingni, Bonaventure Nana, Guy R. Kol, Victor Kamdoum Tamba and Paul Woafo

1 Introduction 23
2 Analysis of Autonomous RC Circuit With Diodes in Antiparallel 25
2.1 Symmetry, Invariance, Dissipative Character, and Linear Stability of the Equilibrium Points 26
2.2 Dynamical Behaviors of the Circuit 29
2.3 Observation of Chaotic Attractors From the Oscilloscope 35
3 Conclusion 36
Acknowledgment 37
References 37

3. Chaos in a System With Parabolic Equilibrium 41

Viet-Thanh Pham, Christos K. Volos, Sundarapandian Vaidyanathan, Sajad Jafari, Gokul P.M. and Tomasz Kapitaniak

1 Introduction 41
2 Model of a System With Infinite Equilibria 43
3 Dynamical Properties of the System With Parabolic Equilibrium 44
3.1 Equilibrium Points and Stability 44
3.2 Dynamical Behavior 45
4 Circuitry Implementation of the System With Parabolic Equilibrium 45
5 Adaptive Control of the System With Parabolic Equilibrium 48
6 Adaptive Synchronization of Two Systems With Parabolic Equilibrium 52
7 Discussion 55
8 Conclusion 58
Acknowledgments 58
References 58

4. A New Four-Dimensional Chaotic System With No Equilibrium Point 63

Shirin Panahi, Viet-Thanh Pham, Karthikeyan Rajagopal, Olfa Boubaker and Sajad Jafari

1 Introduction 63
2 New No Equilibrium Chaotic System 64
3 Fractional Order Hyperjerk System (FOHJS) 65
3.1 Bifurcation 68
4 Circuit Design 70
5 Conclusion 72
References 72

5. A New Five Dimensional Multistable Chaotic System With Hidden Attractors 77

Atefeh Ahmadi, Karthikeyan Rajagopal, Viet-Thanh Pham, Olfa Boubaker and Sajad Jafari

1 Introduction 77
2 New System and Its Dynamical Properties 78
3 Digital Implementation for the System Using Field Programmable Gate Arrays 78
4 Analog Circuit Design 82
5 Conclusion 84
Acknowledgment 84
References 84

6. Extreme Multistability in a Hyperjerk Memristive System With Hidden Attractors 89

Dimitrios A. Prousalis, Christos K. Volos, Bocheng Bao, Efthymia Meletlidou, Ioannis N. Stouboulos and Ioannis M. Kyprianidis

1 Introduction 89
2 The Hyperjerk Memristive System 91
3 System's Extreme Multistability 96
4 Conclusion 101
References 101
Further Reading 103

7. Parameter Estimation of Chaotic Systems Using Density Estimation of Strange Attractors in the State Space 105

Yasser Shekofteh, Shirin Panahi, Olfa Boubaker and Sajad Jafari

1 Introduction 105
2 A New Chaotic System and Its Bifurcation Analysis 106
3 Density Estimation of the Attractor in the State Space Using GMM 107
3.1 Initialization Step 109
3.2 Expectation Step 110
3.3 Maximization Step 110
3.4 Likelihood Score Evaluation 110
3.5 Determining the Appropriate Number of Gaussian Components 111
4 Parameter Estimation of the Chaotic System Using GMM-Based Cost Functions 112
4.1 Learning Phase 112
4.2 Evaluation Phase 112
5 Simulation Results 113
5.1 The GMM of the Chaotic System 113
5.2 1D Parameter Estimation of the Chaotic System 115
5.3 The Effect of the Length of the Evaluation Data 116
5.4 2D Parameter Estimation of the Chaotic System 120
6 Conclusion 122
Acknowledgment 123
References 123

Part II
Real World Applications

8. Virtualization of Chua's Circuit State Space 127

Branislav Sobota and Milan Guzan

1 Introduction 127
2 Chua's Circuit 128

3 **Boundary Surface** 129
4 **Virtual Reality and Other Visualization Forms** 130
4.1 3D Virtualization Sequence and Its Implementation 130
4.2 Input Data Preparation and Acquisition 131
4.3 Modeling, Editing and Verification 131
4.4 Visualization and Working With Virtual World 132
4.5 3D Display Systems 133
4.6 3D Printing and Real Objects Creation 134
5 **Visualization and Immersive Environments** 135
5.1 Mixed Reality System Implementation 136
5.2 CAVE System Implementation 137
6 **Visualization of Objects in State Space** 137
6.1 Visualization of a Chaotic Attractor by Software Means 138
6.2 Visualization of BS 145
6.3 Visualization in Virtual Cave Environment 151
7 **Possibilities of Calculation of C-SBS** 151
7.1 Calculation of C-SBS by a PC With Single or Multicore Processor 154
7.2 GRID Technology or High-Performance Computer Clusters 159
7.3 GPGPU Technology Usage 159
8 **Conclusion** 159
References 160

9. Some New Chaotic Maps With Application in Stochastic 165

Ezzedine Mliki, Navid Hasanzadeh, Fahimeh Nazarimehr, Akif Akgul, Olfa Boubaker and Sajad Jafari

1 **Introduction** 165
2 **One-dimensional White Gaussian Noise Generator** 166
3 **Attention Deficit Disorder Model** 175
4 **Discussion and Conclusion** 181
References 182

10. Chaotic Solutions in a Forced Two-Dimensional Hindmarsh-Rose Neuron 187

Zahra Rostami, Mohsen Mousavi, Karthikeyan Rajagopal, Olfa Boubaker and Sajad Jafari

1 **Introduction** 187
2 **Model and Description** 190
3 **Numerical Results of the Bifurcation Analysis** 191
4 **Numerical Results of Wave Propagation in the Designed Network** 194
5 **Conclusion** 206
References 207

11. PD Bifurcation and Chaos Behavior in a Predator-Prey Model With Allee Effect and Seasonal Perturbation 211

Afef Ben Saad, Olfa Boubaker and Zeraoulia Elhadj

1 Introduction 211
2 Mathematical Modeling 212
3 Preliminaries 213
4 Dynamical Analysis of System for Both Types of Allee Effect 213
5 Numerical Analysis 217
5.1 Predator-Prey System With Strong Allee Effect 217
5.2 Predator-Prey System With Weak Allee Effect Case Study 221
5.3 Seasonally Perturbed System 227
6 Conclusion 230
References 231

12. Chaotic Path Planning for a Two-Link Flexible Robot Manipulator Using a Composite Control Technique 233

Kshetrimayum Lochan, Jay Prakash Singh, Binoy Krishna Roy and Bidyadhar Subudhi

1 Introduction 233
2 Modeling of the Two-Link Flexible Manipulator 236
3 Singular Perturbation Modeling of a TLFM 239
3.1 Dynamic Model of the Slow Subsystem 240
3.2 Dynamic Model of the Fast Subsystem 240
4 Design of a Composite Control 241
4.1 Dynamic Surface Control of the Slow Subsystem 241
4.2 Backstepping Control for the Fast Subsystem 242
5 Chaotic Signal as the Desired Trajectory 244
6 Results and Discussion for the Composite Control 244
6.1 Simulation Results With the Nominal Payload (0.145 kg) 245
6.2 Chaotic Trajectory Tracking With a 0.3 kg Payload 248
7 Conclusions 252
References 252

Part III
New Trends in Chaos Synchronization

13. Robust Synchronization of Master Slave Chaotic Systems: A Continuous Sliding-Mode Control Approach With Experimental Study 261

Hafiz Ahmed, Héctor Ríos and Ivan Salgado

1 Introduction 261
2 Preliminaries 263

3 Problem Statement 263
4 Output-Feedback-Based Continuous Singular Terminal Sliding-Mode (CSTSM) Controller Design 264
4.1 Finite-Time Sliding-Mode Observer 266
5 Numerical Simulation Results 267
6 Experimental Study 270
7 Conclusion 272
References 273

14. A Four-Dimensional Chaotic System With One or Without Equilibrium Points: Dynamical Analysis and Its Application to Text Encryption 277

Victor Kamdoum Tamba, Romanic Kengne, Sifeu Takougang Kingni and Hilaire Bertrand Fotsin

1 Introduction 277
2 Model of Proposed Autonomous System With One or Without Equilibrium Points 279
3 Dynamical Analysis of Proposed Autonomous System With One or Without Equilibrium Points 280
3.1 Self-Excited Attractor in Proposed Autonomous System With Only One Equilibrium Point 280
3.2 Hidden Attractor in Proposed Autonomous System Without Equilibrium Point 281
4 Electronic Circuit Implementation of Proposed Autonomous System With One or Without Equilibrium Points 287
5 Adaptive Finite-Time Synchronization of Proposed Autonomous System With Hidden Attractor 291
5.1 Preliminaries 292
5.2 Main Results 293
5.3 Numerical Verifications 295
6 A Text Encryption Application Using Hidden Chaotic Attractor of Proposed Autonomous System With One or Without Equilibrium Points 296
6.1 Proposed Affine Cipher 296
6.2 Key Generation 296
6.3 Numerical Verifications 297
7 Concluding Remarks 298
References 299

15. FPGA Implementation of Chaotic Oscillators, Their Synchronization, and Application to Secure Communications 301

Esteban Tlelo-Cuautle, Omar Guillén-Fernández, Jose de Jesus Rangel-Magdaleno, Ashley Melendez-Cano, Jose Cruz Nuñez-Perez and Luis Gerardo de la Fraga

1 Introduction 301

2 **Simulation of Chaotic Oscillators Based on PWL Functions** 303
2.1 Numerical Methods to Simulate Chaotic Oscillators 305
2.2 Simulation of the PWL-Function-Based Chaotic Oscillators 306
3 **Cosimulation Between Active-HDL and Simulink** 306
4 **Experimental Observation of Chaotic Attractors and Synchronization of Two Chaotic Oscillators** 311
4.1 Synchronization in a Master-Slave Topology 316
4.2 Synchronization in a Master-Slave Topology for the Three Chaotic Oscillators 319
5 **Application to Image Transmission** 323
6 **Conclusion** 326
Acknowledgments 326
References 326

16. **On Nonidentical Discrete-Time Hyperchaotic Systems Synchronization: Towards Secure Medical Image Transmission** 329

Narjes Khalifa and Mohamed Benrejeb

1 **Introduction** 329
2 **Proposed Method for Coupled Nonidentical Chaotic Systems Synchronization Study** 331
2.1 Proposed Synchronization Method: Basic Idea 331
2.2 Case of the Synchronization of Wang System Coupled to Hénon Hitzl Zele System 333
2.3 Case of the Synchronization of Generalized Hénon 3D System Coupled to Stéfanski System 336
3 **Proposed Cryptosystem to Secure Medical Images Based on Coupled Hyperchaotic Hénon 3D and Stefanski Systems Synchronization** 337
3.1 Cryptosystem Design: Problem Statement 337
3.2 Encryption and Decryption Process and Results 339
3.3 Measurement of Encryption and Decryption Quality 341
3.4 Security Analysis 341
4 **Conclusion** 345
Appendix A 345
Appendix B 346
References 346
Further Reading 349

17. **Fractional-Order Hybrid Synchronization for Multiple Hyperchaotic Systems** 351

Abir Lassoued and Olfa Boubaker

1 **Introduction** 351
2 **Preliminaries and Problem Position** 352
2.1 Preliminaries 352
2.2 Problem Position 353
3 **Main Results** 354

4 Hybrid Synchronization Between Identical FO Hyperchaotic Systems 356
4.1 FO Hyperchaotic Systems 356
4.2 Control Design 357
4.3 Simulation Results 358
5 Hybrid Synchronization Between Nonidentical FO Hyperchaotic Systems 360
5.1 Different FO Hyperchaotic Systems 360
5.2 Control Design 361
5.3 Simulation Results 363
6 Conclusion 363
References 365
Further Reading 366

Index 367

Contributors

Numbers in parantheses indicate the pages on which the authors' contributions begin.

Atefeh Ahmadi (77), Biomedical Engineering Department, Amirkabir University of Technology, Tehran, Iran

Hafiz Ahmed (261), School of Mechanical, Aerospace and Automotive Engineering, Coventry University, Coventry, United Kingdom

Akif Akgul (165), Department of Electrical and Electronics Engineering, Sakarya University, Adapazarı, Turkey

Bocheng Bao (89), School of Information Science and Engineering, Changzhou University, Changzhou, China

Mohamed Benrejeb (329), University of Tunis El Manar, National Engineering School of Tunis, Automatic Control Research Laboratory, Tunis, Tunisia

Olfa Boubaker (3,63,77,105,187,165,211,351), National Institute of Applied Sciences and Technology, Tunis, Tunisia

Jose de Jesus Rangel-Magdaleno (301), Department of Electronics, INAOE, Puebla, Mexico

Luis Gerardo de la Fraga (301), Department of Computer Sciences, CINVESTAV, Gustavo A. Madero, Mexico

Rachid Dhifaoui (3), National Institute of Applied Sciences and Technology, Tunis, Tunisia

Zeraoulia Elhadj (211), Department of Mathematics and Computing, University of Tebessa, Tebessa, Algeria

Hilaire Bertrand Fotsin (277), Laboratory of Condensed Matters, Electronics and Signal Processing (LAMACETS), Department of Physics, Faculty of Science, University of Dschang, Dschang, Cameroon

Omar Guillén-Fernández (301), Department of Electronics, INAOE, Puebla, Mexico

Milan Guzan (127), Department of Theoretical and Industrial Electrical Engineering, Faculty of Electrical Engineering and Informatics, Technical University of Košice, Košice, Slovak Republic

Navid Hasanzadeh (165), Biomedical Engineering Department, Amirkabir University of Technology, Tehran, Iran

Sajad Jafari (3,41,63,77,105,165,187), Biomedical Engineering Department, Amirkabir University of Technology, Tehran, Iran

Tomasz Kapitaniak (41), Division of Dynamics, Lodz University of Technology, Lodz, Poland

Romanic Kengne (277), Laboratory of Condensed Matters, Electronics and Signal Processing (LAMACETS), Department of Physics, Faculty of Science, University of Dschang, Dschang, Cameroon

Narjes Khalifa (329), University of Tunis El Manar, National Engineering School of Tunis, Automatic Control Research Laboratory, Tunis, Tunisia

Sifeu Takougang Kingni (23,277), Department of Mechanical and Electrical Engineering, Faculty of Mines and Petroleum Industries, University of Maroua, Maroua, Cameroon

Guy R. Kol (23), Department of Mechanical, Petroleum and Gas Engineering, Faculty of Mines and Petroleum Industries, University of Maroua, Maroua; School of Geology and Mining Engineering of University of Ngaoundere, Ngaoundere, Cameroon

Ioannis M. Kyprianidis (89), Laboratory of Nonlinear Systems, Circuits and Complexity (LaNSCom), Department of Physics, Aristotle University of Thessaloniki, Thessaloniki, Greece

Abir Lassoued (3,351), National Institute of Applied Sciences and Technology, Tunis, Tunisia

Kshetrimayum Lochan (233), Department of Electrical Engineering, National Institute of Technology Silchar, Silchar; Department of Mechatronics Engineering, Manipal Institute of Technology, Manipal Academy of Higher Education, Manipal, India

Ashley Melendez-Cano (301), Department of Telecommunications, CITEDI-IPN, Tijuana, Mexico

Efthymia Meletlidou (89), Department of Physics, Aristotle University of Thessaloniki, Thessaloniki, Greece

Ezzedine Mliki (165), Department of Mathematics, College of Science, Imam Abdulrahman Bin Faisal University, Dammam, Saudi Arabia

Mohsen Mousavi (187), Biomedical Engineering Department, Amirkabir University of Technology, Tehran, Iran

Bonaventure Nana (23), Department of Physics, Higher Teacher Training College, University of Bamenda, Bamenda, Cameroon

Fahimeh Nazarimehr (165), Biomedical Engineering Department, Amirkabir University of Technology, Tehran, Iran

Jose Cruz Nuñez-Perez (301), Department of Telecommunications, CITEDI-IPN, Tijuana, Mexico

Gokul P.M. (41), Division of Dynamics, Lodz University of Technology, Lodz, Poland

Shirin Panahi (63,105), Biomedical Engineering Department, Amirkabir University of Technology, Tehran, Iran

Viet-Thanh Pham (41,63,77), School of Electronics and Telecommunications, Hanoi University of Science and Technology, Hanoi, Vietnam; Division of Dynamics, Lodz University of Technology, Lodz, Poland

Dimitrios A. Prousalis (89), Laboratory of Nonlinear Systems, Circuits and Complexity (LaNSCom), Department of Physics, Aristotle University of Thessaloniki, Thessaloniki, Greece

Karthikeyan Rajagopal (63,77,187), Center for Nonlinear Dynamics, Defence University, Bishoftu, Ethiopia

Héctor Ríos (261), División de Estudios de Posgrado e Investigación, CONACYT—Tecnológico Nacional de México/I.T. La Laguna, Torreón, Coahuila, México

Zahra Rostami (187), Biomedical Engineering Department, Amirkabir University of Technology, Tehran, Iran

Binoy Krishna Roy (233), Department of Electrical Engineering, National Institute of Technology Silchar, Silchar, India

Afef Ben Saad (211), National Institute of Applied Sciences and Technology, Tunis, Tunisia

Ivan Salgado (261), Centro de Innovación y Desarrollo Tecnológico en Cómputo, Instituto Politécnico Nacional, Mexico City, Mexico

Yasser Shekofteh (105), Faculty of Computer Science and Engineering, Shahid Beheshti University, Tehran, Iran

Jay Prakash Singh (233), Department of Electrical Engineering, National Institute of Technology Silchar, Silchar, India

Branislav Sobota (127), Department of Computers and Informatics, Faculty of Electrical Engineering and Informatics, Technical University of Košice, Košice, Slovak Republic

Ioannis N. Stouboulos (89), Laboratory of Nonlinear Systems, Circuits and Complexity (LaNSCom), Department of Physics, Aristotle University of Thessaloniki, Thessaloniki, Greece

Bidyadhar Subudhi (233), Department of Electrical Engineering, National Institute of Technology Rourkela, Rourkela, India

Victor Kamdoum Tamba (23,277), Department of Telecommunication and Network Engineering, IUT-Fotso Victor of Bandjoun, University of Dschang, Bandjoun; Laboratory of Condensed Matters, Electronics and Signal Processing (LAMACETS), Department of Physics, Faculty of Science, University of Dschang, Dschang, Cameroon

Esteban Tlelo-Cuautle (301), Department of Electronics, INAOE, Puebla, Mexico

Sundarapandian Vaidyanathan (41), Research and Development Center, Vel Tech University, Chennai, India

Christos K. Volos (41,89), Laboratory of Nonlinear Systems, Circuits and Complexity (LaNSCom), Department of Physics, Aristotle University of Thessaloniki, Thessaloniki, Greece

Paul Woafo (23), Laboratory of Modelling and Simulation in Engineering, Biomimetics and Prototypes (LaMSEBP) and TWAS Research Unit, Department of Physics, Faculty of Science, University of Yaoundé I, Yaoundé, Cameroon

Preface

Chaos is a science full of surprises and unpredictability. It teaches us to expect the unexpected. Chaos is also the property of complex systems whose behaviors are so unpredictable as to appear at random, owing to its great sensitivity to small changes in initial conditions. It is believed that chaos exists in realworld systems (all of which are nonlinear) like weather, brain waves, turbulence in fluids, erratic flows of epidemics, arrhythmic heartbeats in the moments before death, and so on.

The main objective of this book is to explore new developments related to chaos theory and its pivotal role in several fields. The selected contributions shed lights on a series of interesting issues related to a range of novel chaotic and hyperchaotic systems, with the aim of demonstrating several novel proprieties as well relevant real-world applications. Experimental results are presented from a series of selected disciplines. A set of advanced chaos synchronization schemes are also included which are supported by relevant applications and experiments.

The book is a timely and comprehensive referencing guide for graduate students, researchers, and practitioners in the area of chaos theory. It presents a clear and concise introduction to the field of chaos theory, suitable not only for researchers in nonlinear systems, control theory, telecommunications, mathematics, and physics, but also in chemistry, medicine, economy, and natural and social sciences. It covers a wide range of topics not usually found in similar books. The motivations of the respective subjects and a clear presentation ease understanding. The book contains worked examples, codes, and videos which make it ideal for an introductory course for students as well as for researchers starting to work in the field. It is also particularly suitable for engineers wishing to enter the field quickly and efficiently.

Written by eminent scientists in the field from 15 countries (Algeria, Cameroon, China, Ethiopia, Greece, India, Iran, México, Poland, Saudi Arabia, the Slovak Republic, Turkey, Tunisia, the United Kingdom, Vietnam), this book offers a concise introduction to the many facets of chaos theory. It covers the latest advances in concise methodologies and key concepts for modeling and analyzing chaos dynamics. A range of new chaotic systems are presented and analyzed over an array of novel proprieties, from circuit design to real-time experiments. New applications highlighting the potential role of chaos in real-world applications are extensively exposed. Last but not least, new trends in chaos synchronization are exposed, where proposed novel synchronization schema are verified for real-world applications, not only via simulation results, but also via some experimental tests.

In total, 17 chapters written by active researchers in the field are compiled in this book to provide an overall picture of the most challenging problems to be solved in chaos theory. The book covers the topic in three parts: Part I, organized in seven chapters, introduces new chaotic systems from design to analysis; Part II, structured in five chapters, describes realworld applications of chaos theory; and finally, Part III, presented in five chapters, introduces new trends in chaos synchronization.

The following researchers are particularly acknowledged for their considerable efforts:

- We are grateful to our collaborators in this artwork, Professor Guanrong Chen from City University of Hong Kong, China, Professor Jun Ma from Lanzhou University of Technology, China, Professor Nikolay Vladimirovich Kuznetsov from Saint Petersburg State University, Russia, and Professor Seyed Mohammad Reza Hashemi Golpayegani from Amirkabir University of Technology, Iran, for their helpful and professional efforts to provide precious comments and reviews.
- We particularity thank Emeritus Professor Mohamed Benrejeb from National Engineering School of Tunis, Tunisia, Professor Sundarapandian Vaidyanathan from Vel Tech University, India, and Professor Tomasz Kapitaniak from Lodz University of Technology, Poland, for the honor they give us by contributing to this artwork.
- Our particular thanks are given to our main collaborators in this work, Professor Viet-Thanh Pham from Hanoi University of Science and Technology, Vietnam, Professor Rachid Dhifaoui from National Institute of Applied Sciences and Technology, Tunisia, and Assistant Professor Christos Volos from Aristotle University of Thessaloniki, Greece.
- The authors of Chapter 1, Abir Lassoued and collaborators, of Chapter 2, Sifeu T. Kingni and collaborators, of Chapter 8, Branislav Sobota and Milan Guzan, and of Chapter 13, Hafiz Ahmed and collaborators, are particularly acknowledged for their efforts to provide experimental data and videos.

Finally, we would like to express our gratefulness to all authors of the book for their valuable contributions and to all reviewers for their helpful and professional efforts to provide precious comments and feedback. In the end, we dedicate this book to the memory of Professor Gennady A. Leonov, who helped and guided us in many stages of the preparation of this book.

Olfa Boubaker
Sajad Jafari

Part I

New Chaotic Systems: Design and Analysis

Chapter 1

Experimental Observations and Circuit Realization of a Jerk Chaotic System With Piecewise Nonlinear Function

Abir Lassoued*, Olfa Boubaker*, Rachid Dhifaoui* and Sajad Jafari†
National Institute of Applied Sciences and Technology, Tunis, Tunisia, †Biomedical Engineering Department, Amirkabir University of Technology, Tehran, Iran

1 INTRODUCTION

Currently, new chaotic systems and circuits have received considerable interest in the research community due to their potential application in nontraditional areas such as secure communication [1, 2], robotics [3], and encryption applications [4]. In order to satisfy the real needs imposed by these technologies, a great number of chaotic circuits have been discovered these last few years [5–7]. In fact, many researchers have attempted to build strange attractors with as simple as possible nonlinear algebraic structures. These systems are strongly recommended for electrical implementations. However, for chaotic circuits, building chaotic attractors is still an open research direction, and further results are expected.

Indeed, one of the most famous chaotic systems with simple algebraic form, in the literature, is described as $\dddot{x} = -a\ddot{x} - \dot{x} + G(x)$, where $G(x)$ is a classical nonlinear function [8]. Remarkably, a variety of conventional nonlinear functions were used to build chaos as well as particularly simple cases. Among these ones, we can cite the sine function [9], the hyperbolic tangent function [10], the polynomials with integer orders [11], and the polynomials with fractional order terms [12]. Usually, the well-known piecewise linear function is still the most used elementary function in jerk systems [13, 14]. In this framework, piecewise nonlinear function could be used as a chaotic generator, and to the best of our knowledge, it has not been harnessed until now. Nevertheless, particular applications (namely, secure communication and encryption) require not only simple

Recent Advances in Chaotic Systems and Synchronization. https://doi.org/10.1016/B978-0-12-815838-8.00001-7

algebraic models but also very complex dynamic behaviors. This compromise between a simple mathematical model and a complex dynamic behavior is very difficult to satisfy and remains infrequently exploited.

From the circuitry point of view, many research works have tried to design electronic circuits generating chaotic behaviors [15, 16]. On the other hand, peculiar nonlinear terms are quite complicated for implementation, such as the Chua circuit, which desires inductance elements [17, 18]. Indeed, inductance components are not recommended for chaos applications because they introduce uncontrollable parameters due to their inherent impedance. Thus, the common analog electronic elements are known in electrical engineering to be solely operational amplifiers, resistors, capacitors, diodes, and transistors. On the other hand, the previous compromise between a simple algebraic structure and complex dynamic behaviors should be satisfied in order to design high performance chaotic circuits.

The main objective of this chapter is to achieve the compromise to build simple jerk model with richer chaotic dynamics than those proposed by related works. Expecting that the piecewise nonlinear function gives us more complex chaotic proprieties than the piecewise linear one, the proposed system is characterized by only one nonlinear term based on the absolute function. The chaotic system presents interesting dynamical behaviors, it can exhibit regular and strange attractors. The corresponding oscillator circuit of the jerk system is designed using MultiSIM software. Experimental investigations also prove the efficiency of the designed circuit.

The remainder of this chapter is arranged as follows. In Section 2, the new jerk system with a piecewise nonlinear function is proposed, and its basic properties are described. In Section 3, the chaotic system is analyzed by focusing on its elementary characteristics such as the Lyapunov exponents, the regular and strange attractor exhibited, and the equilibrium points. In Section 3, the oscillator circuit of the jerk system is designed using simple electronic components. In Section 4, the experimental results of the implemented circuit are presented and compared to simulation results with the MultiSIM software.

2 MATHEMATICAL MODEL AND BASIC PROPERTIES

Let consider the piecewise nonlinear function $G(x) = kx|x|$ which can be also defined by the following expression

$$\begin{cases} kx^2, & \text{if } x \geq 0, \\ -kx^2, & \text{if } x < 0. \end{cases} \tag{1}$$

Thus, the jerk chaotic system with only one nonlinear term is expressed by these three differential equations:

$$\begin{cases} \dot{x} &= y, \\ \dot{y} &= z, \\ \dot{z} &= -az - by - cx + kx|x|, \end{cases} \tag{2}$$

where (x, y, z) are the state variables and (a, b, c, k) are the system parameters.

System (2) can exhibit chaotic behavior only if the general condition of dis-ipativity is satisfied. Hence, let consider V the volume element of the flow of system trajectories given by

$$\nabla V = \frac{\partial \dot{x}}{\partial x} + \frac{\partial \dot{y}}{\partial y} + \frac{\partial \dot{Z}}{\partial z} = -a < 0.$$

As long as $a > 0$ and time goes to infinity, each volume containing the trajectory of system (1) shrinks to zero at an exponential rate. As a result, all system orbits are finally confined to a subset of zero volume, and the asymptotic motion settles onto an attractor in the three-dimensional phase space.

In fact, when the initial conditions are chosen as (1, 1, 1) and the system parameters (a, b, c, k) are equal to (1, 1, – 2.625, – 0.25), system (2) generates a double scrolls chaotic attractor. As shown in Fig. 1, the exhibited attractor is symmetric with respect to the origin point O (0, 0, 0).

The time series of the state variables x, y, and z are described in Fig. 2. These signals represent the chaotification rates of each state variable. More precisely, these curves reflect the variation of the dynamic behaviors of each variable in the course of time.

On the other hand, system (2) is sensitive to initial conditions which is the most visible signature of chaotic behaviors. Indeed, two time series of each state variables of system (2) for neighboring initial conditions are described in Fig. 3. These two curves are started from (1, 1, 1) and (1.001, 1, 1), respectively. It is clear that the two curves are perfectly superimposed in the beginning, but they diverge suddenly after that.

3 DYNAMIC ANALYSIS

In this section, the dynamic analysis of system (2) is focused on the elementary characteristic of chaotic behaviors. All numerical simulations are realized using the two Matlab package MatCont and Matds for the visualization of the attractor forms and the Lyapunov spectrum, respectively.

3.1 Equilibrium and Stability

The equilibrium points of system (2) are obtained by resolving the following equations

$$\dot{x} = 0, \quad \dot{y} = 0, \text{ and } \dot{Z} = 0.$$

When the system parameters (a, b, c, k) are equal to (1, 1, – 2.625, – 0.25), since the parameters c and k are negative constants, then system (2) allows only three equilibrium points: $P_1(0, 0, 0)$, $P_2(\frac{c}{k},0,0)$, and $P_3(\frac{-c}{k},0,0)$. It is clear that the equilibrium points P_2 and P_3 are symmetrical with respect

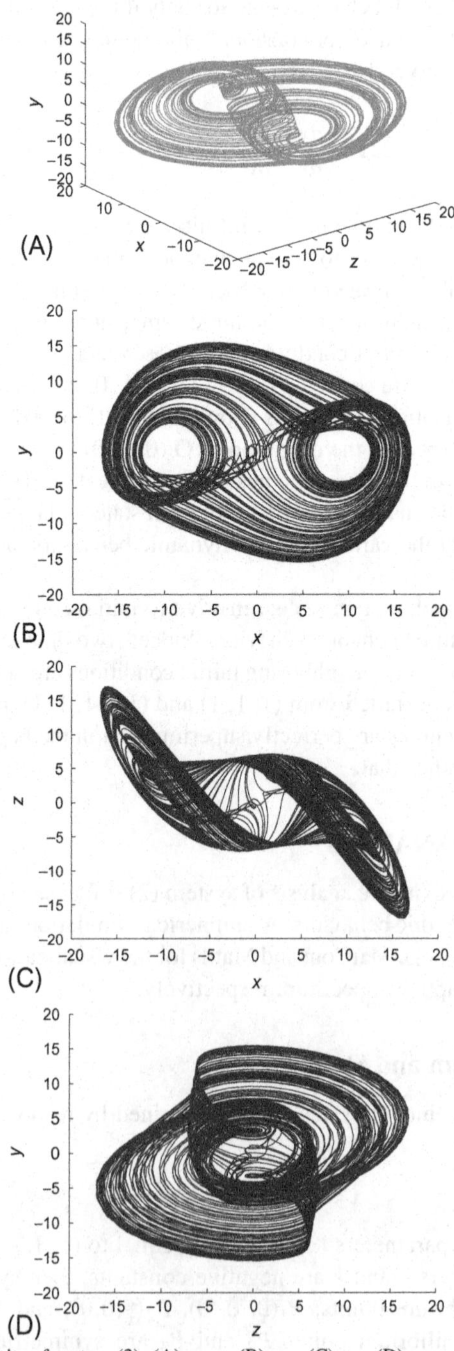

FIG. 1 Phase portraits of system (2): (A) *x*-*y*-*z*; (B) *y*-*x*; (C) *x*-*z*; (D) *y*-*z*.

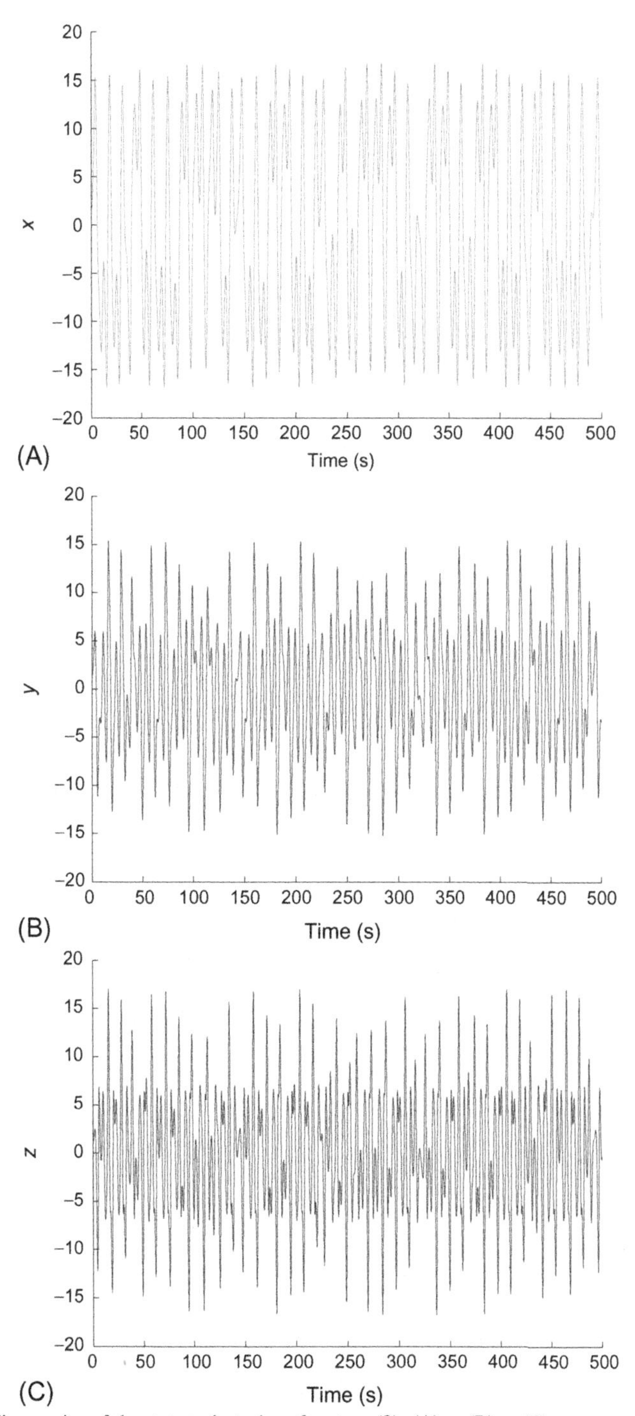

FIG. 2 Time series of the state trajectories of system (2): (A) x; (B) y; (C) z.

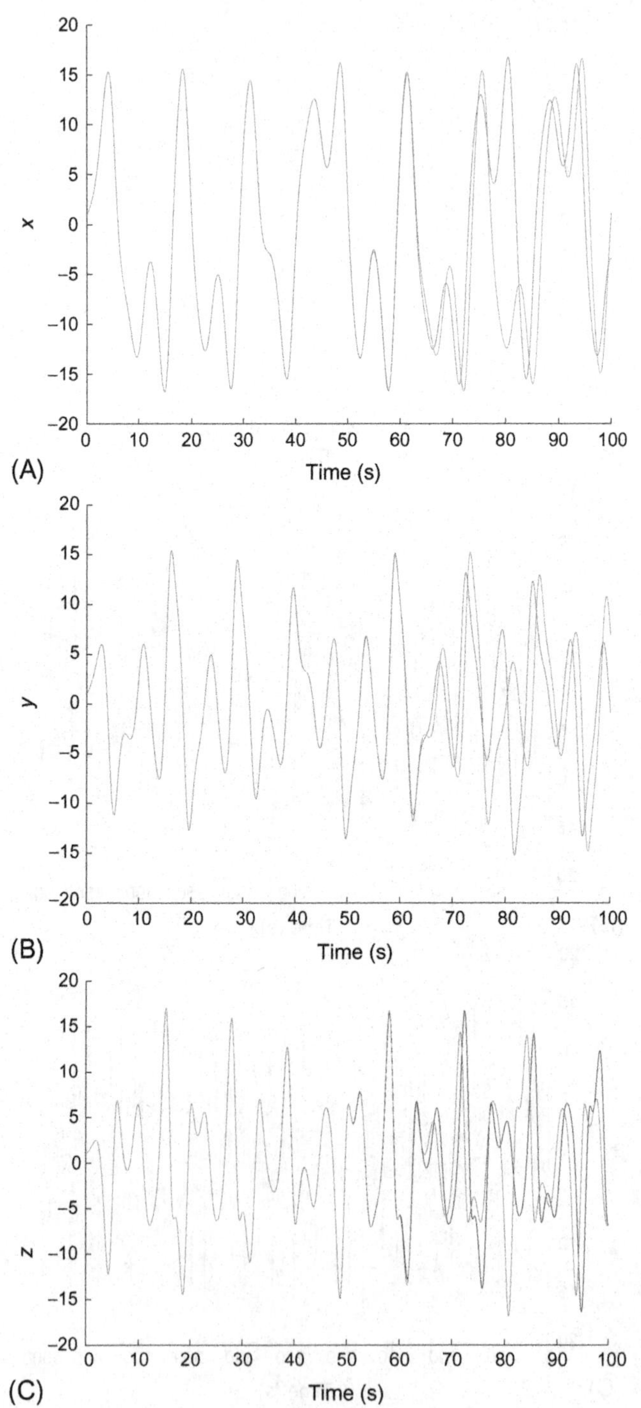

FIG. 3 Sensitive dependence to initial conditions of the state trajectories of system (2): (A) x; (B) y; (C) z.

TABLE 1 Stability Analysis of System (2)

Equilibrium Point	Jacobian Matrix	Corresponding Eigenvalues
P_1	$\begin{pmatrix} 0 & 1 & 0 \\ 0 & 0 & 1 \\ -c & -b & -a \end{pmatrix}$	λ_1=0.934 λ_2=-0.967+1.386i λ_3=-0.967-1.386i
P_2	$\begin{pmatrix} 0 & 1 & 0 \\ 0 & 0 & 1 \\ c & -b & -a \end{pmatrix}$	λ_1=-1.500 λ_2=0.250+1.299i λ_3=0.250-1.299i
P_3	$\begin{pmatrix} 0 & 1 & 0 \\ 0 & 0 & 1 \\ c & -b & -a \end{pmatrix}$	λ_1=-1.500 λ_2=0.250+1.299i λ_3=0.250-1.299i

to the origin point P_1. This reflects the symmetry of the obtained chaotic attractor.

In order to study the stability analysis of system (2), we aim to make explicit in Table 1 the Jacobian matrix and their corresponding eigenvalues for each equilibrium point P_1, P_2, and P_3.

For the equilibrium points, two types of unstable dissipative points (UDS) are defined according to Campos-Cantn et al. [19]. Thus, we introduce below the corresponding definition of each type.

Definition 1 (Campos-Cantn et al. [19]) An equilibrium point, whose eigenvalues are $(\lambda_1, \lambda_2, \lambda_3)$, is said to be an UDS Type I, if the sum of its eigenvalues is negative, λ_1 is a real negative and the other ones are complex conjugates with positive real parts.

Definition 2 (Campos-Cantn et al. [19]) An equilibrium point, whose eigenvalues are $(\lambda_1, \lambda_2, \lambda_3)$, is said to be an UDS Type II, if the sum of its eigenvalues is negative, λ_1 is a real positive and the other ones are complex conjugates with negative real parts.

In strange attractors, only an equilibrium point USD type I can allows the generation of wings. According to Table 1 and Definitions 1 and 2, the stability results of system (2) are defined such as:

- P_1 is an equilibrium USD type II since λ_1 is a positive real and (λ_2, λ_3) are complex conjugates with negative real parts.
- P_2 and P_3 are two symmetrical points. These equilibrium points have the same eigenvalues and are USD type I points. Indeed, the corresponding eigenvalues are $\lambda_1 < 0$ and (λ_2, λ_3) complex conjugates ones with positive real parts.

3.2 Routes to Chaos

When the system parameters are varied, system (2) generates different periodic and chaotic attractors. According to several tests using the MaCont package, we have chosen the parameter c as the only bifurcation parameters in the jerk system. Indeed, as parameter c increases and the parameters (a, b, k) are fixed, system (2) undergoes particular routes as shown in the bifurcation diagram illustrated in Fig. 4.

The obtained behaviors are defined under peculiar conditions as follows:

- If $-1 \leq c \leq -1.1$, then system (2) converges to a fixed point as shown in Fig. 5A.
- If $-1.1 < c < -1.8$, then jerk system exhibits periodic orbit around the equilibrium point P_2. Fig. 5B shows this regular attractor with $c = -1.5$. At $c = -1.59$, a period doubling bifurcation is detected as presented in Fig. 5C.
- If $-1.8 \leq c \leq -1.9$, then another periodic orbit around the two equilibrium points P_2 and P_3 is obtained as described in Fig. 5D with $c = -1.85$.
- If $-1.9 < c \leq -2.7$, system (2) exhibits chaotic attractor. Fig. 5E shows this strange attractor with $c = -2.5$.

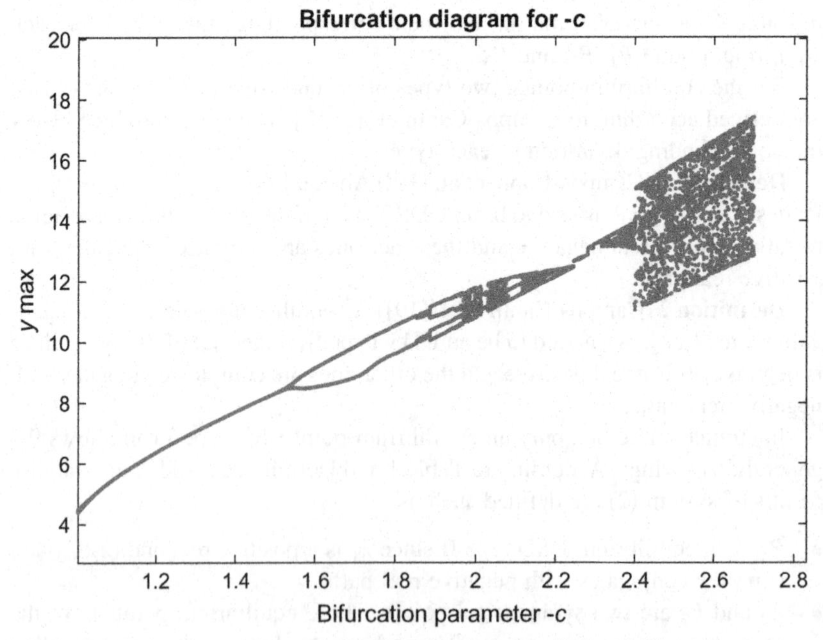

FIG. 4 Bifurcation diagram for $-c \in [1, 2.7]$ while $(a, b, k) = (1, 1, -0.25)$.

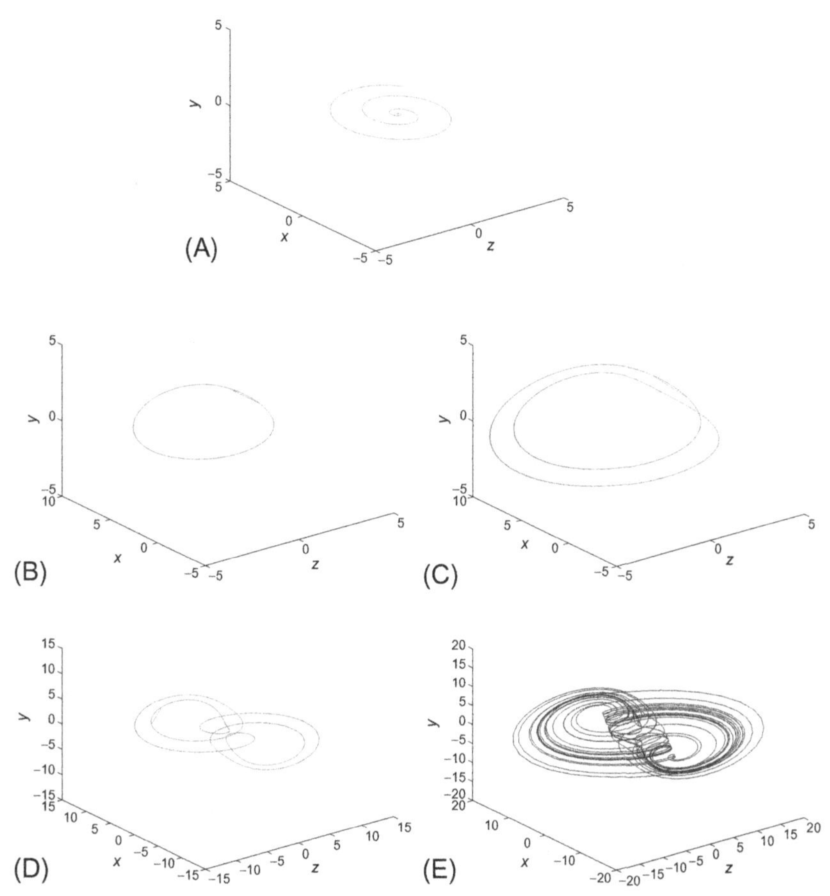

FIG. 5 Different regular and strange attractors exhibited by system (2) when (A) $c = 0.3$; (B) $c = -1.3$; (C) $c = -1.59$; (D) $c = -1.85$; (E) $c = -2.5$.

3.3 Lyapunov Exponents Analysis

The Lyapunov exponent (LE) is the principal criteria of chaos and represents the growth or decline rate of small perturbation along each main axis of the phase space system. For the three-dimensional jerk system (2), three Lyapunov exponents are esteemed using the Wolf algorithm for the system parameters $(a,\ b,\ c,\ k) = (1,\ 1,\ -2.625,\ -0.25)$. The obtained Lyapunov spectrum is described in Fig. 6 by using the Matds software. This Matlab software is based on the ode87. Therefore, the local error normally expected is O (h^9) since the ode87 is an eighth-order accurate integrator.

The largest positive exponent λ_1, which increases the expansion degree of the attractor in the phase space, is equal to 0.130 where $\lambda_3 = -1.136$ increases

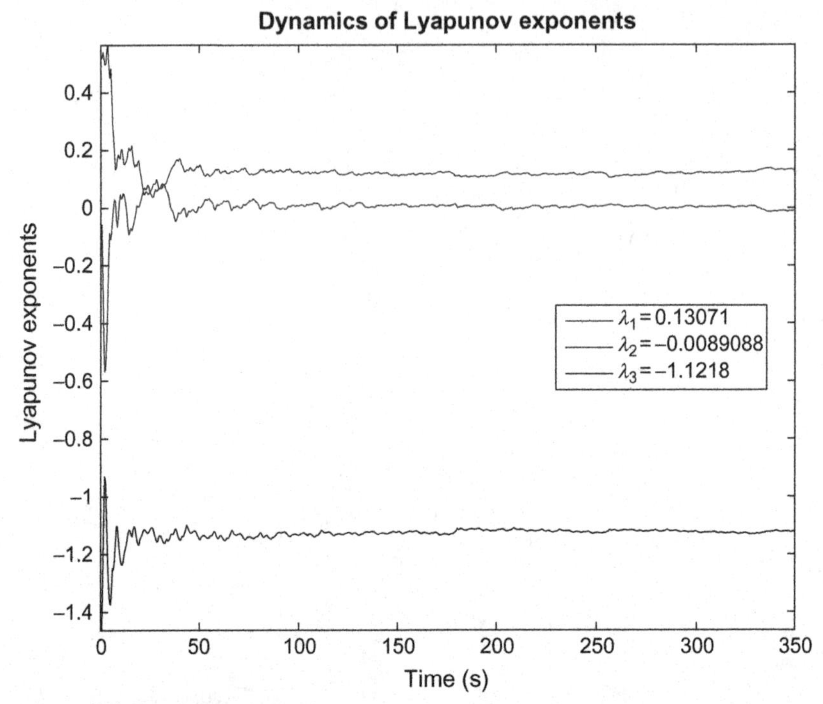

FIG. 6 Lyapunov spectrum of system (2).

the contraction degree of the chaotic attractor. Besides, the λ_2 translates the critical nature of this attractor.

The Kaplan-Yorke dimension of system (2), which presents the complexity of the strange attractor, is given by

$$D_L = j + \frac{1}{|\lambda_{j+1}|}\sum_{i=1}^{j}(\lambda_i) = 2 + \frac{\lambda_1 + \lambda_2}{|\lambda_3|} = 2.124, \tag{3}$$

where j is equal to $n-1$ with n the number of Lyapunov exponents. Thus, system (2) generates chaotic behaviors characterized by fractional-order dimension.

3.4 Comparative Analysis

Based on the review works presented in [20], one of the comparative criteria between chaotic systems is the first Lyapunov exponent. A comparative analysis between system (2) and two related jerk systems using the last criterion is described in Table 2. Identically to the system (2), the selected systems, recently proposed in literature, are based on the jerk equation but they include piecewise linear functions.

TABLE 2 Comparative Analysis With Related Jerk Systems

Jerk System	Lyapunov Exponents
Proposed system	LE_1=0.130
	LE_2=0
	LE_3=-1.136
Campos-Cantn [23]	LE_1=0.121
	LE_2=0
	LE_3<0
Li et al. [7]	LE_1=0.062
	LE_2=0
	LE_3=-1.062

Referring to Table 2, system (2) exhibits more complex dynamics than related ones. Thus, this confirms the highlight potential application of nonlinear piecewise functions with respect to the piecewise linear ones.

For the comparative analysis, we have considered only jerk systems that contain piecewise linear functions. However, there are other jerk systems that contain more than one nonlinear terms and whose Lyapunov exponents are relatively higher. On the other hand, the implementation of these systems is more difficult, such as the jerk system with cross-product terms [21] and a jerk system with tangent function [22].

4 CIRCUIT DESIGN

According to the above dynamic analysis, it is evident that the chaotic system with piecewise nonlinear function generates extremely high behaviors. For this reason, the design and the implementation of its corresponding analog circuit is envisaged. Thus in this section, a chaotic circuit is designed using MultiSIM and numerical simulations are carried out to verify the theoretical analysis previously established.

System (2) can be implement in the form of an electronic circuit following these equations:

$$\begin{cases} \dot{x} = \dfrac{1}{R_1C_1}y, \\ \dot{y} = \dfrac{1}{R_2C_2}z, \\ \dot{z} = -\dfrac{1}{R_3C_3}z - \dfrac{1}{R_4C_3}y + \dfrac{1}{R_5C_3}x - \dfrac{1}{R_6C_3}|x|x, \end{cases} \tag{4}$$

where $R_i(i = 1, ..., 6)$ and $C_j(j = 1, ..., 3)$ are the corresponding resistors and capacitors, respectively. The voltage (x, y, z) could be obtained by applying a simple integration operation from the derivatives variables $(\dot{x}, \dot{y}, \dot{z})$. The phase portraits of system (2) in Fig. 1 show that the maximum value of the state signals is included in the interval $]-15, 15[$. Because ± 15 V could not saturate the classical operational amplifier *LM*741, then we chose this type of amplifier in the designed circuit. The operational amplifier *LM*741 allows to reduce drift due to offset current.

On the one hand, the integrator blocks could be designed by inverting operational amplifier *LM*741, one capacitor 1 nF and one resistance 100 kΩ. Also, the basic of addition and subtraction are implemented using *LM*741. On the other hand, in order to design the absolute value circuit, two diodes, three resistors, and one operational amplifier are recommended.

For future works, a linear transformation for system (2) is essential to reduce the state variables amplitude for microprocessor implementations.

The designed circuit of system (4) obtained with MultiSIM package is illustrated in Fig. 7 where the numerical values of the resistors and capacitors are given by:

$$C_1 = C_2 = C_3 = 1 \text{ nF},$$
$$R_1 = R_2 = R_3 = R_4 = 100 \text{ k}\Omega, R_5 = R_6 = 40 \text{ k}\Omega.$$

The product operation is realized by the *AD*633 multiplier. In addition, all active devices (*LM*741 and *AD*633) are powered by ± 15 V. Some considerations are taken in order to generate the chaotic behaviors. One of them is the adjustment of the components values.

Fig. 8 describes the attractors generated by the proposed circuit using MultiSIM. The outputs of the designed circuit are illustrated on the virtual oscilloscope. The chaotic attractors are symmetric around the origin point and have the same form than the theoretical strange attractors. Therefore, by comparing the chaotic attractors generated with Matlab (Fig. 1) and with MultiSIM (Fig. 8), a good qualitative agreement between the numerical simulations and the electric simulations was observed.

To obtain different regular and strange attractors, the adjustment of the resistance values is recommended. In fact, we found that the chaotic circuit exhibits the following attractors when the resistance values of R_1 and R_2 are fixed and the other ones varied:

- If $R_3 = 100$ kΩ, $R_4 = 20$ kΩ, $R_5 = 40$ kΩ, and $R_6 = 40$ kΩ, then the chaotic circuit converges to a fixed point as shown in Fig. 8A.
- If $R_3 = 100$ kΩ, $R_4 = 85$ kΩ, $R_5 = 55$ kΩ, and $R_6 = 40$ kΩ, then the chaotic exhibits periodic orbit around the equilibrium point P_2. Fig. 8B shows this regular attractor. If $R_5 = 50$ kΩ, a period doubling bifurcation is detected as presented in Fig. 8C.

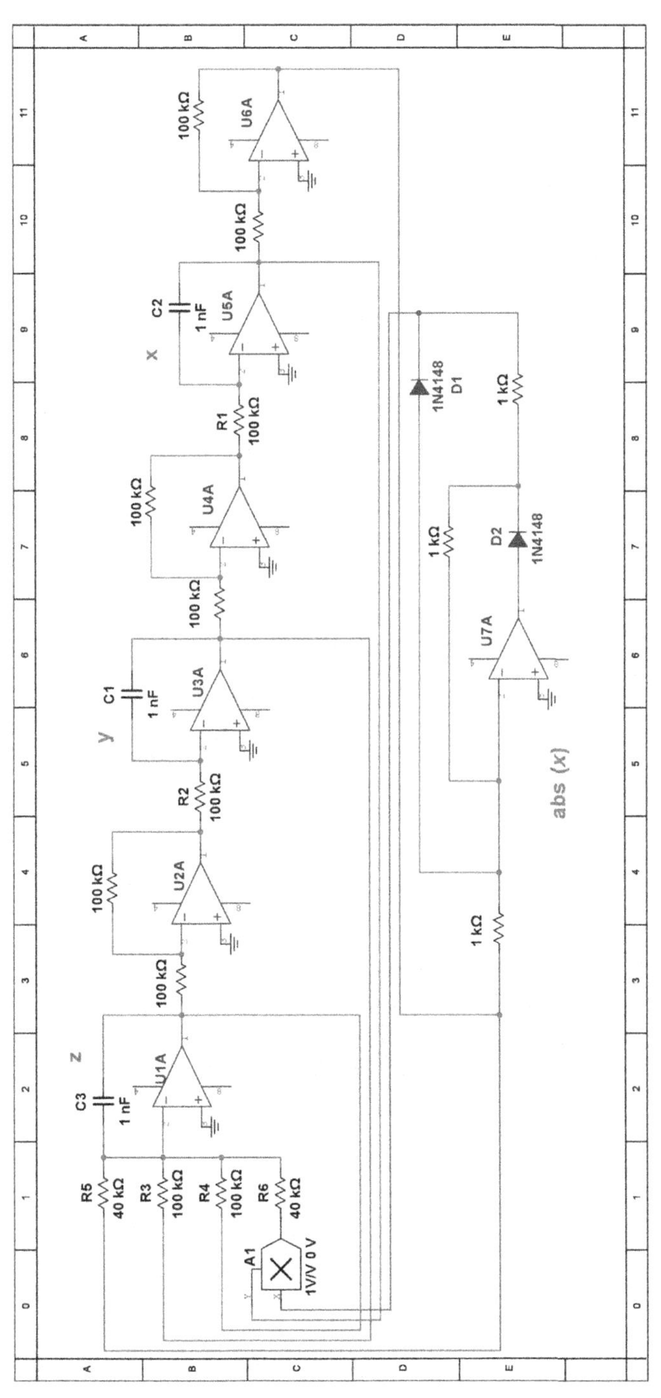

FIG. 7 Designed circuit of system (4) via MultiSIM package.

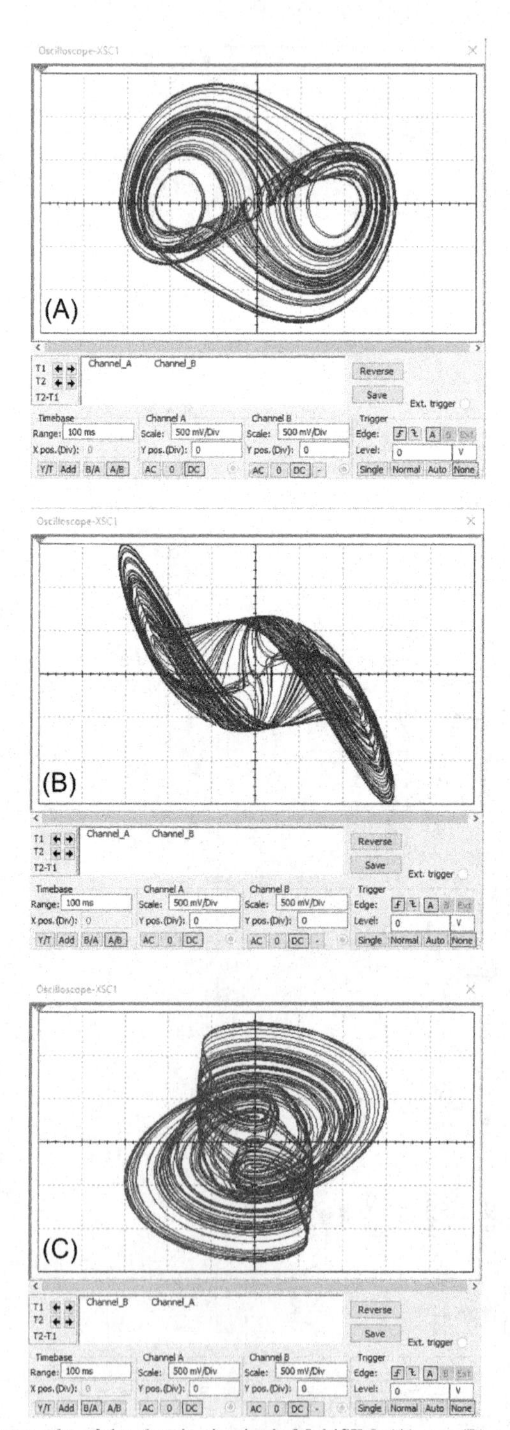

FIG. 8 Simulation results of the chaotic circuit via MultiSIM: (A) x-y; (B) y-z; (C) x-z.

- If $R_3 = 100\ \text{k}\Omega$, $R_4 = 85\ \text{k}\Omega$, $R_5 = 40\ \text{k}\Omega$, and $R_6 = 40\ \text{k}\Omega$, then another periodic orbit around the two equilibrium points P_2 and P_3 is obtained.

According to the previous MultiSIM simulations, it is clear that the resistances (R_3, R_6) are fixed to $(100\ \text{k}\Omega, 40\ \text{k}\Omega)$ for any type of attractor contrary to the resistances R_4 and R_5. For this reason, in order to observe the different dynamic behaviors of the jerk system, the last resistors can be replaced by two potentiometers whose resistances are variables.

In addition, the jerk circuit exhibits chaotic attractor when the value of the resistors R_4 and R_5 are necessary included in the intervals $[110\ \text{k}\Omega, 90\ \text{k}\Omega]$ and $[50\ \text{k}\Omega, 36\ \text{k}\Omega]$, respectively.

5 EXPERIMENTAL RESULTS

In order to verify the theoretical and simulation results, the electronic circuit described previously is implemented as shown in Fig. 9. The same component types and values as those obtained with MultiSIM software are used in the experimental implementation.

The first step was to visualize the time series of the state variables x and y via the oscilloscope as shown in Fig. 10. The obtained results confirm the complex dynamic behavior of system (4). After that the strange attractor is observed as described by Fig. 11. Indeed, the visualized attractor shown in the oscilloscope is symmetrical and characterized by two wings.

To verify the potential application of the proposed system, we changed the value of resistor R_4. Indeed, when $R_4 = 80\ \text{k}\Omega$, regular attractors are visualized as shown in Fig. 12. The obtained periodic orbits are consistent with the MultiSIM simulations as shown in Fig. 13.

FIG. 9 Electrical assembly of the chaotic circuit.

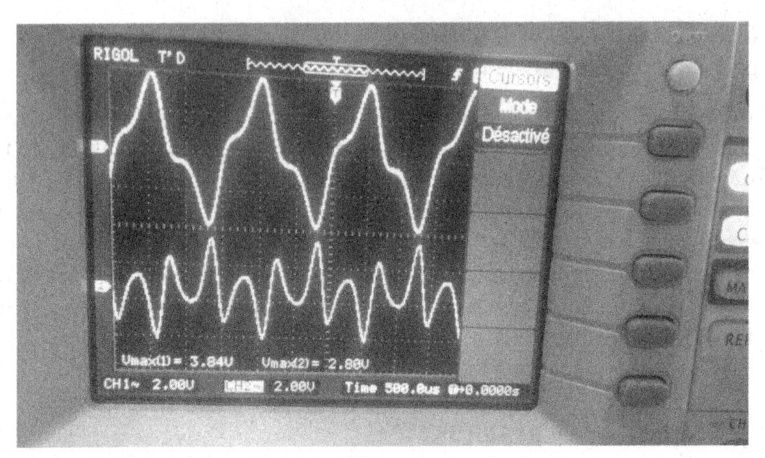

FIG. 10 Experimental time series of the variables x and y.

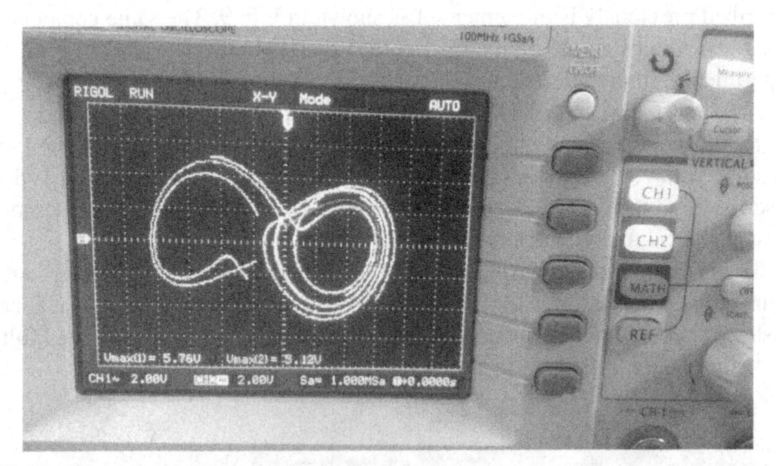

FIG. 11 Experimental chaotic attractor.

By comparing the phase portraits presented in Figs. 8, 11 and also in Figs. 12A and 13D, a good agreement between experimental and numerical simulations was observed. However, slight differences are detected due to the minimal defects in real components which are not always considered in the MultiSIM software.

6 CONCLUSION

The compromise between simplest algebraic structures and complex dynamic behaviors in electronic circuits displaying chaos is recommended for nontraditional applications. In this chapter, we aimed to satisfy such compromise. For that, a new jerk system with only one piecewise nonlinear function is proposed

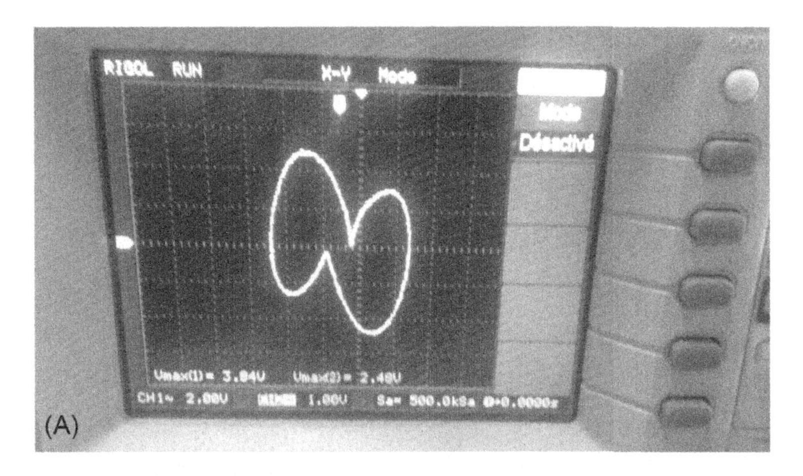

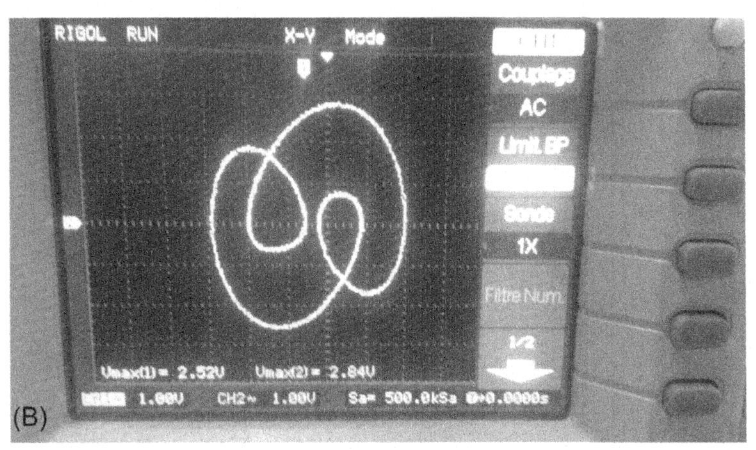

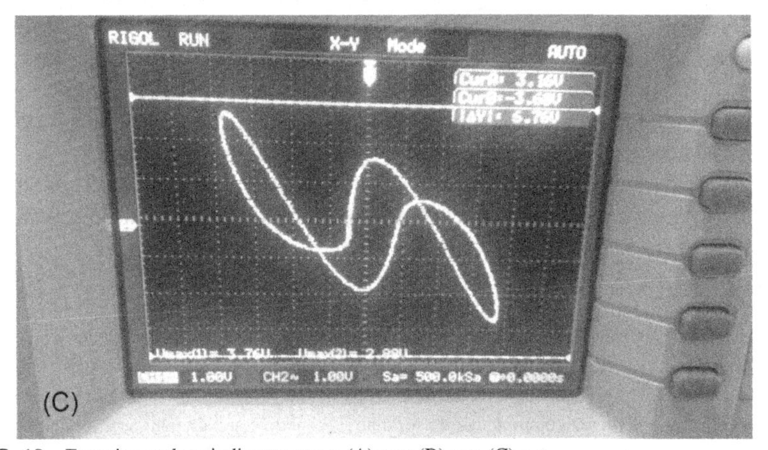

FIG. 12 Experimental periodic attractors: (A) *x*-*y*; (B) *y*-*z*; (C) *x*-*z*.

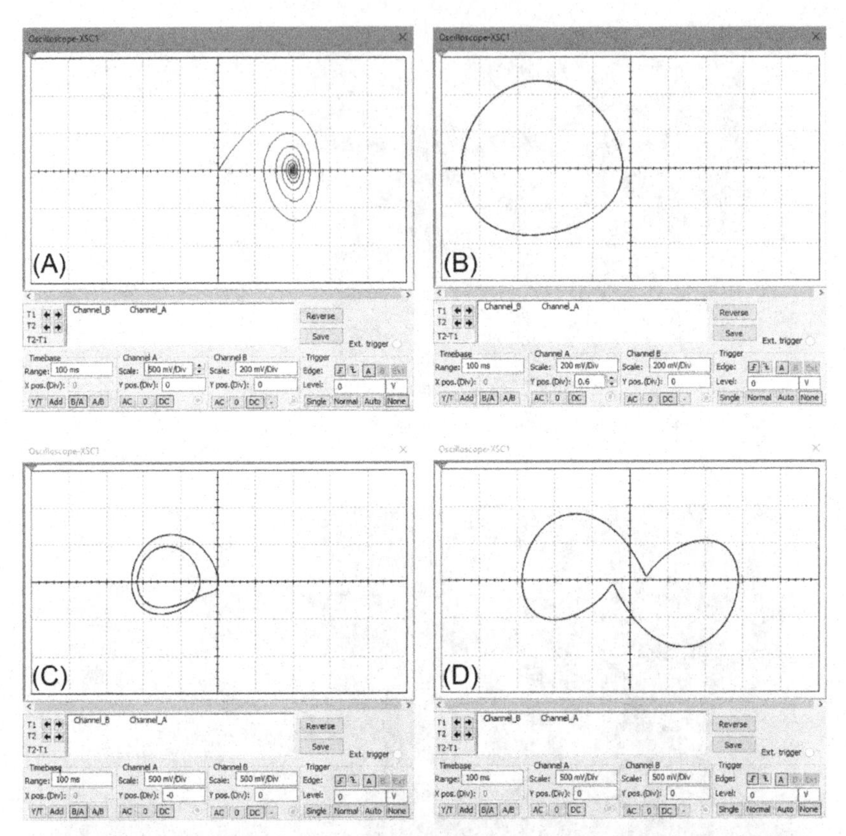

FIG. 13 Different regular and strange attractors exhibited by system (4) when (A) fixed point; (B) period orbit 1; (C) period doubling; (D) period orbit 2.

and analyzed. This system exhibits different regular and strange attractors. The comparative results show that the designed system exhibits more complex behaviors than jerk systems with piecewise linear functions recently proposed in the literature. Furthermore, its analog circuit is modeled and simulated with MultiSIM software. Experimental results confirm the efficiency of the proposed chaotic circuit and their agreement with simulation results. In future works, the proposed circuit will be used for secure communication applications. Thereafter, we aim to realize the electrical implementation of a synchronization schema.

REFERENCES

[1] Kocamaz U, Cicek UE, Uyarolu Y. Secure communication with chaos and electronic circuit design using passivity-based synchronization. J Circuits Syst Comp 2018;27(4):1850057.

[2] Mahmoud EE, Abo-Dahab SM. Dynamical properties and complex anti-synchronization with applications to secure communications for a novel chaotic complex nonlinear model. Chaos Solitons Fractals 2018;106:273–84.

[3] Singh JP, Lochan K, Kuznetsov NV, Roy BK. Coexistence of single- and multi-scroll chaotic orbits in a single-link flexible joint robot manipulator with stable spiral and index-4 spiral repellor types of equilibria. Nonlinear Dyn 2017;90(2):1277–99.
[4] Parvaz R, Zarebnia M. A combination chaotic system and application in color image encryption. Opt Laser Technol 2018;101:30–41.
[5] Lai Q, Nestor T, Nestor J, Zhao XW. Coexisting attractors and circuit implementation of a new 4D chaotic system with two equilibria. Chaos Solitons Fractals 2018;107:92–102.
[6] Wang X, Pham VT, Jafari S, Volos C, Munoz-Pacheco JM, Tlelo-Cuautle E. A new chaotic system with stable equilibrium: from theoretical model to circuit implementation. IEEE Access 2017;5:8851–8.
[7] Li P, Zheng T, Li C, Wang X, Hu W. A unique jerk system with hidden chaotic oscillation. Nonlinear Dyn 2016;86(1):197–203.
[8] Sprott JC. Some simple chaotic flows. Phys Rev E 1994;20(2):R647.
[9] Volos C, Akgul A, Pham VT, Stouboulos I, Kyprianidis I. A simple chaotic circuit with a hyperbolic sine function and its use in a sound encryption scheme. Nonlinear Dyn 2017;89:1047–61.
[10] Kengne J, Njikam SM, Signing VF. A plethora of coexisting strange attractors in a simple jerk system with hyperbolic tangent nonlinearity. Chaos Solitons Fractals 2018;106:201–13.
[11] Elsonbaty AR, El-Sayed AM. Further nonlinear dynamical analysis of simple jerk system with multiple attractors. Nonlinear Dyn 2017;87(2):1169–86.
[12] Lassoued A, Boubaker O. Dynamic analysis and circuit design of a novel hyperchaotic system with fractional-order terms. Complexity 2017;2017:11.
[13] Mendoza J, Araque-Lameda L, Colina-Morles E. Understanding chaos through a jerk circuit. In: Technologies Applied to Electronics Teaching IEEE; 2016. p. 1–5.
[14] Tchitnga R, Nguazon T, Fotso PHL, Gallas JA. Chaos in a single OP-ampbased jerk circuit: experiments and simulations. IEEE Trans Circuits Syst Express Briefs 2016;63(3):239–43.
[15] Akgul A, Li C, Pehlivan I. Amplitude control analysis of a four-wing chaotic attractor, its electronic circuit designs and microcontroller-based random number generator. J Circuits Syst Comp 2017;26(12):1750190.
[16] Lai Q, Nestor T, Kengne J, Zhao XW. Coexisting attractors and circuit implementation of a new 4D chaotic system with two equilibria. Chaos Solitons Fractals 2018;107:92–102.
[17] Stankevich NV, Kuznetsov NV, Leonov GA, Chua LO. Scenario of the birth of hidden attractors in the Chua circuit. Int J Bifurcation Chaos 2017;27(12):1730038.
[18] Mkaouar H, Boubaker O. Chaos synchronization for master slave piecewise linear systems: application to Chuas circuit. Commun Nonlinear Sci Numer Simul 2012;17(3):1292–302.
[19] Campos-Cantn E, Femat R, Chen G. Attractors generated from switching unstable dissipative systems. Chaos 2012;22(3):033121.
[20] Lassoued A, Boubaker O. On new chaotic and hyperchaotic systems: a literature survey. Nonlinear Anal Modell Control 2016;21(6):770–89.
[21] Molaie M, Jafari S, Sprott J, Golpayegani S. Simple chaotic flows with one stable equilibrium. Int J Bifurcation Chaos 2013;23(11):11.
[22] Sprott J. Simple chaotic systems and circuits. Am J Phys 2000;68(8):758–63.
[23] Campos-Cantn E. Chaotic attractors based on unstable dissipative systems via third-order differential equation. Int J Mod Phys C 2016;27(1):11.

Chapter 2

Analytical, Numerical and Experimental Analysis of an RC Autonomous Circuit With Diodes in Antiparallel

Sifeu Takougang Kingni*, Bonaventure Nana†, Guy R. Kol*,¶, Victor Kamdoum Tamba‡ and Paul Woafo§

*Department of Mechanical and Electrical Engineering, Faculty of Mines and Petroleum Industries, University of Maroua, Maroua, Cameroon, †Department of Physics, Higher Teacher Training College, University of Bamenda, Bamenda, Cameroon, ‡Department of Telecommunication and Network Engineering, IUT-Fotso Victor of Bandjoun, University of Dschang, Bandjoun, Cameroon, §Laboratory of Modelling and Simulation in Engineering, Biomimetics and Prototypes (LaMSEBP) and TWAS Research Unit, Department of Physics, Faculty of Science, University of Yaoundé I, Yaoundé, Cameroon, ¶School of Geology and Mining Engineering of University of Ngaoundéré, Ngaoundéré, Cameroon

1 INTRODUCTION

During the last two decades, a large number of electronic circuits and systems exhibiting chaotic behavior have been proposed in the literature [1–13]. The design of chaotic electronic circuits and systems attracted the attention of researchers for various reasons. First of all, one can observe chaos in the laboratory by simply using off-the-shelf components (e.g., op-amps, passive components, diodes, etc.). Second, a chaotic electronic circuit can be controlled by merely varying physically accessible parameters, like resistors, voltage, etc. The third reason lies in its variety of applications in, for example, spacecraft trajectory control, stabilization of the intensity of a laser beam, radar and sonar [14], synchronization [15] and secure communications [14,16]. The fourth reason that motivates researchers to design a chaotic electronic circuit is that we still do not know the sufficient condition(s) for designing a chaotic oscillator [17]. Only necessary conditions for existence of chaotic behavior in an oscillator are known.

One of the best known chaotic circuits is Chua's circuit [18] as well as its variants [19,20] using a Chua's diode. However, an active nonlinear resistor

Recent Advances in Chaotic Systems and Synchronization. https://doi.org/10.1016/B978-0-12-815838-8.00002-9

such as the Chua's diode is not recommended by Ref. [21] because it does not follow the design rules of Ref. [21]. Elwakil and Kennedy [22] proposed a semi-systematic methodology of designing a chaotic oscillator from a sinusoidal oscillator by introducing a suitable passive nonlinearity and a storage element. The authors of Ref. [21] recommended using either a junction field effect transistor or diode as the passive nonlinear component. Using this methodology, many chaotic oscillators have been reported, for example, chaotic Wien-bridge oscillator, Ref. [21] single amplifier bi-quad-based inductor-free Chua's circuit [22], current feedback op amp-based chaotic oscillators [23], three–dimensional chaotic autonomous system with a circular equilibrium [24], chaotic Lu system [25], chaotic dsss-cdma [26], bi-quad-based chaotic oscillator [27], chaotic electronic oscillator from single amplifier bi-quad [17], chaotic RC (resistance and capacitance) with diode [28], a simple chaotic oscillator with a diode [29], a simple chaotic circuit with a light-emitting diode [30], a chaotic jerk oscillator with an exponential nonlinearity diode [31], four simple current tunable chaotic oscillators with two diode-reversible pairs [32], a simple diode-based chaotic oscillator [33], etc. Chaotic RC oscillators such as the Wien bridge chaotic oscillator, for example, have received a great deal of attention mainly because they are more convenient than inductance- and capacitance-based ones, especially at low-to-moderate frequencies. Moreover, capacitors, resistors, and nonlinear devices with sigmoid function can be relatively easily realized, whereas, inductance and nonlinear elements with both positive and negative slope of the sections of nonlinear function cannot be implemented in a monolithic integrated circuit. So, it is important to design chaotic oscillators from elements as capacitors, resistors, and nonlinear devices with sigmoid function, such as hyperbolic sine nonlinearity. A hyperbolic sine nonlinearity could be simply implemented by semiconductor diodes in antiparallel. To the best of our knowledge, studies of chaotic circuit with hyperbolic sine nonlinearity remain scarce [34–36]. In Ref. [34], Kengne et al. proposed a chaotic RC Jerk circuit with diodes in antiparallel described by a hyperbolic sine nonlinearity and demonstrated that the circuit exhibits the coexistence of multiple attractors and a crisis route to chaos. In Ref. [35], Nana and Woafo proposed a chaotic RC circuit with diodes in antiparallel and reported on complete synchronization between two, three, and four almost identical circuits. They investigated the complete synchronization between coupled oscillators both theoretically and experimentally in order to find the synchronization threshold. In Ref. [35], Nana and Woafo theoretically and experimentally demonstrated the transmission of information using synchronization of three chaotic RC circuit with diodes in antiparallel coupled as emitter-relay-receiver system. However, these results (though very interesting) are restricted on the synchronization and chaos-based communications using chaotic oscillators and make no mention of the chaotic mechanism. So, in order to shed more light on the dynamics of this RC circuit with diodes in antiparallel, our objective in this work is threefold: (i) to find the equilibrium points of the system describing the RC circuit with diodes in antiparallel of

Nana and Woafo [35,36] and analyze their stability in order to explain the chaotic mechanism; (ii) to define the region in the parameter space in which the RC circuit with diodes in antiparallel exhibits Hopf bifurcation, antimonotonicity, bistable periodic and chaotic attractors, and periodic and chaotic bubble attractors; and (iii) to carry out an experimental study of the RC circuit with diodes in antiparallel to validate the numerical analysis.

The chapter is organized as follows. Section 2 is devoted to the analysis of stability of the equilibrium points and a systematic numerical and experimental analysis of RC circuit with diodes in antiparallel. The conclusion is given in Sect. 3.

2 ANALYSIS OF AUTONOMOUS RC CIRCUIT WITH DIODES IN ANTIPARALLEL

The autonomous RC circuit with diodes in antiparallel introduced by Nana and Woafo [35] is shown in Fig. 1.

The circuit of Fig. 1 consists of resistors, capacitors, and linear operational amplifiers as gain elements and diodes. The diodes act like a nonlinear component, and its voltage-current characteristic can be modeled with an exponential function, namely: $i = I_0[\exp(v/v_0) - 1]$ where i is the current through the diode, v is the voltage across the diode, $I_0 = 2.682\,\text{nA}$ is the inverse saturation current, and $v_0 \approx 26\,\text{mV}$ at the room temperature. The voltage-current characteristic in diodes in antiparallel is $i_1 = I_0[\exp(v/v_0) - 1] - I_0[\exp(-v/v_0) - 1] = 2I_0 \sinh(v_1/5v_0)$. The dynamics of the circuit is given by the following set of differential equations:

$$\frac{dv_1}{dt'} = \frac{1}{R_6C_1}v_2 - \frac{1}{R_6C_1}v_3 \tag{1a}$$

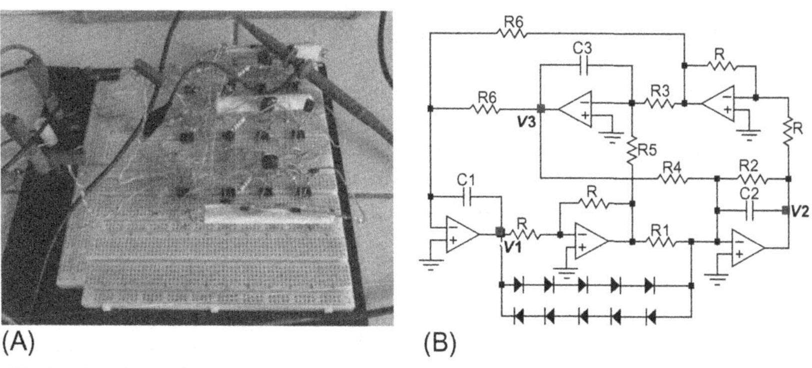

FIG. 1 Experimental set-up (A) and circuit diagram (B) of sinusoidal oscillator with diodes in antiparallel.

$$\frac{dv_2}{dt'} = \frac{1}{R_1C_2}v_1 - \frac{1}{R_2C_2}v_2 - \frac{1}{R_4C_2}v_3 - \frac{2I_0}{C_2}\sinh\left(\frac{v_1}{5v_0}\right) \tag{1b}$$

$$\frac{dv_3}{dt'} = \frac{1}{R_5C_3}v_1 + \frac{1}{R_3C_3}v_2 \tag{1c}$$

The set of Eqs. (1a)–(1c) can be put in the dimensionless form using the dimensionless variables:

$$x = \frac{v_1}{5v_0},\ y = \frac{v_2}{5v_0},\ z = \frac{v_3}{5v_0},\ t = \frac{t'}{\tau},\ a = \frac{\tau}{R_1C_2},\ b = \frac{\tau}{R_2C_2},\ c = \frac{\tau}{R_4C_2},\ d = \frac{\tau}{R_3C_3} \tag{2}$$

where $\tau = R_6C_1, R_6C_1 = R_5C_3$ and $5v_0C_2 = 2I_0R_6C_1$. Introducing Eq. (2) in the set of Eqs. (1a)–(1c), we obtain the following dimensionless equations:

$$\frac{dx}{dt} = y - z \tag{3a}$$

$$\frac{dy}{dt} = ax - by - cz - \sinh(x) \tag{3b}$$

$$\frac{dz}{dt} = x + dy \tag{3c}$$

System (3) is a three-dimensional autonomous system with a nonlinear term in the form of a hyperbolic sine function. Recently, Nana and Woafo reported on the synchronization and chaos-based communications using chaotic oscillators presented in Fig. 1 [35,36]. In this section, we will show that system (3) can exhibit Hopf bifurcation, antimonotonicity, bistability, bubbles, and one-scroll and double-scroll chaotic attractors.

2.1 Symmetry, Invariance, Dissipative Character, and Linear Stability of the Equilibrium Points

System (3) is invariant under the transformation:$(x,y,z) \Leftrightarrow (-x,-y,-z)$. Therefore, if (x,y,z) is a solution of system (3) for a specific set of parameters, then $(-x,-y,-z)$ is also a solution for the same parameters set.

It is noted that if $b > 0$, system (3) is dissipative with an exponential contraction rate $\frac{dV}{dt} = e^{-bt}$ because $\nabla V = \frac{\partial(dx/dt)}{\partial x} + \frac{\partial(dy/dt)}{\partial y} + \frac{\partial(dz/dt)}{\partial z} = -b$.

The equilibrium points of system (3) are obtained by solving $\dot{x} = 0, \dot{y} = 0, \dot{z} = 0$, which gives:

$$z^* = y^* = -x^*/d \tag{4a}$$

$$(a + b/d + c/d)x^* - \sinh(x^*) = 0 \tag{4b}$$

Eq. (4b) cannot be solved analytically. We thus use the Newton-Raphson method to find the value of x^*. The values of y^* *and* z^* are obtained by replacing x^* by its value in Eq. (4a). Depending on the values of parameters (a, b, c, d), Eq. (4b) presents three roots (0 and $\pm x^*$) which correspond to the three equilibrium points of system (3) (see Fig. 2). These three equilibrium points are: $E_0 = (0,0,0)$ and $E_{1,\ 2} = (\pm x^*, \mp x^*/d, \mp x^*/d)$. The characteristic equation evaluated at the equilibrium point $E(x^*, y^*, z^*)$ is:

$$\lambda^3 + b\lambda^2 + [1 - a + dc + \cosh(x^*)]\lambda + c + b + d[a - \cosh(x^*)] = 0 \quad (5)$$

According to the Routh-Hurwitz criterion, the real parts of all the roots λ of Eq. (5) are negative if and only if

$$b > 0 \quad (6a)$$

$$(b + d)[\cosh(x^*) - a] + c(bd - 1) > 0 \quad (6b)$$

$$c + b + d[a - \cosh(x^*)] > 0 \quad (6c)$$

The stability analysis of E_0 and $E_{1,2}$ as function of the parameters of system (3) is depicted in Fig. 2.

From Fig. 2B–D, one can notice that the equilibrium points E_0 and $E_{1,2}$ are always unstable. In Fig. 2A, the equilibrium point E_0 is stable for $a < 0.41$ and unstable for $a > 0.41$ whereas the equilibrium points $E_{1,2}$ are unstable for $0 \leq a \leq 5$ because the equilibrium points $E_{1,2}$ are unstable for $0 \leq a \leq 5$, but

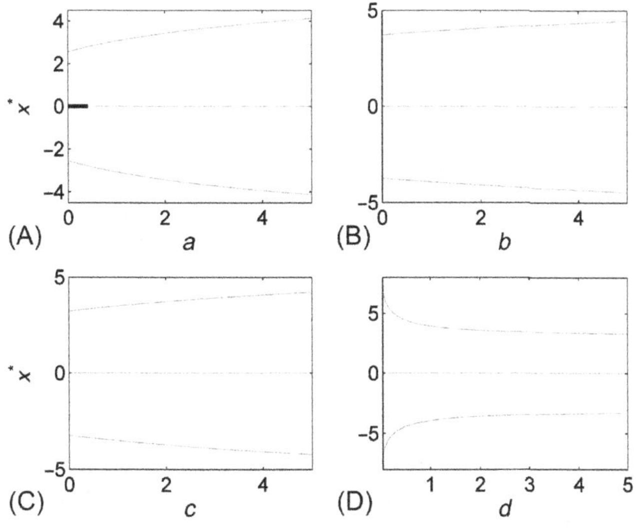

FIG. 2 Stability diagram of the equilibrium points E_0 and $E_{1,\ 2}$ with respect to the parameters of system (3): (A) $b = 0.5$, $c = 2.5$, $d = 1.2$, (B) $a = 3.5$, $c = 2.5$, $d = 1.2$, (C) $a = 3.5$, $b = 0.5$, $d = 1.2$ and (D) $a = 3.5$, $b = 0.5$, $c = 2.5$.

E_0 changes stability properties at $a \approx 0.41$. In the following paragraph, we study the Hopf bifurcation from the equilibrium point E_0 regarding a as the bifurcation parameter.

Theorem. If $1+\frac{c}{b}+\frac{d}{b}(a-1)>0$, *then system (3) undergoes a Hopf bifurcation at the equilibrium point* E_0 *when* a *passes through the critical value* $a_H = 1 + c\frac{db-1}{b+d}$.

Proof. Let $\lambda = i\omega(\omega > 0)$ be a root of Eq. (5) with $x^* = 0$. Substituting $\lambda = i\omega$ into Eq. (5) (for $x^* = 0$) and separating real and imaginary parts, we obtain

$$\omega = \omega_0 = \sqrt{1 + \frac{c}{b} + \frac{d}{b}(a-1)} \tag{7a}$$

$$a = a_H = 1 + c\frac{db-1}{b+d} \tag{7b}$$

Differentiating both sides of Eq. (5) (for $x^* = 0$) with respect to a, we can obtain

$$3\lambda^2\frac{d\lambda}{da} + 2b\lambda\frac{d\lambda}{da} + (2-a+cd)\frac{d\lambda}{da} - \lambda + d = 0 \tag{8}$$

and

$$\frac{d\lambda}{da} = \frac{\lambda - d}{3\lambda^2 + 2b\lambda + 2 - a + cd} \tag{9}$$

then

$$\mathrm{Re}\left(\left.\frac{d\lambda}{da}\right|_{a=a_H,\lambda=i\omega_0}\right) = \frac{2\left(cd^2+b+c+d\right)}{\left(2-a_H+cd-3\omega_0^2\right)^2 + 4b^2\omega_0^2} \neq 0 \tag{10}$$

Because the Jacobian matrix of system (3) at the equilibrium point E_0 have two purely imaginary eigenvalues, and the real parts of eigenvalues satisfy $\mathrm{Re}\left(\left.\frac{d\lambda}{da}\right|_{a=a_H,\lambda=i\omega_0}\right) \neq 0$, all the conditions for Hopf bifurcation to occur are met. Consequently, system (3) undergoes a Hopf bifurcation at E_0 when $a = a_H = 1 + c\frac{db-1}{b+d}$, and periodic solutions will exist in a neighborhood of the point a_H [provided that $1+\frac{c}{b}+\frac{d}{b}(a-1)>0$ holds]. If $b = 0.5$, $c = 2.5$ and $d = 1.2$, then the critical value is $a = a_H \approx 0.41$. In Fig. 3, we plot the time series of the output $x(t)$ and the phase portraits in the plane (x, y) for two specific values of parameter a.

For $a = 0.2 < a_H$, Fig. 3A reveals that the trajectories of system (3) converge to the equilibrium point E_0. The system (3) exhibits a limit cycle for $a = 1 > a_H$ as shown in Fig. 3B.

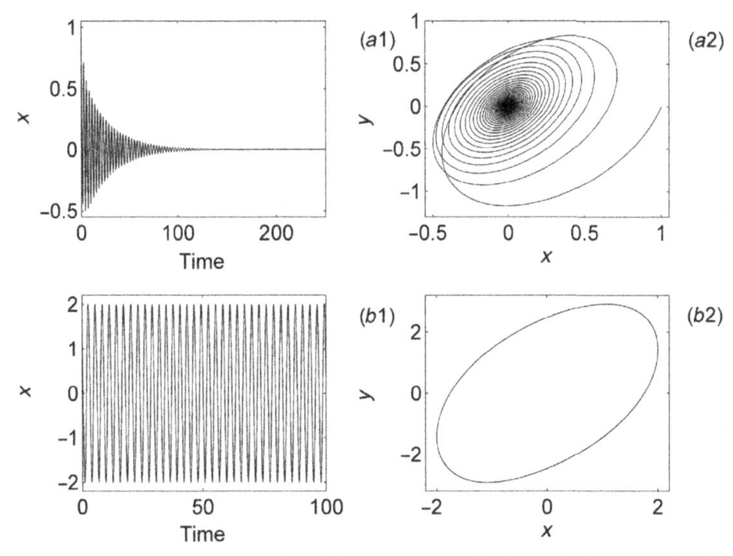

FIG. 3 In panel (1), we plot time series of the output $x(t)$ while in panel (2), we depict the phase portraits in the plane (x, y) for specific values of parameter a: (A) $a = 0.2$ and (B) $a = 1$. The initial conditions are $(x(0), y(0), z(0)) = (1, 0, 0.5)$. The other parameters are $b = 0.5$, $c = 2.5$, and $d = 1.2$.

2.2 Dynamical Behaviors of the Circuit

In Ref. [35], the authors fixed the parameters a, b, d and varied the parameter c in order to find the different dynamical behaviors of system (3). Since the system (3) has four parameters a, b, c, and d. In this section, we construct the two parameters (a, b) and (c, d) bifurcation diagrams by examining the Lyapunov exponents and time series for each cell as shown in Fig. 4. The largest Lyapunov exponent is computed numerically using the algorithm of Wolf et al. described

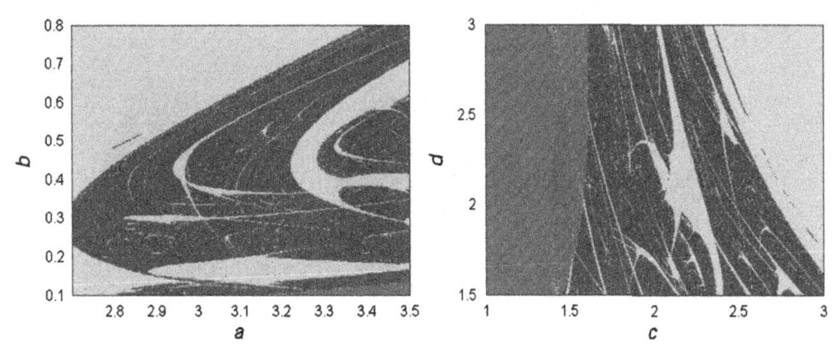

FIG. 4 Regions of dynamical behaviors in the parameters space: (a, b) and (c, d) for the electronic version of the book. Periodic oscillations are in light blue regions, and chaotic behaviors are in red regions. The green regions lead to unbounded orbits.

in Ref. [37]. The dynamics of the circuit is determined by the sign of the largest Lyapunov exponent. The circuit exhibits a fixed point in the state space when the largest Lyapunov exponent is negative. For the zero and positive values of the largest Lyapunov exponent, the circuit experiences periodic and chaotic motion, respectively.

From Fig. 4, we can see that system (3) can display not only periodic and chaotic behaviors but it can also lead to unbounded orbits. For a suitable choice of the parameters, the system (3) can display one- or double-scroll chaotic attractors as shown in Fig. 5.

In Fig. 5A, the trajectories of the chaotic attractor of system (3) are swirling around the equilibrium point E_0. Therefore, the chaotic attractor has the signature of one-scroll chaotic attractor. While in Fig. 5B, the trajectories of the chaotic attractor of system (3) are swirling between the two equilibrium points E_1 and E_2. Therefore, the chaotic attractor is a double-scroll chaotic attractor. The frequency spectra of the outputs $x(t)$ and $y(t)$ of double-scroll chaotic attractor presented in Fig. 5B (not shown) exhibit broad-band power spectrum and lower dimension of strong chaos. The power spectrums show randomly distributed harmonics which confirm the lower dimensions of the strong chaotic nature of the oscillations. The broad frequency spectral bandwidth of RC circuit with diodes in antiparallel indicates its great potential to be used for some relevant engineering applications such as secure communication [36].

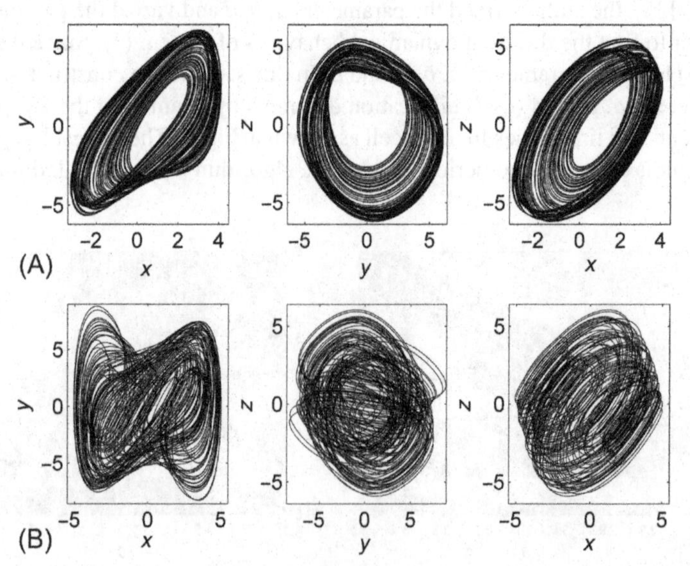

FIG. 5 The phase portraits in plane(x, y), (y, z) and (x, z) for: (A) $b = 0.5, d = 2.3$and (B) $b = 0.15$, $d = 1.2$. The remaining parameters are $a = 3.5$ and $c = 2.5$. The initial conditions are$(x(0), y(0), z(0)) = (1.0, 0, 0.5)$.

In order to know the route to chaotic behavior exhibited by system (3), we plot the bifurcation diagrams depicting the local extrema of $x(t)$ versus one of the parameters a, b and d.

2.2.1 Routes to Chaos With Respect to Parameter a

The bifurcation diagram depicting the local extrema of $x(t)$ and the largest Lyapunov exponent versus the parameter a is plotted in Fig. 6 for $b = 0.5$, $c = 2.5$, and $d = 1.2$.

When the parameter a varies from 0.0 to 4.9, the bifurcation diagram of the output $x(t)$ presented in Fig. 6A displays no oscillation up to $a \approx 0.41$ where a Hopf bifurcation triggers period-1-oscillations. By further increasing the parameter a, the output $x(t)$ reveals a period-doubling route to chaotic oscillations interspersed with periodic windows. The chaotic behavior is confirmed by the largest Lyapunov exponent shown in Fig. 6B.

2.2.2 Routes to Chaos to Parameter b

The bifurcation diagram depicting local maxima of $x(t)$ and the largest Lyapunov versus the parameter b are presented in Fig. 7 for $a = 3.5$, $c = 2.5$, $d = 1.2$.

In Fig. 7A, the system (3) presents a reverse period-doubling to chaos interspersed with periodic windows when the parameter b varies from 0.14 to 1. The chaotic behavior is confirmed by the largest Lyapunov exponent as shown in Fig. 7B.

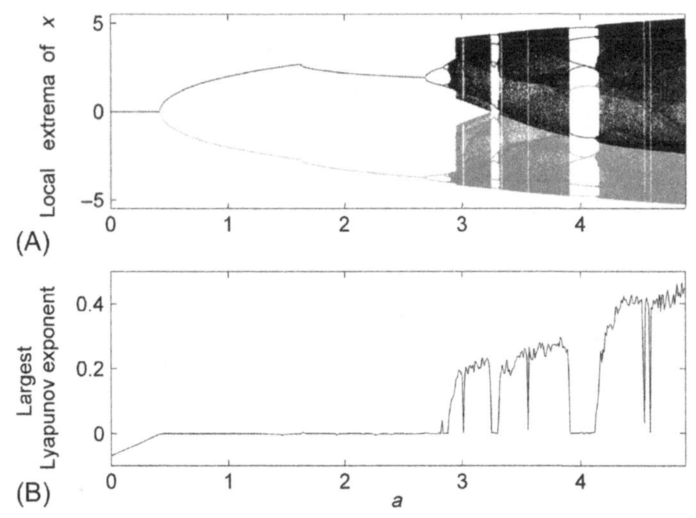

FIG. 6 The bifurcation diagram depicting the local maxima (black dots) and local minima (gray dots) of $x(t)$ (A) and the largest Lyapunov exponent (B) versus the parameter a for $b = 0.5$, $c = 2.5$, and $d = 1.2$.

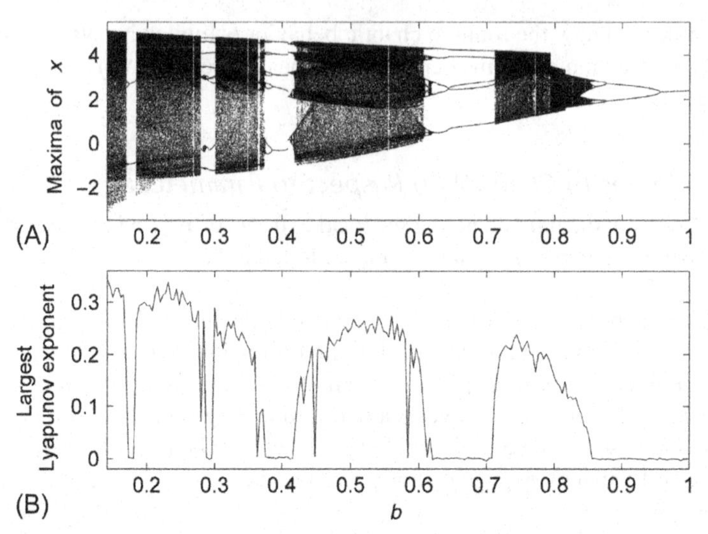

FIG. 7 The bifurcation diagram depicting the maxima of $x(t)$ (A) and the largest Lyapunov exponent (B) versus the parameter b for $a = 3.5$, $c = 2.5$, and $d = 1.2$.

2.2.3 *Routes to Chaos With Respect to Parameter* d

The bifurcation diagram depicting local maxima of $x(t)$ and the largest Lyapunov versus the parameter d are depicted in Fig. 8 for $a = 3.5$, $b = 0.5$, and $c = 2.5$.

The bifurcation diagram of Fig. 8A is obtained by plotting local maxima of the output $x(t)$ versus the parameter d which increased (or decreased) in tiny steps in the range $0.1 \leq d \leq 3$. The final state at each iteration of the control parameter a serves as the initial state for the next iteration. When the parameter d varies from 0.1 to 3 [see black dot in Fig. 8A], the bifurcation diagram of the output $x(t)$ shows a period-doubling bifurcation to chaos for $0.1 < d < 0.5$. Then a period-5-oscillations window is observed followed by a chaotic region interspersed with periodic windows. By further increasing the parameter d, system (3) undergoes a reverse period-doubling bifurcation and a period-1-oscillations is observed for $d > 2.88$. These forward and reverse period-doubling sequences, as a parameter of the system increases in a monotone way, is called antimonotonicity. [38,39] When performing the same analysis by ramping the parameter d [see red dot (gray dot in print versions) in Fig. 8A], the output $x(t)$ displays the same dynamical behaviors as in Fig. 8A (see black dot) but the amplitudes of the output $x(t)$ are not the same. Therefore, one can notice that system (3) shows bistable periodic and chaotic attractors in the ranges $0.1 \leq d \leq 0.357$ and $2.197 \leq d \leq 3$. Besides, at $d \approx 0.357$ and $d \approx 2.197$, system (3) presents symmetry restoring and breaking bifurcations,

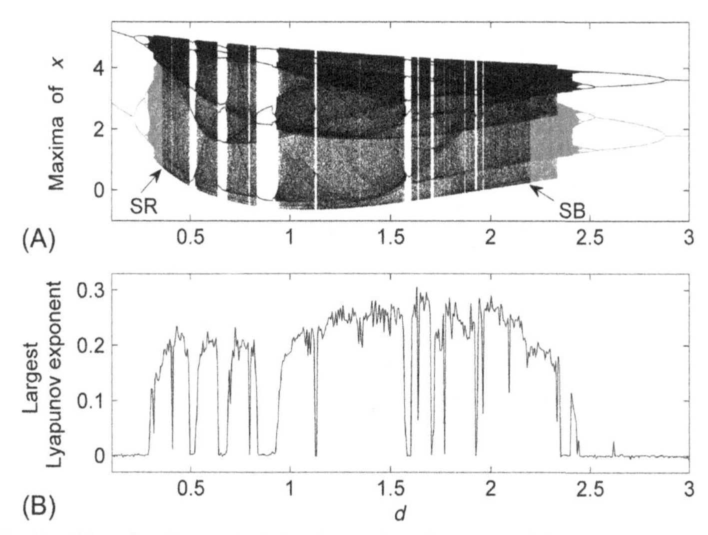

FIG. 8 The bifurcation diagram depicting the maxima of $x(t)$ (A) and the largest Lyapunov exponent (B) versus the parameter d for $a = 3.5$, $b = 0.5$ and $c = 2.5$. Bifurcation diagrams are obtained by scanning the parameter d upwards (black dot) and downwards (red dot; gray dot in print versions). The acronym SB means symmetry-breaking, while SR corresponds to symmetry-restoring.

respectively. The chaotic behavior found in Fig. 5A is confirmed by the largest Lyapunov exponent shown in Fig. 8B. The bistable periodic attractors are depicted in Fig. 9 at $d = 2.95$.

In Fig. 9A, the output $x(t)$ displays period-1-oscillations for the initial conditions $(x(0), y(0), z(0)) = (1, 0, 0.5)$, while in Fig. 9B for the initial conditions$(x(0), y(0), z(0)) = (2, 0, 0)$, the output $x(t)$ exhibits also period-1-oscillations but with higher amplitude than in Fig. 9A.

By varying the value of the parameter c, the bifurcation diagrams of system (3) versus the parameter d are shown in Fig. 10.

In the bifurcation diagrams of Fig. 10A and B chaotic states are observed (chaotic bubbles [39]), while in the bifurcation diagrams of Fig. 10C–F, only periodic states (periodic bubbles). For $c = 4$, system (3) undergoes the sequence: period-1-oscillations $\rightarrow$ period-2-oscillations $\rightarrow$ period-1-oscillations. Bier and Bountis [38] named this sequence "primary bubble." Dawson et al. [40] have proposed a geometric mechanism called dimple formation as a possible means for antimonotonicity in one-dimensional maps that contain two critical points inside the chaotic attractor. Now, we will ascertain whether the mechanism of antimonotonicity found in Figs. 8 and 10 are related to the dimple formation mechanism described in Ref. [40, 41]. For a specific value of parameter d, the phase portraits of chaotic bubble presented in Fig. 10 c and the first-return map of the local maxima of x are illustrated in Fig. 11.

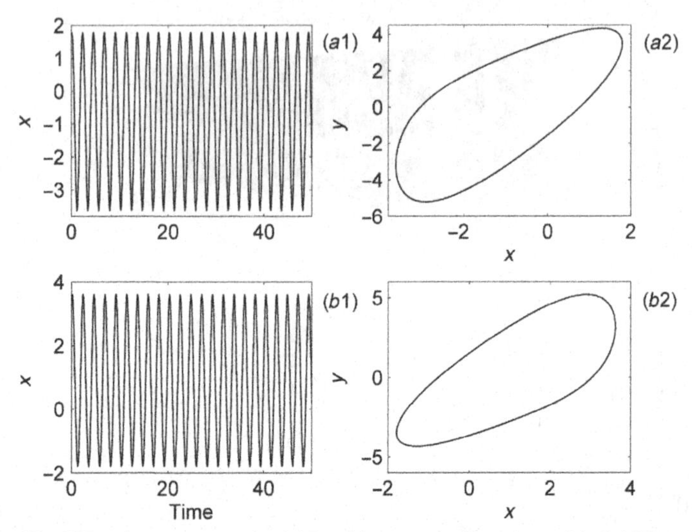

FIG. 9 Bistability phenomenon at $d = 2.95$ and for two specific values of the initial conditions: (A) $(x(0), y(0), z(0)) = (1, 0, 0.5)$ and (B) $(x(0), y(0), z(0)) = (2, 0, 0)$. Time series of the output $x(t)$ is depicted in panel (1) and the phase portrait in the plane (x, y) as shown in panel (2). The remaining parameters are given in Fig. 8.

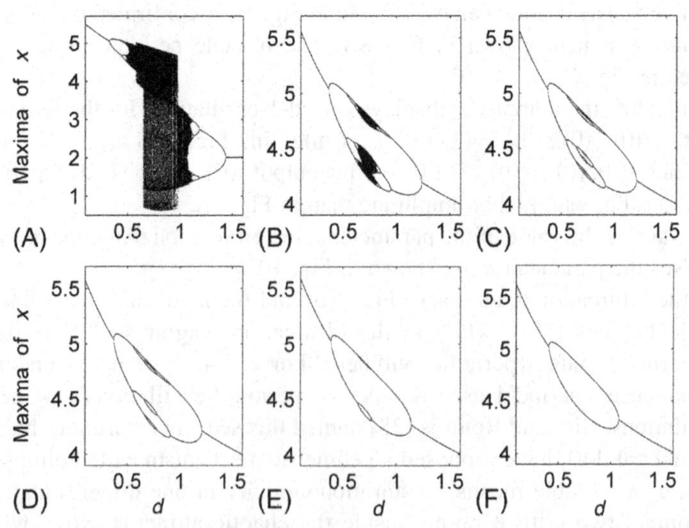

FIG. 10 The bifurcation diagrams depicting the maxima of $x(t)$ versus parameter d for specific values of the parameter c: (A) chaotic bubble at $c = 3.5$, (B) chaotic bubble at$c = 3.76$, (C) bubble of period 16 at $c = 3.79$, (D) bubble of period 8 at $c = 3.8$, (E) bubble of period 4 at $c = 3.9$, and (F) bubble of period-2 at $c = 4$. The others parameters are given in Fig. 8.

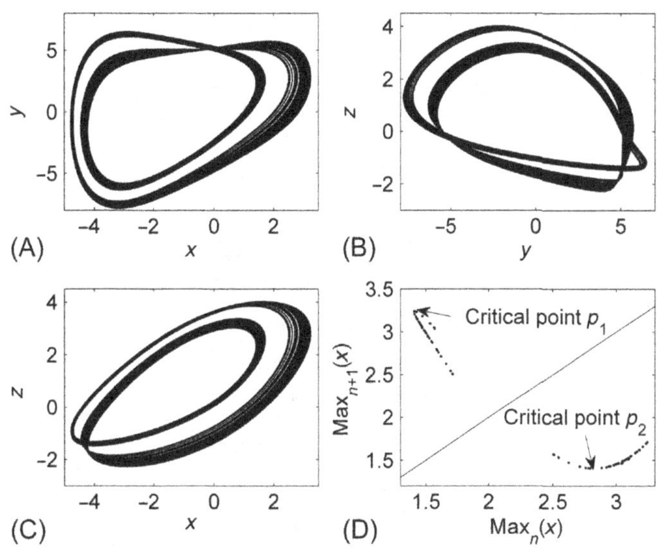

FIG. 11 The phase portraits in plane (x, y), (y, z), (x, z) (A)–(c) and the first-return map $(\mathrm{Max}_{n+1}(x) = f(\mathrm{Max}_n(x)))$ of the maxima of x (D) for $c = 3.76$ and $d = 0.75$. The others parameters are given in Fig. 10. The initial conditions are$(x(0), y(0), z(0)) = (1.0, 0, 0.5)$.

The chaotic bubble attractor presented in Figs. 11A–C is confirmed by the numerical calculation of the Lyapunov exponents which give $LE_1 \approx 0.064$, $LE_2 \approx 0$ and $LE_3 \approx -0.563$. The Lyapunov dimension of the chaotic bubble is $D_L \approx 2.113$. In Fig. 11D, the map is indicative of one-dimensional maps with two critical points p_1 and p_2, which support the occurrence of antimonotonicity phenomenon in RC autonomous circuit with diodes in antiparallel according to the results of Dawson et al. [40,41]

2.3 Observation of Chaotic Attractors From the Oscilloscope

The experimental observations from the oscilloscope of the phase portrait of Figs. 5 and 11A–C obtained by using specific values of resistors and capacitors are shown in Fig. 12.

The occurrence of the one- and double-scroll chaotic attractors and chaotic bubbles attractors can be clearly seen from Fig. 12. By comparing it with Figs. 5 and 12A–C, it can be concluded that a good qualitative agreement between the numerical simulations and the experimental realizations is obtained.

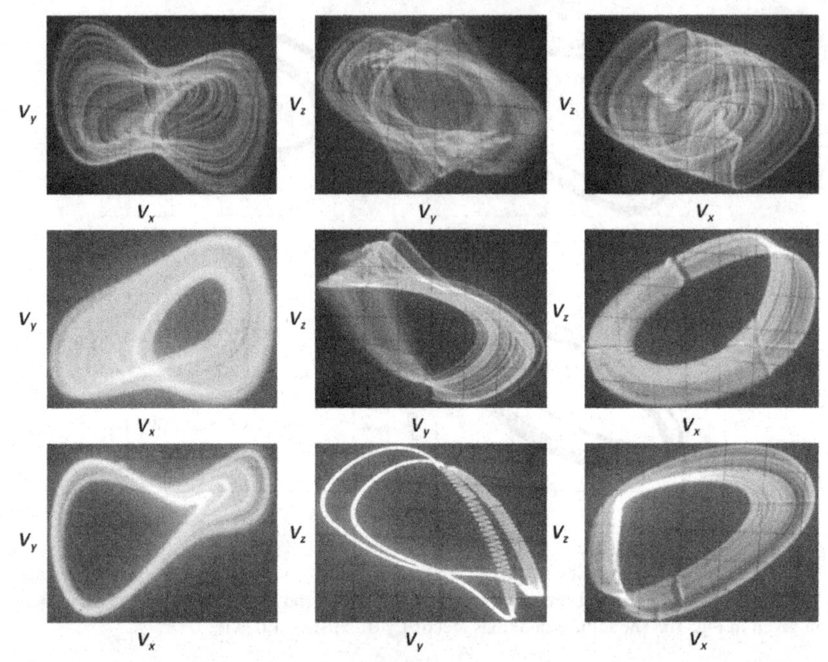

FIG. 12 The phase portraits of the chaotic attractors observed on the oscilloscope. In panels 1, 2, and 3, we depict the phase portraits in planes(V_x, V_y), (V_y, V_z) and (V_x, V_z), respectively. The values of capacitors and resistances are $R_1 = 2.38\,\text{K}\Omega$, $R_2 = 55.5\,\text{K}\Omega$, $R_3 = 6.94\,\text{K}\Omega$, $R_4 = 3.3\,\text{K}\Omega$, $R_5 = R_6 = R = 5\,\text{K}\Omega$, $C_1 = C_2 = C_3 = 12\,\text{nF}$, [first line which reproduces Fig. 5B], $R_1 = 2.38\,\text{K}\Omega$, $R_2 = 16.6\,\text{K}\Omega$, $R_3 = 3.62\,\text{K}\Omega$, $R_4 = 3.3\,\text{K}\Omega$, $R_5 = R_6 = R = 5\,\text{K}\Omega$, $C_1 = C_2 = C_3 = 12\,\text{nF}$, [second line which reproduces Fig. 5A] and $R_1 = 2.38\,\text{K}\Omega$, $R_1 = 2.38\,\text{K}\Omega$, $R_2 = 16.6\,\text{K}\Omega$, $R_3 = 11.1\,\text{K}\Omega$, $R_4 = 2.22\,\text{K}\Omega$, $R_5 = R_6 = R = 5\,\text{K}\Omega$, $C_1 = C_2 = C_3 = 12\,\text{nF}$, [third line which reproduces Fig. 11A–C].

3 CONCLUSION

In this paper, we reported results on the analytical, numerical, and experimental analysis of RC autonomous circuit with diodes in antiparallel. Based on the Routh–Hurwitz stability criterion and linear stability of the equilibrium points, we found that Hopf bifurcation occurs in the circuit at the parameter value $a = a_H$, where a_H is a function of the other model parameters, b, c, and d, provided certain conditions on the model parameters hold. Numerical simulations results have not only verified the mathematical analysis, but also have displayed some other interesting dynamics. For example, bistabilty, antimonotonicity, and periodic and chaotic bubbles emerge in RC autonomous circuits with diodes in antiparallel when the parameter d varies. For specific parameters, we revealed that the circuit exhibits one- and double-scroll chaotic attractors. The one- and double-scroll chaotic attractors and chaotic bubbles attractors obtained during numerical simulations were experimentally confirmed using electronic implementation of the circuit. It is important to note that the RC circuit with diodes in antiparallel could be considered a cell of generalized cellular neural networks [42] because it contains more information with respect to the standard cell and could lead from the point of view of cellular neural networks system with behaviorally interesting aspects.

ACKNOWLEDGMENT

S.T.K. wishes to thank Mr. Andre Chaégé Chamgoué (LaMSEBP, University of Yaounde I, Cameroon) for valuable discussions.

REFERENCES

[1] Special issue on chaos in nonlinear electronic circuits—part A, tutorials and reviews. IEEE Trans Circ Syst II 1993;40:640–786.

[2] Special issue on chaos in nonlinear electronic circuits—part B, bifurcations and chaos. IEEE Trans Circ Syst II 1993;40:792–884.

[3] Kingni ST, Keuninckx L, Woafo P, Van der Sande G, Danckaert J. Dissipative chaos, Shilnikov chaos and bursting oscillations in a three-dimensional autonomous system: theory and electronic implementation. Nonlinear Dyn 2013;73:1111–23.

[4] Ogorzalek MJ. Chaos and complexity in nonlinear electronic circuits. World Sci Ser Nonlinear Sci Ser A 1997;22:.

[5] Ramos J. Introduction to nonlinear dynamics of electronic systems: tutorial. Nonlinear Dyn 2006;44:3–14.

[6] Sprott JC. Elegant chaos: algebraically simple flow. Singapore: World Scientific Publishing; 2010.

[7] Lassoued A, Boubaker O. On new chaotic and hyperchaotic systems: a literature survey. Nonlinear Anal Model Control 2016;21:770–89.

[8] Lassoued A. Dynamic analysis and circuit design of a novel hyperchaotic system with fractional-order terms, Complexity 2017;2017:10 pp. https://doi.org/10.1155/2017/3273408.

[9] Mkaouar H, Boubaker O. Robust control of a class of chaotic and hyperchaotic driven systems. Pramana 2017;88:9–18.

[10] Mkaouar H, Boubaker O. Chaos synchronization for master slave piecewise linear systems: application to Chua's circuit. Commun Nonlinear Sci Numer Simul 2012;17:1292–302.

[11] Boubaker O, Dhifaoui R. Robust chaos synchronization for chua's circuits via active sliding mode control. In: Chaos, complexity and leadership. 2012:Dordrecht: Springer; 2014. p. 141–51.

[12] Mkaouar H. In: On electronic design of the piecewise linear characteristic of the chua's diode: application to chaos synchronization. Electrotechnical conference (MELECON), 16th IEEE mediterranean; 2012. p. 197–200.

[13] Lassoued A, Boubaker O. In: Hybrid synchronization of multiple fractional-order chaotic systems with ring connection. 8th international conference on Modelling, Identification and Control (ICMIC), IEEE; 2016. p. 109–14.

[14] Kilias T, Kelber K, Mogel A, Schwarz W. Electronic chaos generators—design and applications. Int J Electron 1995;79:737–53.

[15] Carroll TL, Pecora LM. Synchronizing chaotic circuits. IEEE Trans Circ Syst I 1991;38:453–6.

[16] Banerjee T, Sarkar BC. Chaos and bifurcation in a third order phase-locked loop. AEU Int J Electron Commun 2006;62:86–91.

[17] Banerjee T, Karmakar B, Sarkar BC. Chaotic electronic oscillator from single amplifier biquad. AEU Int J Electron Commun 2012;66:593–7.

[18] Fortuna L, Frasca M, Xibilia MG. Chua's circuit implementations: yesterday, today and tomorrow. Singapore: World Scientific; 2009.

[19] Bilotta E, Pantano P. A gallery of chua attractors. Singapore: World Scientific; 2008.

[20] Chua LO, Lin G. Canonical realization of Chua's circuit family. IEEE Trans Circ Syst 1990;37:885–902.

[21] Elwakil AS, Kennedy MP. A semi-systematic procedure for producing chaos from sinusoidal oscillators using diode-inductor and FET-Capacitor composites. IEEE Trans Circ Syst 2000;47:582–90.

[22] Banerjee T. Single amplifier biquad based inductor-free Chua's circuit. Nonlinear Dyn 2012;68(2010):565–73.

[23] Elwakil AS, Kennedy MP. Chaotic oscillators derived from sinusoidal oscillators based on the current feedback opamp. Analog Int Circ Signal Process 2000;24:239–51.

[24] Kingni ST, Pham VT, Jafari S, Kol GR, Woafo P. Three-dimensional chaotic autonomous system with a circular equilibrium: Analysis, circuit implementation and its fractional-order form. Circ Syst Signal Process 2016;35:1931–48.

[25] Chang JF, Liao TL, Yan JJ, Chen HC. Implementation of synchronized chaotic Lu systems and its application in secure communication using PSO-based PI controller. Circ Syst Signal Process 2010;29:527–38.

[26] Tayebi A, Berber S, Swain A. Performance analysis of chaotic dsss–cdma synchronization under jamming attack. Circ Syst Signal Process 2016. https://doi.org/10.1007/s00,034-016-0266-y.

[27] Banerjee T, Karmakar B, Sarkar BC. Single amplifier biquad based autonomous electronic oscillators for chaos generation. Nonlinear Dyn 2010;62(2012):859–66.

[28] Bernát P, Baláž I. RC autonomous circuit with chaotic behaviour. Radioengineering 2002;11:1–5.

[29] Tamasevicius A, Mykolaitis G, Pyragas V, Pyragas K. A simple chaotic oscillator for educational purposes. Eur J Phys 2005;26:61–3.

[30] Pham V-T, Volos C, Jafari S, Wang X, Kapitaniak T. A simple chaotic circuit with a light-emitting diode. Optoelectron Adv Mater 2016;10:640–6.

[31] Sprott JC. Elegant chaos: algebraically simple flow. Singapore: World Scientific Publishing; 2010.

[32] Srisuchinwong B, Munmuangsaen B. Four-current-tunable chaotic oscillators in set of two diode-reversible pairs. Electron Lett 2006;48:1051–3.

[33] San-Um W, Suksiri B, Ketthong P. A simple RLCC-diode-opamp chaotic oscillator. Int J Bifurcat Chaos 2014;24:1450. 155.

[34] Kengne J, Njitache ZT, Nguomkam AN, Tsostop MF, Fotsin HB. Coexistence of multiple attractors and crisis route to chaos in a novel chaotic jerk circuit. Int J Bifurcat Chaos 2016;26:1650081–100.

[35] Nana B, Woafo P. Synchronization of diffusively coupled oscillators: theory and experiment. Am J Electr Electron Eng 2015;3:37–43.

[36] Nana B, Woafo P. Chaotic masking of communication in an emitter-relay-receiver electronic setup. Nonlinear Dyn 2015;24:899–908.

[37] Wolf A, Swift JB, Swinney HL, Wastano JA. Determining Lyapunov exponents from a time series. Physica D 1985;16:285–317.

[38] Bier M, Bountis TC. Remerging Feigenbaum trees in dynamical systems. Phys Lett A 1984;104:239–44.

[39] Kyprianidis IM, Haralabidis P, Stouboulos IN, Bountis TC. Antimonotonicity and chaotic dynamics in a fourth order autonomous nonlinear electric circuit. Int J Bifurcat Chaos 2000;10:1903–15.

[40] Dawson SP, Grebogi C, Kan I, Kocak H, Yorke JA. Antimonotonicity: inevitable reversals of period-doubling cascades. Phys Lett A 1992;162:249–54.

[41] Dawson SP, Grebogi C, Kocak H. Geometric mechanism for antimonotonicity in scalar maps with two critical points. Phys Rev E 1993;48:1676–80.

[42] Fortuna L, Arena P, Bâlya D, Zarândy A. Cellular neural networks: a paradigm for nonlinear spatio-temporal processing. IEEE Circ Syst Mag 2001;1:6–21.

Chapter 3

Chaos in a System With Parabolic Equilibrium

Viet-Thanh Pham*,†, Christos K. Volos‡, Sundarapandian Vaidyanathan§, Sajad Jafari¶, Gokul P.M.† and Tomasz Kapitaniak†

**School of Electronics and Telecommunications, Hanoi University of Science and Technology, Hanoi, Vietnam, †Division of Dynamics, Lodz University of Technology, Lodz, Poland, ‡Laboratory of Nonlinear Systems, Circuits and Complexity (LaNSCom), Department of Physics, Aristotle University of Thessaloniki, Thessaloniki, Greece, §Research and Development Center, Vel Tech University, Chennai, India, ¶Biomedical Engineering Department, Amirkabir University of Technology, Tehran, Iran*

1 INTRODUCTION

After the discovery of Lorenz chaotic system [1], chaos theory, chaos-based applications, and new chaotic systems have been studied intensively [2–13]. A novel hyperchaotic system with fractional-order terms has been proposed in [14]. Lassoued and Boubaker have done a literature survey on new chaotic and hyperchaotic systems [15]. Various three-dimensional (3D) chaotic systems have been found and reported in the literature [7]. It is well known that most of the 3D chaotic systems, such as the Lorenz system [1], Rössler system [16], Chen system [17], or Sprott systems (cases B to S) [18], have a countable number of equilibrium points.

It is worth noting that a list of nine chaotic flows with a line of equilibria has been introduced recently by Jafari and Sprott [19]. The authors found these flows by applying a general search routine. These chaotic flows are different from conventional others because of their uncountable equilibria. Their corresponding attractor is hidden from the computing view point [20–23]. On the other word, they are classified as systems with "hidden attractor," which are important in theoretical and engineering areas [24–32]. The properties of hidden attractors can be compared with those of rare attractors [33–35]. Motivated by special features of such chaotic flows with a line equilibrium, other systems with unlimited equilibria were proposed (see Table 1). When studying chaotic flows with a single nonquadratic term, Li and Sprott discovered five chaotic flows with a line equilibrium and an especially complicated one with two infinite parallel lines of equilibrium points [36]. In order to create a line or two

Recent Advances in Chaotic Systems and Synchronization. https://doi.org/10.1016/B978-0-12-815838-8.00003-0

TABLE 1 Three-Dimensional Autonomous Chaotic Systems With an Infinite Number of Equilibrium Points and Their Shapes of Equilibria

Shape of Equilibria	Related Works
A line	[19, 36, 37]
Two parallel lines	[36]
Two perpendicular lines	[37]
A circle	[38, 39]
A square	[39]
An open smooth curve	[40]
A hyperbolic sine curve	[41]
A exponential curve	[42]
A cubic curve	[43]
A parabola	This work

perpendicular lines of equilibrium points, authors modified the Sprott B system by using signum functions and absolute-value functions [37]. The existence of circular equilibrium was investigated by mathematical analysis and circuitry implementation [38, 39]. Gotthans et al. obtained a 3D system with a square equilibrium by modifying a system with a circular equilibrium [39]. A system with an open smooth curve of equilibrium points was presented in [40], while a novel chaotic system with hyperbolic sine equilibrium was reported in [41]. Tlelo-Cuautle realized a system with an exponential curve of equilibria by using FPGA and applied it for transmitting secure images [42]. Pham et al. studied a novel cubic-equilibrium chaotic system [43]. One may wonder whether or not there is a chaotic system with parabolic equilibrium. The aim of this chapter is to give a positive answer for the exciting question.

In this work, we proposed a 3D chaotic system with the presence of parabolic equilibrium. The investigation of such a system will lead to studies of other chaotic systems with different equilibrium curves. Section 2 introduces a general model to construct the system. Dynamical properties, such as stability of equilibrium points, phase portrait, Poincaré map, bifurcation diagram, and Lyapunov exponents of the novel system, are investigated in Section 3. An electronic implementation of such system is studied and PSpice results are reported in Section 4. Adaptive control and adaptive synchronization of such a system with parabolic equilibrium are also reported in Sections 5 and 6, respectively. Additional examples of chaotic systems with parabolic equilibrium are discussed in Section 7. Finally, conclusion remarks are drawn in the last section.

2 MODEL OF A SYSTEM WITH INFINITE EQUILIBRIA

We consider a general model of 3D systems given by

$$\begin{cases} \dot{x} = -z, \\ \dot{y} = xz^2 + a \ \ \text{sgn}(z), \\ \dot{z} = f_1(x,y) + zf_2(x,y,z), \end{cases} \tag{1}$$

where x, y, z are state variables, a is a positive parameter while $f_1(x,y), f_2(x,y,z)$ are two nonlinear functions. Here, the signum function [44, 45] is defined as

$$\text{sgn}(z) = \begin{cases} 1 & \text{for } z > 0, \\ 0 & \text{for } z = 0, \\ -1 & \text{for } z < 0. \end{cases} \tag{2}$$

Note that the signum function (2) can be implemented easily by an operational amplifier [45].

The equilibria of system (1) can be found by solving $\dot{x} = 0, \ \ \dot{y} = 0, \ \ \dot{z} = 0$ or

$$-z = 0, \tag{3}$$

$$xz^2 + a \ \ \text{sgn}(z) = 0, \tag{4}$$

$$f_1(x,y) + zf_2(x,y,z) = 0. \tag{5}$$

Therefore, in the plane $z = 0$, system (1) has many equilibrium points. Furthermore, such points are located on a curve given by

$$f_1(x,y) = 0. \tag{6}$$

Obviously, points of system (1) create different smooth curves when we choose different kinds of nonlinear function $f_1(x,y)$. Moreover, system (1) for appropriate parameters can exhibit a complex behavior like chaos.

In this work, we construct a system with parabolic equilibrium by selecting two nonlinear functions $f_1(x,y), f_2(x,y,z)$ as follows:

$$f_1(x,y) = x - y^2, \tag{7}$$

$$f_2(x,y,z) = by^2 - z^2, \tag{8}$$

in which b is a positive parameter. As a result, system (1) is rewritten in the following form

$$\begin{cases} \dot{x} = -z, \\ \dot{y} = xz^2 + a \ \ \text{sgn}(z), \\ \dot{z} = x - y^2 + z\left(by^2 - z^2\right). \end{cases} \tag{9}$$

Equilibrium of system (9) is located on a parabolic curve as illustrated in Fig. 1.

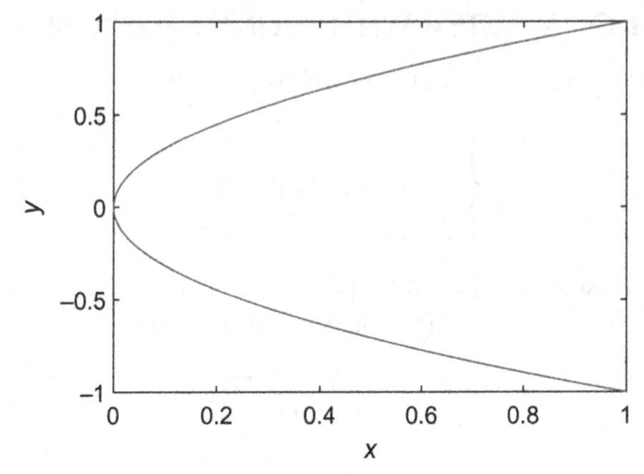

FIG. 1 Equilibrium of system (9) located on a parabolic curve in the x-y plane.

3 DYNAMICAL PROPERTIES OF THE SYSTEM WITH PARABOLIC EQUILIBRIUM

3.1 Equilibrium Points and Stability

From Eqs. (6), (7), we have the set of equilibrium points in system (9):

$$E=\left\{(x,y,z)\in R^3\middle|x=(y^*)^2,y=y^*,z=0\right\}. \tag{10}$$

Interestingly, such equilibrium is located on a parabola. In order to calculate the Jacobian matrix, we have applied the reported approach in [44–47]. The signum function $\text{sgn}(z)$ is approximated to a hyperbolic tangent function $\tanh(kz)$ for $k \gg 1$. At the equilibrium point E, the Jacobian matrix of system (9) is given by

$$\mathbf{J}_E=\begin{bmatrix} -\lambda & 0 & -1 \\ 0 & -\lambda & ak \\ 1 & -2y^* & -\lambda+b(y^*)^2 \end{bmatrix}. \tag{11}$$

As a result, the characteristic equation of system (9) at the equilibrium point E is

$$\lambda\left(\lambda^2-b(y^*)^2\lambda+1+2aky^*\right)=0. \tag{12}$$

From Eq. (12), it is easy to see that the eigenvalues depend on the sign of the discriminant $\Delta=b^2(y^*)^4-4(1+2aky^*)$. If the discriminant is positive $\Delta>0$, there are one zero eigenvalue ($\lambda_1 = 0$) and two nonzero eigenvalues ($\lambda_{2,3}=\frac{b(y^*)^2\pm\sqrt{\Delta}}{2}$). In addition, there is at least one positive real eigenvalue, the equilibrium point E, therefore, is unstable. If the discriminant is zero $\Delta = 0$, there are one zero eigenvalue ($\lambda_1 = 0$) and one real eigenvalue ($\lambda_2=\frac{b(y^*)^2}{2}$). The equilibrium point E is unstable due to the presence of the positive real eigenvalue λ_2. If the discriminant is negative $\Delta<0$ and $y^*\neq 0$, there are one zero eigenvalue ($\lambda_1 = 0$) and a pair of complex conjugate eigenvalues ($\lambda_{2,3} = \frac{b(y^*)^2 \pm i\sqrt{\Delta}}{2}$). These complex eigenvalues have positive real parts (because $b > 0$), the equilibrium point E in this case, therefore, is unstable.

3.2 Dynamical Behavior

It is interesting that system (9) displays chaotic behavior as shown in Fig. 2, for $a = 0.1$, $b = 3$ and initial conditions $(x(0), y(0), z(0)) = (0.1, 0.1, 0.1)$. In this case, the Lyapunov exponents of system (9) are calculated as $L_1 = 0.0797$, $L_2 = 0$, and $L_3 = -0.3415$, while the corresponding Kaplan-Yorke dimension of the system is $D_{KY} = 2.2334$. Here Wolf's algorithm [48] has been applied to calculate Lyapunov exponents. However, in order to achieve precise values, the calculation of Lyapunov exponents should be considered seriously [49–51]. In addition, the Poincaré map of system (9) in Fig. 3 also indicates properties of chaos. It is noted that we have considered and applied the simulation programs in [52].

To better investigate dynamical behavior of system (9), we change the value of the parameter b from 2.7 to 3.7. The bifurcation diagram and the corresponding Lyapunov spectrum are shown in Figs. 4 and 5. As shown in Figs. 4 and 5, there are some windows of limit cycles and of chaotic behavior. Fig. 4 suggests that system (9) generates chaotic oscillations for $2.865 < b < 3.135$. When increasing the value of the parameter b, the system exhibits a reverse period-doubling route to period-1-oscillation.

4 CIRCUITRY IMPLEMENTATION OF THE SYSTEM WITH PARABOLIC EQUILIBRIUM

Physical implementations of mathematical chaotic models have been applied in various engineering applications, for example, in secure communications [53, 54], robotics [55], signal generation [56, 57], multimedia security application [58]. Therefore, the feasibility of theoretical model is often considered when studying new chaotic models [59–61]. In this section, we will show the feasibility of system (9) by presenting its electronic realization. It is noting that the amplitudes of state variables of system (9) are small (see Fig. 2). In order to generate enough larger signals in an electronic circuit, the three state variables of system (9) x, y, z are rescaled with a factor 10. As a result, we transform system (9) into system (13):

$$\begin{cases} \dot{X} = -Z, \\ \dot{Y} = \frac{1}{10^2} XZ^2 + 10a \ \text{sgn}\left(\frac{1}{10} Z\right), \\ \dot{Z} = X - \frac{1}{10} Y^2 + \frac{b}{10^2} Y^2 Z - \frac{1}{10^2} Z^3, \end{cases} \tag{13}$$

in which $X = 10x$, $Y = 10y$, and $Z = 10z$. Fig. 6 presents the schematic of the designed circuit including 13 resistors, 3 capacitors, 7 operational amplifiers

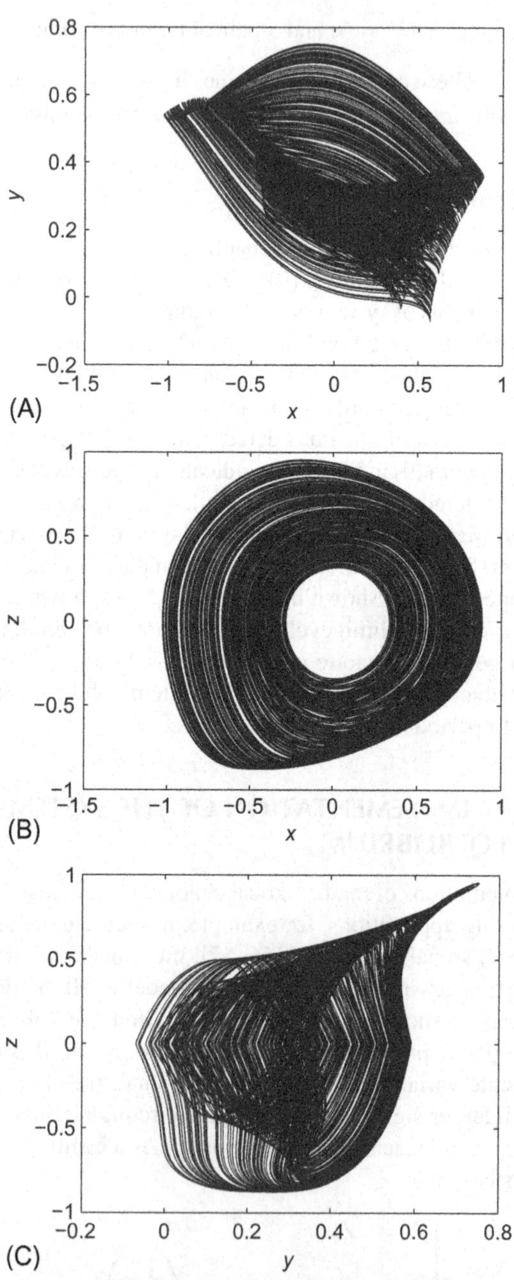

FIG. 2 Phase portraits of system (9) in (A) x-y plane, (B) x-z plane, and (C) y-z plane, for $a = 0.1$ and $b = 3$.

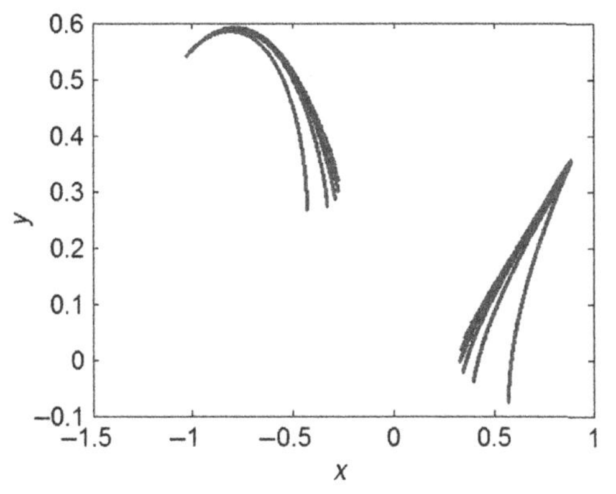

FIG. 3 Poincaré map of system (9) in the x-y plane when $z = 0$, for $a = 0.1$, $b = 3$.

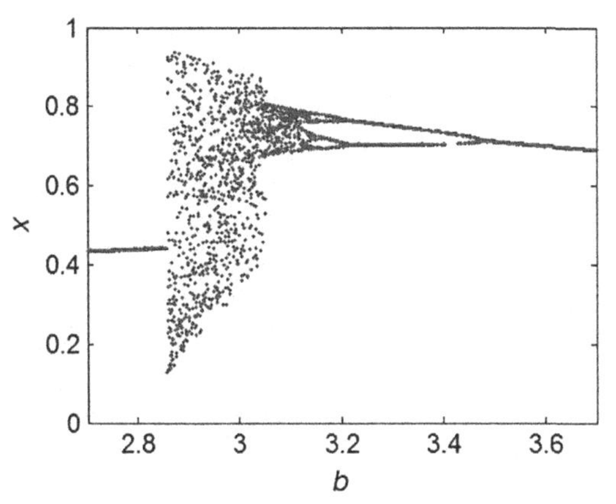

FIG. 4 Bifurcation diagram of x versus b, for $a = 0.1$.

(TL084), 5 analog multipliers (AD633), and the power supplies (± 15 V_{DC}). There are three integrators (U_1–U_3), which are created by the operational amplifiers. The signal $\text{sgn}\left(\frac{1}{10}Z\right)$ is provide by the circuitry which is constituted by two operational amplifiers (U_4, U_5) [45]. By applying Kirchhoff's circuit laws into the circuit of Fig. 6, we get its circuital equations:

$$\begin{cases} \dot{X} = \dfrac{1}{RC}(-Z), \\ \dot{Y} = \dfrac{1}{RC}\left(\dfrac{1}{10^2V^2}XZ^2 + \dfrac{RV_{sat}}{R_a}\text{sgn}\left(\dfrac{1}{10}Z\right)\right), \\ \dot{Z} = \dfrac{1}{RC}\left(X - \dfrac{1}{10V}Y^2 + \dfrac{R}{R_b 10^2 V^2}Y^2 Z - \dfrac{1}{10^2V^2}Z^3\right), \end{cases} \tag{14}$$

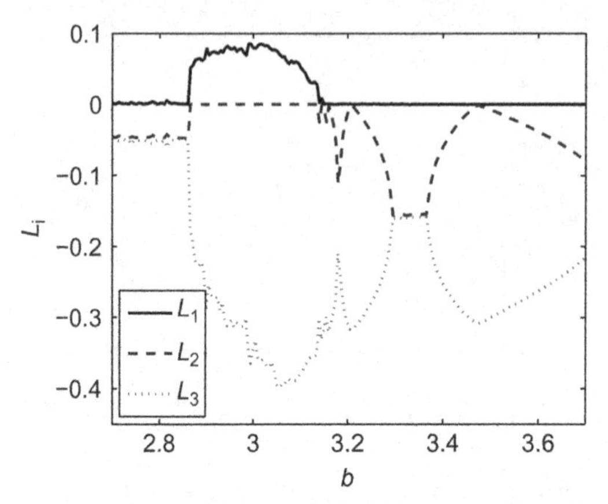

FIG. 5 Lyapunov exponents L_1 (*solid line*), L_2 (*dashed line*), and L_3 (*dotted line*) when changing b for $a = 0.1$.

where X, Y, and Z are correspond to the voltages on the integrators U_1, U_2, and U_3, respectively, while V_{sat} is the saturation voltage of the operational amplifier U_5. In this work, we normalize system (14) by using $\tau = \frac{t}{RC}$. Therefore, system (14) is equivalent to system (13) with $a = \frac{RV_{sat}}{10R_a}$ and $b = \frac{R}{R_b}$. The values of circuit components have been chosen as $R = 10$ kΩ, $R_a = 142.5$ kΩ, $R_b = 3.333$ kΩ, and $C = 4.7$ nF to realize the two system parameters $a = 0.1$ and $b = 3$. The designed circuit in Fig. 6 is implemented by using the electronic simulation package Cadence OrCAD. Obtained PSpice results are reported in Fig. 7. We see that there is a good agreement between the PSpice phase portraits in Fig. 7 and the theoretical ones in Fig. 2.

5 ADAPTIVE CONTROL OF THE SYSTEM WITH PARABOLIC EQUILIBRIUM

Here, we devise an adaptive controller so as to stabilize the trajectories of the new system with parabolic equilibrium.

The main system under consideration is the controlled system with parabolic equilibrium defined by

$$\begin{cases} \dot{x} = -z + u_x, \\ \dot{y} = xz^2 + a \ \text{sgn}(z) + u_y, \\ \dot{z} = x - y^2 + z\left(by^2 - z^2\right) + u_z, \end{cases} \tag{15}$$

where $X = (x, y, z)$ is the state, a, b are unknown parameters, and $\mathbf{u} = (u_x, u_y, u_z)$ is the control.

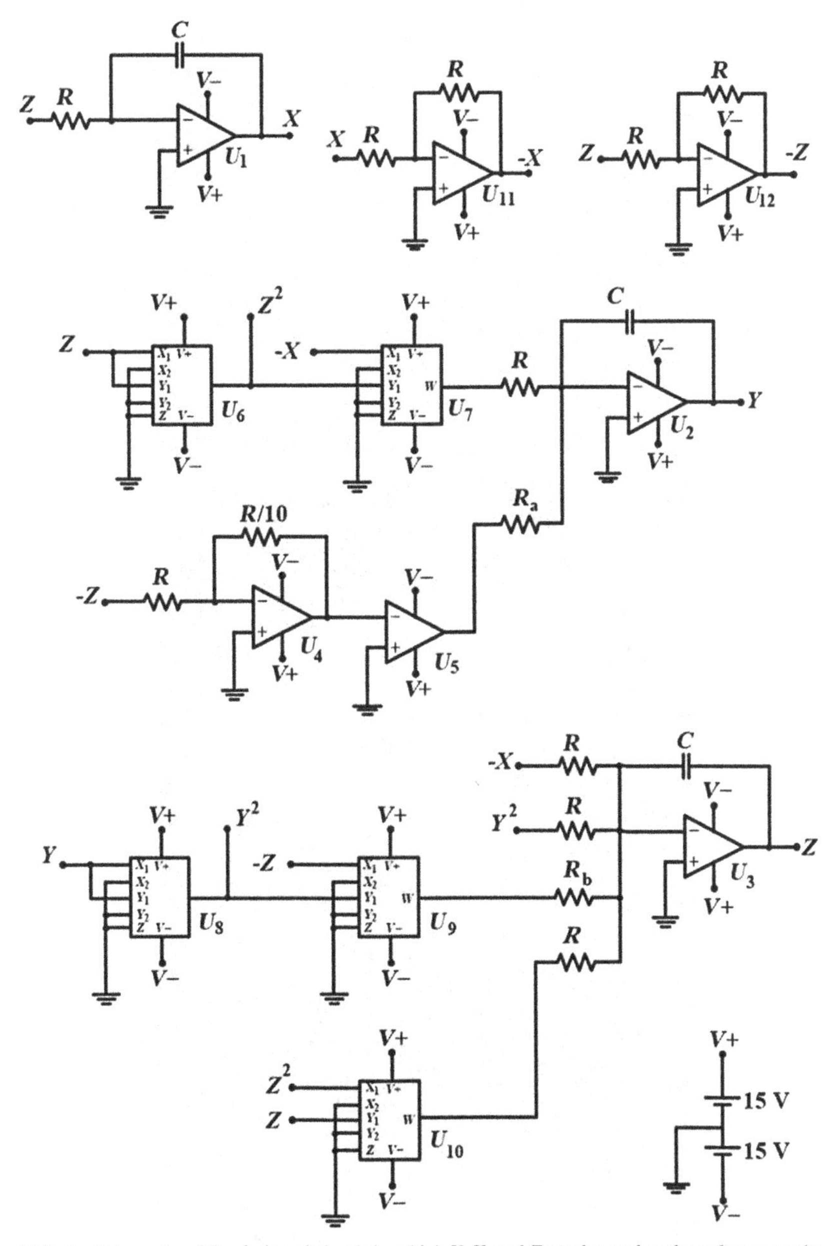

FIG. 6 Schematic of the designed circuit in which X, Y, and Z are denoted as the voltages on the integrators U_1, U_2, and U_3.

Let $A(t)$ and $B(t)$ denote estimates of the unknown parameters a and b, respectively.

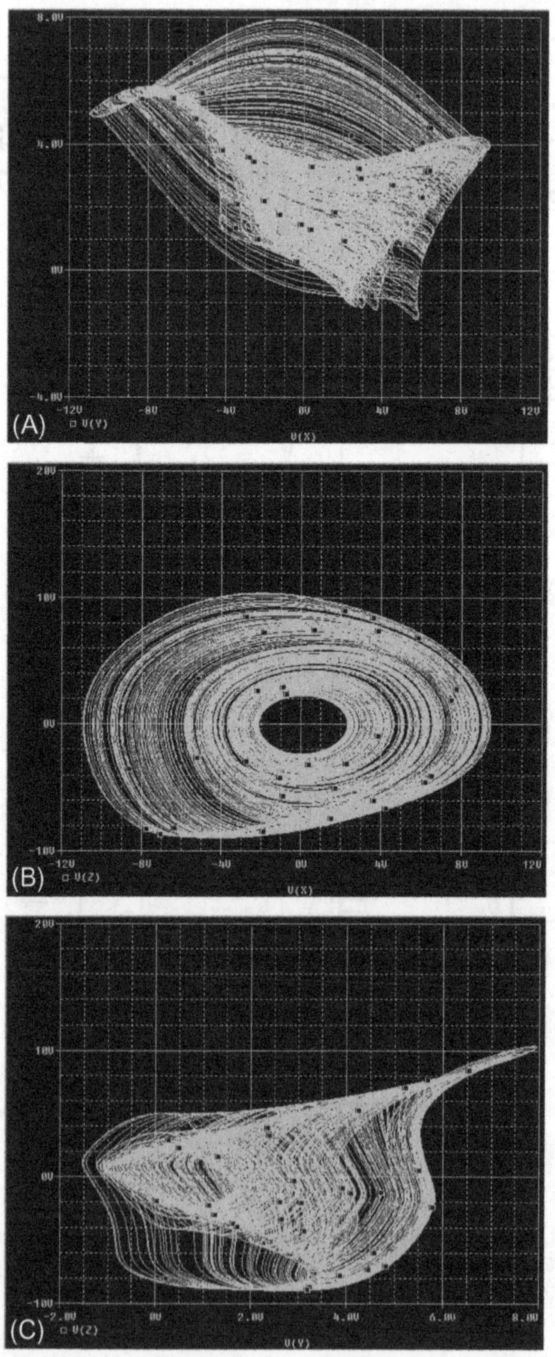

FIG. 7 PSpice phase portraits in different planes from the circuit: (A) *X-Y* plane, (B) *X-Z* plane, and (C) *Y* -*Z* plane.

As an adaptive control scheme for the main system (15), we take

$$\begin{cases} u_x &= z-\alpha_x x, \\ u_y &= -xz^2 - A(t)\ \ \mathrm{sgn}(z) - \alpha_y y, \\ u_z &= -x+y^2 - zB(t)y^2 + z^3 - \alpha_z z, \end{cases} \tag{16}$$

where α_x, α_y, α_z $\in \mathbf{R}^+$ are constants.

Application of the scheme (16) yields the closed-loop plant

$$\begin{cases} \dot{x} = -\alpha_x x, \\ \dot{y} = [a-A(t)]\ \ \mathrm{sgn}(z) - \alpha_y y, \\ \dot{z} = [b-B(t)]zy^2 - \alpha_z z. \end{cases} \tag{17}$$

We define the parameter estimation errors as $e_a = a - A(t)$ and $e_b = b - B(t)$. With this notation, we can simplify Eq. (17) as follows:

$$\begin{cases} \dot{x} = -\alpha_x x, \\ \dot{y} = e_a\ \ \mathrm{sgn}(z) - \alpha_y y, \\ \dot{z} = e_b z y^2 - \alpha_z z. \end{cases} \tag{18}$$

It is also easy to deduce that

$$\dot{e}_a = -\dot{A}, \quad \dot{e}_b = -\dot{B}. \tag{19}$$

As a candidate Lyapunov function, we take

$$V(X, e_a, e_b) = \frac{1}{2}(x^2 + y^2 + z^2) + \frac{1}{2}(e_a^2 + e_b^2). \tag{20}$$

Differentiating V along Eqs. (18), (19), we get

$$\dot{V} = -\alpha_x x^2 - \alpha_y y^2 - \alpha_z z^2 + e_a(y\ \ \mathrm{sgn}(z) - \dot{A}) + e_b(z^2y^2 - \dot{B}). \tag{21}$$

Let us take the update law for parameter estimates $A(t)$ and $B(t)$ as follows:

$$\dot{A} = y\ \ \mathrm{sgn}(z) \quad \text{and} \quad \dot{B} = z^2y^2. \tag{22}$$

The following result is a consequence of Lyapunov stability theory [62].

Theorem 1. *The new system with parabolic equilibrium* (15) *is globally and asymptotically stabilized for all values of $X(0) \in$. $\mathbf{R}^3$ by the adaptive control scheme* (16) *and parameter update scheme* (22), *where $A(t)$, $B(t)$ are updates of the unknown parameters a, b, respectively, and α_x, α_y, α_z are positive constants.*

Proof. The candidate Lyapunov function V defined in Eq. (20) is positive definite on $\mathbf{R}^5$. Substitution of the update scheme (22) into Eq. (21) yields

$$\dot{V} = -\alpha_x x^2 - \alpha_y y^2 - \alpha_z z^2, \tag{23}$$

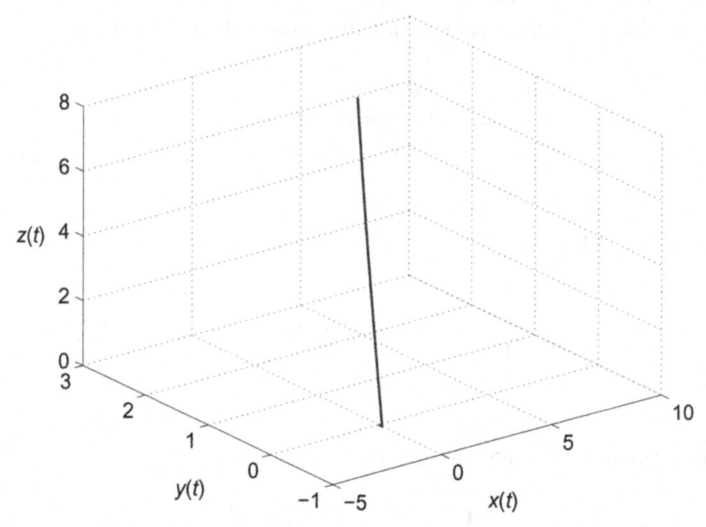

FIG. 8 Numerical simulation of the 3D orbit of the controlled system (15) for $(a, b) = (0.1, 3)$, $(A(0), B(0)) = (2.7, 4.8)$, and $X(0) = (7.3, 2.9, 6.1)$.

which exemplifies that $\dot{V}$ is a negative semidefinite function on $\mathbf{R}^3$.

Barbalat's lemma [62] shows that the closed-loop system (18) is globally asymptotically stable. This completes the proof of the theorem.

For numerical simulations, we take the parameters of the new system (15) as in the chaotic case (i.e., $a = 0.1$ and $b = 3$). We take $A(0) = 2.7$ and $B(0) = 4.8$. We take the gain constants as $\alpha_x = \alpha_y = \alpha_z = 50$. Also, we take $X(0) = (7.3, 2.9, 6.1)$. Fig. 8 shows the asymptotic convergence of the 3D phase orbit of the controlled system (15), while Fig. 9 shows the time-history of the controlled states $X(t) = (x(t), y(t), z(t))$.

6 ADAPTIVE SYNCHRONIZATION OF TWO SYSTEMS WITH PARABOLIC EQUILIBRIUM

Here, we devise an adaptive controller so as to completely synchronize the state trajectories of a pair of systems with parabolic equilibrium which are called as *leader* and *follower* systems.

As the leader system, we take the new system with parabolic equilibrium defined by

$$\begin{cases} \dot{x}_1 = -z_1, \\ \dot{y}_1 = x_1 z_1^2 + a \ \mathrm{sgn}(z_1), \\ \dot{z}_1 = x_1 - y_1^2 + z_1 \left(b y_1^2 - z_1^2 \right), \end{cases} \tag{24}$$

where $X_1 = (x_1, y_1, z_1)$ is the state and a, b are unknown parameters.

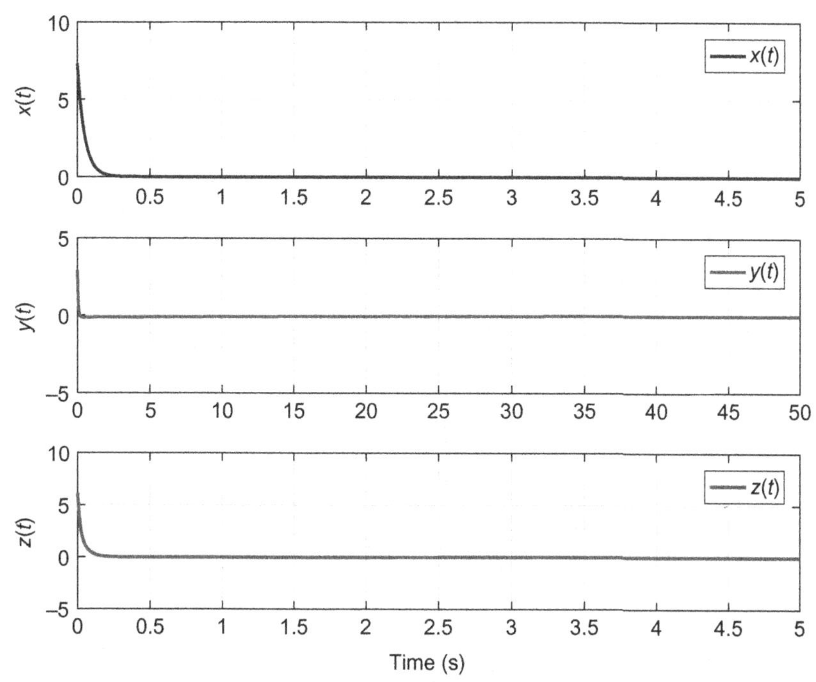

FIG. 9 Numerical simulation of the time-history of the controlled states of the system (15) for $(a, b) = (0.1, 3)$, $(A(0), B(0)) = (2.7, 4.8)$, and $X(0) = (7.3, 2.9, 6.1)$.

As the follower system, we take the controlled new system with parabolic equilibrium defined by

$$\begin{cases} \dot{x}_2 = -z_2 + u_x, \\ \dot{y}_2 = x_2 z_2^2 + a \ \ \mathrm{sgn}(z_2) + u_y, \\ \dot{z}_2 = x_2 - y_2^2 + z_2\left(b y_2^2 - z_2^2\right) + u_z, \end{cases} \tag{25}$$

where $X_2 = (x_2, y_2, z_2)$ is the state and $\mathbf{u} = (u_x, u_y, u_z)$ is the control.

Let $A(t)$ and $B(t)$ denote estimates of the unknown parameters a and b, respectively.

The complete synchronization error between the states X_1 and X_2 is defined by $\mathbf{e} = X_2 - X_1$ or

$$\begin{cases} e_x = x_2 - x_1, \\ e_y = y_2 - y_1, \\ e_z = z_2 - z_1. \end{cases} \tag{26}$$

It is easy to calculate the complete synchronization error dynamics as

$$\begin{cases} \dot{e}_x &= -e_z + u_x, \\ \dot{e}_y &= x_2 z_2^2 - x_1 z_1^2 + a[\mathrm{sgn}(z_2) - \mathrm{sgn}(z_1)] + u_y, \\ \dot{e}_z &= e_x - y_2^2 + y_1^2 + b(z_2 y_2^2 - z_1 y_1^2) - z_2^3 + z_1^3 + u_z. \end{cases} \tag{27}$$

As an adaptive control scheme for the main system (27), we take

$$\begin{cases} u_x &= e_z - \alpha_x e_x, \\ u_y &= -x_2 z_2^2 + x_1 z_1^2 - A(t)[\mathrm{sgn}(z_2) - \mathrm{sgn}(z_1)] - \alpha_y e_y, \\ u_z &= -e_x + y_2^2 - y_1^2 - B(t)(z_2 y_2^2 - z_1 y_1^2) + z_2^3 - z_1^3 - \alpha_z e_z, \end{cases} \tag{28}$$

where α_x, α_y, $\alpha_z \in$. $\mathbf{R}^+$ are constants.

Application of the scheme (28) yields the closed-loop error plant

$$\begin{cases} \dot{e}_x &= -\alpha_x e_x, \\ \dot{e}_y &= [a - A(t)][\mathrm{sgn}(z_2) - \mathrm{sgn}(z_1)] - \alpha_y e_y, \\ \dot{e}_z &= [b - B(t)](z_2 y_2^2 - z_1 y_1^2) - \alpha_z e_z. \end{cases} \tag{29}$$

We define the parameter estimation errors as $e_a = a - A(t)$ and $e_b = b - B(t)$. With this notation, we can simplify Eq. (29) as follows:

$$\begin{cases} \dot{e}_x &= -\alpha_x e_x, \\ \dot{e}_y &= e_a[\mathrm{sgn}(z_2) - \mathrm{sgn}(z_1)] - \alpha_y e_y, \\ \dot{e}_z &= e_b(z_2 y_2^2 - z_1 y_1^2) - \alpha_z e_z. \end{cases} \tag{30}$$

It is also easy to deduce that

$$\dot{e}_a = -\dot{A}, \quad \dot{e}_b = -\dot{B}. \tag{31}$$

As a candidate Lyapunov function, we take

$$V(\mathbf{e}, e_a, e_b) = \frac{1}{2}(e_x^2 + e_y^2 + e_z^2) + \frac{1}{2}(e_a^2 + e_b^2). \tag{32}$$

Differentiating V along Eqs. (30), (31), we get

$$\begin{aligned} \dot{V} = &-\alpha_x e_x^2 - \alpha_y e_y^2 - \alpha_z e_z^2 + e_a\left[e_y[\mathrm{sgn}(z_2) - \mathrm{sgn}(z_1)] - \dot{A}\right] \\ &+ e_b\left[e_z(z_2 y_2^2 - z_1 y_1^2) - \dot{B}\right]. \end{aligned} \tag{33}$$

Let us take the update law for parameter estimates $A(t)$ and $B(t)$ as follows:

$$\dot{A} = e_y[\mathrm{sgn}(z_2) - \mathrm{sgn}(z_1)] \quad \text{and} \quad \dot{B} = e_z\left(z_2 y_2^2 - z_1 y_1^2\right). \tag{34}$$

Theorem 2. *The new systems with parabolic equilibrium* (24), (25) *are globally and asymptotically synchronized for all values of $X_1(0)$, $X_2(0)$ $\in \mathbf{R}^3$ by the adaptive control scheme* (28) *and parameter update scheme* (34), *where $A(t)$, $B(t)$ are updates of the unknown parameters a, b, respectively, and α_x, α_y, α_z are positive constants.*

Proof. The candidate Lyapunov function V defined in Eq. (32) is positive definite on $\mathbf{R}^5$. Substitution of the update scheme (34) into Eq. (33) yields

$$\dot{V} = -\alpha_x e_x^2 - \alpha_y e_y^2 - \alpha_z e_z^2, \tag{35}$$

which exemplifies that $\dot{V}$ is a negative semidefinite function on $\mathbf{R}^3$.

Barbalat's lemma [62] shows that the closed-loop system (30) is globally asymptotically stable. This completes the proof of the theorem.

For numerical simulations, we take the parameters of the new systems (24), (25) as in the chaotic case (i.e., $a = 0.1$ and $b = 3$). We take $A(0) = 5.4$ and $B(0) = 8.3$. We take the gain constants as $\alpha_x = \alpha_y = \alpha_z = 20$.

Also, we take $X_1(0) = (0.6, 2.8, 3.1)$ and $X_2(0) = (2.4, 1.2, 6.5)$.

Figs. 10 and 11 illustrate the asymptotic synchronization of the states of the new systems with parametric equilibrium (24), (25).

7 DISCUSSION

In fact, we can use the general model (1) to find additional cases with parabolic equilibrium. In this section, we select nonlinear functions $f_1(x,y)$, $f_2(x,y,z)$ described by:

$$f_1(x,y) = x - y^2 + by + c, \tag{36}$$

$$f_2(x,y,z) = dy^2 - z^2, \tag{37}$$

where b, c, d are three positive parameters. As a result, system (1) is rewritten as

$$\begin{cases} \dot{x} = -z, \\ \dot{y} = xz^2 + a \;\; \mathrm{sgn}(z), \\ \dot{Z} = x - y^2 + by + c + z(dy^2 - z^2). \end{cases} \tag{38}$$

Obviously, system (38) has parabolic equilibrium $E((y^*)^2 - by^* - c, y^*, 0)$. In more detail we consider two elegant cases. In the first case, when $a = 0.25$, $b = 0.1$, $c = 0$, and $d = 3$, system (38) becomes

$$\begin{cases} \dot{x} = -z, \\ \dot{y} = xz^2 + 0.25 \;\; \mathrm{sgn}(z), \\ \dot{Z} = x - y^2 + 0.1y + z(3y^2 - z^2). \end{cases} \tag{39}$$

In the second case, when $a = 0.25$, $b = c = 0.04$, and $d = 3$, system (38) has the following form

$$\begin{cases} \dot{x} = -z, \\ \dot{y} = xz^2 + 0.25 \;\; \mathrm{sgn}(z), \\ \dot{Z} = x - y^2 + 0.04y + 0.04 + z(3y^2 - z^2). \end{cases} \tag{40}$$

Interestingly, the two new cases generate chaotic behavior as illustrated in Figs. 12 and 13. The Lyapunov exponents, Kaplan-Yorke dimensions, and initial conditions of such cases are summarized in Table 2.

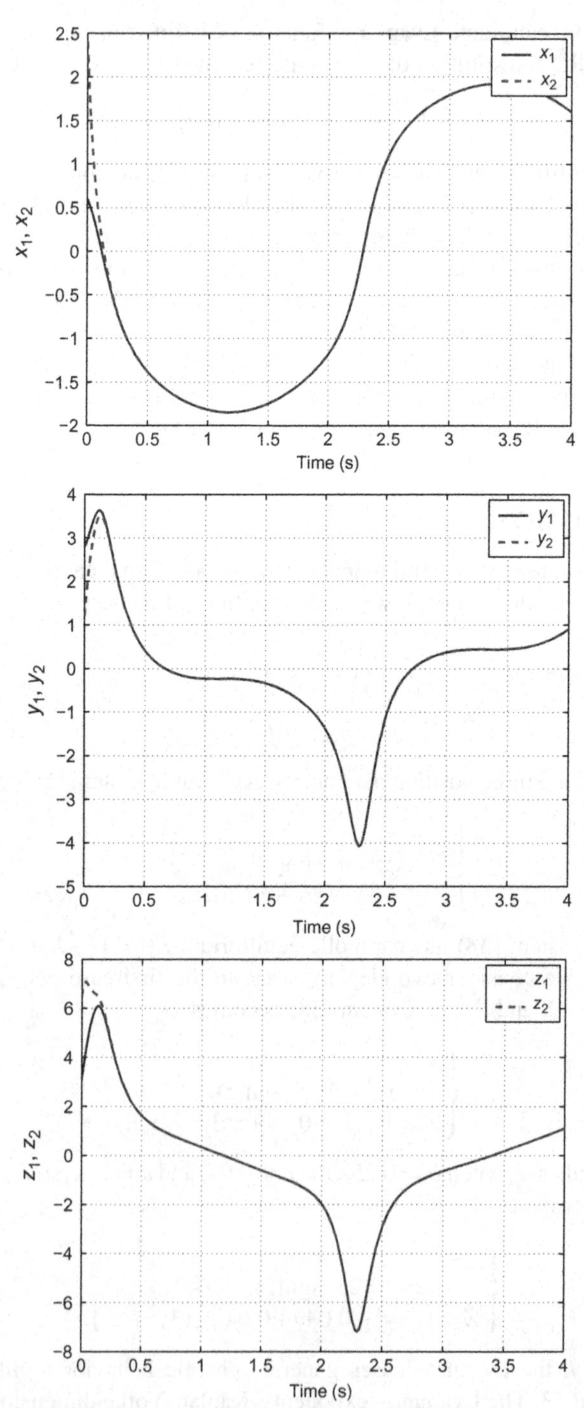

FIG. 10 Numerical simulation showing the complete synchronization of the new systems (24), (25): (A) x_1 and x_2, (B) y_1 and y_2, and (C) z_1 and z_2.

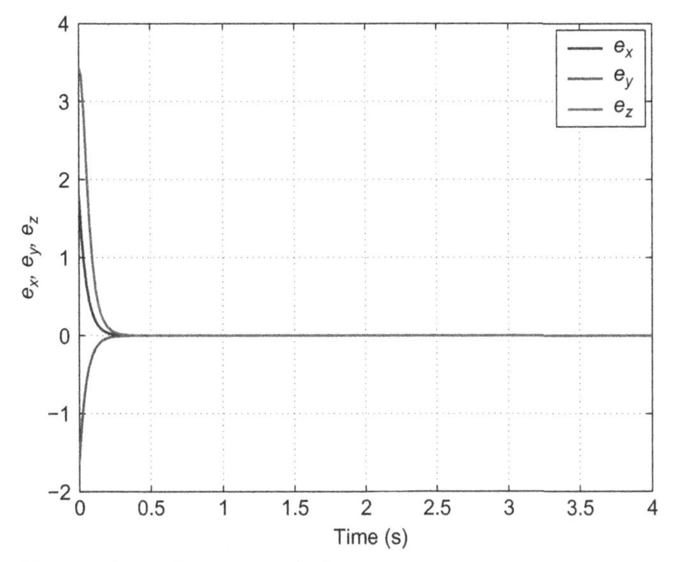

FIG. 11 The time-history of the synchronization errors.

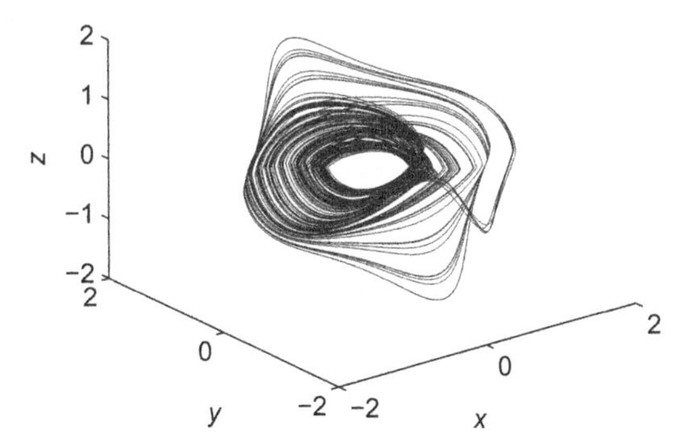

FIG. 12 The strange attractor of the system (39).

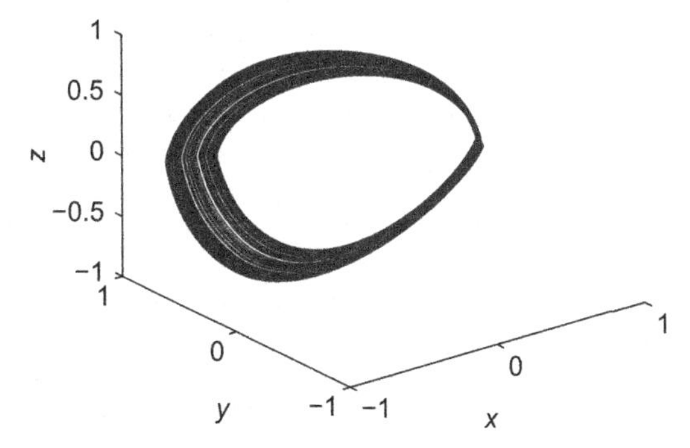

FIG. 13 The strange attractor of the system (40).

TABLE 2 Additional Chaotic Systems With Parabolic Equilibrium

System	LEs	D_{KY}	$(x(0), y(0), z(0))$
Eq. (39)	0.0768, 0, − 0.4624	2.1661	(0.1,0.1,0.1)
Eq. (40)	0.0631, 0, − 0.3624	2.1741	(0.1,0.1,0.1)

8 CONCLUSION

Motivated by recent researches about systems with an infinite number of equilibrium points, a new system with uncountable equilibria is proposed. The major difference from the published works is the presence of parabolic equilibrium. For this reason, our presented system has expanded the list of systems with different shapes of equilibrium points. It is noted that such a new system is able to display chaos, control, and synchronization. Moreover, we are capable of implementing the system by using electronic components, we, therefore, can use it in practical chaos-based applications. We will implement the real circuit of such a system and report the experimental results in our future works.

ACKNOWLEDGMENTS

The authors acknowledge Prof. GuanRong Chen, Department of Electronic Engineering, City University of Hong Kong for suggesting many helpful references. This work has been supported by the Polish National Science Center, MAESTRO Programme—Project No. 2013/08/A/ST8/00/780.

REFERENCES

[1] Lorenz EN. Deterministic nonperiodic flow. J Atmos Sci 1963;20:130–41.
[2] Pecora L, Carroll TL. Synchronization in chaotic systems. Phys Rev Lett 1990;64:821–4.
[3] Kapitaniak T. Chaos in systems with noise. Singapore: World Scientific; 1990.
[4] Strogatz SH. Nonlinear dynamics and chaos: with applications to physics, biology, chemistry, and engineering. Cambridge, MA: Perseus Books; 1990.
[5] Kapitaniak T. Controlling chaos. London: Academic Press; 1996.
[6] Banerjee S. Chaos synchronization and cryptography for secure communication. Hershey, PA: IGI Global; 2010.
[7] Sprott JC. Elegant chaos algebraically simple chaotic flows. Singapore: World Scientific; 2010.
[8] Morel C, Vlad R, Morel JY, Petreus D. Generating chaotic attractors on a surface. Math Comput Simul 2011;81:2549–63.
[9] Torres L, Besancon G, Georges D, Verde C. Exponential nonlinear observer for parametric identification and synchronization of chaotic systems. Math Comput Simul 2012;82:836–46.

[10] Das S, Acharya A, Pan I. Simulation studies on the design of optimum PID controllers to suppress chaotic oscillations in a family of Lorenz-like multi-wing attractors. Math Comput Simul 2014;100:72–87.

[11] Senouci A, Boukabou A. Predictive control and synchronization of chaotic and hyperchaotic systems based on a T-S fuzzy model. Math Comput Simul 2014;105:62–78.

[12] Soriano-Sanchez AG, Posadas-Castillo C, Platas-Garza MA, Diaz-Romero DA. Performance improvement of chaotic encryption via energy and frequency location criteria. Math Comput Simul 2015;112:14–27.

[13] Orlando G. A discrete mathematical model for chaotic dynamics in economics: Kaldor's model on business cycle. Math Comput Simul 2016;125:83–98.

[14] Lassoued A, Boubaker O. Dynamic analysis and circuit design of a novel hyperchaotic system with fractional-order terms. Complexity 2017;2017:3273408.

[15] Lassoued A, Boubaker O. On new chaotic and hyperchaotic systems: a literature survey. Nonlinear Anal Modell Control 2016;21:770–89.

[16] Rössler OE. An equation for continuous chaos. Phys Lett A 1976;57:397–8.

[17] Chen GR, Ueta T. Yet another chaotic attractor. Int J Bifurcation Chaos 1999;9:1465–6.

[18] Sprott JC. Some simple chaotic flows. Phys Rev E 1994;50:R647–50.

[19] Jafari S, Sprott JC. Simple chaotic flows with a line equilibrium. Chaos Solitons Fractals 2013;57:79–84.

[20] Leonov GA, Kuznetsov NV, Kuznetsova OA, Seldedzhi SM, Vagaitsev VI. Hidden oscillations in dynamical systems. Trans Syst Control 2011;6:54–67.

[21] Leonov GA, Kuznetsov NV, Vagaitsev VI. Localization of hidden Chua's attractors. Phys Lett A 2011;375:2230–3.

[22] Leonov GA, Kuznetsov NV, Vagaitsev VI. Hidden attractor in smooth Chua system. Physica D 2012;241:1482–6.

[23] Leonov GA, Kuznetsov NV. Hidden attractors in dynamical systems: from hidden oscillation in Hilbert-Kolmogorov, Aizerman and Kalman problems to hidden chaotic attractor in Chua circuits. Int J Bifurcation Chaos 2013;23:1330002.

[24] Wei Z. Dynamical behaviors of a chaotic system with no equilibria. Phys Lett A 2011;376:102–8.

[25] Jafari S, Sprott JC, Golpayegani SMRH. Elementary quadratic chaotic flows with no equilibria. Phys Lett A 2013;377:699–702.

[26] Wang X, Chen G. A chaotic system with only one stable equilibrium. Commun Nonlinear Sci Numer Simul 2012;17:1264–72.

[27] Wang X, Chen G. Constructing a chaotic system with any number of equilibria. Nonlinear Dyn 2013;71:429–36.

[28] Leonov GA, Kuznetsov NV, Kiseleva MA, Solovyeva EP, Zaretskiy AM. Hidden oscillations in mathematical model of drilling system actuated by induction motor with a wound rotor. Nonlinear Dyn 2014;77:277–88.

[29] Chen Y, Yang Q. A new Lorenz-type hyperchaotic system with a curve of equilibria. Math Comput Simul 2015;112:40–55.

[30] Wei Z, Wang R, Liu A. A new finding of the existence of hidden hyperchaotic attractor with no equilibria. Math Comput Simul 2014;100:13–23.

[31] Leonov GA, Kuznetsov NV, Mokaev TN. Hidden attractor and homoclinic orbit in Lorenz-like system describing convective fluid motion in rotating cavity. Commun Nonlinear Sci Numer Simul 2015;28:166–74.

[32] Zhusubaliyev ZT, Mosekilde E. Multistability and hidden attractors in a multilevel DC/DC converter. Math Comput Simul 2015;109:32–45.

[33] Chudzid A, Perlikowski P, Stefanski A, Kapitaniak T. Multistability and rare attractors in Van der Pol-duffing oscillator. Int J Bifurcation Chaos 2011;21:1907–12.
[34] Kuzma P, Kapitaniak M, Kapitaniak T. Coupling multistable systems: uncertainty due to the initial positions on the attractors. J Theor Appl Mech 2014;52:281–4.
[35] Brezetskyi S, Dudkowski D, Kapitaniak T. Rare and hidden attractors in Van der Pol-duffing oscillators. Eur Phys J Special Topics 2015;224:1459–67.
[36] Li C, Sprott JC. Chaotic flows with a single nonquadratic term. Phys Lett A 2014;378:178–83.
[37] Li C, Sprott JC, Yuan Z, Li H. Constructing chaotic systems with total amplitude control. Int J Bifurcation Chaos 2015;25:1530025.
[38] Gotthans T, Petržela J. New class of chaotic systems with circular equilibrium. Nonlinear Dyn 2015;73:429–36.
[39] Gotthans T, Sportt JC, Petržela J. Simple chaotic flow with circle and square equilibrium. Int J Bifurcation Chaos 2016;26:1650137.
[40] Pham V-T, Volos C, Kapitaniak T, Jafari S, Wang X. Dynamics and circuit of a chaotic system with a curve of equilibrium points. Int J Electron 2018;105:385–97.
[41] Pham V-T, Volos C, Kingni ST, Kapitaniak T, Jafari S. Bistable hidden attractors in a novel chaotic system with hyperbolic sine equilibrium. Circuits Syst Signal Process 2018;37:1028–43.
[42] Tlelo-Cuautle E, de la Fraga LG, Pham V-T, Volos C, Jafari S, de Jesus Quintas-Valles A. Dynamics, FPGA realization and application of a chaotic system with an infinite number of equilibrium points. Nonlinear Dyn 2017;89:1129–39.
[43] Pham V-T, Volos C, Jafari S, Kapitaniak T. A novel cubic-equilibrium chaotic system with coexisting hidden attractors: analysis, and circuit implementation. J Circuits Syst Comp 2018;27:1850066.
[44] Yalcin ME, Ozoguz S. *n*-scroll chaotic attractors from a first-order time-delay differential equation. Chaos 2007;17:033112.
[45] Piper JR, Sprott JC. Simple autonomous chaotic circuits. IEEE Trans Circuits Syst Express Briefs 2010;57:730–4.
[46] Li C, Sprott JC, Thio W, Zhu H. A new piecewise linear hyperchaotic circuit. IEEE Trans Circuits Syst Express Briefs 2014;61:977–81.
[47] Li C, Sprott JC, Thio W. Linearization of the Lorenz system. Phys Lett A 2015;379:888–93.
[48] Wolf A, Swift JB, Swinney HL, Vastano JA. Determining Lyapunov exponents from a time series. Physica D 1985;16:285–317.
[49] Kuznetsov NV. The Lyapunov dimension and its estimation via the Leonov method. Phys Lett A 2016;380:2142–9.
[50] Kuznetsov NV, Alexeeva TA, Leonov GA. Invariance of Lyapunov exponents and Lyapunov dimension for regular and irregular linearizations. Nonlinear Dyn 2016;85:195–201.
[51] Leonov GA, Kuznetsov NV, Korzhemanova NA, Kusakin DV. Lyapunov dimension formula for the global attractor of the Lorenz system. Commun Nonlinear Sci Numer Simul 2016;41:84–103.
[52] Banerjee S, Mitra M, Rondoni L. Applications of chaos and nonlinear dynamics in engineering. vol. 1. Berlin: Springer; 2011.
[53] Volos CK, Kyprianidis IM, Stouboulos IN. Image encryption process based on chaotic synchronization phenomena. Signal Process 2013;93:1328–40.
[54] Cicek S, Ferikoglu A, Pehlivan I. A new 3D chaotic system: dynamical analysis, electronic circuit design, active control synchronization and chaotic masking communication application. Optik 2016;127:4024–30.

[55] Volos CK, Kyprianidis IM, Stouboulos IN. A chaotic path planning generator for autonomous mobile robots. Robot Auton Syst 2012;60:651–6.

[56] Yalcin ME, Suykens JAK, Vandewalle J. True random bit generation from a double-scroll attractor. IEEE Trans Circuits Syst I, Reg Papers 2004;51:1395–404.

[57] Akgul A, Moroz I, Pehlivan I, Vaidyanathan S. A new four-scroll chaotic attractor and its engineering applications. Optik 2016;127:5491–9.

[58] Wang X, Akgul A, Kacar S, Pham V-T. Multimedia security application of a ten-term chaotic system without equilibrium. Complexity 2017;2017:8412093.

[59] Sprott JC. Simple chaotic systems and circuits. Am J Phys 2000;68:758–63.

[60] Bouali S, Buscarino A, Fortuna L, Frasca M, Gambuzza LV. Emulating complex business cycles by using an electronic analogue. Nonlinear Anal Real World Appl 2012;13:2459–65.

[61] Li C, Pehlivan I, Sprott JC, Akgul A. A novel four-wing strange attractor born in bistability. IEICE Electron Lett 2005;12:1–12.

[62] Khalil HK. Nonlinear systems. Upper Saddle River, NJ: Prentice Hall; 2002.

Chapter 4

A New Four-Dimensional Chaotic System With No Equilibrium Point

Shirin Panahi*, Viet-Thanh Pham†, Karthikeyan Rajagopal‡, Olfa Boubaker§ and Sajad Jafari*
**Biomedical Engineering Department, Amirkabir University of Technology, Tehran, Iran, †School of Electronics and Telecommunications, Hanoi University of Science and Technology, Hanoi, Vietnam, ‡Center for Nonlinear Dynamics, Defence University, Bishoftu, Ethiopia, §National Institute of Applied Sciences and Technology, Tunis, Tunisia*

1 INTRODUCTION

The word "chaos," which means a state of disorder and confusion, entered into science atmosphere, when Lorenz observed some unusual features in usual differential equations during his investigations on weather dynamics [1]. Due to the strangeness and attractiveness of this subject, many scientists choose this field as their research topic and explored the different aspects of chaotic systems [2–4]. Until then, the general thought about the generation of chaos in mathematics centered around the equilibrium points which could help the scientists to expound this mystery [5,6]. They believed unstable equilibria might be the reason of the generation of chaotic attractors. It seems that a point near the unstable fixed point on unstable manifold is needed to forward the trajectory to the attractors. Therefore, they named these attractors self-excited when the equilibria of the chaotic system were involved in forming the strange attractors. But all of these hypotheses were rejected by the discovery of chaotic systems without fixed points. The strange attractors of these systems are called hidden [7–12]. The mathematical procedure of finding these attractors is hard because the basin of attraction for a hidden attractor is not connected with any equilibrium [13–15]. On the other hand, it seems they can be useful in engineering applications [16,17]. One of the main branches in chaos theory is to build up the paradigm of the design of chaotic electronic circuits [18–25] in order to pick up the performance of strange attractors of chaotic systems for application purposes.

Recent Advances in Chaotic Systems and Synchronization. https://doi.org/10.1016/B978-0-12-815838-8.00004-2

Fractional order dynamic systems are the other method to improve the mathematical models for some actual physical and engineering systems.

In this paper, a novel, no-equilibrium [26–34] chaotic system with quadratic nonlinearity is proposed. Such systems are in close relation with other rare chaotic systems which have been introduced recently [35,36]. From them we can point out systems with surfaces of equilibria [37–40], curves of equilibria [41–46], stable equilibria [47–49], nonhyperbolic equilibria [50], infinite number of equilibria [51–54], multiscroll attractors [55,56], different families of hidden attractors [57–63], multistability [64–67], megastability [68], extreme multistability [69–73], and some other specific properties [74–78].

The paper is structured as follows: Section 2 is the introduction of the new proposed system and describes its dynamical properties. Section 3 is designing the fractional order differential equation of the proposed system. In Section 4, we present an electronic design for the system. Finally, Section 5 is the conclusion.

2 NEW NO EQUILIBRIUM CHAOTIC SYSTEM

The new 4D chaotic flow with no equilibria is described as follows:

$$\begin{aligned} \dot{x} &= y \\ \dot{y} &= z \\ \dot{z} &= w \\ \dot{w} &= a(1)w + a(2)x^2 + a(3)y^2 + a(4)xy + a(5)xz + a(6) \end{aligned} \tag{1}$$

The model parameters are available in Table 1.

Eq. (1) has no equilibria which can be investigated mathematically as follows:

$$\begin{aligned} &\dot{x} = 0 \to y = 0 \\ &\dot{y} = 0 \to z = 0 \\ &\dot{z} = 0 \to w = 0 \\ &\dot{w} = 0 \to -1.02w + 1.64x^2 - 1.36y^2 + 0.28xy + 2.42xz + 1.45 \\ &\quad = 0 \to 1.64x^2 + 1.45 = 0 \end{aligned} \tag{2}$$

Fig. 1 is plotted to take a look at the strange behavior of the system which is the different projection of the strange attractors.

TABLE 1 The Model Parameter

Parameter	Value	Parameter	Value
a(1)	−1.02	a(4)	0.28
a(2)	1.64	a(5)	2.42
a(3)	−1.36	a(6)	1.45

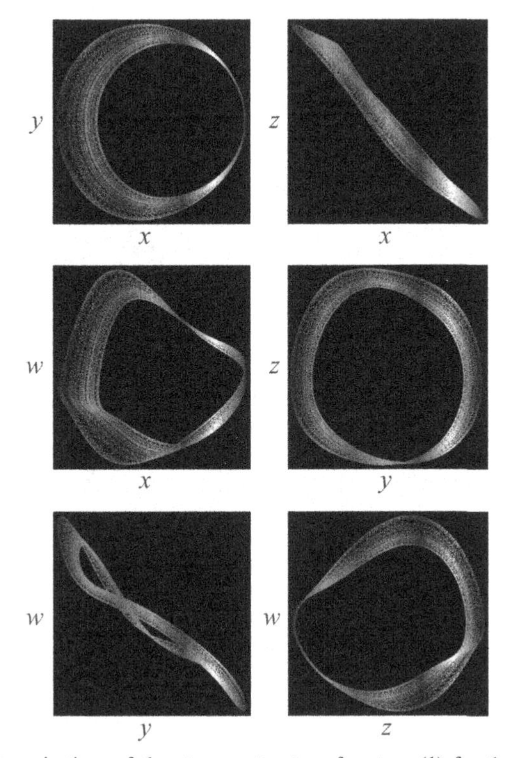

FIG. 1 Different projections of the strange attractor of system (1) for the initial conditions $(-2.77,-0.53,2.7,-0.34)$.

Dynamical behavior of the proposed system with respect to variations of parameters $a(2)$ and $a(3)$ are presented as follows: Fig. 2A shows the bifurcation diagram, and Fig. 2B exposes the Lyapunov exponents of System (1) with respect to variations of parameter $a(2)$. Fig. 3 is plotted for parameter $a(3)$. A bifurcation diagram with respect to the other parameters is shown in Fig. 4.

3 FRACTIONAL ORDER HYPERJERK SYSTEM (FOHJS)

Fractional order calculus and its applications have been of greater interest in the recent years, and fractional order controls with different approaches have been achieved in Refs. [79–81]. Various fractional order systems which can show chaotic and hyperchaotic oscillations are proposed [82–87]. A novel fractional order no equilibrium chaotic system is investigated in Ref. [88], and a fractional order hyperchaotic system without equilibrium points is investigated in Ref. [89]. Memristor-based fractional order system with a capacitor and an inductor is discussed in Ref. [90]. Numerical analysis and methods for simulating fractional order nonlinear system is proposed in Ref. [91], and Matlab solutions for fractional order chaotic systems is discussed in Ref. [92].

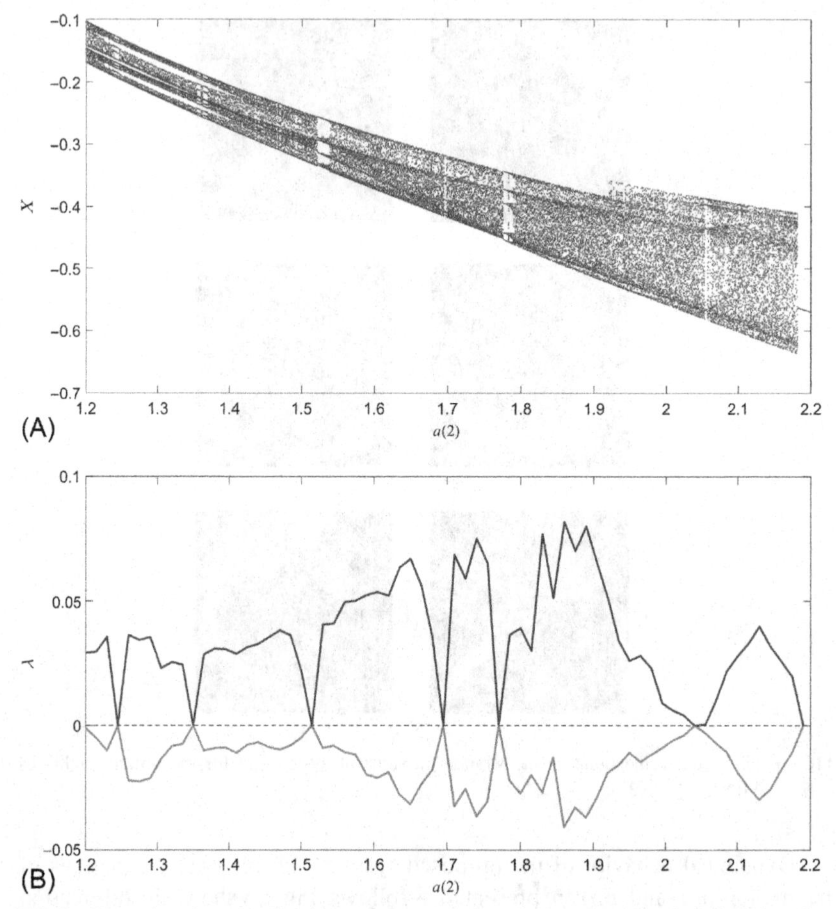

FIG. 2 Dynamics of system (1) with respect to variation of parameter $a(2)$: (A) bifurcation diagram and (B) Lyapunov exponents.

The fractional order hyperjerk system (FOHJS) model can be derived from the integer order model as,

$$\begin{aligned}
\frac{d^{q_x}x}{dt^{q_x}} &= y \\
\frac{d^{q_y}y}{dt^{q_y}} &= z \\
\frac{d^{q_z}z}{dt^{q_z}} &= w \\
\frac{d^{q_w}w}{dt^{q_w}} &= a_1w + a_2x^2 + a_3y^2 + a_4xy + a_5xz + a_6
\end{aligned} \tag{3}$$

The system shows chaotic oscillations for the parameters

$a_1 = -1.02;\ \ a_2 = 1.64;\ \ a_3 = -1.36;\ \ a_4 = 0.28;\ \ a_5 = 2.42;\ \ a_6 = 1.45$

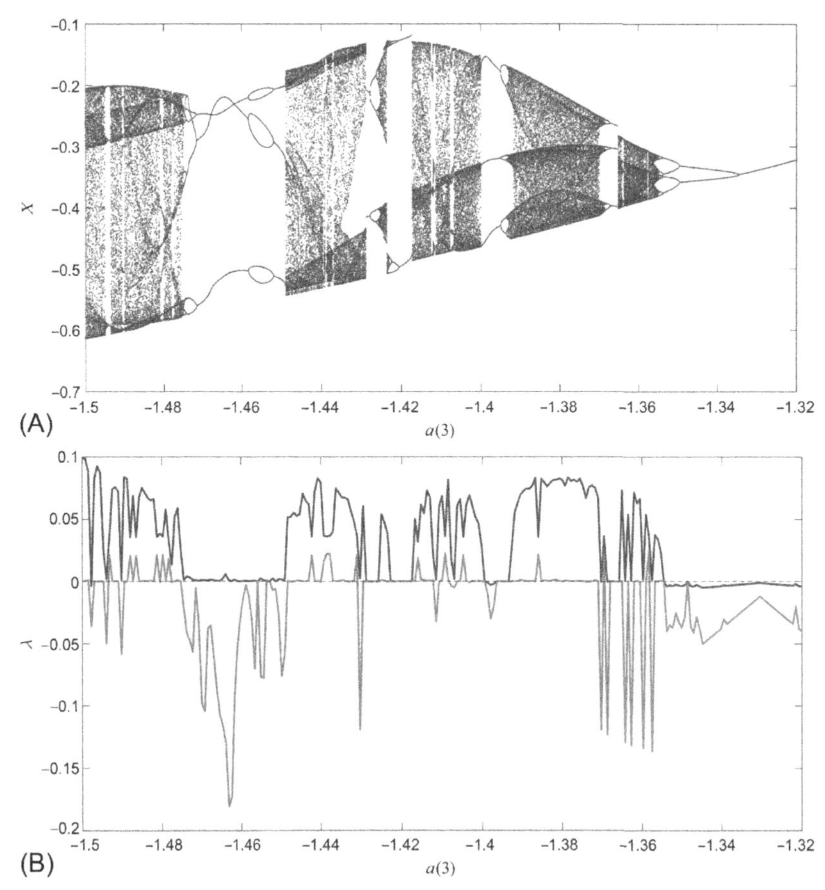

FIG. 3 Dynamics of system (1) with respect to variation of parameter $a(3)$: (A) bifurcation diagram and (B) Lyapunov exponents.

We use the ADM method [93,94] to numerically solve the FOHJS system. The fractional order discrete form of the dimensionless state equations for the FOHJS system can be given as,

$$
\begin{aligned}
x_{n+1} &= \sum_{j=0}^{6} A_1^j \frac{h^{jq}}{\Gamma(jq+1)} \\
y_{n+1} &= \sum_{j=0}^{6} A_2^j \frac{h^{jq}}{\Gamma(jq+1)} \\
z_{n+1} &= \sum_{j=0}^{6} A_3^j \frac{h^{jq}}{\Gamma(jq+1)} \\
w_{n+1} &= \sum_{j=0}^{6} A_4^j \frac{h^{jq}}{\Gamma(jq+1)}
\end{aligned}
\tag{4}
$$

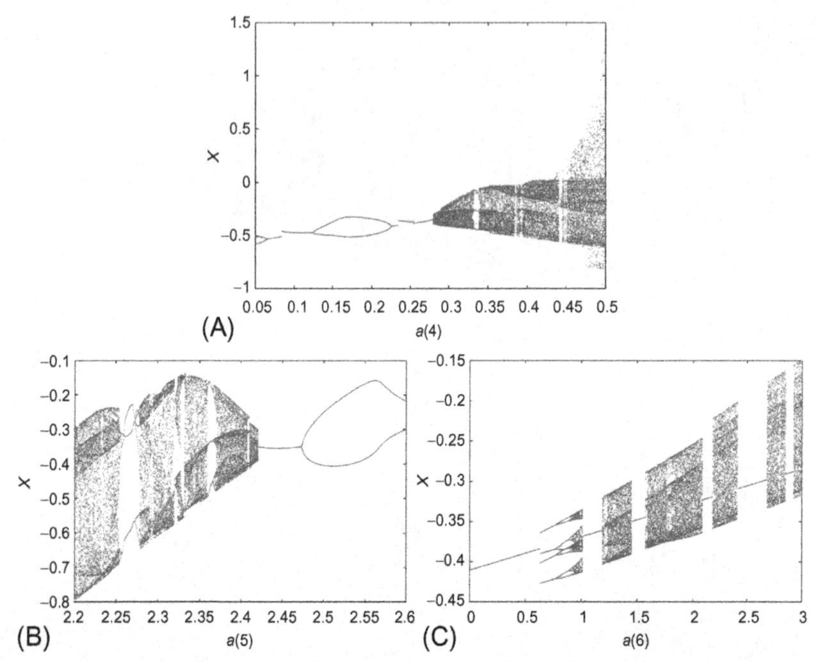

FIG. 4 Bifurcation diagram with respect to variations of different parameters of system (A) $a(4)$, (B)$a(5)$, and (C) $a(6)$.

where A_i^j are the Adomian polynomials with $i = 1, 2, 3, 4$ and $A_1^0 = x_n$, $A_2^0 = y_n$, $A_3^0 = z_n$, $A_4^0 = w_n$. As can be seen from the FOHJS (3), there are four nonlinear terms x^2, y^2, xz and xy. The Adomian polynomials for the linear terms can be calculated as$A_1^{j+1} = A_2^j$; $A_2^{j+1} = A_3^j$; $A_3^{j+1} = A_4^j$, and the first six Adomian polynomials for the nonlinear terms can be derived as in Table 2.

where $h = t_{n+1} - t_n$ and $\Gamma(\bullet)$ is the gamma function. Similarly the other two nonlinear terms y^2, xz can be calculated. The 2D phase portraits of the system are given in Fig. 5.

3.1 Bifurcation

To investigate the impact of fractional orders on the FOHJS we derive the bifurcation plots. First, we consider the system to be a commensurate FOHJS and fix the order $q = 0.995$ and vary the parameter a_1 of the system from -1.2 to-0.7. The initial conditions for the first iteration are taken as $[-2.77; -0.53; 2.7; -0.34]$ and is reinitialized to the last values of the state variables at every iteration, the local maximum values of the state variables are then plotted. Fig. 6 shows the bifurcation of the FOHJS with respect to the parameter a_1. As can be observed from Fig. 6, the FOHJS takes period doubling route to chaotic region. We can see regions of periodic limit cycles like period 1

TABLE 2 Adomian Polynomials of the Nonlinear Terms in FOHJS

Nonlinear Term	Adomian Polynomials
x^2	$A_1^1=(A_1^0)^2$ $A_1^2=2A_1^0A_1^1$ $A_1^3=\left[2A_1^0A_1^2+(A_1^1)^2\right]\frac{\Gamma(2q+1)}{\Gamma^2(q+1)}$ $A_1^4=\left[2A_1^0A_1^3+2A_1^1A_1^2\right]\frac{\Gamma(3q+1)}{\Gamma(q+1)\Gamma(2q+1)}$ $A_1^5=\left[(A_1^2)^2+2A_1^1A_1^3+2A_1^0A_1^4\right]\frac{\Gamma(4q+1)}{\Gamma(q+1)\Gamma(3q+1)}$ $A_1^6=\left[2A_1^2A_1^3+2A_1^0A_1^5+2A_1^2A_1^4+(A_1^3)^3\right]\frac{\Gamma(5q+1)}{\Gamma(q+1)\Gamma(4q+1)}$
xy	$A_1^1=A_1^0A_2^0$ $A_1^2=A_1^1A_2^0+A_1^0A_2^1$ $A_1^3=\left[A_1^2A_2^0+A_1^0A_2^2+A_1^1A_2^1\right]\frac{\Gamma(2q+1)}{\Gamma^2(q+1)}$ $A_1^4=\left[A_1^3A_2^0+A_1^0A_2^3+A_1^1A_2^2+A_1^2A_2^1\right]\frac{\Gamma(3q+1)}{\Gamma(q+1)\Gamma(2q+1)}$ $A_1^5=\left[A_1^4A_2^0+A_1^0A_2^4+A_1^3A_2^1+A_1^1A_2^3+A_1^2A_2^2\right]\frac{\Gamma(4q+1)}{\Gamma(q+1)\Gamma(3q+1)}$ $A_1^6=\left[A_1^5A_2^0+A_1^0A_2^5+A_1^3A_2^2+A_1^2A_2^3+A_1^4A_2^1+A_1^1A_2^4\right]\frac{\Gamma(5q+1)}{\Gamma(q+1)\Gamma(4q+1)}$

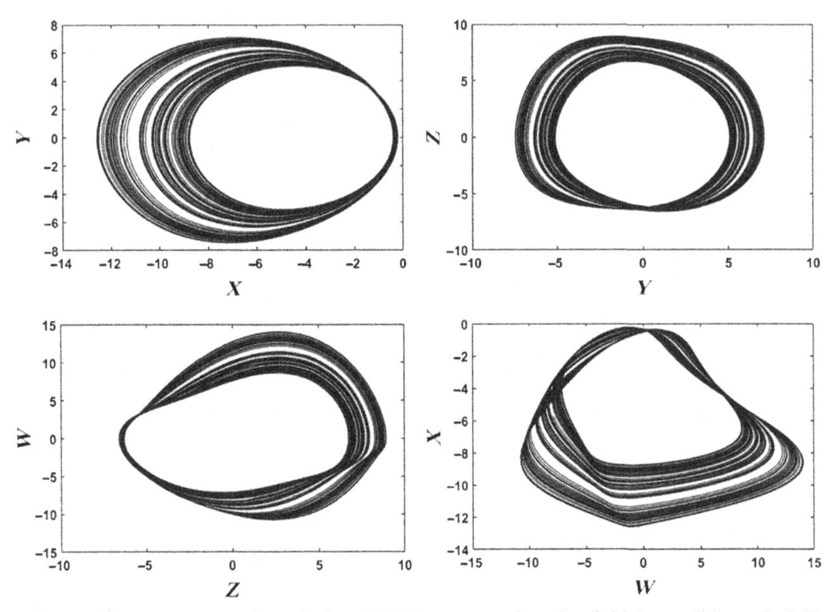

FIG. 5 2D phase portraits of the FOHJS system for the initial conditions $[-2.77; -0.53; 2.7; -0.34]$ and commensurate fractional order $q=0.995$ using the ADM discretization method.

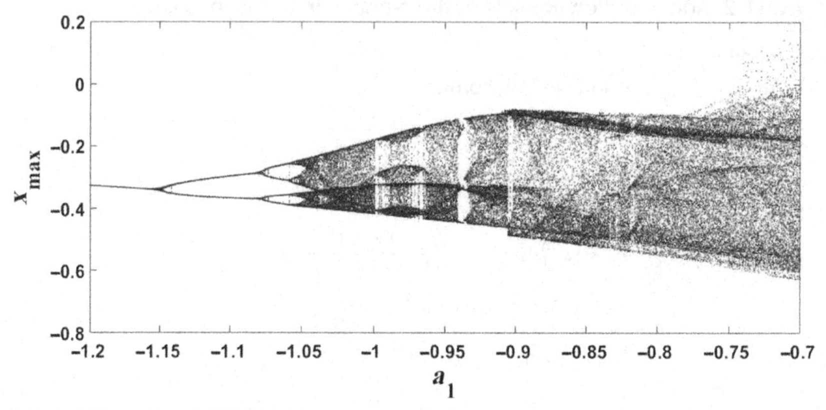

FIG. 6 Bifurcation of FOHJS with respect to a_1 for commensurate fractional order.

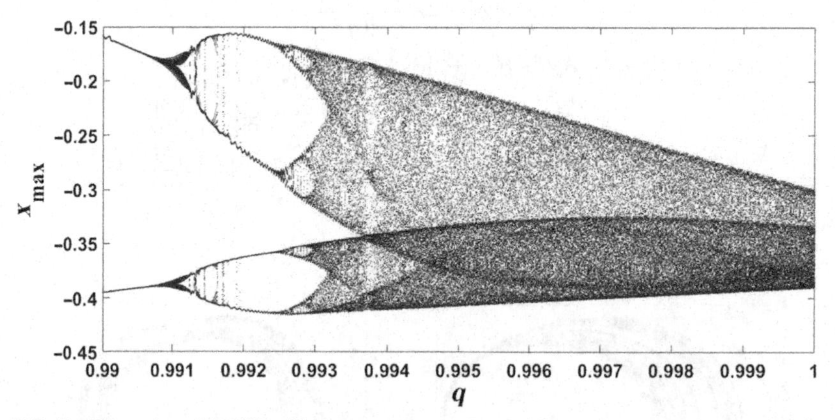

FIG. 7 Bifurcation of FOHJS with respect to the parameter q for commensurate fractional order with reinitializing of the initial conditions.

oscillations from -1.2 to -1.153, period 2 oscillations from -1.153 to -1.078 and period 4 oscillations for -1.078 to -1.054.

By the similar method of bifurcation, we derive the plots for fractional order q as shown in Fig. 7. The FOHJS take period doubling route to chaos. We can see regions of periodic limit cycles like period 1 oscillations from 0.99 to 0.991, period 2 oscillations from 0.991 to 0.9925 and period 4 oscillations for 0.9925 to 0.9929.

4 CIRCUIT DESIGN

Circuit implementation of chaotic systems has been of much interest [50,76,95–99]. This section presents a circuit based on the theoretical 4D system. By applying the operational-amplifier approach [100–102], we have designed the circuit as presented in Fig. 8, in which X, Y, Z, W are the output

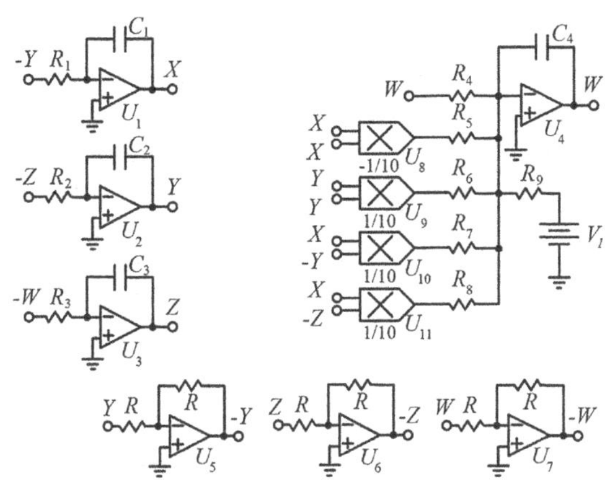

FIG. 8 Designed circuit for the 4D system. It is noted that the power suppliers for the circuit are $\pm 16V_{DC}$.

voltages of four operational amplifiers (U_1, U_2, U_3, U_4). Therefore, the circuit's equation is given by

$$\begin{cases} \dot{X} = \dfrac{1}{R_1C_1}Y \\ \dot{Y} = \dfrac{1}{R_2C_2}Z \\ \dot{Z} = \dfrac{1}{R_3C_3}W \\ \dot{W} = -\dfrac{1}{R_4C_4}W + \dfrac{1}{10R_5C_4}X^2 - \dfrac{1}{10R_6C_4}Y^2 + \dfrac{1}{10R_7C_4}XY + \dfrac{1}{10R_8C_4}XZ - \dfrac{1}{R_9C_4}V_1 \end{cases} \tag{5}$$

By comparing circuital Eq. (5) and the 4D system, it is easy to confirm that the voltages (X,Y,Z,W) correspond to the state variables (x,y,z,w). As a result, we can select electronic components in Fig. 8 as follows: $R_1 = R_2 = R_3 = R_9 = R = 100$ kΩ, $R_4 = 98.039$ kΩ, $R_5 = 6.098$ kΩ, $R_6 = 7.353$ kΩ, $R_7 = 35.714$ kΩ, $R_8 = 4.132$ kΩ, $C = 2.2$ nF, and $V_1 = -1.45V_{DC}$. PSpice phase portraits in Fig. 9 verify the chaotic behavior of the circuit.

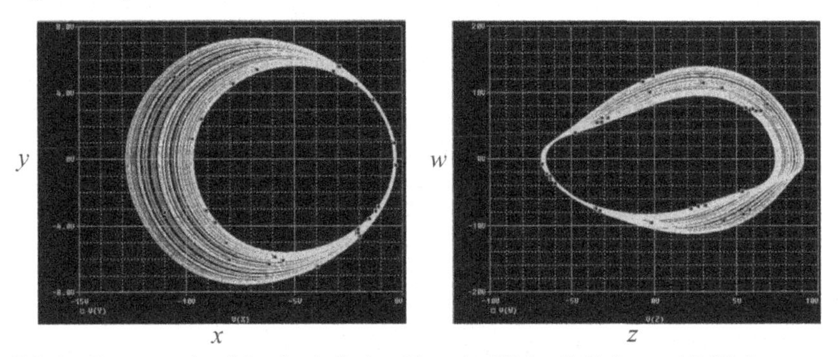

FIG. 9 Phase portraits of the circuit displayed by using PSpice: X–Y plane and Z–W plane.

5 CONCLUSION

A new 4D hyperjerk chaotic system is proposed which has no equilibria and belongs then to the class of chaotic systems with hidden attractors. Dynamical features of this new system such as state-space portraits, bifurcation diagrams, and Lyapunov exponents are investigated. A fractional model of this system is also introduced. A related electronic circuit is finally designed.

REFERENCES

[1] Lorenz EN. Deterministic nonperiodic flow. J Atmos Sci 1963;20(2):130–41.
[2] Chen G, Ueta T. Yet another chaotic attractor. Int J Bifurcat Chaos 1999;9(07):1465–6.
[3] Sprott JC. Some simple chaotic flows. Phys Rev E 1994;50(2):R647.
[4] El Naschie M, Kapitaniak T. Soliton chaos models for mechanical and biological elastic chains. Phys Lett A 1990;147(5–6):275–81.
[5] Shil'nikov LP. Methods of qualitative theory in nonlinear dynamics. vol. 5. World Scientific; 2001.
[6] Lassoued A, Boubaker O. On new chaotic and hyperchaotic systems: a literature survey. Nonlinear Anal: Model Control 2016;21(6):770–89.
[7] Lao S-K, et al. Cost function based on gaussian mixture model for parameter estimation of a chaotic circuit with a hidden attractor. Int J Bifurcat Chaos 2014;24(01):1450010.
[8] Pham V-T, et al. Hidden hyperchaotic attractor in a novel simple memristive neural network. Optoelectron Adv Mater—Rapid Commun 2014;8(11 – 12):1157–63.
[9] Pham V-T, et al. Hidden attractors in a chaotic system with an exponential nonlinear term. Eur Phys J Spec Top 2015;224(8):1507–17.
[10] Shahzad M, et al. Synchronization and circuit design of a chaotic system with coexisting hidden attractors. Eur Phys J Spec Top 2015;224(8):1637–52.
[11] Dudkowski D, et al. Hidden attractors in dynamical systems. Phys Rep 2016;637:1–50.
[12] Jafari S, et al. The relationship between chaotic maps and some chaotic systems with hidden attractors. Int J Bifurcat Chaos 2016;26(13):1650211.
[13] Jafari S, et al. Limitation of perpetual points for confirming conservation in dynamical systems. Int J Bifurcat Chaos 2015;25(13):1550182.
[14] Nazarimehr F, et al. Categorizing chaotic flows from the viewpoint of fixed points and perpetual points. Int J Bifurcat Chaos 2017;27(02):1750023.
[15] Nazarimehr F, et al. Are perpetual points sufficient for locating hidden attractors? Int J Bifurcat Chaos 2017;28(3):1750037.
[16] Jafari S, Sprott JC, Nazarimehr F. Recent new examples of hidden attractors. Eur Phys J Spec Top 2015;224(8):1469–76.
[17] Panahi S, et al. Parameter identification of a chaotic circuit with a hidden attractor using krill herd optimization. Int J Bifurcat Chaos 2016;26(13):1650221.
[18] Pham V-T, et al. A novel memristive neural network with hidden attractors and its circuitry implementation. Sci China Technol Sci 2016;59(3):358–63.
[19] Pham V-T, et al. A chaotic system with equilibria located on the rounded square loop and its circuit implementation. IEEE Trans Circ Syst II: Express Briefs 2016;63(9):878–82.
[20] Pham V-T, et al. A chaotic hyperjerk system based on memristive device. In: Advances and applications in chaotic systems. Springer International Publishing; 2016. p. 39–58.
[21] Kengne J, et al. Dynamic analysis and electronic circuit implementation of a novel 3D autonomous system without linear terms. Commun Nonlinear Sci Numer Simul 2017;52:62–76.

[22] Pham V-T, et al. Dynamics and circuit realization of a no-equilibrium chaotic system with a boostable variable. AEU Int J Electron Commun 2017;78:134–40.

[23] Pham V-T, Jafari S, Volos C. A novel chaotic system with heart-shaped equilibrium and its circuital implementation. Optik 2017;131:343–9.

[24] Pham V-T, et al. A simple three-dimensional fractional-order chaotic system without equilibrium: dynamics, circuitry implementation, chaos control and synchronization. AEU Int J Electron Commun 2017;78:220–7.

[25] Pham V-T, et al. A novel cubic-equilibrium chaotic system with coexisting hidden attractors: analysis, and circuit implementation. J Circ Syst Comput 2017;1850066.

[26] Jafari S, Sprott JC, Hashemi Golpayegani SMR. Elementary quadratic chaotic flows with no equilibria. Phys Lett A 2013;377(9):699–702.

[27] Jafari S, et al. A new cost function for parameter estimation of chaotic systems using return maps as fingerprints. Int J Bifurcat Chaos 2014;24(10):1450134.

[28] Pham V-T, et al. Is that really hidden? The presence of complex fixed-points in chaotic flows with no equilibria. Int J Bifurcat Chaos 2014;24(11):1450146.

[29] Pham V-T, et al. Generating a novel hyperchaotic system out of equilibrium. Optoelectron Adv Mater-Rapid Commun 2014;8(5–6):535–9.

[30] Pham V-T, et al. Constructing a novel no-equilibrium chaotic system. Int J Bifurcat Chaos 2014;24(05):1450073.

[31] Pham V-T, et al. A no-equilibrium hyperchaotic system with a cubic nonlinear term. Optik 2016;127(6):3259–65.

[32] Pham V-T, et al. A novel memristive time-delay chaotic system without equilibrium points. Eur Phys J Spec Top 2016;225(1):127–36.

[33] Pham V-T, et al. Coexistence of hidden chaotic attractors in a novel no-equilibrium system. Nonlinear Dyn 2017;87(3):2001–10.

[34] Ren S, et al. A new chaotic flow with hidden attractor: the first hyperjerk system with no equilibrium. Zeitsch Naturforsch A 2018;73(3):239–49.

[35] Sprott JC, et al. A chaotic system with a single unstable node. Phys Lett A 2015;379 (36):2030–6.

[36] Goudarzi S, et al. NARX prediction of some rare chaotic flows: recurrent fuzzy functions approach. Phys Lett A 2016;380(5):696–706.

[37] Jafari S, Sprott JC, Molaie M. A simple chaotic flow with a plane of equilibria. Int J Bifurcat Chaos 2016;26(06):1650098.

[38] Jafari S, et al. Simple chaotic 3D flows with surfaces of equilibria. Nonlinear Dyn 2016;86 (2):1349–58.

[39] Panahi S, et al. A new transiently chaotic flow with ellipsoid equilibria. Pramana 2018;90 (3):31.

[40] Singh JP, Roy BK, Jafari S. New family of 4-D hyperchaotic and chaotic systems with quadric surfaces of equilibria. Chaos Solitons Fractals 2018;106:243–57.

[41] Jafari S, Sprott JC. Simple chaotic flows with a line equilibrium. Chaos Solitons Fractals 2013;57:79–84.

[42] Jafari S, Sprott JC. Erratum to: "simple chaotic flows with a line equilibrium" [Chaos, Solitons and Fractals 57 (2013) 79–84]. Chaos Solitons Fractals 2015;77:341–2.

[43] Barati K, et al. Simple chaotic flows with a curve of equilibria. Int J Bifurcat Chaos 2016;26 (12):1630034.

[44] Kingni ST, et al. Three-dimensional chaotic autonomous system with a circular equilibrium: analysis, circuit implementation and its fractional-order form. Circ Syst Signal Process 2016;35(6):1933–48.

[45] Pham V-T, et al. A chaotic system with infinite equilibria located on a piecewise linear curve. Optik 2016;127(20):9111–7.
[46] Pham V-T, et al. A chaotic system with different shapes of equilibria. Int J Bifurcat Chaos 2016;26(04):1650069.
[47] Molaie M, et al. Simple chaotic flows with one stable equilibrium. Int J Bifurcat Chaos 2013;23(11):1350188.
[48] Kingni ST, et al. Three-dimensional chaotic autonomous system with only one stable equilibrium: analysis, circuit design, parameter estimation, control, synchronization and its fractional-order form. Eur Phys J Plus 2014;129(5):1–16.
[49] Pham V-T, et al. Generating a chaotic system with one stable equilibrium. Int J Bifurcat Chaos 2017;27(04):1750053.
[50] Rajagopal K, et al. A chaotic jerk system with non-hyperbolic equilibrium: dynamics, effect of time delay and circuit realisation. Pramana 2018;90(4):52.
[51] Pham VT, et al. A gallery of chaotic systems with an infinite number of equilibrium points. Chaos Solitons Fractals 2016;93:58–63.
[52] Pham V-T, Jafari S, Kapitaniak T. Constructing a chaotic system with an infinite number of equilibrium points. Int J Bifurcat Chaos 2016;26(13):1650225.
[53] Kingni ST, et al. A chaotic system with an infinite number of equilibrium points located on a line and on a hyperbola and its fractional-order form. Chaos Solitons Fractals 2017;99: 209–18.
[54] Tlelo-Cuautle E, et al. Dynamics, FPGA realization and application of a chaotic system with an infinite number of equilibrium points. Nonlinear Dyn 2017;1–11.
[55] Tahir FR, et al. A novel no-equilibrium chaotic system with multiwing butterfly attractors. Int J Bifurcat Chaos 2015;25(04):1550056.
[56] Jafari S, Pham V-T, Kapitaniak T. Multiscroll chaotic sea obtained from a simple 3D system without equilibrium. Int J Bifurcat Chaos 2016;26(02):1650031.
[57] Pham V-T, et al. A chaotic system with different families of hidden attractors. Int J Bifurcat Chaos 2016;26(08):1650139.
[58] Jafari MA, et al. Chameleon: the most hidden chaotic flow. Nonlinear Dyn 2017;1–15.
[59] Kingni ST, et al. Constructing and analyzing of a unique three-dimensional chaotic autonomous system exhibiting three families of hidden attractors. Math Comput Simul 2017;132:172–82.
[60] Pham V-T, et al. A chaotic system with rounded square equilibrium and with no-equilibrium. Optik 2017;130:365–71.
[61] Pham V-T, et al. Different families of hidden attractors in a new chaotic system with variable equilibrium. Int J Bifurcat Chaos 2017;27(09):1750138.
[62] Pham V-T, et al. From Wang-Chen system with only one stable equilibrium to a new chaotic system without equilibrium. Int J Bifurcat Chaos 2017;27(06):1750097.
[63] Rajagopal K, Jafari S, Laarem G. Time-delayed chameleon: analysis, synchronization and FPGA implementation. Pramana 2017;89(6):92.
[64] Lai Q, Chen S. Coexisting attractors generated from a new 4D smooth chaotic system. Int J Control Autom Syst 2016;14(4):1124–31.
[65] Lai Q, Wang L. Chaos, bifurcation, coexisting attractors and circuit design of a three-dimensional continuous autonomous system. Optik 2016;127(13):5400–6.
[66] Chudzik A, et al. Multistability and rare attractors in Van Der Pol-Duffing oscillator. Int J Bifurcat Chaos 2011;21(07):1907–12.
[67] Yanchuk S, Kapitaniak T. Symmetry-increasing bifurcation as a predictor of a chaos-hyperchaos transition in coupled systems. Phys Rev E 2001;64(5):056235.

[68] Sprott JC, et al. Megastability: coexistence of a countable infinity of nested attractors in a periodically-forced oscillator with spatially-periodic damping. Eur Phys J Spec Top 2017;226(9):1979–85.

[69] Jafari S, et al. A new hidden chaotic attractor with extreme multi-stability. AEU—Int J Electron Commun 2018;.

[70] Jafari S, et al. Extreme multi-stability: when imperfection changes quality. Chaos Solitons Fractals 2018;108:182–6.

[71] Bao B, et al. Two-memristor-based Chua's hyperchaotic circuit with plane equilibrium and its extreme multistability. Nonlinear Dyn 2017;1–15.

[72] Bao H, et al. Initial condition-dependent dynamics and transient period in memristor-based hypogenetic jerk system with four line equilibria. Commun Nonlinear Sci Numer Simul 2018;57:264–75.

[73] Chen M, et al. Controlling extreme multistability of memristor emulator-based dynamical circuit in flux-charge domain. Nonlinear Dyn 2018;91(2):1395–412.

[74] Wang Z, et al. A new chaotic attractor around a pre-located ring. Int J Bifurcat Chaos 2017;27 (10):1750152.

[75] Chen H, et al. A flexible chaotic system with adjustable amplitude, largest Lyapunov exponent, and local Kaplan-Yorke dimension and its usage in engineering applications. Nonlinear Dyn 2018;1–10.

[76] Pham V-T, et al. Bistable hidden attractors in a novel chaotic system with hyperbolic sine equilibrium. Circ Syst Signal Process 2018;37(3):1028–43.

[77] Nazarimehr F, et al. Fuzzy predictive controller for chaotic flows based on continuous signals. Chaos Solitons Fractals 2018;106:349–54.

[78] Shekofteh Y, Jafari S, Rajagopal K. Cost function based on hidden Markov models for parameter estimation of chaotic systems. Soft Comput 2018;1–12.

[79] Rajagopal K, et al. Fractional order synchronous reluctance motor: analysis, chaos control and FPGA implementation. Asian J Control 2018;.

[80] Rajagopal K, et al. A chaotic memcapacitor oscillator with two unstable equilibriums and its fractional form with engineering applications. Nonlinear Dyn 2018;91(2):957–74.

[81] Rajagopal K, et al. Chaotic chameleon: dynamic analyses, circuit implementation, FPGA design and fractional-order form with basic analyses. Chaos Solitons Fractals 2017;103:476–87.

[82] Rajagopal K, Karthikeyan A, Srinivasan A. Bifurcation and chaos in time delayed fractional order chaotic memfractor oscillator and its sliding mode synchronization with uncertainties. Chaos Solitons Fractals 2017;103:347–56.

[83] Karthikeyan A, Rajagopal K. FPGA implementation of fractional-order discrete memristor chaotic system and its commensurate and incommensurate synchronisations. Pramana 2018;90(1):14.

[84] Rajagopal K, Karthikeyan A, Srinivasan A. Dynamical analysis and FPGA implementation of a chaotic oscillator with fractional-order memristor components. Nonlinear Dyn 2018;91 (3):1491–512.

[85] Zhou P, Huang K. A new 4-D non-equilibrium fractional-order chaotic system and its circuit implementation. Commun Nonlinear Sci Numer Simul 2014;19(6):2005–11.

[86] Zhou P, Bai R. The adaptive synchronization of fractional-order chaotic system with fractional-order $$\varvec {1}<\varvec {q}<\varvec {2}$$ 1< q< 2 via linear parameter update law. Nonlinear Dyn 2015;80(1–2):753–65.

[87] Zhou P, Zhu P. A practical synchronization approach for fractional-order chaotic systems. Nonlinear Dyn 2017;89(3):1719–26.

[88] Li R, Chen W. Fractional order systems without equilibria. Chin Phys B 2013;22:040503.
[89] Cafagna D, Grassi G. Fractional-order systems without equilibria: the first example of hyperchaos and its application to synchronization. Chin Phys B 2015;24(8):080502.
[90] Danca M-F, Tang WK, Chen G. Suppressing chaos in a simplest autonomous memristor-based circuit of fractional order by periodic impulses. Chaos Solitons Fractals 2016;84:31–40.
[91] Petráš I. Method for simulation of the fractional order chaotic systems. Acta Montan Slovaca 2006;11(4):273–7.
[92] Trzaska Z. Matlab solutions of chaotic fractional order circuits. In: All Assi. Engineering Education and Research using MATLAB. Rijeka: Intech; 2011. Chapter 19.
[93] He S, Sun K, Wang H. Complexity analysis and DSP implementation of the fractional-order Lorenz hyperchaotic system. Entropy 2015;17(12):8299–311.
[94] Adomian G. A review of the decomposition method and some recent results for nonlinear equations. Math Comput Model 1990;13(7):17–43.
[95] Mkaouar, H., & Boubaker, O. On electronic design of the piecewise linear characteristic of the chua's diode: application to chaos synchronization. In Electrotechnical conference (MELECON), 2012 16th IEEE mediterranean, pp. 197–200. IEEE.
[96] Pham V-T, et al. A novel cubic-equilibrium chaotic system with coexisting hidden attractors: analysis, and circuit implementation. J Circ Syst Comput 2018;27(04):1850066.
[97] Rajagopal K, et al. Hyperchaotic memcapacitor oscillator with infinite equilibria and coexisting attractors. Circ Syst Signal Process 2018;1–3.
[98] Lassoued A, Boubaker O. Dynamic analysis and circuit design of a novel hyperchaotic system with fractional-order terms. Complexity 2017;.
[99] Rajagopal K, Pham VT, Tahir FR, et al. A chaotic jerk system with non-hyperbolic equilibrium: Dynamics, effect of time delay and circuit realisation. Pramana J Phys 2018;90:52. https://doi.org/10.1007/s12043-018-1545-x.
[100] Bao H, et al. Bi-stability in an improved memristor-based third-order wien-bridge oscillator. IETE Tech Rev 2018;1–8.
[101] Buscarino A, et al. A concise guide to chaotic electronic circuits. Springer; 2014.
[102] Buscarino A, et al. Nonideal behavior of analog multipliers for Chaos generation. IEEE Trans Circ Syst II: Express Briefs 2016;63(4):396–400.

Chapter 5

A New Five Dimensional Multistable Chaotic System With Hidden Attractors

Atefeh Ahmadi*, Karthikeyan Rajagopal†, Viet-Thanh Pham‡, Olfa Boubaker§ and Sajad Jafari*

**Biomedical Engineering Department, Amirkabir University of Technology, Tehran, Iran, †Center for Nonlinear Dynamics, Defence University, Bishoftu, Ethiopia, ‡School of Electronics and Telecommunications, Hanoi University of Science and Technology, Hanoi, Vietnam, §National Institute of Applied Sciences and Technology, Tunis, Tunisia*

1 INTRODUCTION

In the field of nonlinear dynamics, chaos is a major area of interest [1–6]. After the invention of the three-dimensional Lorenz system [1], numerous chaotic systems were introduced [7–12]. Extensive research has shown that chaotic system is the subject of many studies in chaos-based applications such as chaos synchronization, motion control, secure communications, improved encryption algorithms, and robust embedded biometric authentication [13–21]. Several attempts have been made to generate higher-dimensional chaotic systems because of their complicated dynamics [22–26]. Tang and Zhang proposed a novel bounded 4D chaotic system [22]. Li et al. were constructed a four-dimensional memristive circuit with infinitely many stable equilibria [23]. Hyperchaos and horseshoe were studied in a 4D memristive system with a line of equilibria [24]. Interestingly, a 5D Rikitake dynamo system was reported in Ref. [25]. By modifying a generalized Lorenz system, a new 5D hyperchaotic system was presented by Yang and Bai [26]. A 5D system with three positive Lyapunov exponents was found by Yang and Chen [27]. When considering a five-dimensional self-exciting homopolar disc dynamo, Wei et al. reported its hidden hyperchaos and electronic circuit application [28]. In addition, Yang et al. designed a novel six-dimensional hyperchaotic system having four positive Lyapunov exponents [29]. Mahmoud added a linear controller into the complex Lorenz model to obtain a new hyperchaotic complex Lorenz system,

Recent Advances in Chaotic Systems and Synchronization. https://doi.org/10.1016/B978-0-12-815838-8.00005-4

which corresponds to a 7-dimensional (7D) continuous real autonomous hyperchaotic system [30]. Synchronization and control of time-delayed chaotic systems were investigated in [31,32].

Recently, Leonov and Kuznetsov have proposed the definition of "hidden attractors," which has received considerable critical attention [33,34]. It is worth noting that conventional numerical procedures cannot be applied for identifying hidden attractors [35,36]. Extensive research has shown that multistability and hidden attractors appeared in different nonlinear systems, ranging from an impulsive Goodwin oscillator with time delay [37], a multilevel DC/DC converter [38], to no-equilibrium chaotic system [39]. Motivated by the complex behavior and undiscovered features of systems with multistability, we study a new five-dimensional multistable chaotic system in this chapter.

2 NEW SYSTEM AND ITS DYNAMICAL PROPERTIES

Consider the following five-dimensional autonomous system.

$$\begin{aligned} \dot{x} &= y \\ \dot{y} &= z \\ \dot{z} &= w \\ \dot{w} &= -1.08\,w + v \\ \dot{v} &= 1.6\,xy + 1.1\,xw - 0.38\,yz \end{aligned} \tag{1}$$

The equilibria of this system can be obtained by setting the right-hand equations equal to zero. Thus, $(x,0,0,0,0)$ is the equilibrium which means this system has a line of equilibria (which is x-axis in the five-dimensional state-space). Fig. 1 shows projections of the strange attractor for initial conditions $(0,-1,0,0,0)$. Interestingly, changing v_0 results in different dynamical solutions in the system. Fig. 2 shows the bifurcation diagram of the proposed system when changing v_0. This proves the existence of extreme multistability in the proposed chaotic system.

3 DIGITAL IMPLEMENTATION FOR THE SYSTEM USING FIELD PROGRAMMABLE GATE ARRAYS

Field programmable gate arrays (FPGAs) have been the choice of implementing chaotic systems in recent years. Hardware and software co-simulation have been the most commonly used implementation method for FPGA [40–47]. There have been various discussions on FPGA implementation of chaotic systems, like the multiscroll attractors discussed in [41,43,45], discretized chaotic systems as discussed in [42,47], and implementation of chaos-based applications such as in image cryptography [44]. Even memristor-based chaotic systems [46] and complex higher-order chaotic systems [48,49] are implemented in FPGAs. Various synchronization and control algorithms such as

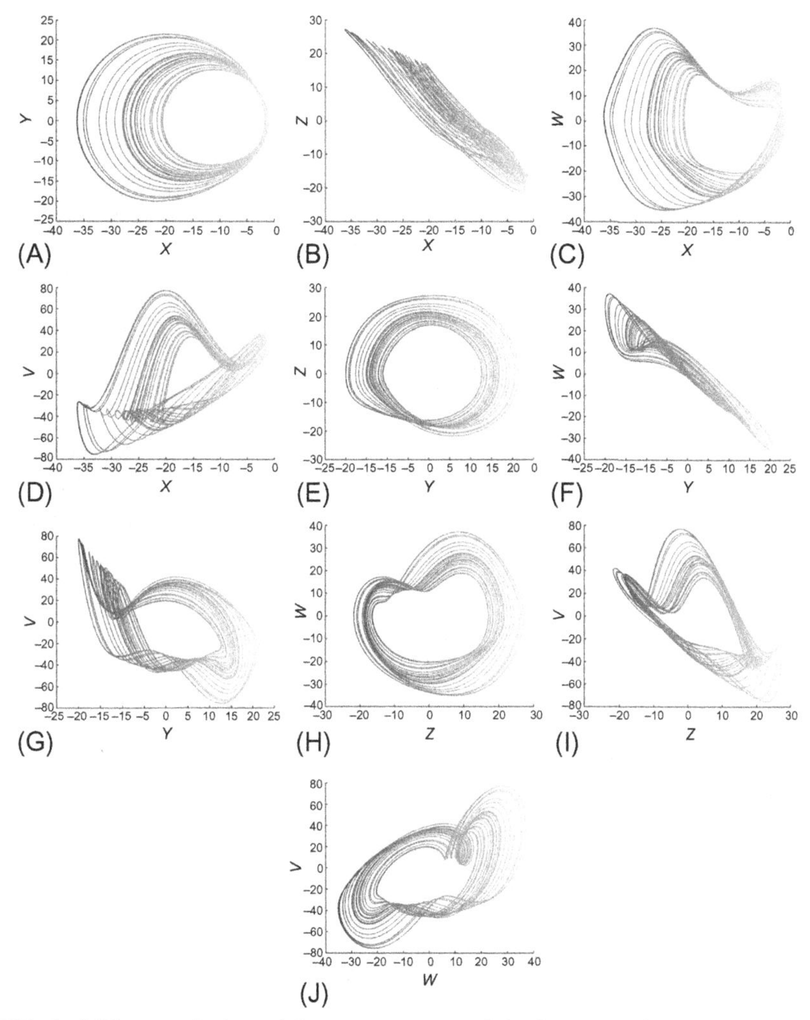

FIG. 1 Different projections of the strange attractor of the System (1) for initial conditions $(0, -1, 0, 0, 0)$.

synchronization of Ikeda systems [50], genetically optimized PID and sliding mode control based synchronizations [51,52], and adaptive sliding mode synchronization [53] are realized using FPGAs. Implementing fractional order chaotic systems efficiently in FPGAs have shown a new way of applying chaotic systems with complex behavior for cryptography [53–59]. Power efficiency of fractional order and integer order nonlinear controllers are analyzed using their equivalent implementations in FPGAs [55–60].

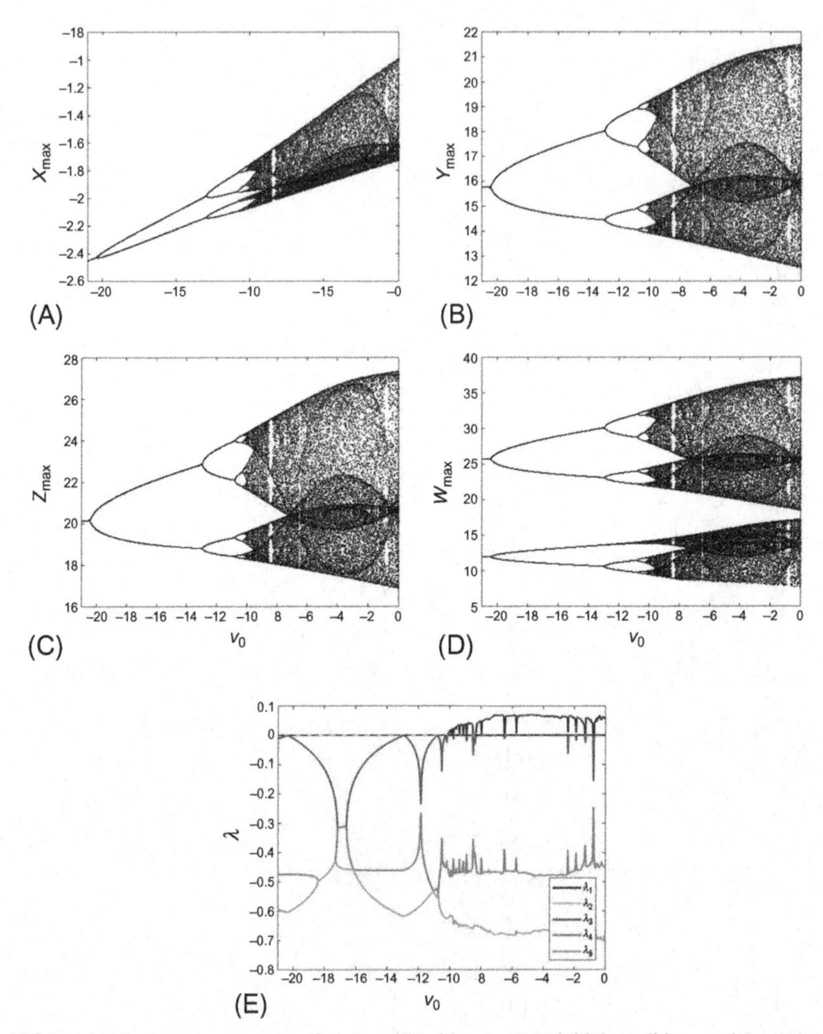

FIG. 2 (A) Lyapunov exponents of system (1) with respect to initial condition v_0. A common period-doubling route to chaos can be observed, while the positive largest Lyapunov exponent indicates the existence of chaos for some ranges of v_0. (B) Bifurcation diagram of System (1) with respect to initial condition v_0 (x_{max} are the local maxima of x). (C) Bifurcation diagram of System (1) with respect to initial condition v_0 (y_{max} are the local maxima of y). (D) Bifurcation diagram of System (1) with respect to initial condition v_0 (z_{max} are the local maxima of z). (E) Bifurcation diagram of System (1) with respect to initial condition v_0 (w_{max} are the local maxima of w), and the initial conditions are $(0,-1,0,0,v_0)$.

The proposed 5D system is implemented in FPGAs using hardware-software co-simulation. Spartan 7 processors are selected for implementation with the clock period kept as 10 ns. We used the forward Euler's method to derive the discrete form of the 5D system (2).

$$
\begin{aligned}
x_{k+1} &= x_k + h[y_{k-1}] \\
y_{k+1} &= y_k + h[z_{k-1}] \\
z_{k+1} &= z_k + h[w_{k-1}] \\
w_{k+1} &= w_k + h[-1.08\, w_{k-1} + v_{k-1}] \\
v_{k+1} &= v_k + h[1.6\, x_{k-1} y_{k-1} + 1.1\, x_{k-1} w_{k-1} - 0.38\, y_{k-1} z_{k-1}]
\end{aligned}
\tag{2}
$$

where h is the step size and in this realization taken as $h = 0.001$.

The discrete state equations are implemented using hardware and software co-simulation, and the needed basic arithmetic operators are implemented using the Xilinx system generator toolkit. We used the Spartan 7 chipset (xc7s50csga324-1) for the co-simulation with the Simulink used to plot the phase plots. Fig. 3A shows the main RTL block of the system (2) with a clock driver. Fig. 3B shows the RTL schematics of the state variables, and Fig. 4 shows the 2D state portraits plotted using system (2) with the required state variables x, y, z, w, v used from the outputs of the FPGA implemented system (2). Table 1 shows the resources consumed by the system (2) for implementing in FPGA.

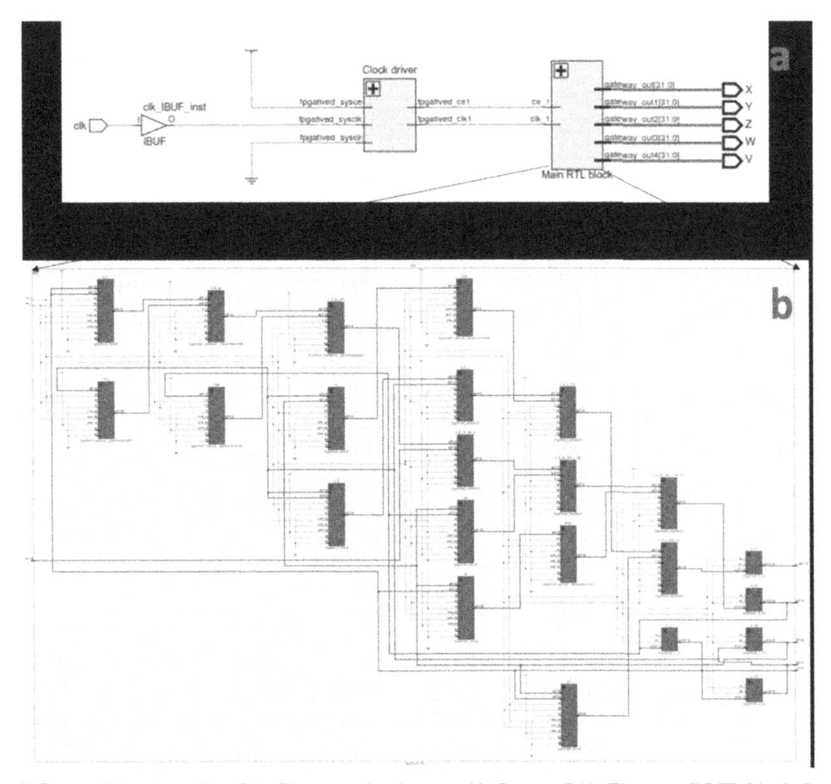

FIG. 3 RTL schematics of the 5D system implemented in Spartan 7 (A: The overall RTL block. B: The RTL of the state equations).

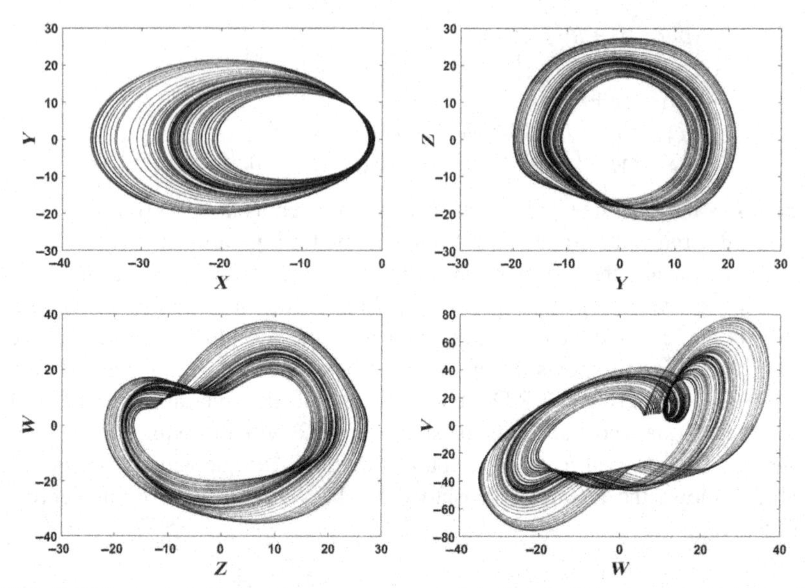

FIG. 4 Different projections of the strange attractor of the System (2) implemented in Spartan 7 processor and the outputs seen in Simulink using hardware—software cosimulation for initial conditions $(0,-1,0,0,0)$.

TABLE 1 Resources Utilized by the 5D System

Resource	Utilization	Available	Utilization %
LUT	803	32,600	2.46
FF	320	65,200	0.49
DSP	24	120	20.00
IO	161	210	76.67
BUFG	1	32	3.13

4 ANALOG CIRCUIT DESIGN

This section introduces an electronic circuit for emulating the five-dimensional autonomous system (1). We used the operational-amplifier approach [61–66] to design the circuit, therefore the output signals must belong to the operational voltage range of operational amplifiers. As a result, the five-dimensional autonomous system (1) has to be rescaled by using $X=\frac{x}{20}, Y=\frac{y}{20}, Z=\frac{z}{20}, W=\frac{w}{20}, V=\frac{v}{40}$.

We obtained the following five-dimensional autonomous system:

$$\begin{aligned}
\dot{X} &= Y \\
\dot{Y} &= Z \\
\dot{Z} &= W \\
\dot{W} &= -1.08\,W + 2V \\
\dot{V} &= 16XY + 11\,XW - 3.8YZ
\end{aligned} \tag{3}$$

The designed circuit is presented in Fig. 5, where X, Y, Z, W, V are the voltages at the operational amplifiers. From Fig. 5, we get the circuit's equation given by:

$$\begin{aligned}
\dot{X} &= \frac{1}{R_1C}Y \\
\dot{Y} &= \frac{1}{R_2C}Z \\
\dot{Z} &= \frac{1}{R_3C}W \\
\dot{W} &= -\frac{1}{R_4C}W + \frac{1}{R_5C}V \\
\dot{V} &= \frac{1}{R_6C}XY + \frac{1}{R_7C}XW - \frac{1}{R_8C}YZ
\end{aligned} \tag{4}$$

We can see that the circuit's equation (4) corresponds to the five-dimensional autonomous system (4). Hence in order to emulate system (4), the selected components in Fig. 5 are: $C = 47$ nF, $R_1 = R_2 = R_3 = R = 40$ kΩ, $R_4 = 37.037$ kΩ, $R_5 = 20$ kΩ, $R_6 = 2.5$ kΩ, $R_7 = 3.643$ kΩ, and $R_8 = 10.526$ kΩ. In Fig. 6, we have reported the PSpice results, which show the chaotic attractors of the circuit.

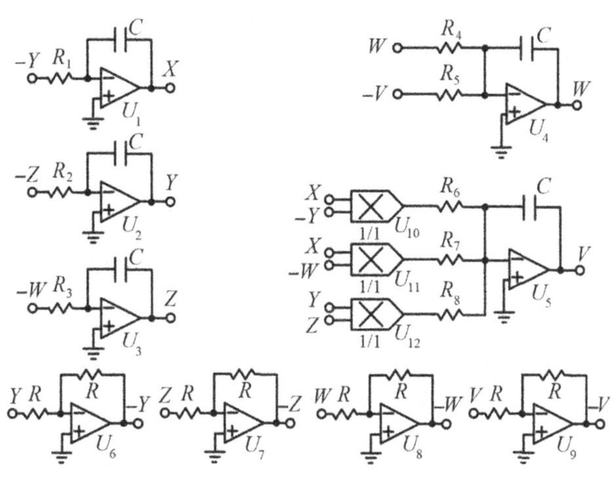

FIG. 5 The circuit based on operational amplifier approach.

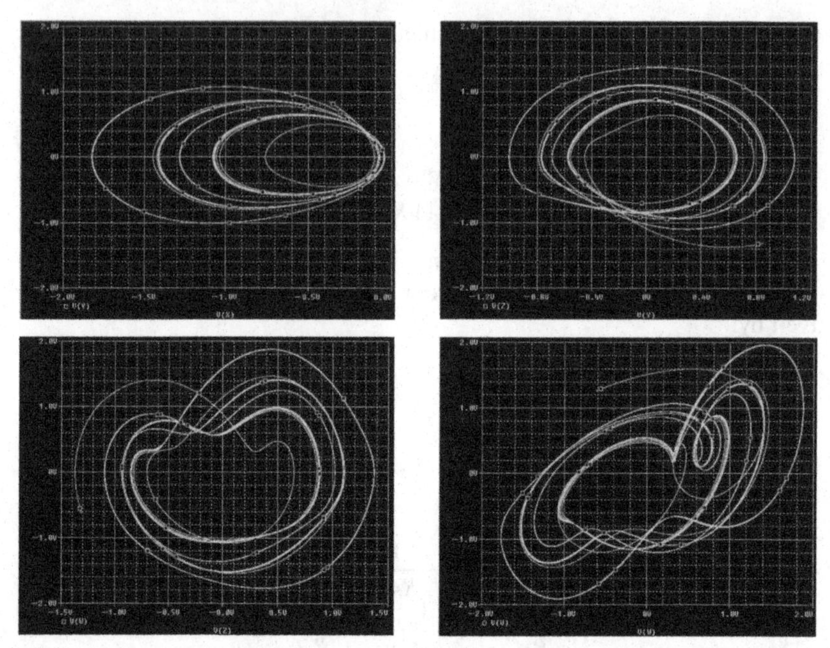

FIG. 6 PSpice projections of the strange attractor observed in the circuit.

5 CONCLUSION

This chapter introduced a novel five-dimensional chaotic system with special characteristics. By investigating system's dynamical properties, we found that the system is multistable. Extreme multistability was displayed when changing the initial condition v0, which was illustrated via Lyapunov exponents and bifurcation diagrams. By using hardware-software co-simulation, the system has been implemented in the FPGA Spartan 7 chipset. We also applied the operational-amplifier approach to design an electronic circuit for the five-dimensional system and reported the PSpice results. In our future work, we will present the detailed studies of the system's chaos-based applications.

ACKNOWLEDGMENT

Sajad Jafari was supported by Iran National Science Foundation (No. 96000815).

REFERENCES

[1] Lorenz EN. Deterministic nonperiodic flow. J Atmos Sci 1963;20:130–41.
[2] Dong E, Liang Z, Du S, Chen Z. Topological horseshoe analysis on a four-wing chaotic attractor and its FPGA implement. Nonlinear Dyn 2016;83(1–2):623–30.
[3] Chen G, Yu X. Chaos control: theory and applications. Berlin: Springer; 2003.
[4] Chua K, Zheng W. Chaos in electric drive systems analysis. Control and applications. Berlin: Wiley; 2011.

[5] Khan M, Shah T, Gondal MA. An efficient technique for the construction of substitution box with chaotic partial differential equation. Nonlinear Dyn 2013;73:1795–801.

[6] Lassoued A, Boubaker O. Dynamic analysis and circuit design of a novel hyperchaotic system with fractional-order terms, Complexity 2017;2017:10 pp. https://doi.org/10.1155/2017/3273408.

[7] Matsumoto T. A chaotic attractor from Chua's circuit. IEEE Trans Circ Syst 1984;31 (12):1055–8.

[8] Zhong G-Q, Ayrom F. Experimental confirmation of chaos from Chua's circuit. Int J Circ Theory Appl 1985;13(1):93–8.

[9] Chua LO. Chua's circuit 10 years later. Int J Circ Theory Appl 1994;22(4):279–305.

[10] Chen G, Ueta T. Yet another chaotic attractor. Int J Bifurc Chaos 1999;9:1465–6.

[11] Lü J, Chen G. A new chaotic attractor coined. Int J Bifurc Chaos 2002;12:659–61.

[12] Lassoued A, Boubaker O. On new chaotic and hyperchaotic systems: a literature survey. Nonlinear Analysis: Model Control 2016;21(6):770–89.

[13] Mkaouar H, Boubaker O. Chaos synchronization for master slave piecewise linear systems: application to Chua's circuit. Commun Nonlinear Sci Numer Simul 2012;17(3):1292–302.

[14] Lassoued, A. Boubaker, O.: Hybrid synchronization of multiple fractional-order chaotic systems with ring connection. In: 2016 8th international conference on modelling, identification and control (ICMIC), pp. 109–114. IEEE.

[15] Lassoued, A. Boubaker, O. A new fractional-order jerk system and its hybrid synchronization. Fractional order control and synchronization of chaotic systems, pp. 699–718. Springer.

[16] Mkaouar H, Boubaker O. Robust control of a class of chaotic and hyperchaotic driven systems. Pramana 2017;88(1):9.

[17] Buscarino A, Fortuna L, Frasca M, Muscato G. Chaos does help motion control. Int J Bifurc Chaos 2007;17:3577–81.

[18] De la Hoz MZ, Acho L, Vidal Y. A modified Chua chaotic oscillator and its application to secure communications. Appl Math Comput 2014;247:712–22.

[19] Kanso A, Ghebleh M. An efficient and robust image encryption scheme for medical applications. Commun Nonlinear Sci Numer Simul 2015;24(1):98–116.

[20] Murillo-Escobar MA, Cruz-Hernández C, Abundiz-Pérez F, López-Gutiérrez RM. Implementation of an improved chaotic encryption algorithm for real-time embedded systems by using a 32-bit microcontroller. Microprocess Microsyst 2016;45:297–309.

[21] Murillo-Escobar MA, Cruz-Hernández C, Abundiz-Pérez F, López-Gutiérrez RM. A robust embedded biometric authentication system based on fingerprint and chaotic encryption. Expert Syst Appl 2015;42(21):8198–211.

[22] Tang W, Zhang J. A novel bounded 4D chaotic system. Nonlinear Dyn 2012;67:2455–65.

[23] Li QD, Zeng HZ, Li J. Hyperchaos in a 4D memristive circuit with infinitely many stable equilibria. Nonlinear Dyn 2015;79:2295–308.

[24] Li QD, Hu SY, Tang S, Zeng G. Hyperchaos and horseshoe in 4D memristive system with a line of equilibria and its implementation. Int J Circ Theory Appl 2014;42:1172–88.

[25] Vaidyanathan S, Pham V-T, Volos CK. A 5-D hyperchaotic Rikitake dynamo system with hidden attractors. Eur Phys J Spec Top 2015;224:1575–92.

[26] Yang Q, Bai M. A new 5D hyperchaotic system based on modified generalized Lorenz system. Nonlinear Dyn 2017;88:189–221.

[27] Yang QG, Chen CT. A 5D hyperchaotic system with three positive Lyapunov exponents coined. Int J Bifurc Chaos 2013;23:1350109-1–1350109-24.

[28] Wei Z, Moroz I, Sprott JC, Akgul A, Zhang W. Hidden hyperchaos and electronic circuit application in a 5D self-exciting homopolar disc dynamo. Chaos 2017;27:033101.

[29] Yang Q, Osman WM, Chen C. A new 6D hyperchaotic system with four positive Lyapunov exponents coined. Int J Bifurc Chaos 2015;25.

[30] Mahmoud EE. Dynamics and synchronization of new hyperchaotic complex Lorenz system. Math Comput Model 2012;55:1951–62.
[31] Shahverdiev EM, Sivaprakasam S, Shore KA. Lag synchronization in time-delayed systems. Phys Lett A 2002;292:320–4.
[32] Ghosh D, Chowdhury R, Saha P. Multiple delay Rossler system—bifurcation and chaos control. Chaos Solitons Fractals 2008;35:472–85.
[33] Leonov GA, Kuznetsov NV, Kuznetsova OA, Seldedzhi SM, Vagaitsev VI. Hidden oscillations in dynamical systems. Trans Syst Control 2011;6:54–67.
[34] Leonov GA, Kuznetsov NV, Vagaitsev VI. Localization of hidden Chua's attractors. Phys Lett A 2011;375:2230–3.
[35] Leonov GA, Kuznetsov NV, Vagaitsev VI. Hidden attractor in smooth Chua system. Physica D 2012;241:1482–6.
[36] Leonov GA, Kuznetsov NV. Hidden attractors in dynamical systems: from hidden oscillation in Hilbert-Kolmogorov, Aizerman and Kalman problems to hidden chaotic attractor in Chua circuits. Int J Bifurc Chaos 2013;23:1330002.
[37] Zhusubaliyev ZT, Mosekilde E, Churilov AN, Medvedev A. Multistability and hidden attractors in an impulsive Goodwin oscillator with time delay. Eur Phys J Spec Top 2015;224:1519–39.
[38] Zhusubaliyev ZT, Mosekilde E. Multistability and hidden attractors in a multilevel DC/DC converter. Math Comput Simulat 2015;109:32–45.
[39] Pham V-T, Volos C, Jafari S, Kapitaniak T. Coexistence of hidden chaotic attractors in a novel no-equilibrium system. Nonlinear Dyn 2017;87:2001–10.
[40] Muthuswamy B, Banerjee S. A route to chaos using FPGAs. Berlin: Springer; 2015.
[41] Tlelo-Cuautle E, Pano-Azucena AD, Rangel-Magdaleno JJ, Carbajal-Gomez VH, Rodriguez-Gomez G. Generating a 50-scroll chaotic attractor at 66 MHz by using FPGAs. Nonlinear Dyn 2016;85(4):2143–57.
[42] Rajagopal K, Akgul A, Jafari S, Karthikeyan A, Koyuncu I. Chaotic chameleon: dynamic analyses, circuit implementation, FPGA design and fractional-order form with basic analyses. Chaos Solitons Fractals 2017;103:476–87. https://doi.org/10.1016/j.chaos.2017.07.007.
[43] Enzeng D, Zhihan L, Shengzhi D. Topological horseshoe analysis on a four-wing chaotic attractor and its FPGA implementation. Nonlinear Dyn 2016;83(1–2):623–30.
[44] Tlelo-Cuautle E, Carbajal-Gomez VH, Obeso-Rodelo PJ. FPGA realization of a chaotic communication system applied to image processing. Nonlinear Dyn 2015;82(4):1879–92.
[45] Tlelo-Cuautle E, Rangel-Magdaleno J, Pano-Azucena J. FPGA realization of multi-scroll chaotic oscillators. Commun Nonlinear Sci Numer Simul 2015;27(1–3):66–80.
[46] Ya-Ming X, Li-Dan W, Shu-Kai D. A memristor-based chaotic system and its field programmable gate array implementation. Acta Phys Sinica 2016;(12):65.
[47] Tlelo-Cuautle E, de la Fraga LG, Pham VT, Volos C, Jafari S, de Jesus Quintas-Valles A. Dynamics, FPGA realization and application of a chaotic system with an infinite number of equilibrium points. Nonlinear Dyn 2017;89(2):1129–39.
[48] Guang-Yi W, Xu-Lei B, Zhong-Lin W. Design and FPGA implementation of a new hyperchaotic system. Chin Phys B 2008;17(10):3596.
[49] Rajagopal K, Karthikeyan A, Duraisamy P. Hyperchaotic chameleon: fractional order FPGA implementation. Complexity 2017;2017:16. https://doi.org/10.1155/2017/8979408. Article ID 8979408.
[50] Valli D, Muthuswamy B, Banerjee S, et al. Synchronization in coupled Ikeda delay systems-experimental observations using field programmable gate arrays. Eur Phys J Spec Top 2014;223:1465. https://doi.org/10.1140/epjst/e2014-02144-8.

[51] Rajagopal K, Guessas L, Vaidyanathan S, Karthikeyan A, Srinivasan A. Dynamical analysis and FPGA implementation of a novel hyperchaotic system and its synchronization using adaptive sliding mode control and genetically optimized PID control, In: Mathematical problems in engineering. vol. 2017. 201714 pp. https://doi.org/10.1155/2017/7307452.
[52] Rajagopal K, Guessas L, Karthikeyan A, Srinivasan A, Adam G. Fractional order memristor no equilibrium chaotic system with its adaptive sliding mode synchronization and genetically optimized fractional order PID synchronization, Complexity 2017;2017:19 pp. https://doi.org/10.1155/2017/1892618.
[53] Rajagopal K, Karthikeyan A, Srinivasan A. FPGA implementation of novel fractional order chaotic system with two equilibriums and no equilibrium and its adaptive sliding mode synchronization. Nonlinear Dyn 2017;https://doi.org/10.1007/s11071-016-3189-z.
[54] Rajagopal K, Laarem G, Karthikeyan A, Srinivasan A. FPGA implementation of adaptive sliding mode control and genetically optimized PID control for fractional-order induction motor system with uncertain load. Adv Differ Equat 2017;273:https://doi.org/10.1186/s13662-017-1341-9.
[55] Karthikeyan A, Rajagopal K. Chaos control in fractional order smart grid with adaptive sliding mode control and genetically optimized PID control and its FPGA implementation. Complexity 2017;2017:18 pp. https://doi.org/10.1155/2017/3815146.
[56] Rajagopal K, Nazarimehr F, Karthikeyan A, Srinivasan A, Jafari S. Fractional order synchronous reluctance motor: analysis, chaos control and FPGA implementation. Asian J Control 2017. https://doi.org/10.1002/asjc.1690.
[57] Rajagopal K, Karthikeyan A, Srinivasan A. Dynamical analysis and FPGA implementation of a chaotic oscillator with fractional-order memristor components. Nonlinear Dyn 2018;91:1491. https://doi.org/10.1007/s11071-017-3960-9.
[58] Rajagopal K, Akgul A, Jafari S, et al. A chaotic memcapacitor oscillator with two unstable equilibriums and its fractional form with engineering applications. Nonlinear Dyn 2018;91:957. https://doi.org/10.1007/s11071-017-3921-3.
[59] Tlelo-Cuautle E, de la Fraga LG, Pham VT, et al. Dynamics, FPGA realization and application of a chaotic system with an infinite number of equilibrium points. Nonlinear Dyn 2017;89:1129. https://doi.org/10.1007/s11071-017-3505-2.
[60] Benrejeb W, Boubaker O. FPGA modeling and real-time embedded control design via LabVIEW software: application for swinging-up a pendulum. Int J Smart Sens Intell Syst 2012;5(3):576–91.
[61] Mkaouar H, Boubaker O. On electronic design of the piecewise linear characteristic of the chua's diode: application to chaos synchronization. Electrotechnical conference (MELECON), 2012 16th IEEE Mediterranean, IEEE; 2012. p. 197–200.
[62] Li CB, Pehlivan I, Sprott JC, Akgul A. A novel four-wing strange attractor born in bistability. IEICE Electron Lett 2015;12:1–12.
[63] Akgul A, Shafqat H, Pehlivan I. A new three-dimensional chaotic system, its dynamical analysis and electronic circuit applications. Optik 2016;127(18):7062–71.
[64] Wei Z, Moroz I, Sprott JC, Akgul A, Zhang W. Hidden hyperchaos and electronic circuit application in a 5D self-exciting homopolar disc dynamo. Chaos 2017;27:033–101.
[65] Lassoued A, Boubaker O, Dhifaoui R. In: A simple chaotic system with only one nonlinear term: Circuit design and experimental validation. 2018 15th international multi-conference on Systems, Signals & Devices (SSD), March 19–22, Hammamet, Tunisia; 2018.
[66] Lassoued A, Boubaker O, Dhifaoui R, Jafari S. Experimental observations and circuit realization of a jerk chaotic system with piecewise nonlinear function. In: Boubaker O, Jafari S, editors. Recent advances in chaotic systems and synchronization: from theory to real world applications. 2018.

Chapter 6

Extreme Multistability in a Hyperjerk Memristive System With Hidden Attractors

Dimitrios A. Prousalis*, **Christos K. Volos***, **Bocheng Bao**[†], **Efthymia Meletlidou**[‡], **Ioannis N. Stouboulos*** and **Ioannis M. Kyprianidis***

[*]Laboratory of Nonlinear Systems, Circuits and Complexity (LaNSCom), Department of Physics, Aristotle University of Thessaloniki, Thessaloniki, Greece, [†]*School of Information Science and Engineering, Changzhou University, Changzhou, China,* [‡]*Department of Physics, Aristotle University of Thessaloniki, Thessaloniki, Greece*

1 INTRODUCTION

In 1971, Chua [1] proposed that the memristor was the missing fourth-circuit element, following the theory of electrical circuits. The memristor combines the charge (q) and the flux (ϕ). Due to the unconventional attributes of the memristors, Chua and Kang in 1976 inferred that there can be an interesting class of nonlinear dynamical systems, the memristive systems [2]. In 2008, the realization of a two-terminal memristor was announced [3]. This announcement influenced many researchers and widened the horizons in various scientific fields. In 2009, other elements with memory from the nano-world, such as memcapacitor and meminductor, were introduced [4].

Many researchers found the survey upon the applications of memristors, based on their properties, such as memristor-based neural networks, memristor-based chaotic oscillators, and memristor-based Charge-Pump Phase-Lock Loops, etc., interesting [5–7]. The memristor-based chaotic systems are considered to be a research topic focused in both the technological and the application domains [8–16]. In addition, the replacement of the nonlinear part of chaotic dynamical systems with memristors supported the development of the design of memristor-based chaotic oscillators [17–23].

The periodic and chaotic attractors were categorized as either self-excited or hidden only in the last decade by Leonov and Kuznetsov [19, 20, 24–28]. As far as self-excited attractors are concerned, they retain a basin of attraction that is

Recent Advances in Chaotic Systems and Synchronization. https://doi.org/10.1016/B978-0-12-815838-8.00006-6

associated with an unstable equilibrium. On the other hand, the basin of attraction of a hidden attractor does not intersect with small neighborhoods of any equilibrium points. This method cannot be used for the hidden attractors. Furthermore, hidden attractors are important in engineering applications.

Also, there are systems in which many different kinds of attractors coexist. This is the phenomenon of multistability. In those systems, the trajectory converges on the coexisting attracting sets and depends on its initial conditions. In a nonlinear dynamical system with an infinite number of equilibrium points, the coexistence of infinitely many attractors is called extreme multistability. This phenomenon was first encountered in nonlinear dynamical coupled systems [29, 30]. This phenomenon can be found also in high-order systems [31, 32]. Recently, the researchers started to study the phenomenon of extreme multistability in memristive systems [21]. Bao et al. [22] present an active band-pass filter-based memristive circuit with a line of equilibrium. Multistability exposes a great diversity on the behavior of a system with many stable states and offers the system a great flexibility. Nonlinear systems with infinite attractors present research interest due to the fact that multistability can be used as an additional source of randomness especially applied in information engineering [33]. Actually, in a memristor-based chaotic circuit the complex dynamical behaviors depend on the initial states of the memristor, which just reflect the emergences of extreme multistability in these memory systems [34, 35]. Multistable systems are very sensitive to noise, initial conditions, and system parameters. Therefore, in order to keep the system on the desired attractor, one needs to apply an appropriate controlling scheme.

During the last decade, a special class of dynamical systems, known as jerk, has attracted the attention of the research community. The jerk systems are third-order differential equations of the form:

$$\dddot{x} = J(\ddot{x}, \dot{x}, x), \tag{1}$$

where J is a nonlinear function, called the "jerk," and denotes the third derivative of x, which corresponds with the derivative of acceleration in a system [36–41]. In the last decade, research interest in jerk systems has been grown beyond three-order systems. Hyperjerk systems are dynamical systems described by an nth order of ordinary differential equation with $n > 3$, depicting the time evolution of a single-scalar variable. The form of the system is:

$$d^{(n)}x/dt^{(n)} = f(d^{(n-1)}x/dt^{(n-1)}, \ldots, x). \tag{2}$$

Such systems excuse the interest of their research because of their wide generality and elegant simplicity. The transformation of dynamical systems to jerk systems is an area with various applications [42].

In this chapter an extended study of the extreme multistability of a recently new proposed memristive hyperjerk system is presented. For the purposes of this work, bifurcation-like diagrams with regard to system variables' initial conditions are investigated. Extreme multistability as well as other interesting phenomena related with chaos theory have been observed as the system's initial conditions are varied.

This research work was organized with a certain procedure, as presented below. Section 2 presents the model of the memristive device as well as the proposed system. Section 3 presents the phenomenon of extreme multistability, while Section 4 concludes this work with a summary of the main results.

2 THE HYPERJERK MEMRISTIVE SYSTEM

Pham et al. [43] proposed the following form of a memristive device:

$$\dot{x}_m = u, \tag{3a}$$

$$g = (1-x)u, \tag{3b}$$

where x, g, and u denote the state of memristive device, output, and input, respectively. The investigation of the behavior of the memristive device demands the application of an external bipolar periodic signal u. The form of u is:

$$u = A\sin(2\pi ft), \tag{4}$$

where A is the amplitude and f is the frequency. Fig. 1 presents the hysteresis loops of the memristive device (3) when a periodic signal (4) for different frequencies, $f=0.1$ (solid line with squares), $f=0.2$ (solid line with triangles), and $f=0.5$ (solid line with circles) is applied. The proposed memristive device displays a pinched hysteresis loop in the input-output plane.

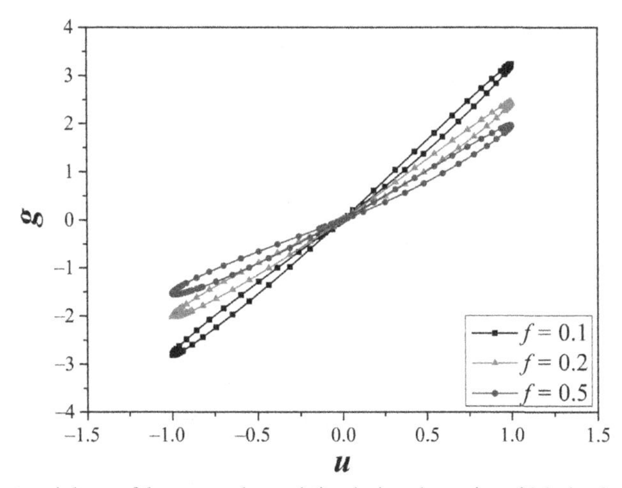

FIG. 1 Hysteresis loops of the proposed memristive device when a sinusoidal stimulus with $f=0.1$ and $w_0=0$, for different frequencies.

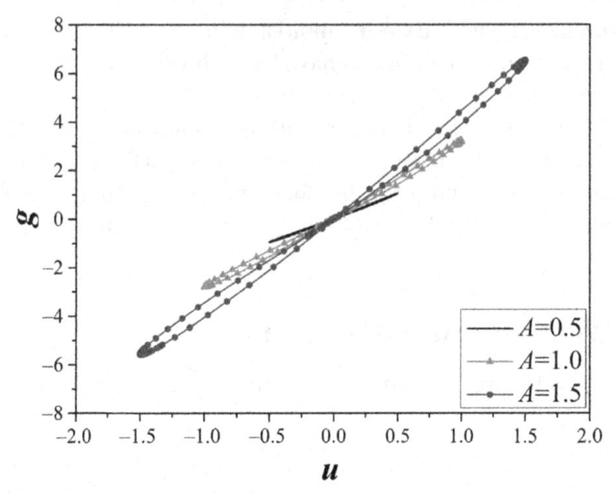

FIG. 2 Hysteresis loops of the proposed memristive device driven by a sinusoidal stimulus with $f = 0.1$ and $w_0 = 0$ for different amplitudes.

Fig. 2 shows the hysteresis loops of the memristive device (3) when a periodic signal (4) for different amplitude $A = 0.5$ (solid line with squares), $A = 1$ (solid line with triangles), and $A = 1.5$ (solid line with circles) is applied. It is obvious that the proposed memristive device displays a pinched hysteresis loop in the input-output plane.

This section presents the proposed 4D memristive system in detail. For this reasons the features of the system, the equilibrium points, their stability, and the dynamical behavior of it have been investigated by using well-known tools of the theory of dynamical systems, such as phase portraits, diagrams of Lyapunov exponents, and bifurcation diagrams. The proposed memristive system belongs to the category of hyperjerk dynamical system [42], which is described as:

$$\frac{d^{(4)}x}{dt^{(4)}} + a\frac{d^{(3)}x}{dt^{(3)}} + \frac{d^{(2)}x}{dt^{(2)}} + b\frac{d^{(2)}x}{dt^{(2)}}\frac{d^{(3)}x}{dt^{(3)}} + (1-x)\frac{dx}{dt} = 0 \tag{5}$$

or in system's form as:

$$\begin{aligned} \dot{x} &= y \\ \dot{y} &= z \\ \dot{z} &= w \\ \dot{w} &= -z - aw - bzw - g \end{aligned} \tag{6}$$

where a, b are two positive parameters and $g = (1 - x)y$ is the output of the memristive device. As it has been reported by Pham et al. [43–45], system (6) is chaotic when the parameters a and b take the values $a = 0.5$, $b = 0.4$. For the selected values of the parameters a and b and the initial conditions have kept as $(x(0), y(0), z(0), w(0)) = (0, 10^{-6}, 0, 0)$ the finite-time local Lyapunov exponents [46] of the hyperjerk memristive system (6) are found as $L_1 = 0.0730$, $L_2 = 0.0018$, $L_3 = 0$, $L_4 = -0.5755$, while the Kaplan-Yorke dimension [47] is:

$$D_{KY} = 3 + \frac{L_1 + L_2 + L_3}{|L_4|} = 3.12299. \tag{7}$$

By solving the equations

$$\begin{aligned} y &= 0 \\ z &= 0 \\ w &= 0 \\ -z - aw - bzw - (1-x)y &= 0 \end{aligned} \tag{8}$$

the equilibrium points of system (6) are obtained as: $(x, y, z, w) = (c, 0, 0, 0)$, where c is a real constant. Due to the fact that system (6) presents infinite equilibrium points, depending of the value of c, it belongs to the recently new proposed category of dynamical systems with hidden attractors [26, 48].

The Jacobian of the system (6), $\mathbf{J}$ at any point is calculated as:

$$\mathbf{J} = \begin{pmatrix} 0 & 1 & 0 & 0 \\ 0 & 0 & 1 & 0 \\ 0 & 0 & 0 & 1 \\ y & -1+x & -1-bw & -a-bz \end{pmatrix}. \tag{9}$$

According to theory of dynamical systems the characteristic equation of the Jacobian matrix for $E(c, 0, 0, 0)$ is:

$$\lambda + \lambda^2 + a\lambda^3 + \lambda^4 - \lambda c = 0. \tag{10}$$

The eigenvalues of the characteristic equation, for $a = 0.5$, $b = 0.4$, are:

$$\begin{aligned} \lambda_1 &= 0, \\ \lambda_2 &= -0.166667 + \frac{0.916667 - 1.58771i}{Q} - (0.0833333 + 0.144338i)Q, \\ \lambda_3 &= -0.166667 + \frac{0.916667 + 1.58771i}{Q} - (0.0833333 - 0.144338i)Q, \\ \lambda_4 &= -0.166667 - 1.83333/Q + 0.0166667Q, \end{aligned} \tag{11}$$

where

$$Q = \left(-91 + 108c + 10.3923\sqrt{89 - 182c + 108c^2}\right)^{(1/3)}. \tag{12}$$

As it is clear the $\lambda_1 = 0$ as expected because the system is degenerative and has a line equilibrium. The rest eigenvalues λ_2, λ_3, λ_4 of the Jacobian matrix depend on the parameter c. So, it is difficult to determine the stability of the equilibrium points. Also, the eigenvalues λ_2 and λ_3 of the Jacobian matrix are complex conjugate and λ_4 is real number for all values of c.

In Fig. 3 the real parts of the eigenvalues of the Jacobian matrix (9) for $a = 0.5$, $b = 0.4$ in respect of the parameter c are depicted.

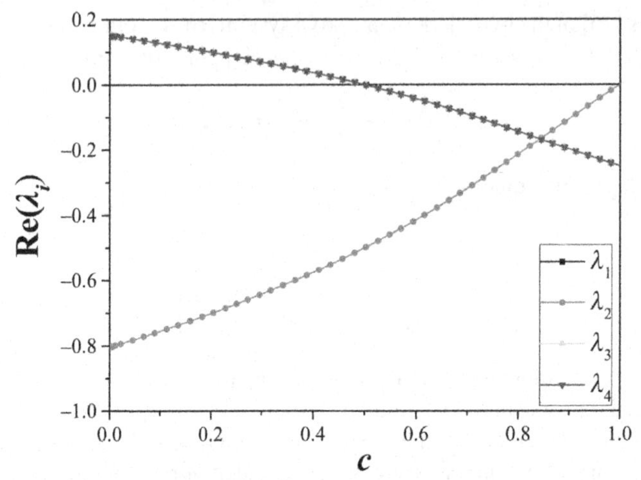

FIG. 3 Real part of the eigenvalues for $a = 0.5$ and $b = 0.4$ versus variable c.

In more details, the equilibrium points of the system (6) can be grouped as follows, in respect to c:

- A region with two unstable foci and a stable linear subspace for $c < 0.5$.
- A region with two stable foci and a stable linear subspace for $1 > c > 0.5$.
- A region with two stable foci and a unstable linear subspace for $c > 1$.

Furthermore, in this chapter, a more complete study of system's (6) dynamical behavior is presented. Consequently, the effect of the parameters a and b in system's dynamics is investigated numerically by taking the bifurcation diagrams of the variable x versus the parameters a and b, respectively, as well as the spectrum of systems' Lyapunov exponents (LE_i, $i = 1, 2, 3, 4$) versus the parameters a and b. In more detail, the bifurcation diagrams are produced when the trajectories cut the plane $y = 0$ with $dy/dt < 0$, by changing the value of the parameter a (or b) in order to investigate the dynamics of system (6), while the initial conditions keeping as $(x(0), y(0), z(0), w(0)) = (0.1, 10^{-6}, 0, 0)$. Also, the proposed system (6) is integrated numerically using the classical fourth-order Runge-Kutta integration algorithm. For each set of parameters used in this work, the calculations are performed using variables and parameters in extended precision mode. Furthermore, the Lyapunov exponents are calculated by using the Wolf's algorithm [49].

As can be seen from the bifurcation diagram of Fig. 4A, the system has a rich dynamical behavior. There are some windows of limit cycles and chaos, when varying the parameter a, as well as other interesting phenomena related with chaos, such as period doubling routes to chaos, period-3 windows and crisis phenomena. For example, system (6) exhibits limit cycles (Fig. 5A–C and E) and chaotic attractors (Fig. 5D and F), when the value of variable b remains

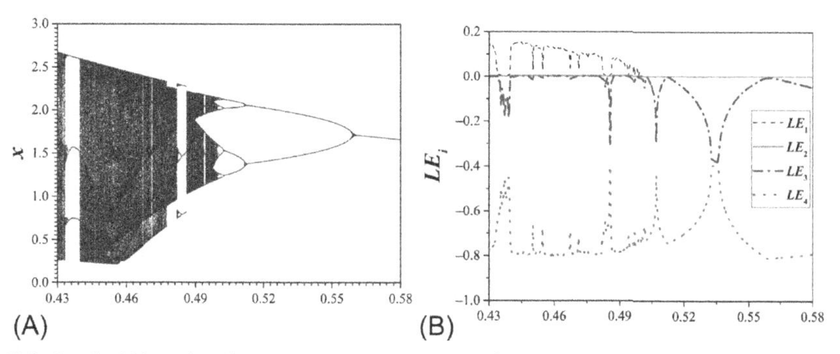

FIG. 4 (A) Bifurcation diagram and (B) the spectrum of Lyapunov exponents of the system (6), when varying the value of the bifurcation parameter a from 0.43 to 0.58, and $b = 0.4$.

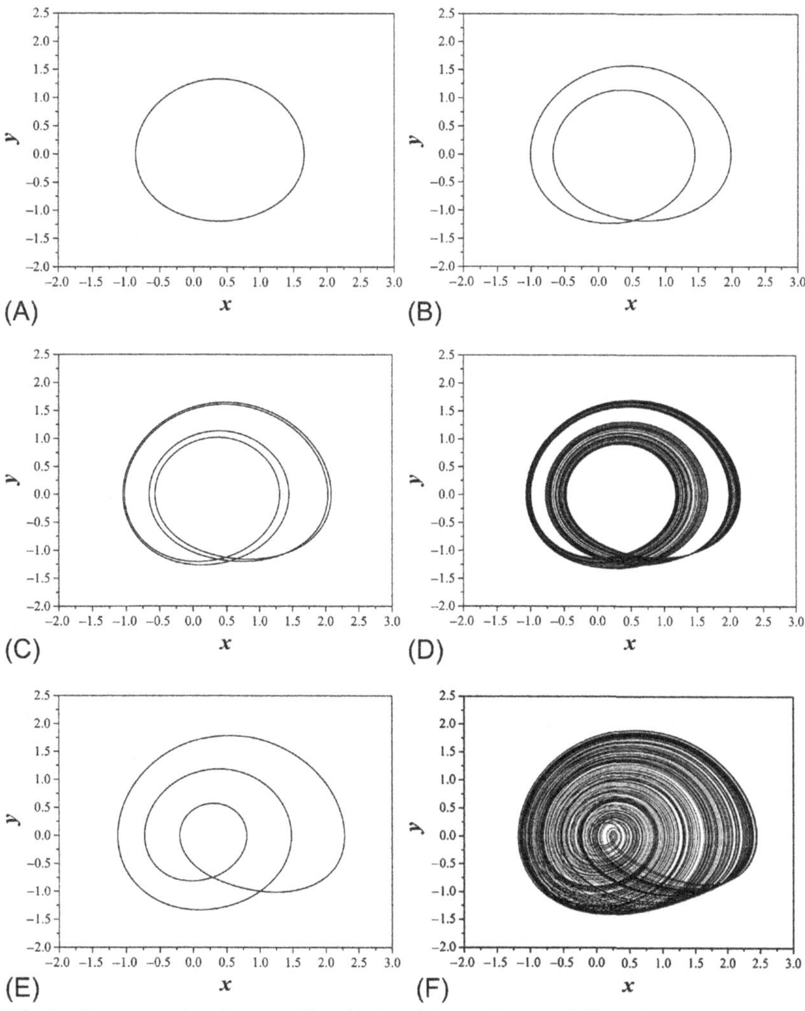

FIG. 5 Phase portraits of system (6), with $b = 0.4$, and (A) $a = 0.58$ (period-1), (B) $a = 0.53$ (period-2), (C) $a = 0.51$ (period-4), (D) $a = 0.497$ (chaos), (E) $a = 0.485$ (period-3), (F) $a = 0.45$ (chaos).

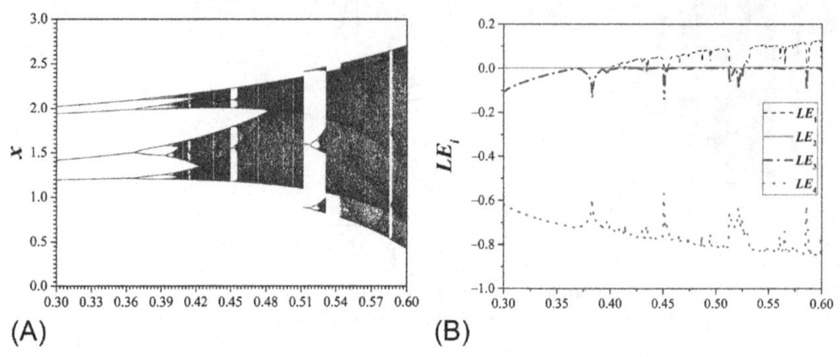

FIG. 6 (A) Bifurcation diagram and (B) the spectrum of Lyapunov exponents of the system (6), when varying the value of the bifurcation parameter b from 0.3 to 0.6, with $a = 0.5$.

the same as $b = 0.4$. The spectrum of Lyapunov exponents in Fig. 4B confirms the system's dynamical behavior as it was described from the bifurcation diagram.

The system (6) exhibits also strange behavior, as in the previous case, when the bifurcation parameter is the parameter b (Fig. 6A), while $a = 0.5$. In the selected window of parameter values the system (6) is initially in period-4 steady state for $b = 0.3$, while b is increased, through a period doubling route, inserts to chaos. There are also windows of limit cycles (i.e., period-3), as well as period doubling routes to chaos and crisis phenomena. As depicted in Fig. 7, system (6) exhibits a chaotic attractor, for $b = 0.43$ (two bands) and $b = 0.58$, as well as limit cycles of different period. The spectrum of Lyapunov exponents in Fig. 6B confirms the system's dynamical behavior as it was discovered from the bifurcation diagram of Fig. 6A.

3 SYSTEM'S EXTREME MULTISTABILITY

The system's dynamics, which are extremely relied on its initial conditions, have been investigated in this section. Memristive hyperjerk system's initial condition-dependent dynamics shows the extreme multistability phenomenon. For this reason, typical system parameters, as they have used in the previous section ($a = 0.5$, $b = 0.4$), are determined. Also, the initial conditions have remained $(x(0), y(0), z(0), w(0)) = (0, 10^{-6}, 0, 0)$. However, in the following study, each one of the variables' initial values varies in a region while the other three remain as they are.

As a beginning, the initial value of the variable x is varied in the range [0, 0.2]. So, when $x(0)$ is gradually decreased, a bifurcation-like diagram of the state variable x versus $x(0)$, as well as the system's Lyapunov exponents versus $x(0)$ have been plotted (Fig. 8B). These diagrams reveal that with the variation of the initial condition $x(0)$, system (6) shows infinite number of coexisting periodic or chaotic attractors, as well as other interesting phenomena concerning

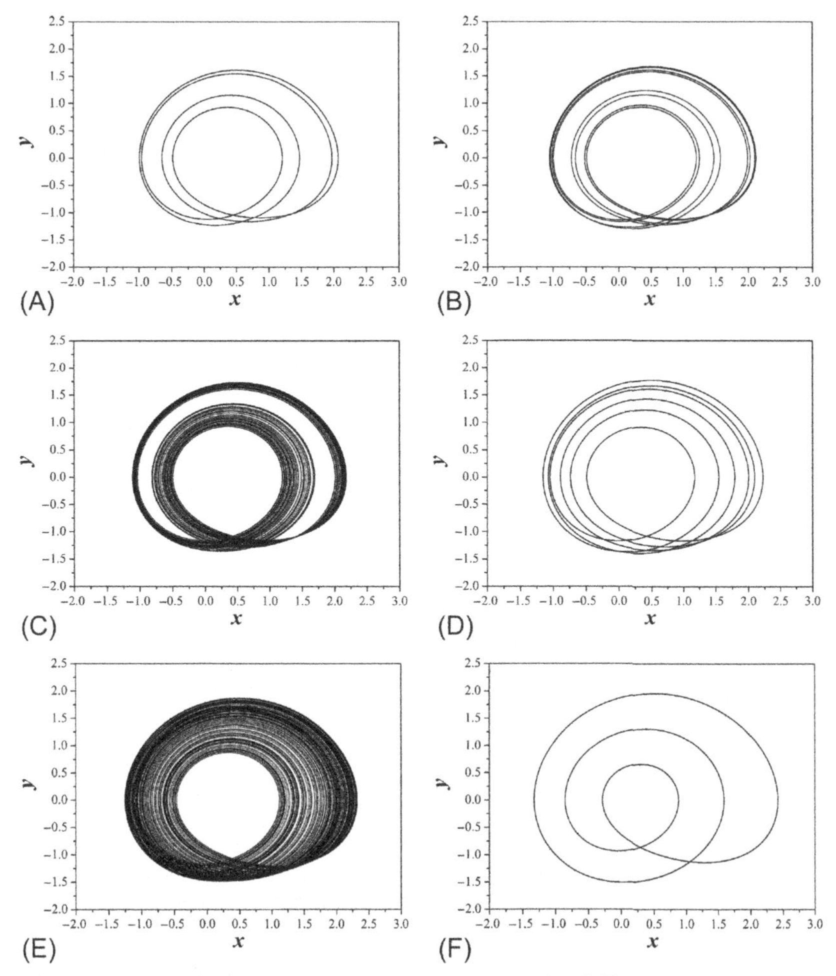

FIG. 7 Phase portraits of system (6), with $a = 0.5$, and (A) $b = 0.35$ (period-4), (B) $b = 0.39$ (period-8), (C) $b = 0.43$ (chaos), (D) $b = 0.451$ (period-6), (E) $b = 0.5$ (chaos), (F) $b = 0.515$ (period-3).

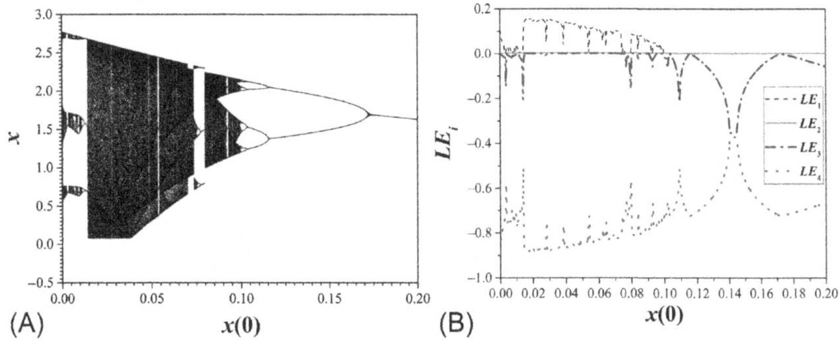

FIG. 8 (A) Bifurcation diagram and (B) the spectrum of Lyapunov exponents of the system (6), when varying the initial value $x(0)$, for $a = 0.5$ and $b = 0.4$.

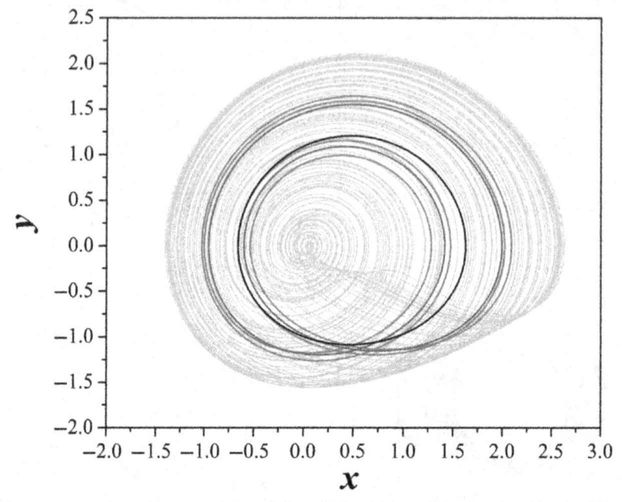

FIG. 9 Different attractors for $a = 0.5$ and $b = 0.4$ and for different initial conditions $x(0)$.

chaos theory, such as period-doubling route to chaos and crisis phenomena. In Fig. 9 some typical phase portraits of coexisting infinitely many attractors are depicted.

Next, the value of initial condition $y(0)$, as a bifurcation parameter, has been used, while the values of system's (6) parameters remain as $a = 0.5$, $b = 0.4$. In the same way, as the initial condition $y(0)$ is decreased starting from the value of $y(0) = 0.1$, a bifurcation diagram and the spectrum of Lyapunov exponents, when varying the initial value $y(0)$, have been plotted (Fig. 10). It can also be concluded that with the variation of the initial condition $y(0)$ an infinite number of coexisting attractors are produced. Fig. 11 displays four of the attractors that have been produced as the value of initial condition $y(0)$ is decreased.

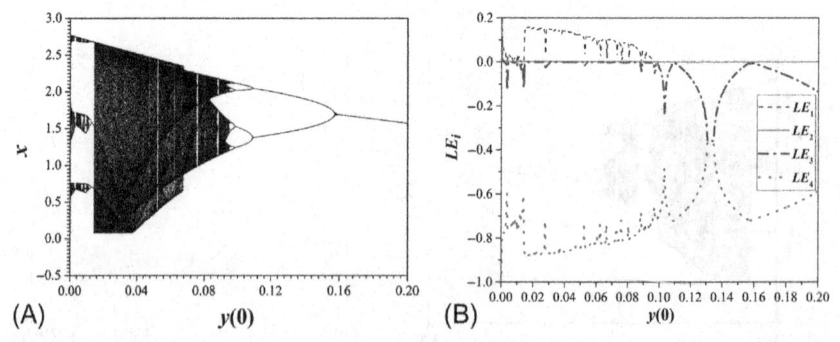

FIG. 10 (A) Bifurcation diagram and (B) the spectrum of Lyapunov exponents of the system (6), when varying the initial value $y(0)$, for $a = 0.5$ and $b = 0.4$.

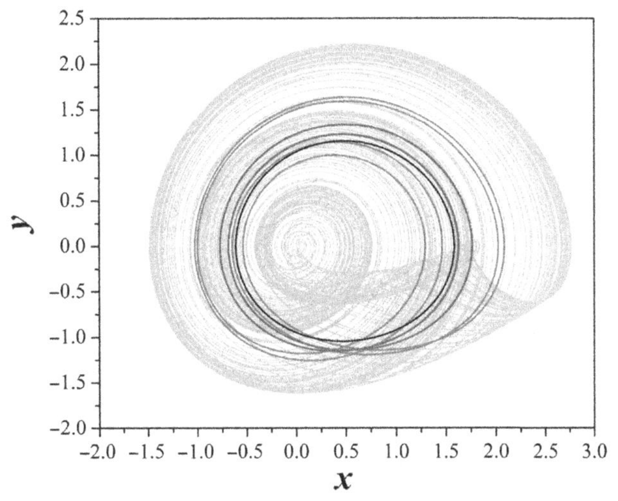

FIG. 11 Different attractors for $a = 0.5$ and $b = 0.4$ and for different initial conditions $y(0)$, where $y(0) = 0.10$ —period-1, $y(0) = 0.06$ —period-2, $y(0) = 0.04$ —period-4, and $y(0) = 0.02$ —chaos.

By choosing the initial value of z as a bifurcation parameter, with the same values of parameters a, b and the rest of initial conditions $(x(0), y(0), w(0)) = (0, 10^{-6}, 0)$, the system's (6) analysis have been done. So, a bifurcation-like diagram (Fig. 12) of the state variable x versus $z(0)$, as well as the system's Lyapunov exponents versus $z(0)$ have been plotted again. These diagrams display that with the variation of the initial condition $z(0)$ (Fig. 12), system (6) has also an infinite number of coexisting periodic or chaotic attractors, while well-known phenomena, such as period-doubling route to chaos, as $z(0)$ is decreased, a wide period-3 window, and crisis phenomenon, has been risen. In Fig. 13, various attractors (periodic or chaotic), depending on the values of initial condition $z(0)$, are depicted.

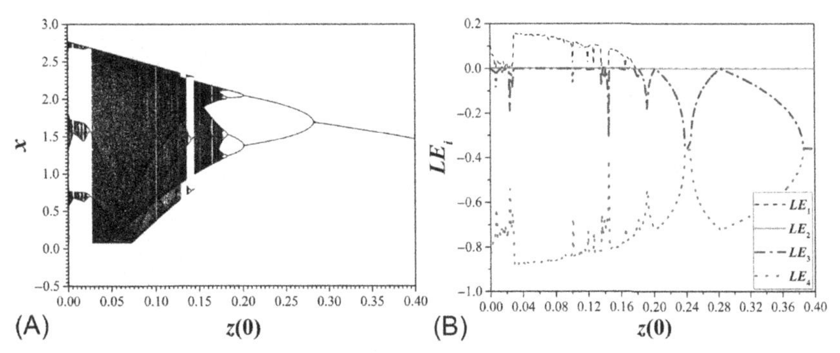

FIG. 12 (A) Bifurcation diagram and (B) the spectrum of Lyapunov exponents of the system (6), when varying the initial value $z(0)$, for $a = 0.5$ and $b = 0.4$.

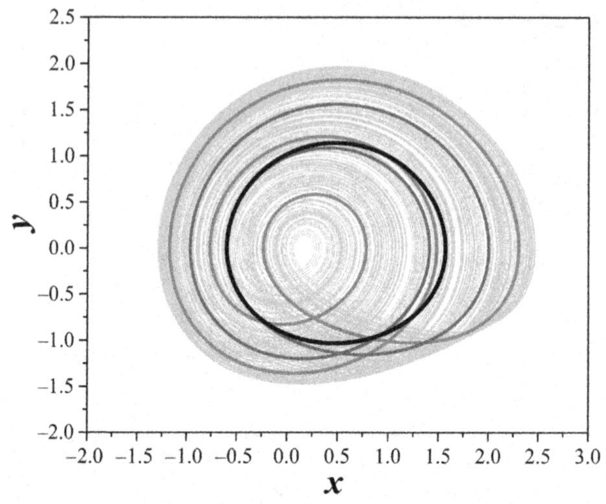

FIG. 13 Different attractors for $a = 0.5$ and $b = 0.4$ and for different initial conditions $z(0)$, where $z(0) = 0.35$ —period-1, $z(0) = 0.22$ —period-2, $z(0) = 0.143$ —period-3, and $z(0) = 0.09$ —chaos.

Finally, the value of initial condition $w(0)$, as a bifurcation parameter, has been used, for the same set of system's (6) parameters. In the same way, as the initial condition $w(0)$ is decreased starting from the value of $w(0) = 0.2$, a bifurcation diagram and the spectrum of Lyapunov exponents, when varying the initial value $w(0)$, have been plotted (Fig. 14). An infinite number of coexisting attractors are also produced in the case of a variation of the initial condition $w(0)$. Fig. 15 displays four of the attractors that have been produced as the value of initial condition $w(0)$ is varied.

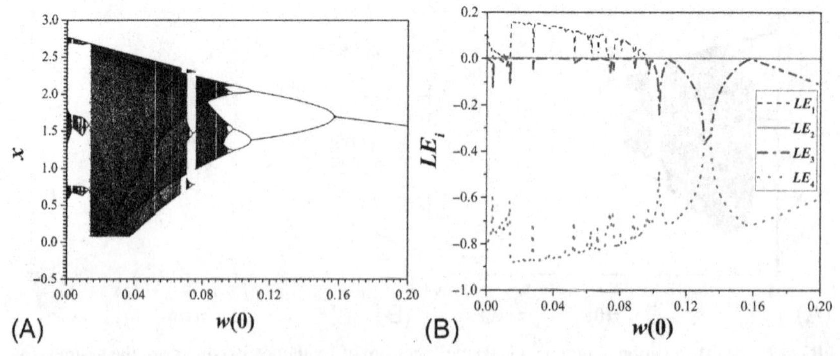

FIG. 14 (A) Bifurcation diagram and (B) the spectrum of Lyapunov exponents of the system (6), when varying the initial value $w(0)$, for $a = 0.5$ and $b = 0.4$.

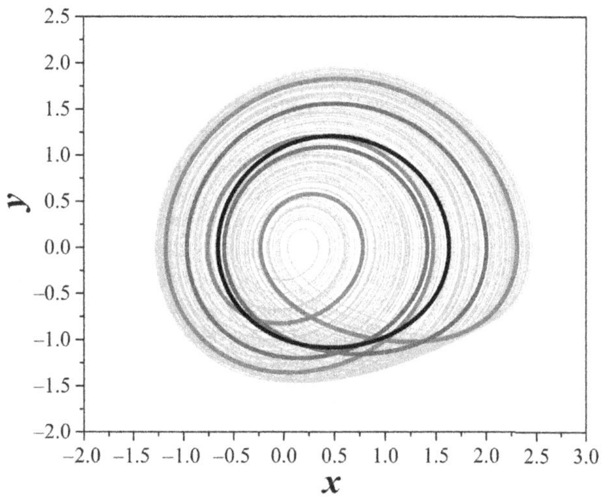

FIG. 15 Different attractors for $a = 0.5$ and $b = 0.4$ and for different initial conditions $w(0)$, where $w(0) = 0.18$ —period-1, $w(0) = 0.12$ —period-2, $w(0) = 0.075$ —period-3, and $w(0) = 0.05$ —chaos.

4 CONCLUSION

A memristor-based chaotic hyperjerk system with line of equibria has been studied in this chapter. The rich dynamical behavior of the system is confirmed by the presented numerical bifurcation diagrams and Lyapunov spectrums. Also, this work proves that the behavior of coexisting infinite attractors and the extreme multistability phenomenon depend on the system's initial conditions. Due to our limited knowledge about the extreme multistability of these systems, we will continue our work focusing on such systems and their applications.

REFERENCES

[1] Chua LO. Memristors—the missing circuit element. IEEE Trans Circuit Theory 1971; 18(5):507–19.

[2] Chua LO, Kang SM. Memristive devices and systems. In: Proceeding of the IEEE no 23; 1976. p. 209–23.

[3] Strukov DB, Snider GS, Stewart GR, Williams RS. The missing memristor found. Nature 2008;453:80–3.

[4] Ventra MD, Pershin YV, Chua LO. Circuit elements with memory: memristors, memcapacitors, and meminductors. Proc IEEE 2009;97(10):1717–24.

[5] Itoh M, Chua LO. Memristor oscillators. Int J Bifurcation Chaos 2008;18:3183–206.

[6] Zhao YB, Tse CK, Feng JC, Guo YC. Application of memristor-based controller for loop filter design in charge-pump phase-locked loop. Circuits Syst Signal Process 2013;32:1013–23.

[7] Wu AL, Zhang JN, Zeng ZG. Dynamical behaviors of a class of memristor-based Hopfield networks. Phys Lett A 2011;375:1661–5.

[8] Volos CK, Kyprianidis IM, Stouboulos IN. The memristor as an electric synapse synchronization phenomena. In: Proceedings of international conference DSP2011 (Confu, Greece); 2011. p. 1–6.

[9] Yang JJ, Strukov DB, Stewart DR. Memristive devices for computing. Nat Nanotechnol 2013;8:13–24.

[10] Driscoll T, Quinn J, Klein S, Kim HT, Kim BJ, Pershin YV, et al. Memristive adaptive filters. Appl Phys Lett 2010;97:093502-1-3.

[11] Wang L, Zhang C, Chen L, Lai J, Tong J. A novel memristor-based rSRAM structure for multiple-bit upsets immunity. IEICE Electron Express 2012;9:861–7.

[12] Shang Y, Fei W, Yu H. Analysis and modeling of internal state variables for dynamic effects of nonvolatile memory devices. IEEE Trans Circuits Syst I Reg Pap 2012;59:1906–18.

[13] Cepisca C, Grigorescu SD, Ganatsios S, Bardis NG. Passive and active compensations for current transformers. Metrologie 2008;4:5–310.

[14] Bogdan M, Buga M, Medianu R, Cepisca C, Bardis N. Obtaining a model of photovoltaic cell with optimized quantum efficiency. Sci Bull Electr Eng Fac 2011;2(16):36–41.

[15] Corinto F, Ascoli A. Memristor based elements for chaotic circuits. PIEICE Nonlinear Theory Its Appl 2012;3(3):336–56.

[16] Corinto F, Ascoli A. Analysis of current-voltage characteristics for memristive elements in pattern recognition systems. Int J Circuit Theory Appl 2012;40(12):1277–320.

[17] Sabarathinam S, Volos CK, Thamilmaran K. Implementation and study of the nonlinear dynamics of a memristor-based Duffing oscillator. Nonlinear Dyn 2016;87:1–13.

[18] Chen M, Li M, Yu Q, Bao B, Xu Q, Wang J. Dynamics of self-excited attractors and hidden attractors in generalized memristor-based Chuas circuit. Nonlinear Dyn 2015;81(1–2):215–26.

[19] Chen M, Xu Q, Lin Y, Bao B. Multistability induced by two symmetric stable nodi-foci in modified Canonical Chuas's circuit. Nonlinear Dyn 2017;87(2):789–802.

[20] Chen G, Leonov GA, Kuznetsov NV, Mokaev TN. Hidden attractors on one path: Glukhovsky-Dolzhansky, Lorenz, and Rabinovich systems. Int J Bifurcation Chaos 2017;27(8). https://doi.org/10.1142/80218127417501152.

[21] Bao B, Li Q, Wang N, Chen M, Xu Q. Multistability in Chua's circuit with two stable nodi-foci. Chaos 2016;26(4):043111.

[22] Bao B, Jiang T, Xu Q, Chen M, Wu H, Hu Y. Coexisting infinitely many attractors in active band-pass filter-based memristive circuit. Nonlinear Dyn 2016;86:1711–23.

[23] Wu H, Bao B, Liu Z, Xu Q, Jiang P. Chaotic and periodic bursting phenomena in a memristive Wien-bridge oscillator. Nonlinear Dyn 2016;83(1–2):893–903.

[24] Kuznetsov NV, Leonov GA, Vagaitsev VI. Analytical-numerical method for attractor localization of generalized Chua's system. IFAC Proc Vol (IFAC-Papers Online) 2010;43 (11):29–33.

[25] Leonov GA, Kuznetsov NV, Vagaytsev VI. Localization of hidden Chua's attractors. Phys Lett A 2011;375(23):2230–3.

[26] Leonov GA, Kuznetsov NV. Hidden attractors in dynamical systems. From hidden oscillations in Hilbert-Kolmogorov, Aizerman, and Kalman problems to hidden chaotic attractor in Chua circuits. Int J Bifurcation Chaos 2013;23(1). https://doi.org/10.1142/S0218127413300024.

[27] Leonov GA, Kuznetsov NV, Mokaev TN. Homoclinic orbits, and self-excited and hidden attractors in a Lorenz-like system describing convective fluid motion. Eur Phys J Spec Top 2015;224:1421–58. https://doi.org/10.1140/epjst/e2015-02470-3.

[28] Stankevich NV, Leonov GA, Kuznetsov NV, Chua L. Scenario of the birth of hidden attractors in the Chua circuit. Int J Bifurcation Chaos 2017;27(12). https://doi.org/10.1142/S0218127417300385.

[29] Ngonghala CN, Feudel U, Showalter K. Extreme multistability in a chemical model system. Phys Rev E 2011;83(5):056206.
[30] Hens C, Dana SK, Feudel U. Extreme multistability: attractor manipulation and robustness. Chaos 2015;25(5):053112.
[31] Patel MS, Patel U, Sen A, Sethia GC, Hens C, Dana SK, et al. Experimental observation of extreme multistability in an electronic system of two coupled Rssler oscillators. Phys Rev E Stat Nonlin Soft Matter Phys 2014;89(2):022918.
[32] Hens CR, Banerjee R, Feudel U, Dana SK. How to obtain extreme multistability in coupled dynamical systems. Phys Rev E 2012;85(3):035202.
[33] Morfu S, Nofiele B, Marqui P. On the use of multistability for image processing. Phys Lett A 2007;367(3):192–8.
[34] Bao B, Bao H, Wang N, Chen M, Xu Q. Hidden extreme multistability in memristive hyperchaotic system. Chaos Solitons Fractals 2017;94:102–11.
[35] Bo-Chen B, Jian-Ping BX, Zhong L. Initial state dependent dynamical behaviors in a memristor based chaotic circuit. Chin Phys Lett 2010;27(7):070504.
[36] Sprott JC. Some simple chaotic flows. Phys Rev E 1994;50:647–50.
[37] Sprott JC. Chaos and time-series analysis. Oxford: Oxford University Press; 2003.
[38] Sprott JC, Linz SJ. Algebraically simple chaotic flows. Int J Chaos Theory Appl 2000;5:3–22.
[39] Linz SJ. Nonlinear dynamical models and jerky motion. Am J Phys 1997;65:523–6.
[40] Eichhorn R, Linz SJ, Hänggi P. Transformations of nonlinear dynamical systems to jerky motion and its application to minimal chaotic flows. Phys Rev E 1998;58:7154–64.
[41] Schot SH. Jerk: the time rate of change of acceleration. Am J Phys 1978;46:1090–4.
[42] Chlouverakis KE, Sprott JC. Chaotic hyperjerk systems. Chaos Solitons Fractals 2006;8:739–46.
[43] Pham VT, Vaidyanathan S, Volos CK, Jafari S, Wang X. A chaotic hyperjerk system based on memristive device. In: Vaidyanathan S, Volos C, editors. Advances and applications in chaotic systems, vol. 636. Studies in computational intelligence. Cham: Springer; 2016.
[44] Pham VT, Vaidyanathan S, Volos CK, Jafari S, Wang X. A chaotic hyperjerk system based on memristive device. In: Advances and applications in chaotic systems. Switzerland: Springer International Publishing; 2016.
[45] Pham VT, Vaidyanathan S, Volos CK, Hoang TM, Van Yem V. Dynamics, synchronization and SPICE implementation of a memristive system with hidden hyperchaotic attractor; 2016, Switzerland: Springer International Publishing; 2016, p. 35–52.
[46] Kuznetsov NV, Leonov GA, Mokaev TN, Prasad A, Shrimali MD. Finite-time Lyapunov dimension and hidden attractor of the Rabinovich system. Nonlinear Dyn 2018;92:267–85. https://doi.org/10.1007/s11071-018-4054-z.
[47] Strogatz SH. Nonlinear dynamics and chaos: with applications to physics, biology, chemistry, and engineering. Cambridge, MA: Perseus Books; 1994.
[48] Pham VT, Volos CK, Kapitaniak T. Systems with hidden attractors: from theory to realization in circuits. Switzerland: Springer International Publishing; 2017. p. 35–52.
[49] Wolf A, Swift JB, Swinney HL, Vastano J. Determining Lyapunov exponents from a time-series. Physica D 1985;16(3):285–317.

FURTHER READING

[50] Lassoued A, Boubaker O. On new chaotic and hyperchaotic systems: a literature survey. Nonlinear Anal Modell Control 2016;21(6):770–89.

FURTHER READING

Chapter 7

Parameter Estimation of Chaotic Systems Using Density Estimation of Strange Attractors in the State Space

Yasser Shekofteh*, Shirin Panahi†, Olfa Boubaker‡ and Sajad Jafari†
**Faculty of Computer Science and Engineering, Shahid Beheshti University, Tehran, Iran,*
†*Biomedical Engineering Department, Amirkabir University of Technology, Tehran, Iran,*
‡*National Institute of Applied Sciences and Technology, Tunis, Tunisia*

1 INTRODUCTION

Parameter estimation methods of chaotic systems have been investigated in a variety of fields [1–4]. The estimation of unknown parameters of chaotic systems is essential because they are very sensitive both to parameters and initial conditions. On the other hand, proving structural identifiability propriety for non-linear systems is still considered a very difficult problem [5]. It is shown that a slight change in the parameters of chaotic systems causes important bifurcation in the time-domain [1,6]. Because of their butterfly effect and random-like behavior in the time domain, the parameter estimation methods of chaotic systems are not easy [7,8].

Defining a proper cost function is a crucial part of a parameter estimation method. Until recently, most parameter estimation methods involving chaotic systems have been carried out using what we may call simple mean-square-error (MSE)-based cost functions. However, these approaches seem to have major limitations, such as random-like behavior of states variables, because these approaches define the cost function in the time-domain [7–9].

In this chapter, we will report the usage of a non-conventional parameter estimation method. Its cost function is based on the density estimation method to extract characteristics of the chaotic system by strange attractors in the state space. As mentioned, the butterfly effect and random-like behavior in the time domain leads us to choose the state space as a suitable domain for analyzing the chaotic system. As we know, the time series generated by the chaotic systems

Recent Advances in Chaotic Systems and Synchronization. https://doi.org/10.1016/B978-0-12-815838-8.00007-8

are ordered in the state space and usually have specific topologies named strange attractors. Therefore, the state space may help the process of the parameter estimation of the chaotic systems. We have considered and reported the efficiency of the new cost function in the parameter estimation of chaotic circuits [10,11] and chaotic biological systems [12]. This method can model the strange attractor of the system in the state space by a parametric density estimation method so-called Gaussian mixture model (GMM).

The GMM is a flexible parametric and probabilistic model. It can model the strange attractor generated from the measured data of the chaotic system properly. The GMM is also a commonly used model in the pattern recognition and machine learning domain [13,14]. For example, in the field of the speech recognition, a set of GMMs was introduced to model phone attractors in a reconstructed phase space (RPS) which is a time-independent domain similar to the state space [15–18]. The phone classification results had shown that the GMM could be a useful model to capture the structure and topology of the speech attractors in the RPS.

The purpose of this chapter is to provide a complete survey of the GMM-based cost function and its one-dimensional and two-dimensional parameter estimation method that can cover the large number of chaotic systems. This may help the researchers who are newcomers to the area of system identification using the strange attractor. They can obtain an overview of the proposed method along with all its mathematical formulas. In addition, some information criteria are also introduced to help them to obtain the best model of the attractor. The length effect of simulation data in the process of the parameter estimation method will also be considered.

The structure of this chapter is organized as follows: in the next section, we introduce and analyze a new chaotic system that we have used in our experiments. In Section 3, we investigate the concept and usage of the GMM carefully as our density estimation method. Section 4 deals with the cost function based on Gaussian mixture model. The results related to the cost function, discussions, and parameter estimation of the new chaotic system are reported in Section 5.

2 A NEW CHAOTIC SYSTEM AND ITS BIFURCATION ANALYSIS

In this section, we first introduce a new three-dimensional (3D) system with chaotic behaviors. We have used this system in a 1D and 2D parameter estimation problem. Consider the system described with the following ordinary differential equations:

$$\begin{aligned} \dot{x} &= az \\ \dot{y} &= -bx + z \\ \dot{z} &= -x + 0.7y + 0.3z + 0.88z^2 - 0.5yz - 6.94 \end{aligned} \tag{1}$$

where (x, y, z) refers to the state variables of the system (1), a and b are constants. They are the unknown parameters of the system that need to be estimated. It is worthy to note that the true parameters of the system $a = 1.0$ and $b = 1.0$ ensure chaotic behavior of the system (1). Different projections of the phase portrait for this system are plotted in Fig. 1, which shows its strange attractor in the 2D state spaces. Here, the initial conditions are taken as (−0.10, −5.05, −6.00).

In this part, we investigate the behaviors of the system (1) with variation of parameters a and b. In Figs. 2 and 3 bifurcation diagrams of the system (1) using one of the state variables, here x, are shown with variation of parameters a and b, respectively. As can be seen in Figs. 2 and 3, changing the parameters of the system causes a familiar reverse period doubling route to chaos. Therefore, the system (1) can be a chaotic system.

3 DENSITY ESTIMATION OF THE ATTRACTOR IN THE STATE SPACE USING GMM

State variables of a chaotic system could show a semi-random-like behavior in the time-domain, but they are ordered in the state space so-called strange attractors. The state space can represent all possible states of a system [6,16,19]. The density of the strange attractor points in the state space can be modeled with probabilistic models. The density estimation of the strange attractor provides

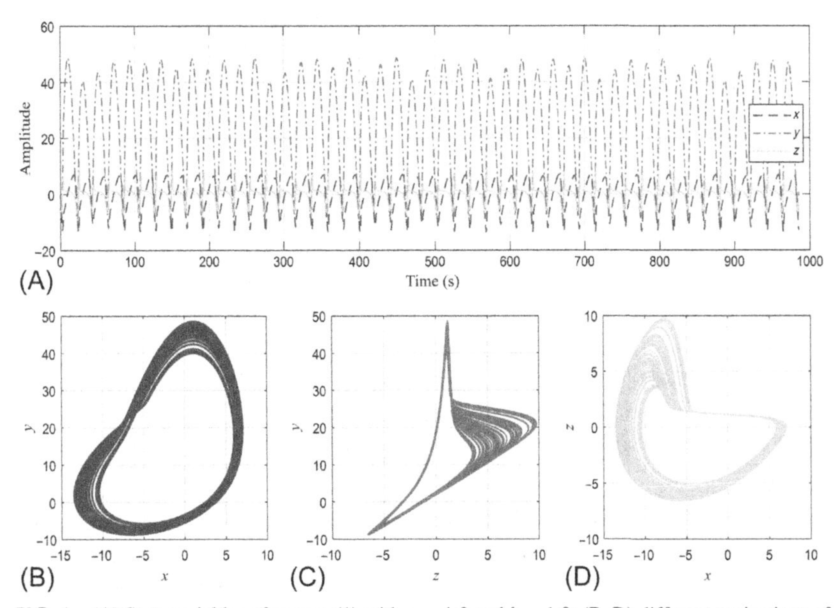

FIG. 1 (A) State variables of system (1) with $a = 1.0$ and $b = 1.0$; (B–D) different projections of the chaotic attractor of system (1) with $a = 1.0$ and $b = 1.0$.

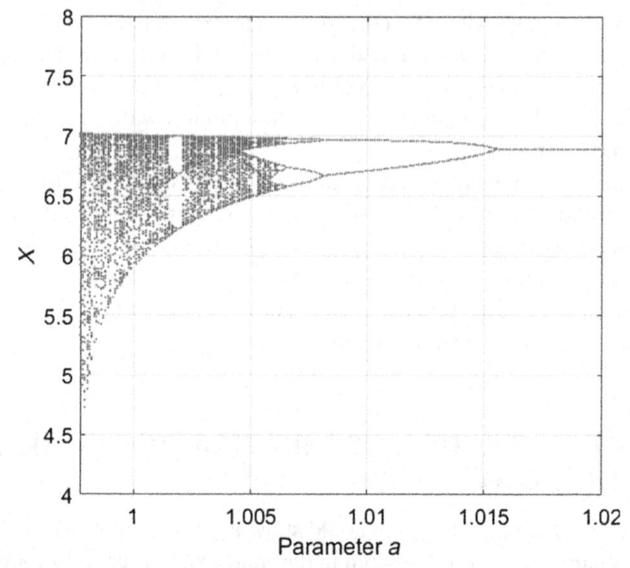

FIG. 2 Bifurcation diagram with variation of parameter a of system (1).

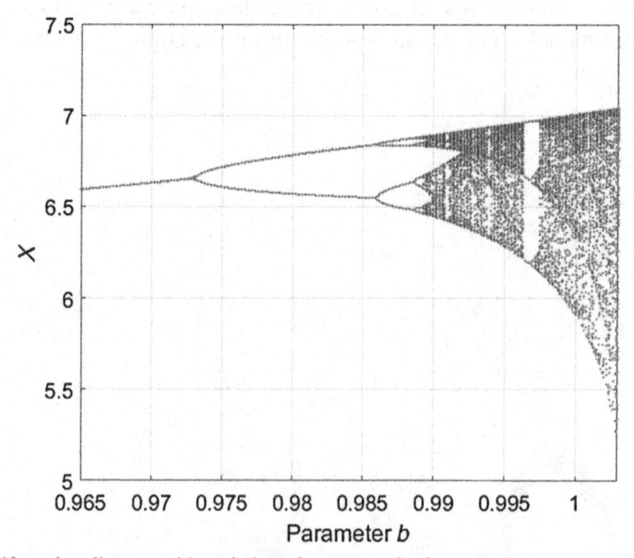

FIG. 3 Bifurcation diagram with variation of parameter b of system (1).

a direct way to obtain its smooth model and interpolates missing values in the time series generated by the strange attractor points. Gaussian mixture model (GMM) as a most-used probabilistic model is one of the best parametric models for representing geometry and density of the strange attractor in the state space. This idea was recently successfully used as a phone classification method by the

GMM modeling of the speech signal in a high-dimension domain named reconstructed phase space (RPS) [15–18].

A GMM with M mixtures is a weighted sum of M individual Gaussian densities. Each Gaussian density as a component of the GMM is represented by three main parts: a mixture weight, a mean vector, and a covariance matrix. A complete GMM is fully defined by all its component densities. So, it can be shown by a set of parameters, λ, as follows:

$$\lambda = \{w_m, \mu_m, \Sigma_m\}, \quad m = 1, \ldots, M$$
$$p(\mathbf{v}|\lambda) = \sum_{m=1}^{M} w_m \frac{1}{(2\pi)^{D/2}} \frac{1}{|\Sigma_m|^{1/2}} \exp\left\{\frac{-1}{2}(\mathbf{v} - \mu_m)^T \Sigma_m^{-1} (\mathbf{v} - \mu_m)\right\} \tag{2}$$

and $p(\mathbf{v} \mid \lambda)$ is a conditional probability of a D-dimensional observation vector $\mathbf{v}$ given the GMM of λ. Here, the observation vector $\mathbf{v}$ is equal to the state vector of the chaotic system. The value of $p(\mathbf{v} \mid \lambda)$ is a so-called likelihood score. It expresses how probable the observed vector $\mathbf{v}$ is for the GMM of λ. In Eq. (2) $|.|$ is the determinant operator, exp.(.) denotes the exponential function, and M is the number of Gaussian components (mixtures). In addition, for mth component, $w_m \in [0, 1]$ is an scalar and named mth mixing coefficient or mixture weight, μ_m is the mth D-dimensional mean vector, and Σ_m is the mth $D \times D$ covariance matrix. The mean vector and covariance matrix of a Gaussian component can show the center and the shape of the state space points around that component. It should be noted that the mixing coefficients w_m are nonnegative and constrained to sum up to 1, that is, $\sum_{m=1}^{M} w_m = 1$.

Now, suppose we have a strange attractor of a chaotic system in a D-dimensional state space. Our problem is how to model this attractor by a GMM. As a popular method, a maximum likelihood estimation (MLE) is commonly used to compute the GMM parameters [20]. In other words, the GMM learns the characteristics of the distribution of the attractor by its parameters. Let $S = \{s_1, s_2, \ldots, s_N\}$ be an $N \times D$ matrix consisted N-samples of the time series of the chaotic system in its D-dimensional state space. To find MLE solutions of the GMM parameters, an iterative expectation-maximization (EM) algorithm as an elegant and powerful method is usually utilized [14] as described in the following sections.

Suppose we have a GMM with M mixtures. The goal is to maximize the likelihood function over the training data, S, with respect to the parameters of the GMM.

3.1 Initialization Step

Choose an initial setting for the parameters of the GMM: the mean vector μ_m, covariance matrix Σ_m and mixing coefficients w_m in Eq. (2).

Then, evaluate the initial value of the logarithm of the total likelihood score obtained from all points of the given time series as follows:

$$\log p(S|\lambda) = \log \prod_{n=1}^{N} p(s_n|\lambda) = \sum_{n=1}^{N} \log (p(s_n|\lambda)) \tag{3}$$

Here, we assumed that N samples of the training data come from a set of independent identically distributed (i.i.d.) data, so, we must use the independence assumption for this method [14].

3.2 Expectation Step

Evaluate values of $r(s_i \cdot m)$, named responsibility of ith sample of the training data S given mth Gaussian component, using the current values of the GMM parameters:

$$r(s_i \cdot m) = \frac{w_m \frac{1}{(2\pi)^{D/2}} \frac{1}{|\Sigma_m|^{1/2}} \exp\left\{\frac{-1}{2}(s_i - \mu_m)^T \Sigma_m^{-1} (s_i - \mu_m)\right\}}{\sum_{j=1}^{M} w_j \frac{1}{(2\pi)^{D/2}} \frac{1}{|\Sigma_j|^{1/2}} \exp\left\{\frac{-1}{2}(s_i - \mu_j)^T \Sigma_j^{-1} (s_i - \mu_j)\right\}} \tag{4}$$

The responsibility $r\ (s_i,\ m)$ illustrates how the mth Gaussian component GMM takes in explaining the observed s_i [14].

3.3 Maximization Step

Re-estimate the parameters of the GMM utilizing the estimated values of the responsibilities as follows:

$$N_m = \sum_{i=1}^{N} r(s_i, m) \tag{5}$$

$$\mu_m = \frac{1}{N_m} \sum_{i=1}^{N} r(s_i, m) s_i \tag{6}$$

$$\Sigma_m = \frac{1}{N_m} \sum_{i=1}^{N} r(s_i, m)(s_i - \mu_m)(s_i - \mu_m)^T \tag{7}$$

$$w_m = \frac{N_m}{N} \tag{8}$$

3.4 Likelihood Score Evaluation

Evaluate the logarithm of the total likelihood score in Eq. (3) and check for convergence criterion. If the convergence criterion is not satisfied, return to step 1.3.2.

3.5 Determining the Appropriate Number of Gaussian Components

As mentioned, a GMM with M mixtures is a weighted sum of M individual Gaussian densities. The selection of the number of components M is crucial and has a significant effect on the performance and accuracy of the parameter estimation method. Generally, it depends on the complexity of the attractor distribution in the state space. There are too few components of the GMM which are not able to model the distribution of the attractor suitably, while too many components of the GMM are the cause of overfitting problems; therefore, the learned model cannot be generalized to use as a global model.

Suppose we have used the GMM for the attractor modeling with some different number of Gaussian components (M). Now, our question is, what model is good for our problem? In fact, there are information criteria to estimate the relative quality of statistical models such as GMMs. For example, the negative of the log-likelihood function (−Log Likelihood), Akaike information criterion (AIC), and Bayesian information criterion (BIC) can be used as the model selection criteria [21–24]. Both BIC and AIC attempt to resolve the model selection problem by introducing a penalty term for the number of parameters in the GMM with a set of N samples of the training data. The AIC deals with the tradeoff between the quality of fit of the GMM (mostly calculated by the likelihood score) and the simplicity of the GMM. It tells nothing about the absolute quality of a GMM, but only illustrates the quality relative to other the GMMs. The penalty discourages overfitting problems because increasing the number of parameters in the model always improves the quality of the fit. AIC and BIC could be expressed as a function of the log-likelihood of the converged mixture model:

$$\mathrm{AIC} = -2\log\left(p(S|\,\lambda)\right) + 2P \tag{9}$$

$$\mathrm{BIC} = -2\,\log\left(p(S|\,\lambda)\right) + P\,\log\left(N\right) \tag{10}$$

where $P = M(d + d(d + 1)/2 + 1)$ is the number of free parameters for the GMM λ with full covariance matrices, and D shows the size of the feature vector in the GMM modeling. In BIC, the penalty term is larger than in AIC, because the penalty is $2P$ for AIC, whereas with BIC the penalty is $P\,\log(N)$and $\log(N) > 2$ in most cases. Therefore, BIC generally results in a smaller choice for M than by AIC [25,26].

So, the number of components can be selected according to these information criteria. The EM algorithm of the GMM learning is then run for several different values of M, and the model which minimizes the chosen criterion is selected. In addition, it is common to utilize a trial-and-error method to choose an adequate value of M for a special task [12]. For example, in our attractor modeling problem, to get a proper GMM model of the attractor in the state space, we can evaluate some values of M and evaluate the performance of the parameter estimation method.

4 PARAMETER ESTIMATION OF THE CHAOTIC SYSTEM USING GMM-BASED COST FUNCTIONS

In this section, we will completely introduce the parameter estimation method based on the GMM modeling for chaotic systems such as the system (1). The time-independent property of the state space, which can show complex behaviors of a strange attractor, is a sufficient reason to use it in the cost function of the parameter estimation process [6]. As mentioned before, the utilized cost function is based on the attractor modeling using a GMM.

Now, we consider the parameter estimation problem. Suppose there is a deterministic chaotic system with some unknown parameters. At first, we must model the attractor of that system by a GMM over the recorded trajectory of the system. To find the unknown parameters, we can then simulate the known model structure of the system with some acceptable values for parameters and generate some new trajectories. Finally, using the similarity score, based on the likelihood score, between the GMM model of the chaotic system and the time series obtained by the stimulation of the model, we can estimate unknown parameters by searching the maximum value of the similarity scores. So, the parameter estimation of a known chaotic system with unknown parameters can be done by the following two phases; a learning phase, which includes fitting the GMM to the attractor of the system, and an evaluation phase to select the best values of parameters which cause the maximum similarity score over the learned GMM. Following are these phases in detail.

4.1 Learning Phase

The first phase of the parameter estimation approach is the learning phase to find the GMM parameters, λ in Eq. (2). The GMM learns the attractor's distribution of the system (1) by an iterative expectation-maximization (EM) algorithm. The GMM is used as density estimation in the state space. In addition, determining the appropriate number of GMM components is crucial for this phase.

4.2 Evaluation Phase

The second phase of the parameter estimation approach is finding the best parameters of the known model structure of the chaotic system (with unknown parameters) using the learned GMM in the first phase. Here, the search space will be formed from a set of acceptable values of the model parameters. For example, as a 2D parameter estimation method, we suppose that the true values of the parameters a & b of the system (1) are unknown. Then, for each pair of parameters (a,b), the chaotic system (1) will be simulated, and a trajectory $T(a,b) = (t_1, t_2, t_3, \ldots, t_K | a, b)$ with K samples will be obtained where each of t_K is a three-dimensional measured data $(D = 3)$ in the state space.

Finally, using an average point-by-point log-likelihood score obtained from the learned GMM, λ, a similarity-based score can be computed as follows:

$$\log\left(p\left(T^{(a,b)}|\lambda\right)\right) = \log\prod_{k=1}^{K} p(t_n|\lambda) = \sum_{k=1}^{K} \log\left(p(t_k|\lambda)\right) \qquad (11)$$

where $\boldsymbol{T}^{(a,b)}$ is a matrix whose rows are composed from the state space vectors of the system trajectory with the model's parameters (a,b), and K is the number of the state space points. Here, we assumed that K samples of the evaluation data come from a set of independent identically distributed (i.i.d.) data, so we can use the independence assumption again.

The parameter estimation method of the model is accomplished by computing Eq. (11) and selecting the parameters of the model which can gain the best similarity-based score, here meaning the maximum score. If we use the negative of the similarity-based score, then the parameter estimation becomes a cost function minimization. Finally, the best selection of the unknown parameters, $(a,b)^*$, would be conducted by the following criteria, $J(.)$, based on the negative of mean log-likelihood score:

$$(a,b)^* = \text{argmin}\{J((a,b))\},\ J((a,b)) = -\frac{1}{K}p\left(T^{(a,b)}|\lambda\right) \qquad (12)$$

where K is the number of the state space points in the evaluation phase. Eq. (12) shows the utilized cost function. Here, λ is the learned GMM of the system attractor obtained from the phase A, and (a,b) is the set of the acceptable pair values for the parameters of system (1).

The objective of the 2D parameter estimation method will be to determine the parameters (a,b) of the system when the cost function is minimized. The minimum value of the cost function guarantees the best solution of the parameter estimation method. Fig. 4 summarizes the usage of the GMM as the introduced cost function for the parameter estimation problem.

5 SIMULATION RESULTS

In this section, a set of simulation results are reported to show the performance of the GMM-based parameter estimation method. They are considered 1D and 2D estimation problems for the system (1). To do this, first, we need to get a proper GMM for the strange attractor modeling of the system (1). Then, some simulations are done to investigate the acceptability of the parameter estimation method of the chaotic system (1).

5.1 The GMM of the Chaotic System

The chaotic system (1) has three variables in the state space. So, the observation vector of its attractor will be formed as $v = [x, y, z]$, and we must select $D = 3$ as

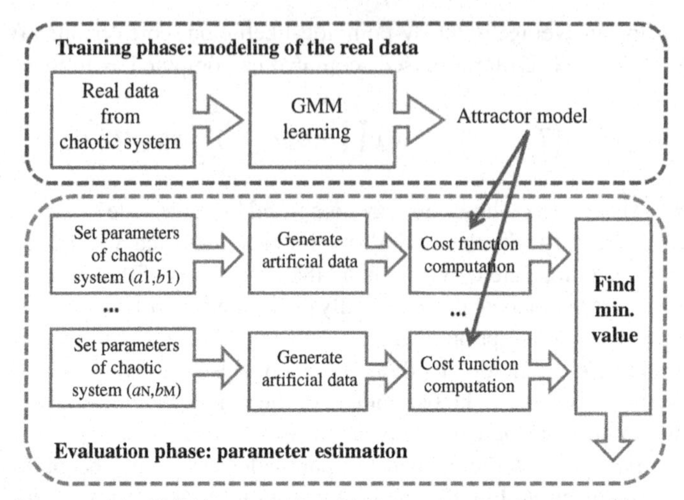

FIG. 4 Diagram of the GMM-based cost function for parameter estimation problem.

the state space dimension in Eq. (2). To generate the attractor points of the chaotic system (1) as real data, its model has been simulated with parameters $a = 1.0$ and $b = 1.0$ by a fourth-order Runge-Kutta method with a step size of 10 ms [27,28]. For training data at the first phase, a set of sequential samples of the system (1) including 100,000 samples (equal to 1000 s time length) has been recorded. The initial conditions were set to (−0.10, −5.05, −6.00) as initial conditions of the system (1). Here, we assume that this recorded training data must lead us to estimate unknown parameters of the chaotic system (1), a and b, by minimization of the GMM-based cost function.

Using obtained training data from the chaotic system, we can learn a GMM in order to model the geometry of the attractor in the state space. In other words, the GMM computation fits a parametric model to the distribution of the attractor in the state space. Fig. 5 shows the attractor of the chaotic system (1) in a three-dimensional state space along with its GMM modeling using $M = 64$ Gaussian components. In this figure, every three-dimensional ellipsoid corresponds to one of the Gaussian components.

As can be seen from Fig. 5, the Gaussian components attempt to cover the attractor in the state space. To show the effect of the number of Gaussian mixtures, in Fig. 6, the attractor of the chaotic system (1) and its GMM models are shown for different values of $M = 16$, 32, 48, and 64.

As can be seen from Fig. 6, when we increase the number of Gaussian components, more details of the trajectory of the chaotic attractor can be covered by the added Gaussian components. Therefore, in these experiments, the best GMM modeling of the attractor can be obtained by $M = 64$, which shows a precise model of the chaotic attractor. Therefore, by increasing the number of Gaussian components in the GMM, it can cover more complexity of the given

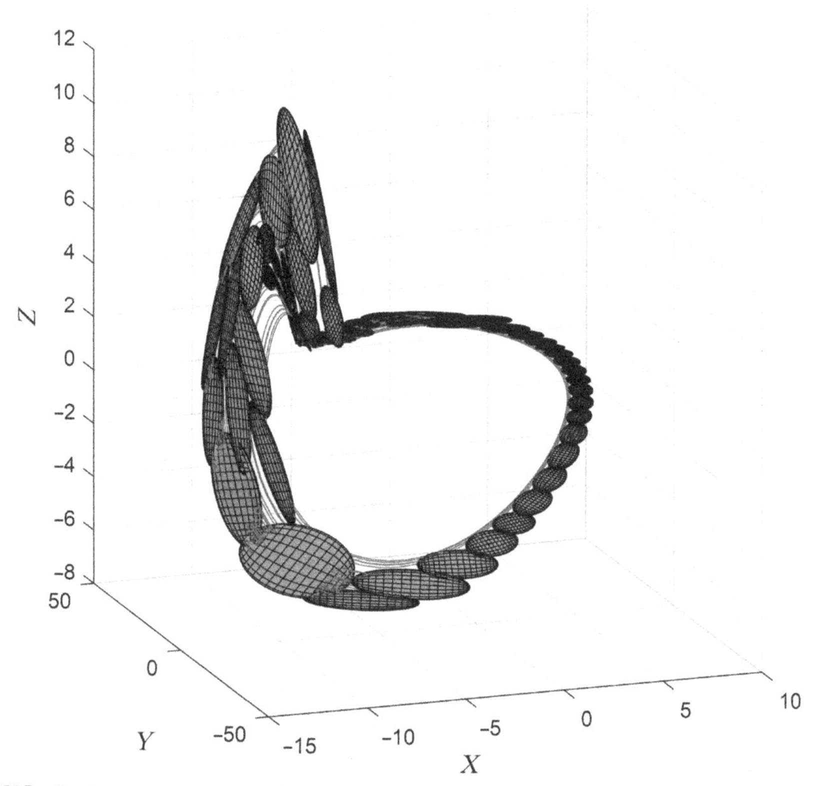

FIG. 5 Plot of the chaotic attractor of the system (1) and its GMM modeling with $M = 64$ components in the 3D state space. Here, the parameters of the system (1) are set to $a = 1.0$ and $b = 1.0$.

time series in its model. The higher value of M can improve the performance of the cost function, but it also increases the computational cost and may be lead to overfitting problems.

In Fig. 7, the information criteria such as AIC, BIC, and the negative of the log-likelihood are considered for the GMM selection problem. It shows that $M = 64$ is a good choice for the number of GMM components, because of minimization of the criteria.

5.2 1D Parameter Estimation of the Chaotic System

In this section, we have used a similar method to generate data for evaluation phase. A fourth-order Runge-Kutta method with a step size of 10 ms and a total of 10,000 samples corresponding to a time of 100 s is used for this phase. As 1D parameter estimation methods, we must evaluate the cost function to estimate true values of the unknown parameter a or b. Here, different numbers of the

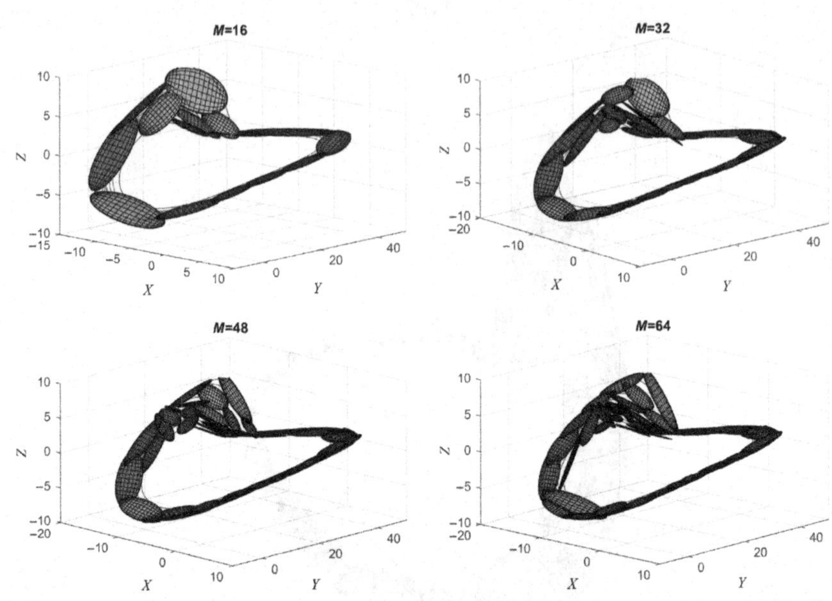

FIG. 6 Plot of the chaotic attractor of the system (1) and its GMM modeling with $M = 16, 32, 48$, and 64 components in the 3D state space.

GMM's components, $M = 16, 32, 48$, and 64, are used to show the sufficiency of the cost function. The experimental results of the cost function versus the values of the parameters a and b are depicted in Figs. 8 and 9, respectively.

As can be seen in Figs. 8 and 9, all of the cost functions can show convex functions around the desired points $a = 1.0$ and $b = 1.0$, respectively. So, they are acceptable for the parameter estimation methods. In particular, the effect of changing the parameter of the system is shown as a monotonically trend along with a global minimum at the exact expected value of the desired parameters $(a = 1.00,\ b = 1.00)$. Therefore, the GMM-based cost function has the desired ideal properties for the parameter estimation problem. Moreover, Figs. 8 and 9 show the effect of increasing the number of GMM components, M, used in the GMM modeling. In this case, $M = 64$ represents the best performance to identify the parameters a & b.

5.3 The Effect of the Length of the Evaluation Data

In this section, we consider the effect of the length of the evaluation data on the performance of the GMM-based cost function for 1D parameter estimation method. As could be seen in Eqs. (12) and (11), for cost function computation of the evaluation data, it is not directly dependent on the number of the samples of the evaluation data, K, or equally the length of the evaluation data, because of the mean operator. But as we know, if we have a longer length of the evaluation

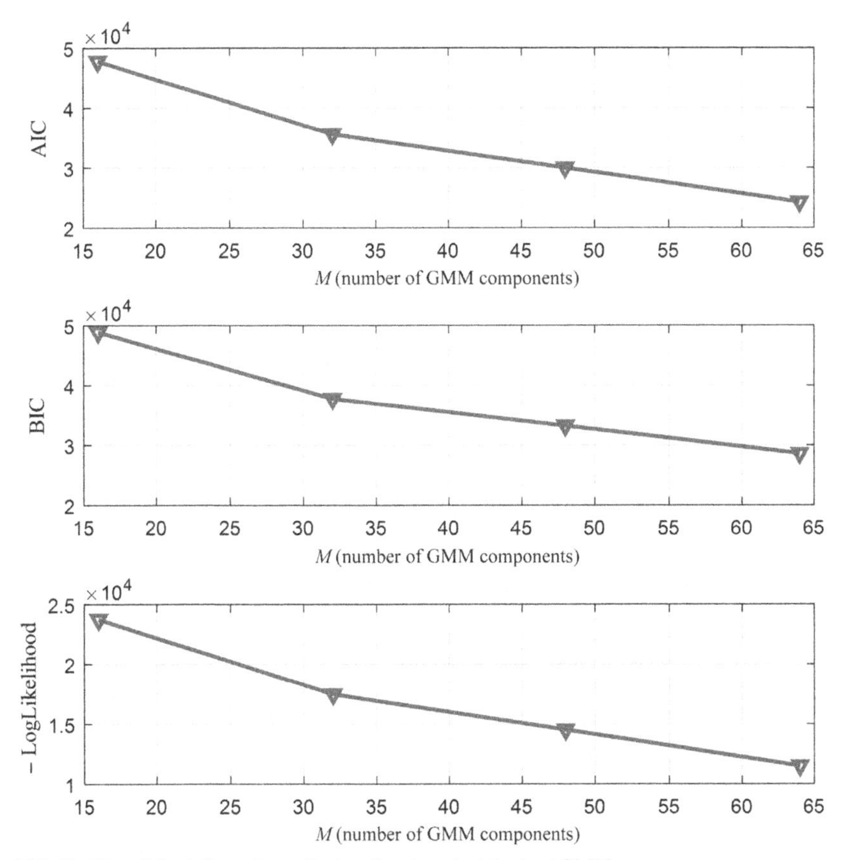

FIG. 7 Plot of the information criteria values to select the best GMM.

data, we can predict better attractor of the chaotic system. Therefore, the similarity computation using the likelihood function will be done truly. In other words, decreasing the length of time series may lead to inadequate or incorrect estimation of the parameters, and increasing the length of time series lead to better estimation of parameters.

In Figs. 10 and 11, we show the effect of reducing the value of K in our parameter estimation problem. The parameter L here represents the used percentages of the samples. For example, $L = 25\%$ means that the first 14 samples of the evaluation data are used. As noted earlier, the initial length of the evaluation data was considered 100 s. So, $L = 25\%$ equals 25 s for the evaluation data length.

As shown in Figs. 10 and 11, when length parameter L is too small, the results are poor because the attractor in the state space will not be formed enough. As length parameter L increases, the results of cost function improve. So, we can find a threshold beyond which the results will not be significantly

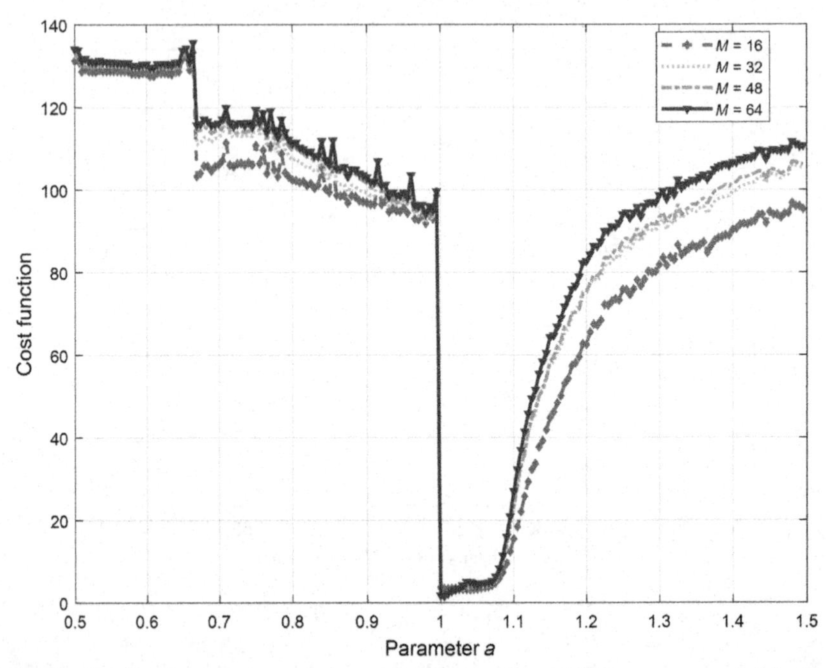

FIG. 8 Cost function versus parameter a, with different number of GMM components (M) for the 1D parameter estimation method.

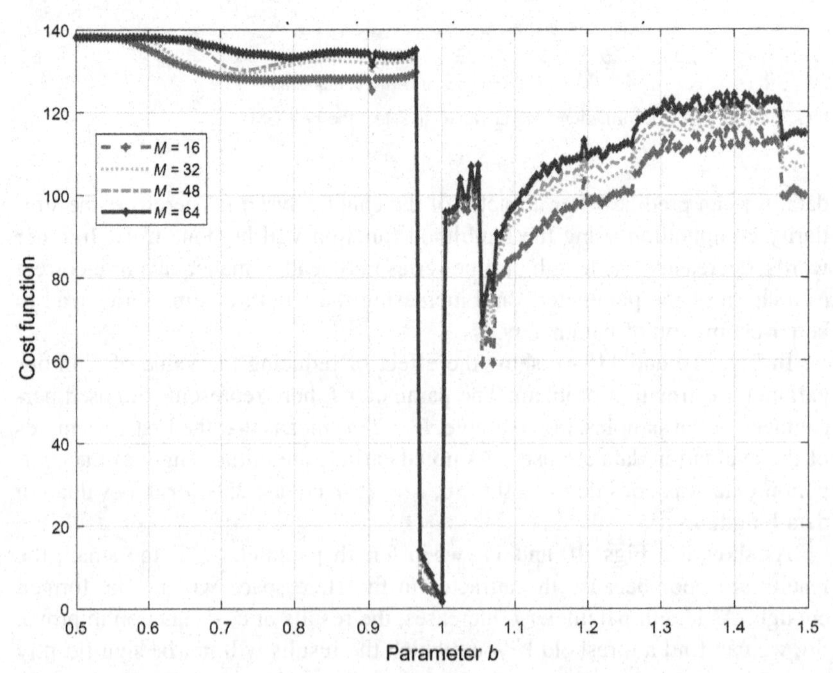

FIG. 9 Cost function versus parameter b, with different number of GMM components (M) for 1D parameter estimation method.

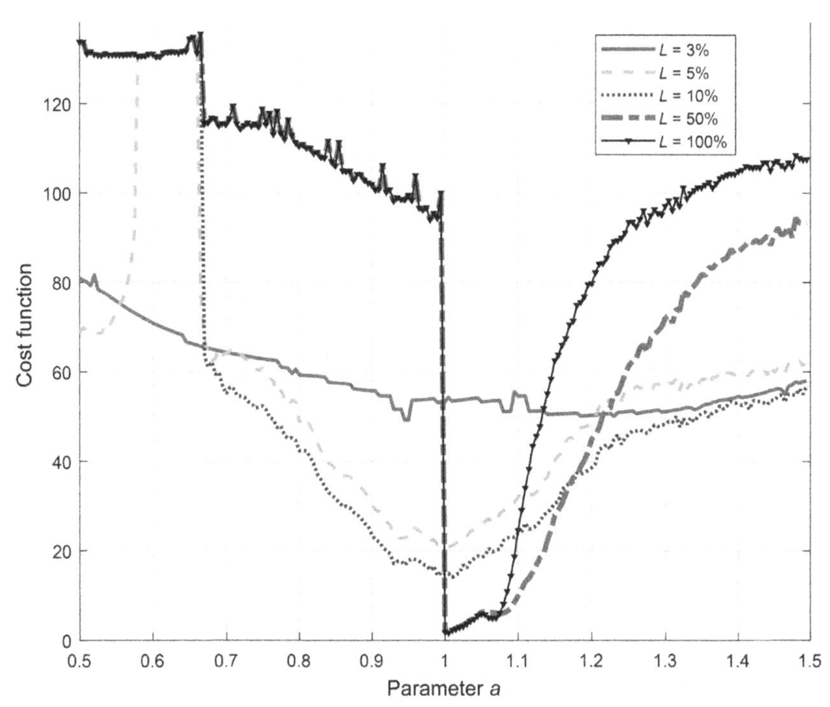

FIG. 10 Effects of reducing the evaluation data samples by parameter L on the cost function versus parameter a ($M = 64$).

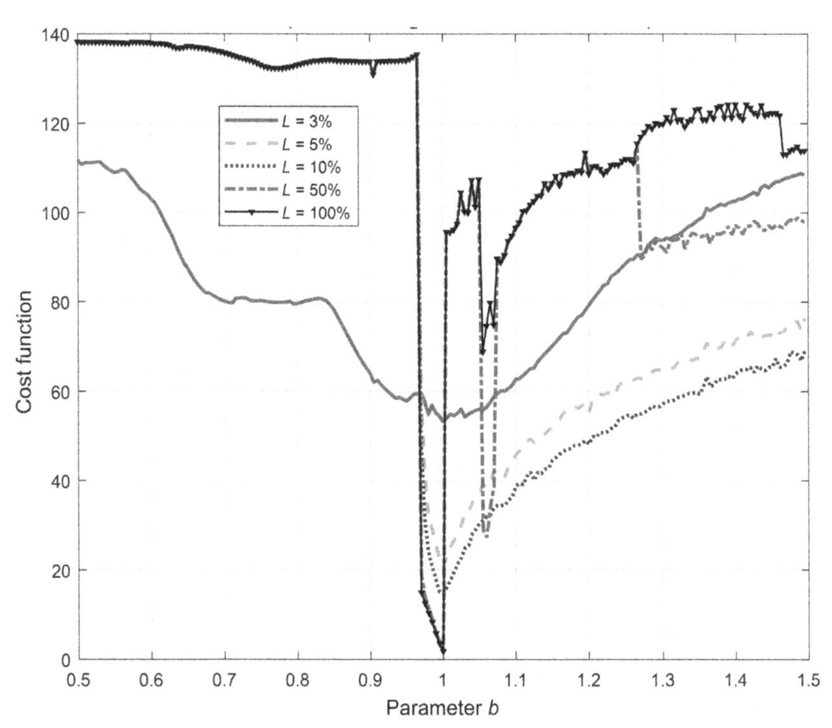

FIG. 11 Effects of reducing the evaluation data samples by parameter L on the cost function versus parameter b ($M = 64$).

affected. In this case, it is recommended to choose the length of the evaluation data between 50 and 100 s, considering good search quality.

In addition, in Figs. 12 and 13, we show the effects of increasing the length of the evaluation data in our parameter estimation problem. Here, the length parameter L is utilized again. For example, $L = 200\%$ means that we have used twice the number of samples for the evaluation data than the length of initial test. So, $L = 200\%$ equals 200 s for the data length.

As shown in Figs. 12 and 13, when length parameter L is increased, the performance of the parameter estimation method is not affected. So, $L = 100\%$ equals 100 s for the evaluation data length is sufficient for our experiments.

5.4 2D Parameter Estimation of the Chaotic System

In this section, we provide the results of a two-dimensional parameter estimation experiments. In Figs. 14 and 15, a contour plot of the cost function and its "cost surface" are respectively shown for the chaotic system (1) with $M = 64$ Gaussian components along with variation in the parameters, a & b. The minimum value of the point on those plots gives the parameters for the best model.

As can be seen in Figs. 14 and 15, the global minimum of the cost function is in the right place ($a = 1.00$ and $b = 1.00$). More than that, the surface of the cost function is almost convex near the best parameters (which makes it an easy case for any optimization approach, such as particle swarm optimization (PSO),

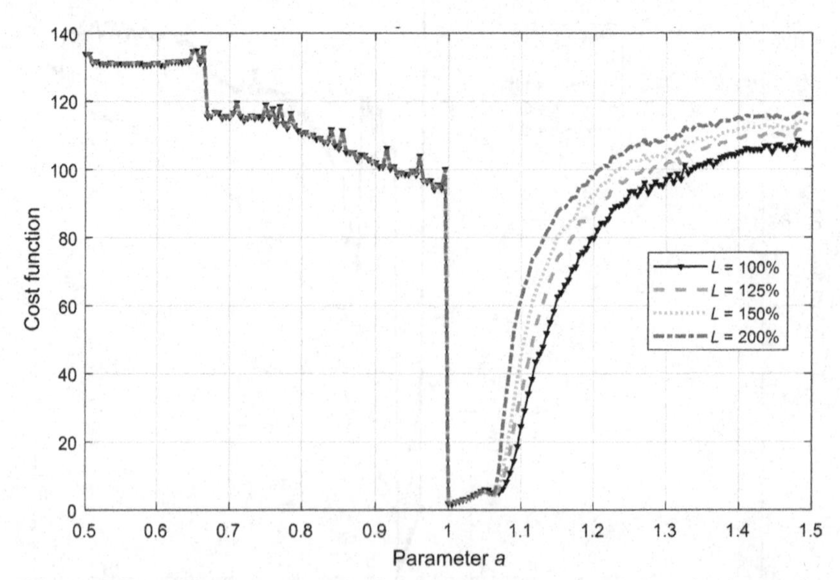

FIG. 12 Effects of increasing the evaluation data samples by parameter L on the cost function versus parameter a ($M = 64$).

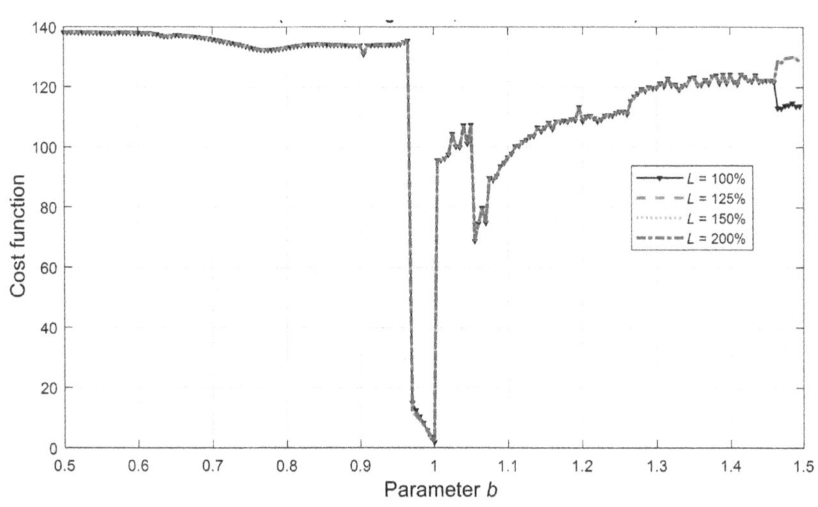

FIG. 13 Effects of increasing the evaluation data samples by parameter L on the cost function versus parameter b ($M = 64$).

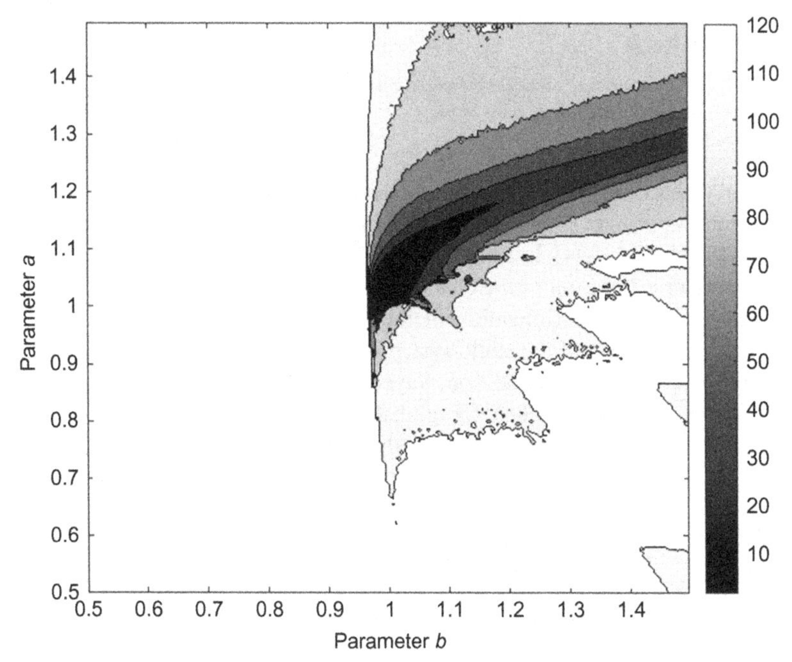

FIG. 14 The "contour plot" of the GMM-based cost function for the chaotic system (1) along with variations in the parameters a and b ($M = 64$).

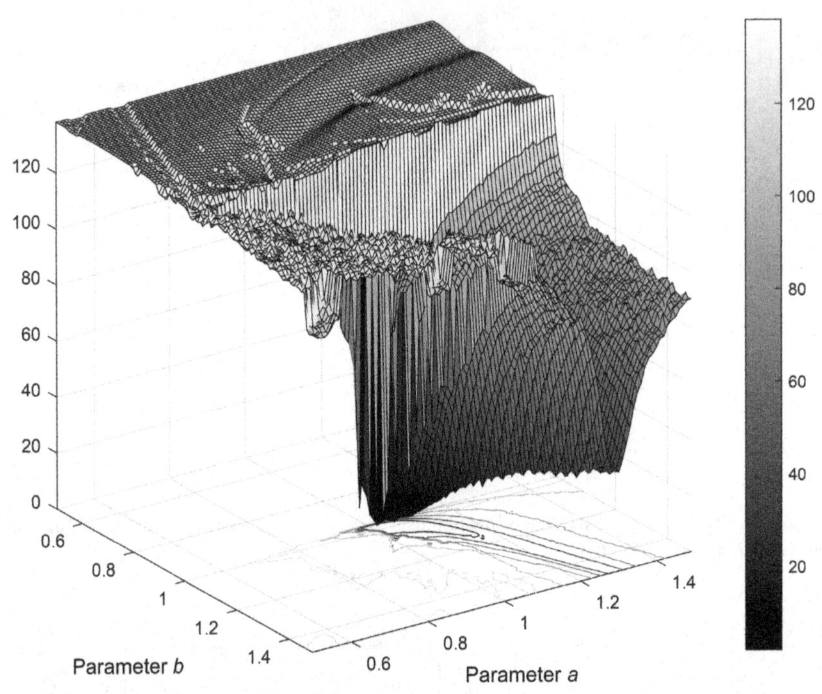

FIG. 15 The "cost surface" of the GMM-based cost function for the chaotic system (1) along with variations in the parameters, a & b ($M = 64$).

artificial bee colony (ABC), teaching–learning-based optimization (TLBO), multi-verse optimizer (MVO), and whale optimization algorithm (WOA)), then moves downhill [29–34]. It is usual that the optimization approaches are used in the parameter estimation of the chaotic systems [35–39]. For example, in [11], two recent efficient optimization methods (WOA and MVO) were used for parameter estimation of a chaotic system with a self-excited attractor in signal encryption application. In addition, Sayed et al. [40] proposed a novel chaotic salp swarm algorithm (SSA) for global optimization and feature selection driven by the simulation behavior of salps.

6 CONCLUSION

In this chapter, an appropriate cost function has been considered for use in the parameter estimation of chaotic systems. The method is based on a density estimation approach of the real system attractor in the state space. Because the GMM is a proper tool as a parametric model, it is used for density estimation approaches. In addition, some information criteria, such as -Log Likelihood, AIC, and BIC were utilized to select the best model of the chaotic attractor, in this case, to choose the number of Gaussian components. Also, the effect of the length of the evaluation data was studied on the performance of the GMM-based cost function, which indicated its importance over the final results of the system identification. Overall results of 1D and 2D parameter estimation experiments showed that the global minimum of the GMM-based cost function was the true value of the model's parameters of the system.

ACKNOWLEDGMENT

The authors would like to thank the anonymous reviewers for their valuable comments and suggestions to improve the quality of the chapter.

REFERENCES

[1] Hilborn RC. Chaos and nonlinear dynamics: an introduction for scientists and engineers. Oxford: Oxford University Press; 2000.
[2] Hubler A. Adaptive-control of chaotic systems. Helvetica Physica Acta 1989;62(2–3):343–6.
[3] Ott E, Grebogi C, Yorke JA. Controlling chaos. Phys Rev Lett 1990;64(11):1196.
[4] Vaidyanathan S. Global chaos synchronisation of identical Li-Wu chaotic systems via sliding mode control. Int J Model Ident Control 2014;22(2):170–7.
[5] Boubaker O, Fourati A. In: Structural identifiability of non linear systems: an overview. IEEE International Conference on Industrial Technology, 2004. IEEE ICIT'04. vol. 3; 2004. p. 1244–8.
[6] Kantz H, Schreiber T. Nonlinear time series analysis. vol. 7. Cambridge: Cambridge University Press; 2004.
[7] Jafari S, Hashemi Golpayegani SMR, Daliri A. Comment on 'parameters identification of chaotic systems by quantum-behaved particle swarm optimization'. Int J Comput Math 2013;90 (5):903–5. [Int J Comput Math 86 (12)(2009), pp. 2225–2235].
[8] Jafari S, Hashemi Golpayegani SMR, Rasoulzadeh Darabad M. Comment on "parameter identification and synchronization of fractional-order chaotic systems" Commun Nonlinear Sci Numer Simulat 2013;18(3):811–4. [Commun Nonlinear Sci Numer Simulat 2012; 17: 305–16].
[9] Jafari S, et al. Some remarks on chaotic systems. Int J Gen Syst 2012;41(3):329–30.
[10] Lao S-K, et al. Cost function based on gaussian mixture model for parameter estimation of a chaotic circuit with a hidden attractor. Int J Bifurcation Chaos 2014;24(01):1450010.
[11] Xu G, et al. A new chaotic system with a self-excited attractor: entropy measurement, signal encryption, and parameter estimation. Entropy 2018;20(2):86.
[12] Shekofteh Y, et al. A gaussian mixture model based cost function for parameter estimation of chaotic biological systems. Commun Nonlinear Sci Numer Simul 2015;20(2):469–81.
[13] Arı Ç, Aksoy S, Arıkan O. Maximum likelihood estimation of Gaussian mixture models using stochastic search. Pattern Recognit 2012;45(7):2804–16.
[14] Bishop CM. Pattern recognition and machine learning. vol. 128. Springer-Verlag New York Inc; 2006. 1–58.
[15] Johnson MT, et al. Time-domain isolated phoneme classification using reconstructed phase spaces. IEEE Trans Speech Audio Process 2005;13(4):458–66.
[16] Povinelli RJ, et al. Statistical models of reconstructed phase spaces for signal classification. IEEE Trans Signal Process 2006;54(6):2178–86.

[17] Shekofteh Y, Almasganj F. Feature extraction based on speech attractors in the reconstructed phase space for automatic speech recognition systems. ETRI J 2013;35(1):100–8.

[18] Shekofteh Y, Almasganj F, Daliri A. MLP-based isolated phoneme classification using likelihood features extracted from reconstructed phase space. Eng Appl Artif Intell 2015;44:1–9.

[19] Shekofteh Y, Gharibzadeh S. Are chaotic models of EEG signals useful in diagnosing attention-deficit/hyperactivity disorder? Clin EEG Neurosci 2014;45(1):57.

[20] Wu X, et al. Top 10 algorithms in data mining. Knowl Inform Syst 2008;14:1–37.

[21] Aho K, Derryberry D, Peterson T. Model selection for ecologists: the worldviews of AIC and BIC. Ecology 2014;95(3):631–6.

[22] Akaike H. Factor analysis and AIC. Psychometrika 1987;52(3):317–32.

[23] Kuha J. AIC and BIC: comparisons of assumptions and performance. Sociol Methods Res 2004;33(2):188–229.

[24] Schwarz G. Estimating the dimension of a model. Ann Stat 1978;6(2):461–4.

[25] Celeux G, Soromenho G. An entropy criterion for assessing the number of clusters in a mixture model. J Classif 1996;13(2):195–212.

[26] Steele RJ, Raftery AE. Performance of Bayesian model selection criteria for Gaussian mixture models. Dept. Stat., Univ; September 2009. Washington, Washington, DC, Tech. Rep. 559.

[27] Cartwright JH, Piro O. The dynamics of Runge–Kutta methods. Int J Bifurcation Chaos 1992;2 (03):427–49.

[28] Mehdi SA, Kareem RS. Using fourth-order Runge-Kutta method to solve Lü chaotic system. Am J Eng Res 2017;6(1):72–7.

[29] Clerc M. Particle swarm optimization. vol. 93. New York: John Wiley & Sons; 2010.

[30] Karaboga D, Basturk B. A powerful and efficient algorithm for numerical function optimization: artificial bee colony (ABC) algorithm. J Global Optim 2007;39(3):459–71.

[31] Mehdi H, Boubaker O. In: Position/force control optimized by particle swarm intelligence for constrained robotic manipulators. *11th International Conference* in *Intelligent Systems Design and Applications (ISDA)*, 2011; 2011. p. 190–5.

[32] Mirjalili S, Lewis A. The whale optimization algorithm. Adv Eng Softw 2016;95:51–67.

[33] Mirjalili S, Mirjalili SM, Hatamlou A. Multi-verse optimizer: a nature-inspired algorithm for global optimization. Neural Comput Applic 2016;27(2):495–513.

[34] Zhang H, et al. Parameter estimation of nonlinear chaotic system by improved TLBO strategy. Soft Comput 2016;20(12):4965–80.

[35] Huang Y, et al. Parameter estimation of fractional-order chaotic systems by using quantum parallel particle swarm optimization algorithm. PLoS One 2015;10(1).

[36] Li X, Yin M. Parameter estimation for chaotic systems by hybrid differential evolution algorithm and artificial bee colony algorithm. Nonlinear Dynam 2014;77(1–2):61–71.

[37] Lin J. Parameter estimation for time-delay chaotic systems by hybrid biogeography-based optimization. Nonlinear Dynam 2014;77(3):983–92.

[38] Lin J, Chen C. Parameter estimation of chaotic systems by an oppositional seeker optimization algorithm. Nonlinear Dynam 2014;76(1):509–17.

[39] Mehdi H, Boubaker O. PSO-Lyapunov motion/force control of robot arms with model uncertainties. Robotica 2016;34(3):634–51.

[40] Sayed GI, Khoriba G, Haggag MH. A novel chaotic salp swarm algorithm for global optimization and feature selection. Appl Intell 2018;1–20. https://doi.org/10.1007/s10489-018-1158-6.

Part II

Real World Applications

Chapter 8

Virtualization of Chua's Circuit State Space

Branislav Sobota* and Milan Guzan†
**Department of Computers and Informatics, Faculty of Electrical Engineering and Informatics, Technical University of Košice, Košice, Slovak Republic,* †*Department of Theoretical and Industrial Electrical Engineering, Faculty of Electrical Engineering and Informatics, Technical University of Košice, Košice, Slovak Republic*

1 INTRODUCTION

The first mention of the Chua's circuit [1] was the global impetus not only for chaos research, but also for the use of chaos in practice. At the same time, Chua's circuit became an impulse for searching for new systems of differential equations generating chaos, see for example, [2–9], and for the creation of illustrative monographs [10–13]. Currently, chaos is being used in many areas such as health, noise generation, cryptography, neural networks, memristors, chaos synchronization, control of chaos, differential equations of fractional order, economics, cancer research, true random number generators, and secret communication [10–23]. In the abovementioned areas, there is still a great potential for utilization of chaos in various applications. Therefore, on the next pages we will focus on visualization of 3D state space not only by conventional 2D visualization, but also by means of software tools (visualization of system dynamics and evolution of representative point (RP) movement), by 3D printing, and by using modern techniques such as 3D visualization with the use of modern and progressive virtual reality technologies—still unpublished results. At the same time, we would like to mention the possibilities and the efficiency of the calculation of boundary surface.

In this chapter, we will deal with the well-known Chua's circuit, which is popular in the academic environment for its simple implementation and its ability to generate hundreds of attractors [10]. However, the solving procedures and visualization related to the Chua's circuit presented here can be extended for any other autonomous systems described by a set of differential equations and characterized by at least two attractors.

Recent Advances in Chaotic Systems and Synchronization. https://doi.org/10.1016/B978-0-12-815838-8.00008-X

2 CHUA'S CIRCUIT

Chua's circuit is shown in Fig. 1A. This autonomous circuit is described by the system (1). Chua's circuit consists of four linear elements and one nonlinear resistor (*NR*)—the so-called Chua's diode, whose possible five-segmented *IV* characteristic is given by the formula (2) and shown in Fig. 1B. The resistor ρ in Fig. 1A represents the DC resistance of coil L and the current I indicates the possibility of circuit control. Chua's circuit is the first physical circuit characterized by three unstable singularities. Until the discovery of Chua's circuit, both stable and unstable singularities were present in the sequence circuits, with stable singularities always outnumbering the unstable singularities by one. When analyzing the sequence circuits using the graphic methods without the use of computers, the author of the works [24, 25] often asked following questions: Is it possible to implement such a nonlinear autonomous circuit that would have all the singularities unstable? If so, what would then be the attractor in such a special circuit? The discovery of Chua's circuit has given answers to these questions, and therefore, the work [26] perceives the uniqueness of the presence of three unstable singularities in the physical circuit as a miracle in circuit theory.

$$\begin{aligned}
C_1\frac{du_1}{dt} &= \frac{1}{R}(v_2 - v_1) - g(v_1) - I \\
C_2\frac{du_2}{dt} &= \frac{1}{R}(v_1 - v_2) + i \\
L\frac{di}{dt} &= -v_2 - \rho i
\end{aligned} \tag{1}$$

$$\begin{aligned}
g(v_1) = m_2 v^2 1 &+ \frac{1}{2}(m_1 - m_0)(|v_1 - B_P| - |v_1 + B_P|) \\
&+ \frac{1}{2}(m_2 - m_1)(|v_1 - B_0| - |v_1 + B_0|)
\end{aligned} \tag{2}$$

It is known that, in dependence to the initial conditions (ICs), two attractors are present in the Chua's circuit: chaotic attractor (CHA) and stable limit cycle

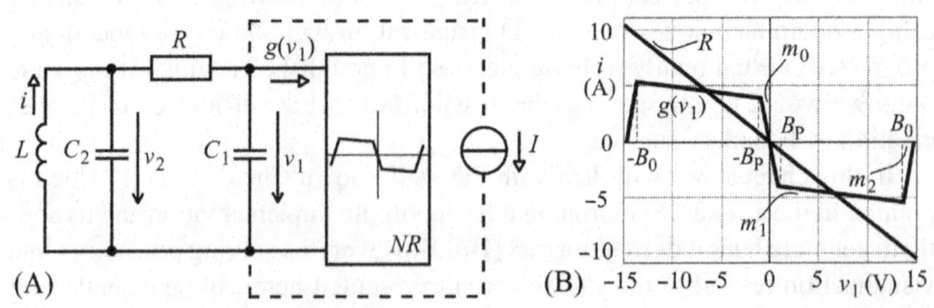

FIG. 1 (A) Chua's circuit, (B) *IV* characteristic of Chua's diode (*NR*)—$g(v_1)$ for parameters (3), together with load R in (i,u_1) plane.

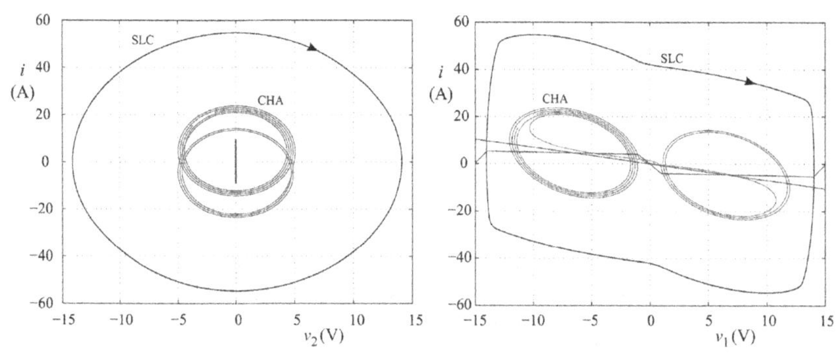

FIG. 2 Monge projection of CHA and SLC for parameters (3).

(SLC). Fig. 2 shows both attractors [26]—CHA attractor and SLC in the Monge projection for parameters (3) adapted from [1]. (In Ref. [27] parameters (4) can be found. We will use them later.) Naturally, the authors of the work [27] have asked similar questions as the author dealing with the sequence circuits [24, 25]. They ask: What boundary separates the domain of attraction for the chaotic attractor and the domain of attraction for the periodic attractor? The answer was given in Ref. [28]. It was so-called *boundary surface* (BS), in principle, the same as the surface discussed in Ref. [24, 25] and the following [29–31]. Trajectories have already been calculated using large mainframes, or personal computers (PC).

$$\begin{array}{l} C_1=0.1\text{F} \;\; C_2=2\text{F} \;\; L=0.142857\text{H} \;\; R=1.42857\Omega \;\; m_0=-4\text{S} \\ m_1=-0.1\text{S} \;\; m_2=5\text{S} \;\; B_P=1\text{V} \;\; B_0=14\text{V} \;\; \rho=0 \end{array} \tag{3}$$

$$\begin{array}{l} C_1=0.11111\text{F} \;\; C_2=1\text{F} \;\; L=0.142857\text{H} \;\; R=1.42857\Omega \;\; m_0=-0.8\text{S} \\ m_1=-0.5\text{S} \;\; m_2=5\text{S} \;\; B_P=1\text{V} \;\; B_0=14\text{V} \;\; \rho=0 \end{array} \tag{4}$$

3 BOUNDARY SURFACE

If there is a need to keep the signal secret by using chaos, or to design new binary or multiple-valued memory structures, it is necessary to know BS (for R^3) or the boundary lines—separatrix (for R^2) for trouble-free and unambiguous memory control [32–34]. Those are geometric formations that divide the state space into regions of attraction (RAs) corresponding to individual attractors. It should be noted that the existence of RA is not the exclusive domain of electrical circuits. Similar RAs can be seen both in space (first, second cosmic velocity, attraction of planets, etc.) as well as on Earth (gravitation, doldrums, isotherms, high- or low-pressure areas, etc.). In general, we can say that the rules of attractivity we see in the microchip of the implemented circuit are also valid in the universe. The natural question that arises is: What is the best and most obvious way to visualize CHA and BS so that it would be possible to show not only BS and CHA itself but also to illustrate the rules of chaotic movement of RP in the state space? In the following section, we will present possible techniques for visualization of BS, CHA, and RP movement in 3D state space, in 3D printing, except that Section 6 will include yet unpublished outputs of visualization from "virtual cave."

4 VIRTUAL REALITY AND OTHER VISUALIZATION FORMS

As it was written earlier, the visualization of Chua's circuit space is already possible in several ways, not just in 2D. It is possible to visualize in 3D mainly by using the program tools described below including 3D printing at present time. The use of the virtual reality (VR) technologies in this field is rapidly moving forward. Some examples or photos used in this chapter were taken in VR laboratory LIRKIS at Department of Computers and Informatics Faculty Electrical Engineering and Informatics Technical university of Košice (LIRKIS DCI FEEI TU Košice).

VR systems provide a better experience and they are more interactive, but their implementation is far more complex [35]. VR subsystems are divided according to the senses they affect: visualization subsystems, acoustic subsystems, kinetic and statokinetic subsystems, touch and contact subsystems, and other senses (e.g., smell, taste, pheromone sensitivity). The VR system represents an interactive computer system operating in close connection with a human computer. In addition to the phrase, "virtual reality," it is possible to work with the phrase, "Fuzzy Reality" (FR, the number of a frames and the quality is lower most often due to technological complexity and price) and Mixed/Augmented Reality (MR/AR, real world view is enhanced/supplemented by synthesized/generated computer elements). The observer or the user of VR system is also called a cybernaut, and the virtual agent (representative) of the user in the virtual world is called an avatar.

In order to work with state space (including simple 3D visualization only), the entire space and its objects must be virtualized. This whole process covers the so-called 3D virtualization sequence.

4.1 3D Virtualization Sequence and Its Implementation

The whole process of object processing and virtualization is called the 3D virtualization sequence [36, 37]. A simplified 3D virtualization sequence scheme can be seen in Fig. 3. Of course, each of these steps contains a number of other substeps, naturally. Not all of these substeps will be described in detail. Many subsequences of this sequence are already standard practice and have their steady conventions and standards.

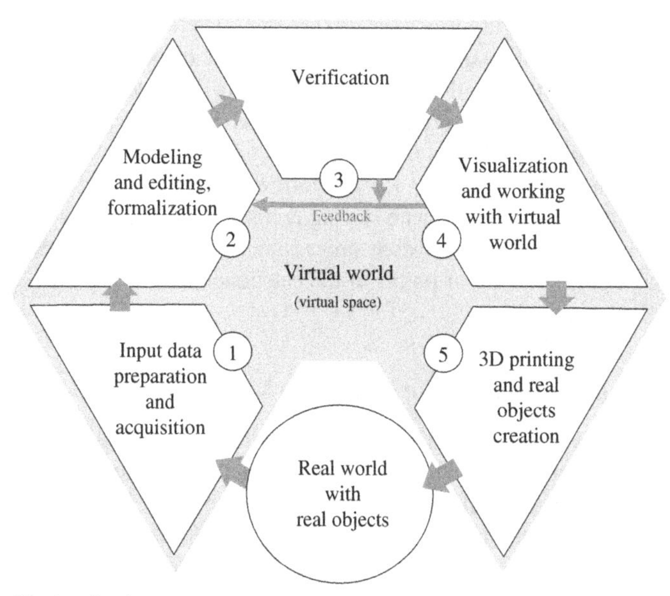

FIG. 3 3D virtualization sequence.

4.2 Input Data Preparation and Acquisition

All possible inputs for future modeling are the basis before of the modeling step. Of course, the number and form of the inputs is dependent on the acquired object (e.g., CHA is obtained as a point cloud). The 3D scanner use is the best way to determine the status of the 3D object at present [36]. However, in terms of nonlinear circuits or the Chua's circuit, the input data are generally obtained as outputs of the calculation of the corresponding equation system. The polygonization, most often in the form of triangulation, is another challenging step. This step is highly suitable for parallelization. More precise models (e.g., BS) can be further transformed into higher surface types, for example, NURBS. The last step is archiving or export to modeling and editing tools.

4.3 Modeling, Editing and Verification

3D computer modeling is the creation or modification of a 3D graphical virtual model using a specific software tool—3D modeling application (e.g., Blender, Autodesk 3D Studio max, Maya, Sketchup, etc.). The modeling process of preparing/editing graphical data is similar to sculpture creation. Modeling of 3D objects is not an exception in this case, and it is based on a four-step principle: model/object preparation, polygon modeling, texture preparation, and texture application and filtering. The properties of these steps depend on the model/object type (wire model, surface model, volume model). The created model/object can be used for visualization, simulation, or for 3D printing.

One of the challenging processes in this step, where parallelization can be advantageous, is model simplification. The number of polygons defining the model/object reduction so that the output model is as close as possible to the original is the role of this process, in principle. From the point of view of the overall virtual scene, when using hundreds of objects, it can be a considerable reduction in the difficulty for the following steps of the virtualization sequence and for the virtual scene storing/archiving. This process was advantageously used, for example, when processing cross-section of BS (C-SBS) or eliminating the number of points of the calculated and then modeled trajectory of RP motion.

4.4 Visualization and Working With Virtual World

We have already spoken about visualization in the introduction to this chapter. Technically, visualization is possible with different technologies. Mainly, the user interface and its implementation are very important [38]. There is also the possibility of manipulation with virtual objects (CHA or BS, in our case). The interactive manipulation of virtual objects can be done either with conventional input devices such as a keyboard and mouse or with various other devices (data glove, MS Kinect, MYO or others) used in VR systems.

We can also use simpler computer graphics technologies to visualize nonlinear circuits, BS or CHA. It is necessary to distinguish the type of visualization that is appropriate and at the same time achievable with the used technical facilities. In principle, visualization can be done using two different methods. Real-time data visualization is the first method. Real-time data rendering means that the data (objects) are processed immediately, so the objects are instantly rendered/shown based on the input, and the user can interact with the objects. An example of real-time visualization for direct user interaction is for example interactive work in CAD design tools or, at the top, work in the VR system. Off-line data visualization is the second method. This type of visualization is based on two steps. The first step is data collection and processing. In the second step, the already processed data are sent to the graphical output. It is used especially in cases where a large amount of data is expected at the input that could not be rendered in real time without processing.

The visualization subsystem of VR system belongs to real-time systems and it provides the main part of the observer information in the virtual environment [35]. This subsystem calculates the image for an observer in the virtual world. The photorealistic visualization is very interesting for the observer. But today, with existing hardware, it cannot be done in real time and in the required range.

World/trajectory/BS model is the main input. It is based on convex polygons (this will be a parallel implementation of NURBS calculations as a core part of the visualization engine used in the future). If necessary, especially in dynamic visualization processes, a descriptive scripting language, used in this case, too, is often used to implement the top layer of the visualization system.

4.5 3D Display Systems

The main part of the VR system is its visualization (display) part. It is optimal, if 3D display systems are used in this part, both on a tightly-coupled displaying system (e.g., data helmets) but also on a loosely coupled system (e.g., virtual caves) [37]. These displaying systems can be divided into:

- Holographic displays
- Volumetric displays
- Stereoscopic displays—passive, active, and autostereoscopic

Classic display (passive or active stereoscopic display) can be usable at the lowest level of visualization. A higher level is the use of an autostereoscopic 3D display (Fig. 4) (e.g., Philips WOWvx working with a depth map). Anaglyphic projection (Fig. 5A) or horizontally split left/right view (Fig. 5B) can be used

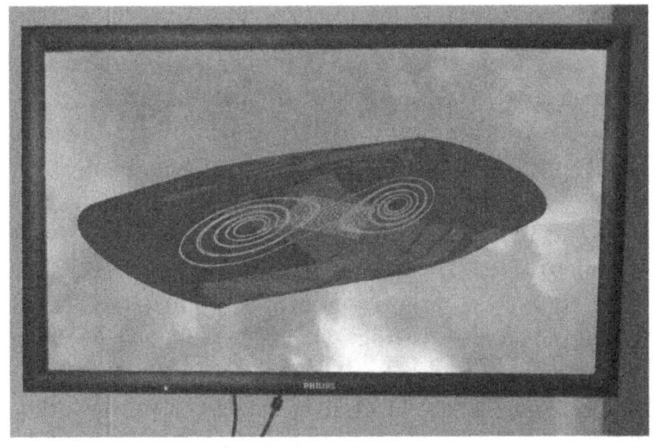

FIG. 4 BS virtual scene visualization on autostereoscopic display (LIRKIS, DCI FEEI TU Košice).

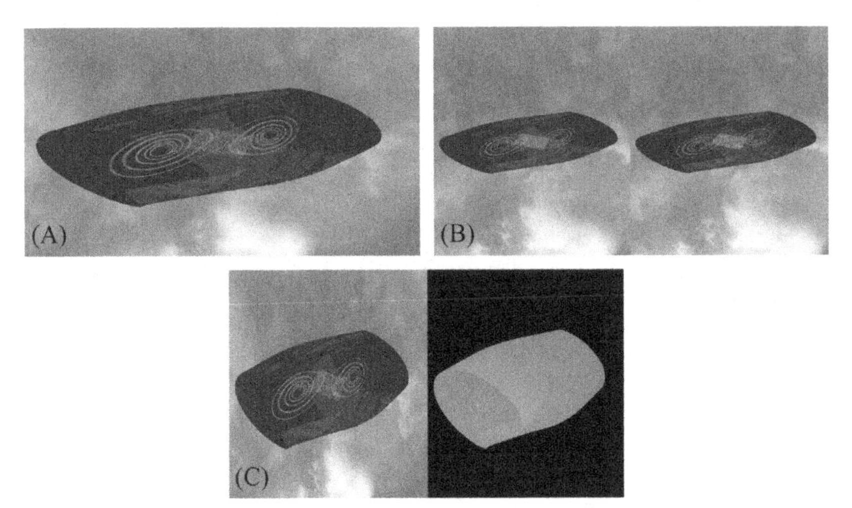

FIG. 5 BS virtual scene visualization: (A) Anaglyphic projection (B) Horizontally divided left-right projection c) Depth map projection (LIRKIS, DCI FEEI TU Košice).

also as standard at present. The best percept, without any aid, is to display 3D content using the 2D-plus-depth method (depth map) (Fig. 5C). This method creates a 3D perception using the 2D image and the depth information image. Another possibility in this mode of rendering is parallelization usage, because calculating the depth from two left/right images is, again, a time-consuming process. Given that real-time computing is required, the use of GPGPU technology is also a good way of implementation.

4.6 3D Printing and Real Objects Creation

3D printing is the last step in the virtualization sequence. It is the creation of real objects based on the virtual models (CHA or BS in this case). The basic 3D printing technology used in the authors' laboratory (LIRKIS Laboratory) uses powder (plaster ($CaSO_4$)) as building material. The binder (water-based) is then applied to the powder. This creates one layer of the real object. The binder blending uses a device similar to the inkjet printer head (some devices use modified printheads of regular inkjet printers, tri-color HP cartridge in this case).

The 3D printer was also used for CHA or BS 3D prints presented in this chapter (e.g., ZPrinter 450, LIRKIS laboratory, Fig. 6 middle). Objects were printed with two print heads, one for binder application, and the other for painting (coloring). The 3D printer enables color 3D printing of 3D objects at 300 × 450 dpi up to size 203 × 254 × 203 mm. Print speed is 2–4 layers per minute, depending on the printed model. The thickness of one layer is 0.089–0.102 mm. The materials from which 3D models were printed are nontoxic. This type of printing is practically waste-free, and it is another advantage of this 3D printing type. The object has to be finalized to obtain the desired hardness, strength, or flexibility after printing. A wide range of finishing materials is offered to help achieve the various final properties of the printed object.

The schematic representation of a 3D model acquisition sequence using 3D printing is shown in Fig. 6. A printed 3D model of Chua's BS generating chaos is on the right. The use of 3D printing is a possible basis for futuristic

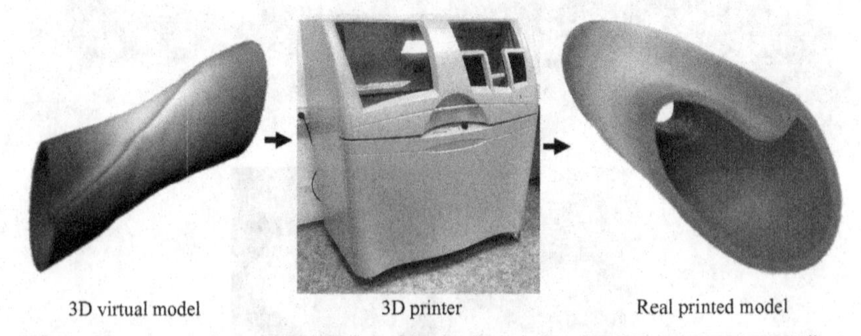

FIG. 6 The process of acquiring the BS model using 3D printing (LIRKIS, DCI FEEI TU Košice).

technologies such as food printing, 3D faxing or even the printing of entire bases on the moon, Mars, or other planets in a future perspective. Presently, it is mainly used in fields such as advertising, architecture, jewellery (printing of gold jewelery), archeology, restoration, engineering, or health (e.g., dental and bone prostheses or artificial heart valves) [39–41].

5 VISUALIZATION AND IMMERSIVE ENVIRONMENTS

Realization of such an interface is possible by involvement of known input and output devices which are widely used for navigation in a variety of virtual worlds. VR technologies are supposed to be used to simulate BS or CHA virtual space to evaluate new ways of interaction. There are two widely used techniques to deliver immersive and semiimmersive experience of VR with satisfying predictions of desired results for BS or CHA virtualization: HMD and CAVE.

HMDs (head-mounted displays) are devices suitable for bringing a virtual environment to one individual user at a time. They are becoming highly portable and available at a steadily decreasing price. The most popular and available HMDs include *Microsoft Hololens* (Fig. 7), *Oculus Rift*, *HTC Vive*, and *PlayStation VR*. The cheapest solution is to use simple HMD with smartphone display. Sensing of user's movement could be provided by cameras and motion capture systems. Recorded outputs have to be mapped to virtual objects representing human body parts. This is a base for mixed reality (MR) systems [42].

CAVE (cave automatic virtual environment) is a room-sized space consisting of several projection walls. Despite the fact that the stereoscopic projection

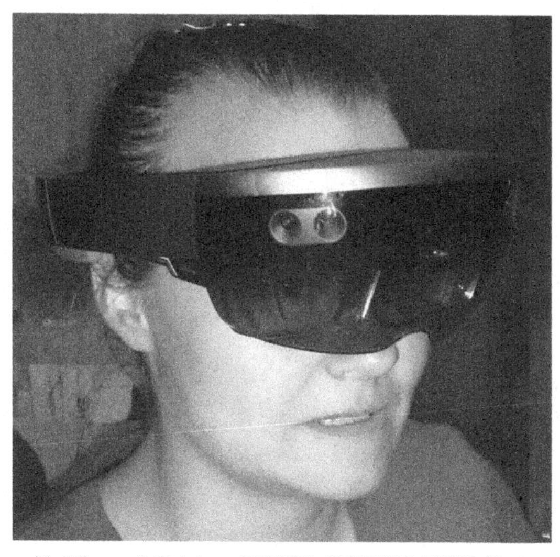

FIG. 7 The user with Microsoft Hololens (LIRKIS, DCI FEEI TU Košice).

is adjusted according to one user's perspective, others could share the same virtual environment without additional costs for perspective change synchronization. The user can move freely in the space of a CAVE and experience his/her body in close interaction with virtual scene, mostly through a contactless motion capture system [43].

5.1 Mixed Reality System Implementation

Mixed reality is one of the ways we can experience immersion visualization of virtual worlds, [42] including virtual scenes with CHA or BS. Implemantation of a mixed reality system is quite a difficult task. The main part is the real-time two-way recalculations between the real-world coordinate systems and virtual world coordinate systems. A lot of attention from the perspective of LIRKIS laboratory is devoted to the mixed reality phenomenon.

Practically speaking, the MR system is already a very flexible tool, and it provides a very immersive experience in working with it. The use of a data helmet with transparent displays (e.g., MS Hololens) is an excellent extension of this system. Increased system dynamics can be achieved by using devices that allow, in addition, a dynamic change of the markers/objects such as interactive table and data gloves or contactless sensors such as Leap Motion or MYO Armband. This makes it much more immersive to explore, for example, 3D models of CHA or BSs. Through the use of mixed reality technologies, the observer is then able to interactively manipulate these whole objects—inserting other objects or combining them with other CHAs or BSs. The observer feels as if he is actually holding the object with his own hands. This again increases the visualization quality. Due to time-consuming computations, the possibility of interactive manipulation with CHA or BS objects and their editing and calculation in real time cannot be expected sooner than in the near future. The experimental LIRKIS mixed reality workplace with data helmets with transparent displays (left) and an example of an observer's view of the BS model (right) is shown in Fig. 8.

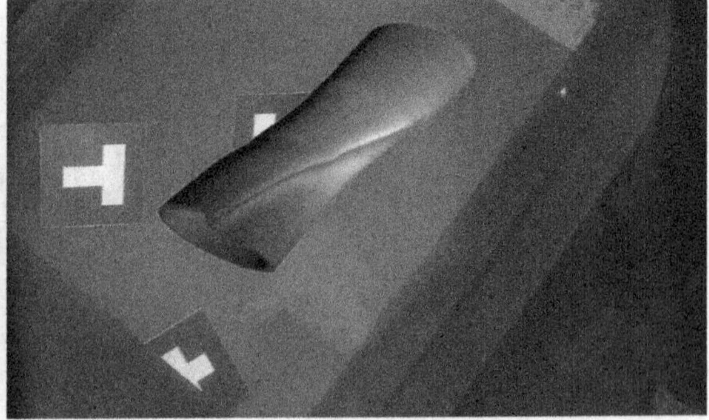

FIG. 8 Experimental MR workplace with data helmets with transparent displays (left) and an example of an observer's view of the BS model (right) (LIRKIS, DCI FEEI TU Košice).

5.2 CAVE System Implementation

In order to provide the experience similar to the real laboratory physical space, it is appropriate to virtualize BS or CHA using the LIRKIS CAVE [43] system shown in Fig. 9. Requirements include the possibility of rapid prototyping and sharing a virtual scene with other people. Another advantage is usage of real electrical appliances and other devices inside of the physical space of CAVE.

The system includes 20 stereoscopic screens (Fig. 9A, around and up and down) in a noncubic layout (a decagon) and spatial sound system. User's movement is captured using a markerless Optitrack system. The Optitrack system consists of eight cameras (red triangles, gray triangles in print versions, in Fig. 9A) with seven situated in the top corners of the screens and one behind the user on a steel frame (dashed gray square Fig. 9A). The LIRKIS CAVE engine consists of three parts: control center, Java console, and video renderer. The control center is the core of the system, and it mediates communication between the other parts. The Java console is used for remote control of the whole system and provides the opportunity to interact with the currently loaded virtual environment. The video renderer is responsible for 3D scene rendering, and it is based on the OpenSG toolkit. Scene packages include textures and 3D models. Additional logic and interaction with enhanced peripheral devices is implemented using Ruby script, which is included in the package. Thus, the system requires only on-site changes at scene level. A distributed computer system is used for computation. It consists of 7 nodes based on an Intel i7 processor, an nVidia Quadra graphics adapter, and 112GB total memory. However, real-time BS or CHA calculation is not possible yet, and the manipulation with BS or CHA is possible only with whole objects.

6 VISUALIZATION OF OBJECTS IN STATE SPACE

The basis for any initial visualization is the plane—2D. In principle, any 3D system can be clearly represented/visualized in two projections—front view

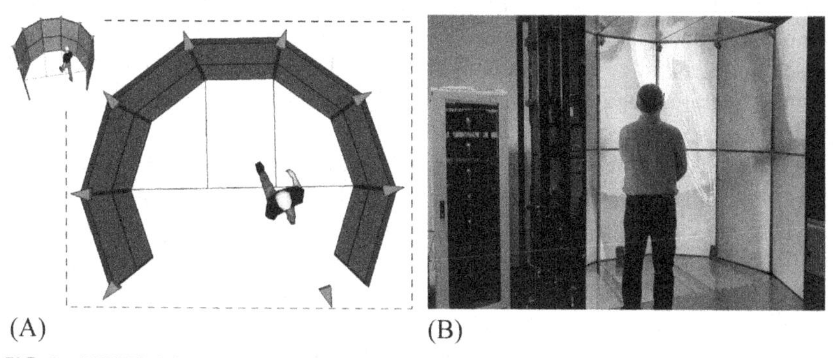

FIG. 9 LIRKIS CAVE (LIRKIS DCI FEEI Košice): A schematic showing (A) the spatial configuration of the CAVE and (B) a photo of a user inside the LIRKIS CAVE.

and side view in Monge projection (Fig. 2). However, the work [44] has shown that a third projection (top view) is also very useful for a clear visualization of the state space of a nonlinear dynamic circuit. Therefore, two Monge projections were used in one figure. Still, there was no possibility of zooming in, rotating, or manipulating with a given attractor.

The progress in computer hardware performance over the past 10 years has had a positive impact not only on the speed of computing, but also on change of simulation outputs. As we have already mentioned, the simulation outputs had only a 2D dimension meanwhile. However, with the increasing performance of graphics cards, the development of VR technologies, and with the new 3D printing options, it was possible to visualize processes taking place in the simulated circuit not only in 3D on a (3D) monitor, but also in VR or 3D printing. Of course, there was no change in simulation results. The calculated trajectories or cross-sections of space remained the same, but by visualization of the original 2D simulation results, it was possible not only to increase the clarity of the presented simulation activities, but even to "attract" the observer into the process running in the integrated circuit or directly on the chip.

6.1 Visualization of a Chaotic Attractor by Software Means

It is known that the possibility of computer programming enabled the development of programs for calculating trajectories [28, 45, 46]. There was a period of time with almost no possibilities for using computer graphics. Computers were mainly used to calculate trajectories, eigenvalues, or Lyapunov exponents. Computer screen views did not allow manipulation of the observed object. The performance of computer graphics was poor; but gradually, in Matlab, it has become possible to display a 3D trajectory, manipulate with it, etc. With the increasing performance of processors and graphics cards, it became possible to access the visualization interactively by the programs created. If we would like to indicate RP movement in the state space on a typical 2D image, it is possible to indicate the direction of RP movement to CS using arrows, like in Fig. 2, but it is not possible to perceive the dynamics of the movement. Using the program created, [47, 48] it was possible to monitor the RP movement in the state space in the so-called continuous and sequential mode (using the comet effect). In this program, it was possible to change the length of the comet's tail to allow visualization of any complicated CHA.

The visualization of RP movement in continuous mode for the parameters (3) along with the drawn *IV* characteristics can be seen in Fig. 10. When rotating a scene, any view on the plotted CHA is possible, as illustrated by the subfigures in 10 (The sequence of images is from left to right, from up to down). The effect of the comet is shown in Fig. 11 as the so-called sequential mode of RP movement visualization.

Another possibility for visualizing CHA or trajectories calculated when changing parameters is represented by the software described in Ref. [49].

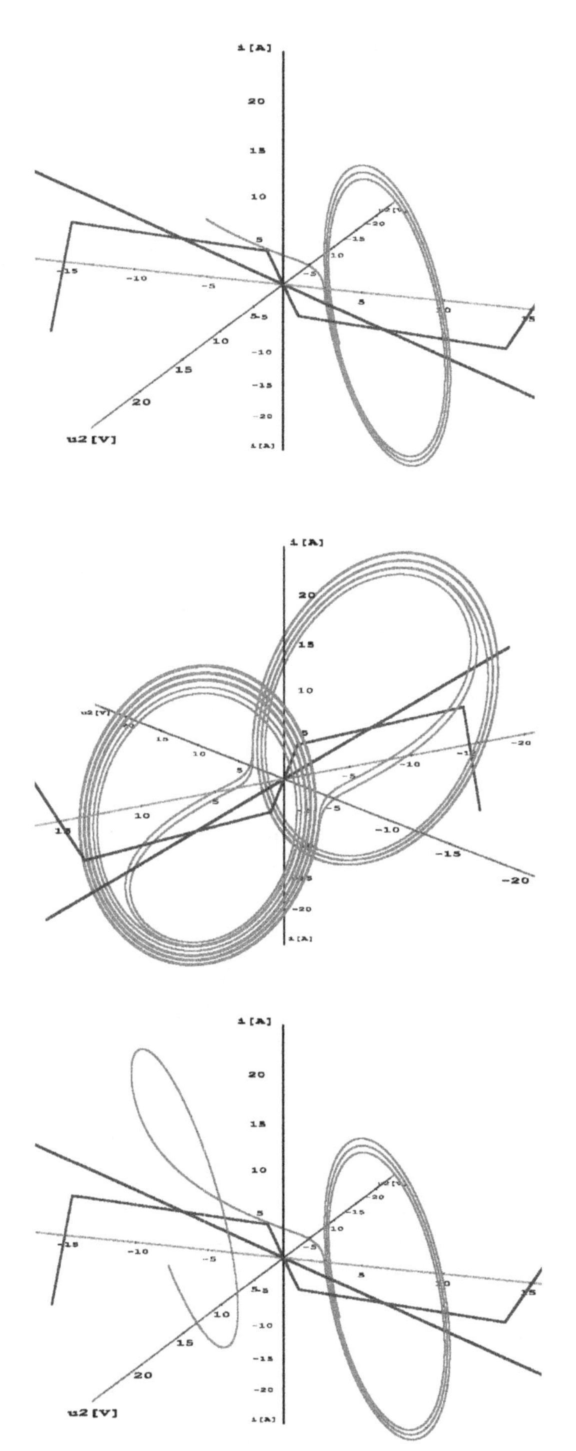

FIG. 10 The CHA corresponding to the parameters (3) in 3D in continuous mode together with *IV* characteristics at different scene rotation (left → right, up → down).

continued

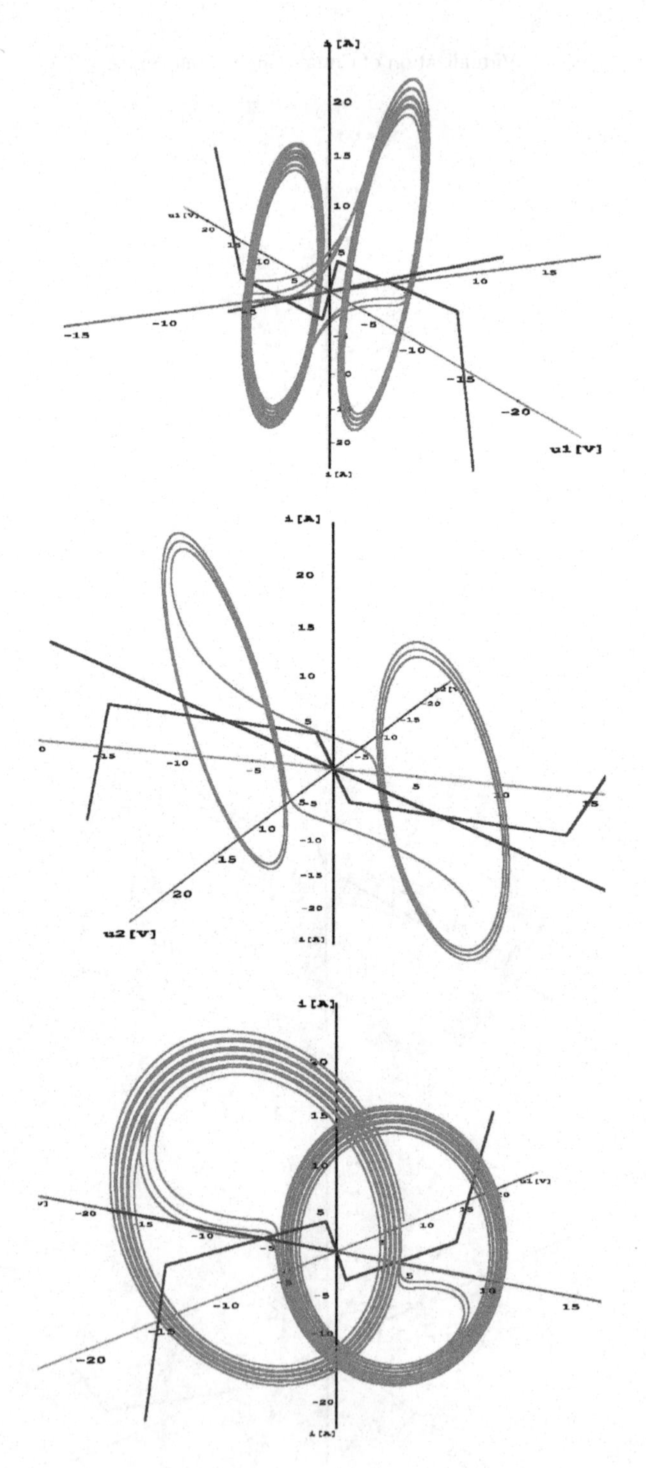

FIG. 10, Cont'd

continued

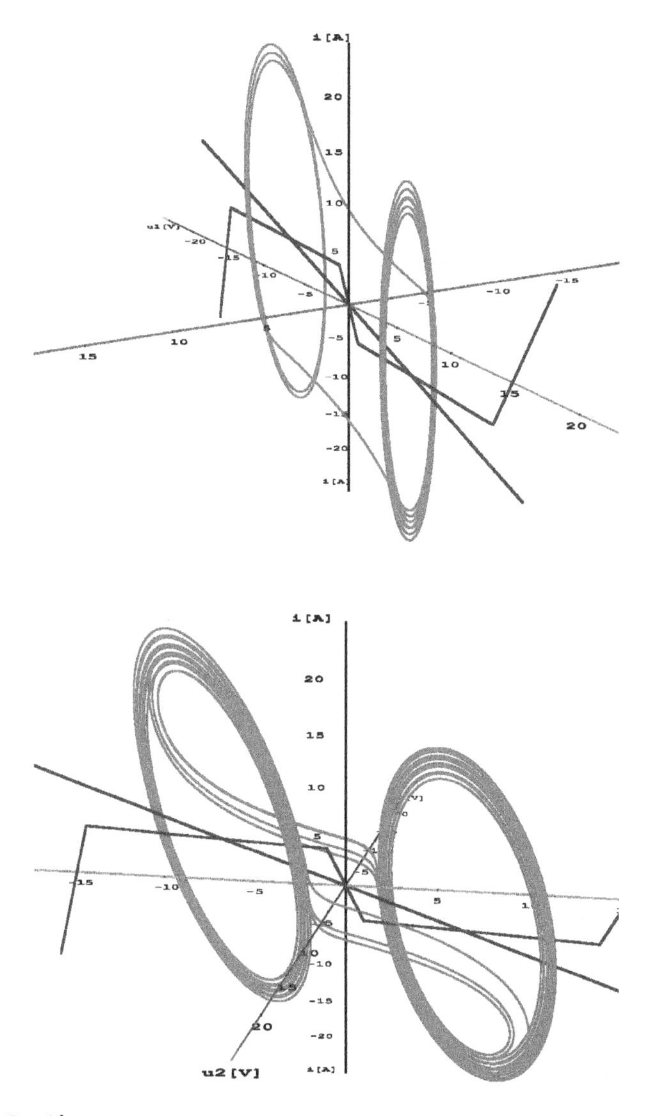

FIG. 10, Cont'd,

Although the software does not allow you to monitor the RP movement, it allows a detailed view of the calculated trajectory in two Monge projections and one 3D view that rotates in the software window. The software allows you to calculate (and thus preedit) input files such as *IV* characteristics or trajectories for later visualization. A view of CHA for parameters (3) shows Fig. 12 (left) and for the modified *IV* characteristic and the corresponding CHA Fig. 12 (right).

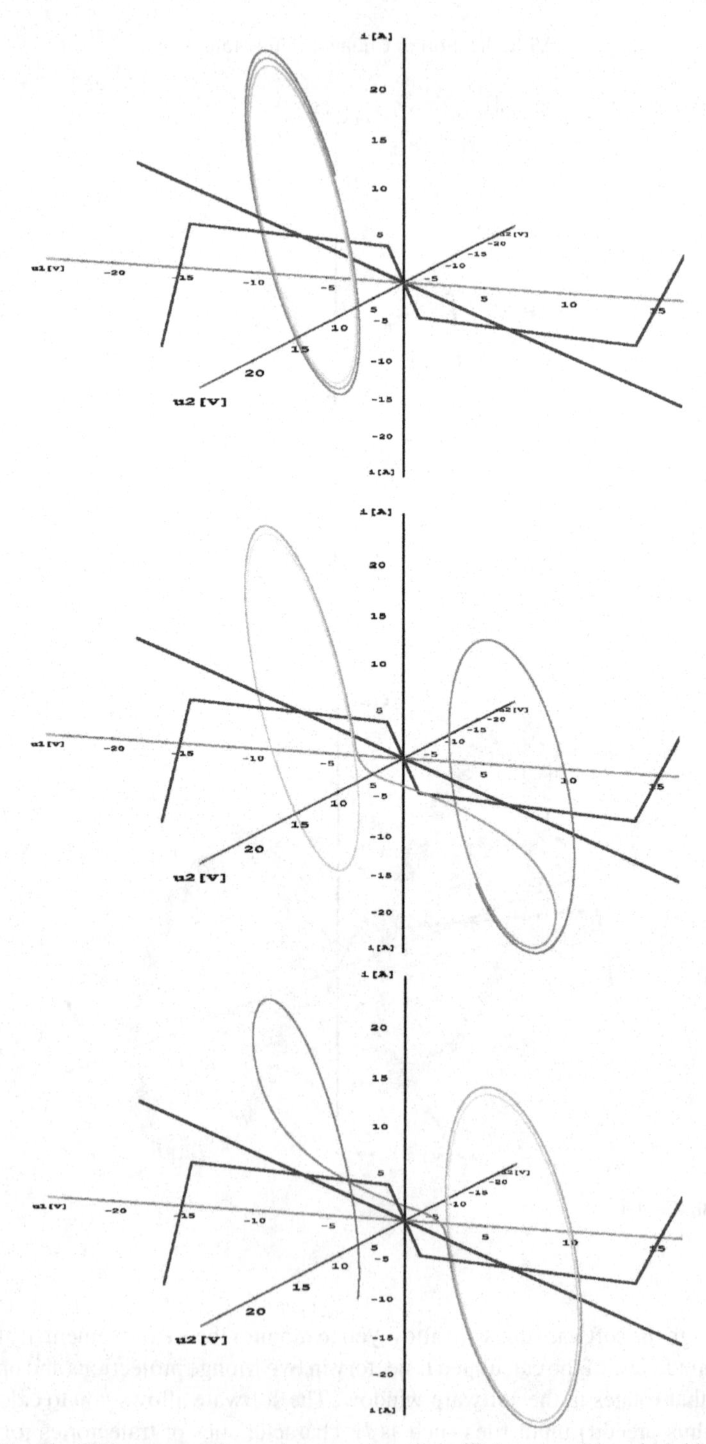

FIG. 11 The CHA corresponding to the parameters (3) in 3D in sequential mode (with comet effect) together with *IV* characteristics at different scene rotation (left → right, up → down).

continued

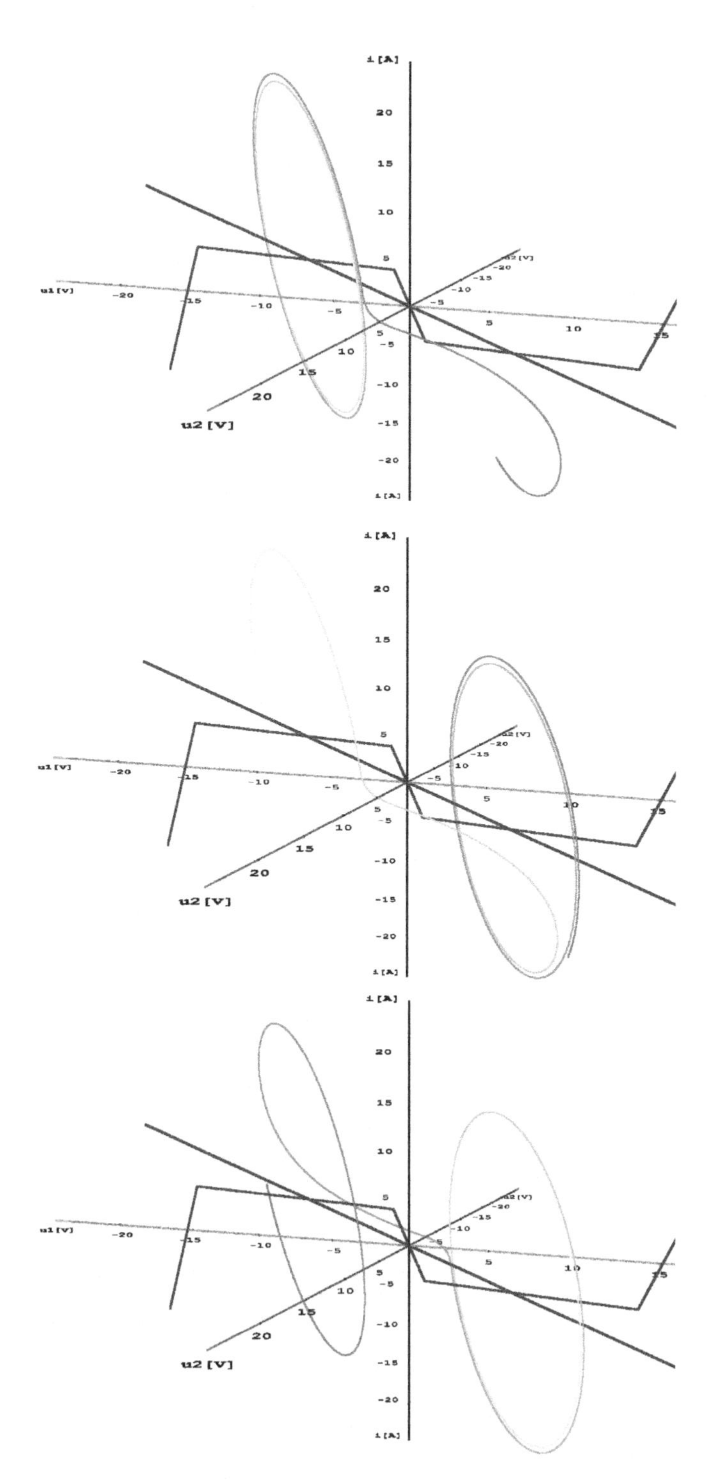

FIG. 11, Cont'd

continued

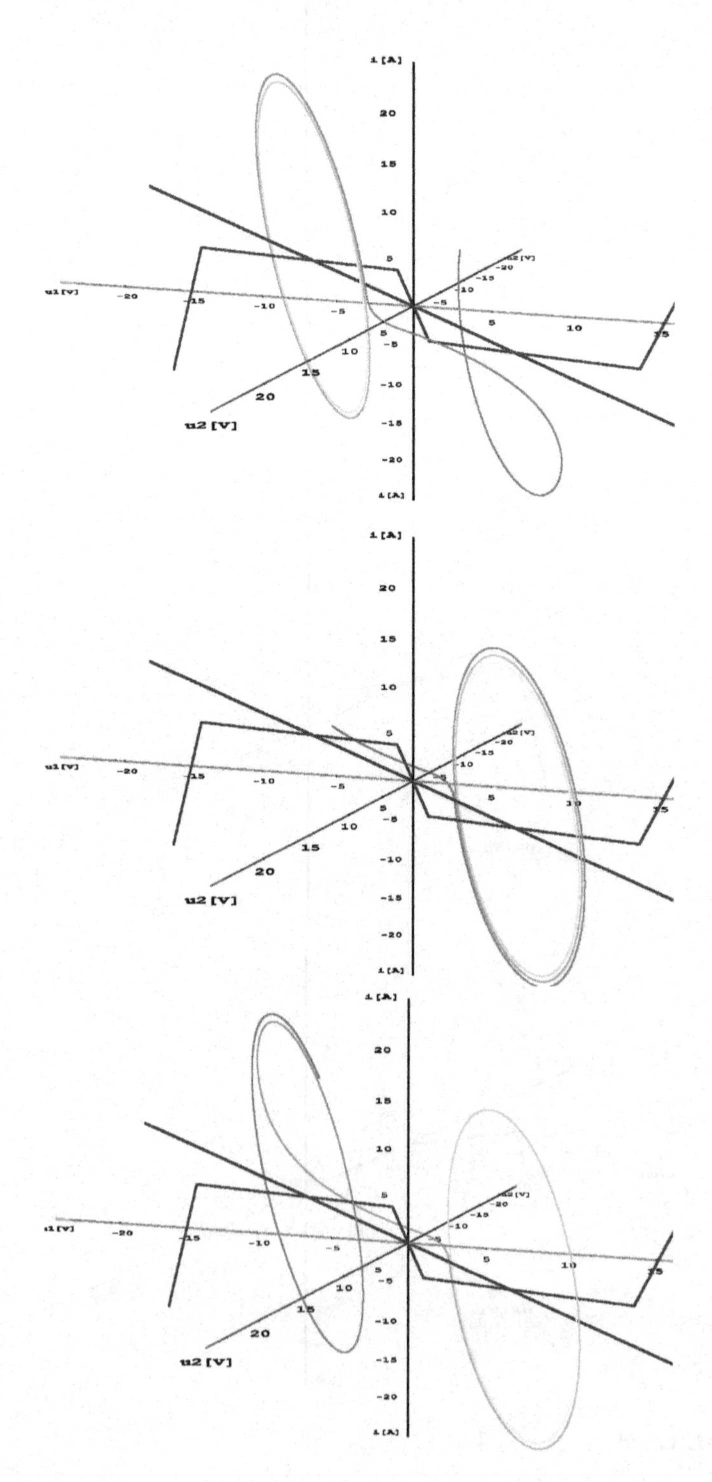
i [A]
20
15
10
5
-5
-10
-15
-20
u1 [V]
u2 [V]
i [A]
20
15
10
5
-5
-10
-15
-20
u1 [V]
u2 [V]
i [A]
20
15
10
5
-5
-10
-15
-20
u2 [V]

FIG. 11, Cont'd

What has been illustrated by Figs. 10–12 (software written in C++, C#, or Java) can also be implemented using web technologies [50, 51]. The advantage is that anyone who joins the website can change the input parameters. It does not need to include the comfort of the sequential mode of CHA display, but will be accessible to the general public similar to the possibility of downloading Matlab source files [51]. Similarly, it is possible to proceed with other software, for example, MathCad, Octave, Mathematica, etc.

An even more clear form of visualization is printing the entire CHA using a 3D printer. Of course, this is not an instant visualization as when using the software. Before 3D printing, it is necessary to filter the calculated file containing the excess points and then edit by the appropriate software for the 3D print. At the same time, it is also necessary to take into consideration the thickness of the "wall" so that no deformation or destruction of the 3D object occurs after printing. The result of such a process is illustrated in Fig. 13A—printed CHA corresponding to parameters (3).

In case it is necessary to create a summary of several CHAs at the same time (in the work [10] the authors refer to up to 195 different CHAs for real Chua circuit values), it is appropriate to use VR for this presentation. The resulting figure depicting several CHAs simultaneously for the parameters listed in Ref. [10] is shown in Fig. 13B. If it were only such a view of the set of CHA as illustrated by Fig. 13B, it would not be a good illustration of the attractors, so in this case, we can use the VR means to view each CHA separately as illustrated by the video to Fig. 13B [52]. A visit to the CHA virtual gallery would also be available at http://galileo.cincom.unical.it/esg/chua, [10] but this website has not worked for a long time. Virtual reality, specifically "virtual cave," is an attractive and modern way of visualization.

6.2 Visualization of BS

As we mentioned in Section 3, BS separates individual attractors from each other. Knowledge of size or morphology of BS gives answers to some questions [53–57] for example:

- Why did memory fail when changing technology or changing the size of parasitic elements on the chip?
- Why did memory fail and generate oscillations instead of some of the logic voltage levels?
- How can we control the newly-built elementary memory to allow reliable information writing?
- What is the principle of operation of the so-called statistical sensor?
- How big should the amplitude of the signal to be hidden using a chaotic signal be?

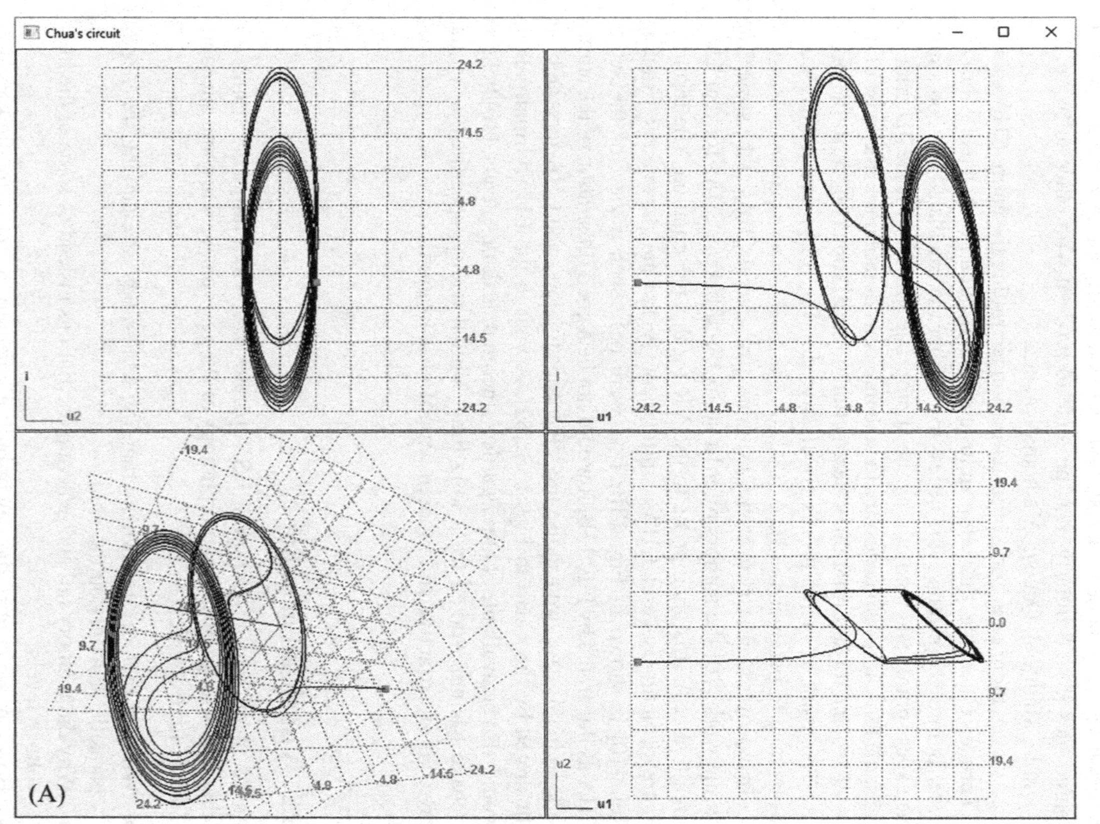

FIG. 12 Visualization of CHA for parameters (3) in three plane projections and in 3D with *IV* characteristics (left), preview of new CHA after change of *IV* characteristic (right).

continued

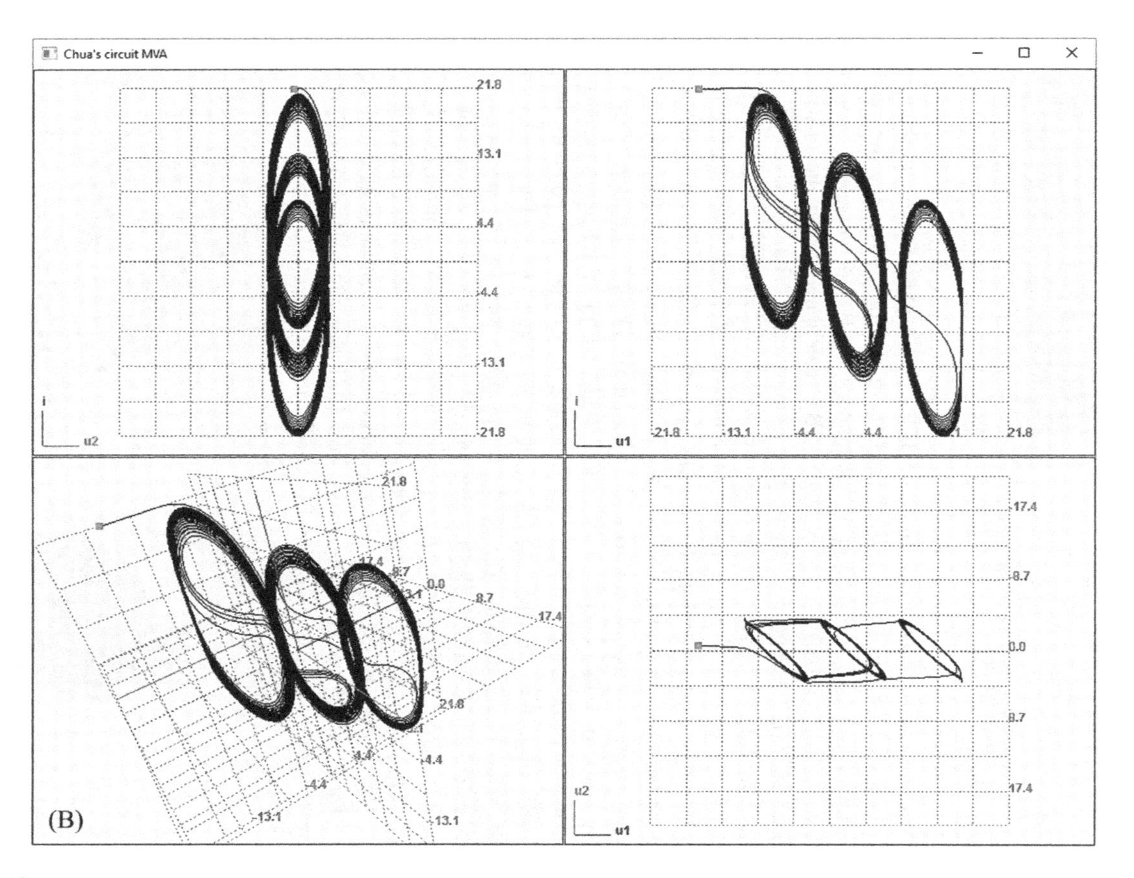

FIG. 12, Cont'd

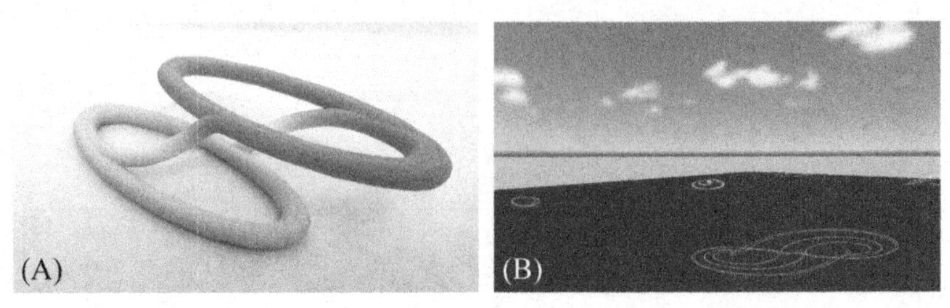

FIG. 13 (A) CHA printed using ZPrinter 450 3D printer, (B) a set of CHAs placed on a virtual field.

The Chua circuit is described by three differential equations (1) and therefore, the BS will be three-dimensional and more demonstrative for the reader. Twenty years ago, the method of (calculated) C-SBS for visualization of the BS was still used. By a set of 2D C-SBS (calculated using the grid technique—[28]), it was possible to imagine its morphology, but it was not possible to see the BS as a single 3D object. One C-SBS for $v_1 = 0$ and parameters (3) is shown in Fig. 14A, wherein the C-SBS represents a contour between the black (RA for CHA) and the white (RA for SLC) interface. This is how we will further understand the concept of C-SBS. One C-SBS consists of M × N points. The points represent the IC for solution of system (1), and situation of M = N (Fig. 14B) can also occur. Later, when visualization in 3D using the Matlab became a common practice, it was possible to display the contours of the C-SBS of ternary memory in the axonometric view [57]. The reader may ask: What is the morphology of BS for parameters (3)? The selected set of 300 calculated C-SBS is illustrated in Fig. 15 [56]. From Fig. 15, it is obvious that it is an object of cylindrical shape with a "sharp tooth" symmetrically moving from the upper right corner (for $v_1 = -15$ V) to the lower left corner (for $v_1 = 15$ V) [56]. BS of Chua's circuit obtained using mentioned grid technique was for the first time presented in the work [28]. But, the question remaining is how to merge, for example, 300 2D C-SBS into a single 3D object.

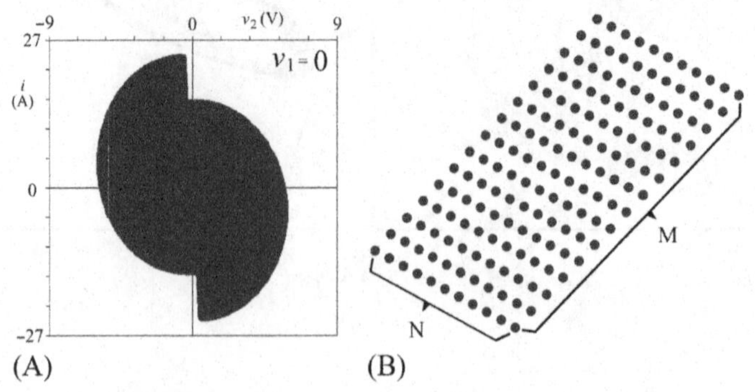

FIG. 14 (A) C-SBS in plane i,u_2, for $v_1 = 0$ and parameters (3), (B) grid of M × N initial conditions for calculating trajectories.

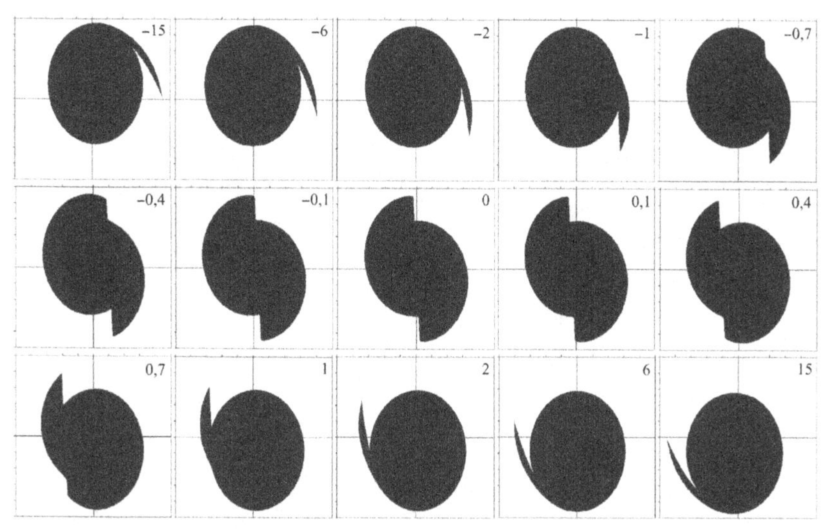

FIG. 15 Gallery of some of the 300 calculated C-SBS demonstrating morphology of BS in projection to plane i,v_2 for $v_1 = \text{const}$, given in the upper right corner. The description of the axes is identical to Fig. 14A.

Since the mid-1970s, researchers have been trying to find a way to reconstruct BS from 2D sections, mostly for medical purposes (MRI, CT, ultrasound). The work [58] was the first work that brought a comprehensive solution to the problem, although only for C-SBS with one closed contour and adjacent contours that did not differ too much. BS in 3D for the parameters (3) was reconstructed from the calculated 300C-SBS [56, 59] using the knowledge coming from [60, 61] (dealing with the problem of more complicated contours). The procedure of BS reconstruction can be briefly summarized into the following items:

1. Detection of contour for each calculated 2D C-SBS,
2. determining the order of points along the border of contours in one direction, [62]
3. creating closed oriented contours, represented by a polygon,
4. finding the triangulation region for each vertex of the contour,
5. determining the optimal triangulation vertex, and
6. triangulation (polygonal design—wire model of BS).

Fig. 16A shows BS visualized in 3D like a continuous model as a result of the procedure mentioned above. Fig. 16B illustrates BS visualized in 3D using contours. At the same time, *IV* characteristics and CHA are also displayed. Fig. 16C is similar like Fig. 16B, but this time the wire model is shown. Fig. 16D shows together BS, CHA, and SLC (the outer part of the spiral around BS). From the figure is clear that CHA is inside of BS and SLC is around BS. The models of BS printed on the 3D printer corresponding to parameters Eqs. (3), (4) can be seen in Fig. 17. The schematic procedure of BS 3D printing for parameters (4) is depicted in Fig. 6. At the same time, Fig. 18 shows an anaglyphic form of

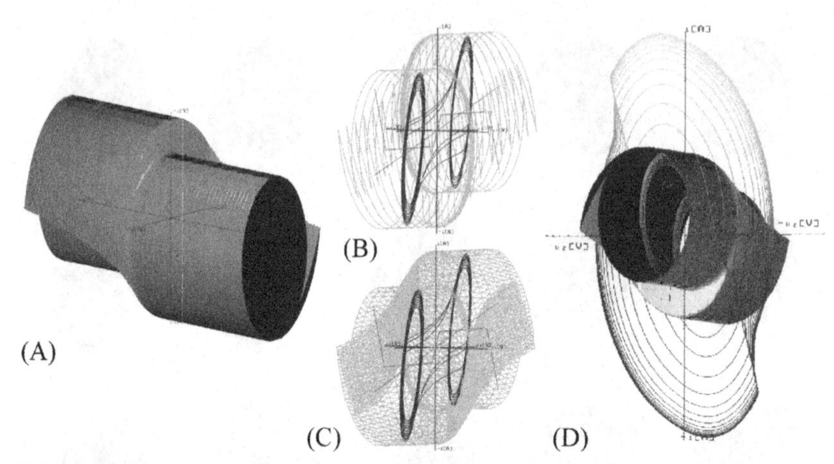

FIG. 16 Different views on the reconstructed BS for parameters (3) (A) reconstructed BS z 300C-SBS, (B) plotted contours C-SBS with *IV* characteristic and CHA, (C) wire model of the BS with *IV* characteristic and CHA, (D) BS with CHA (inside BS) and with SLC (the outer part of the spiral around BS).

FIG. 17 Printed 3D models of BS corresponding to the parameter (3)—left and (4)—right.

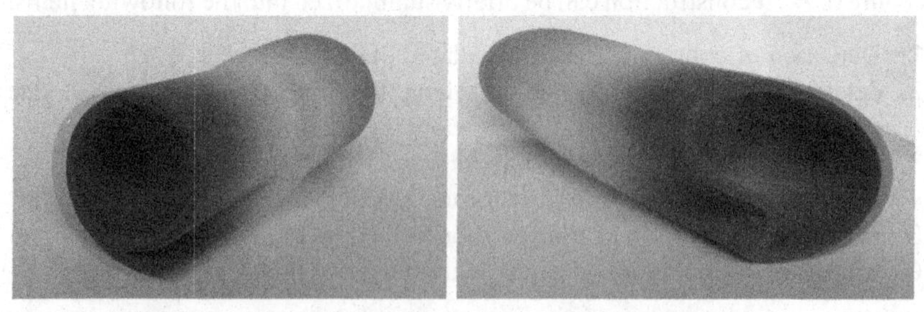

FIG. 18 Printed 3D models of BS in anaglyphic form of display.

display BS created from the Fig. 17. Using anaglyphic glasses, both BSs can be perceived as 3D objects, even if they are printed as 2D figures on paper.

The final visualizations in 3D could not be obtained without the large computational power required for calculation several tens to hundreds of C-SBS in 2D. Therefore, in the following section, we will point out the possibilities of C-SBS calculations, which also include calculations of individual trajectories.

6.3 Visualization in Virtual Cave Environment

The virtual cave solution itself has been described in Section 5.2. Given that virtual cave utilization provides a high degree of immersion in visualization, it is possible to compare the importance of virtual cave usage to transition from Monge projection to using computer 3D modeling systems. Given that the cave environment is projected beyond the viewer's peripheral vision and at the same time is of a size equivalent to the observer's size, the observer's perception level is truly fascinating. The observer can gain not only an amazing visual touch, but also a feeling of, for example, "staying inside BS or CHA environment." It is also interesting to use it in learning process to help students understand the CHA or BS environment better, especially if the work with virtual CHA or BS is set up as game, similarly to [63].

The position, whole object transformations, and stereoscopy for depth vision in this virtual cave environment are computed in real-time. It is true that with respect to the current calculation power, it is possible to manipulate the BS respectively with CHA only as a whole. The time-consuming calculation does not allow for real-time dynamic change yet, so it is not possible to fully interact with the BS or CHA virtual scenes. Nevertheless, this feeling is amazing, and dynamism can be achieved by at least some form of animation. Until now, without the CAVE usage, it was virtually not possible to look at, nor to manipulate BS or CHA. Virtual environments with BS or CHA can be shared by multiple users. The following photos (Figs. 19–24) are previews from BS or CHA observing in the virtual CAVE environment in the LIRKIS Laboratory at DCI FEEI TU Košice (nonsharp projection on the screens is not an error; it is stereoscopic projection). The parameters used in the following figures are taken from [10], chapter *Physical circuit* (parameters designation PCXYZ) with the difference that the *IV* characteristic for the Chua's diode (2) is only three-segments (without the parameters B_0 and m_2 in Ref. [2]) and the parameters given in nF, mS, and mH we used in F, S, and H.

Box 1. Video support

Short videos are available in this chapter for a better reader imagination. Therefore, you can download the video files in the following link:

http://ktpe.fei.tuke.sk/guzan/FromTheory2RealWorldApplications and watch videos relating to figures marked with symbol (🎬) in the figure caption.

---

7 POSSIBILITIES OF CALCULATION OF C-SBS

Reconstruction of BS in 3D is not possible without the calculation of 2D C-SBS. You cannot calculate and display in 2D without calculating individual M × N trajectories. Therefore, it is necessary to optimize the speed of trajectory calculation. In general, the following technological resources can be used to calculate trajectories and C-SBS:

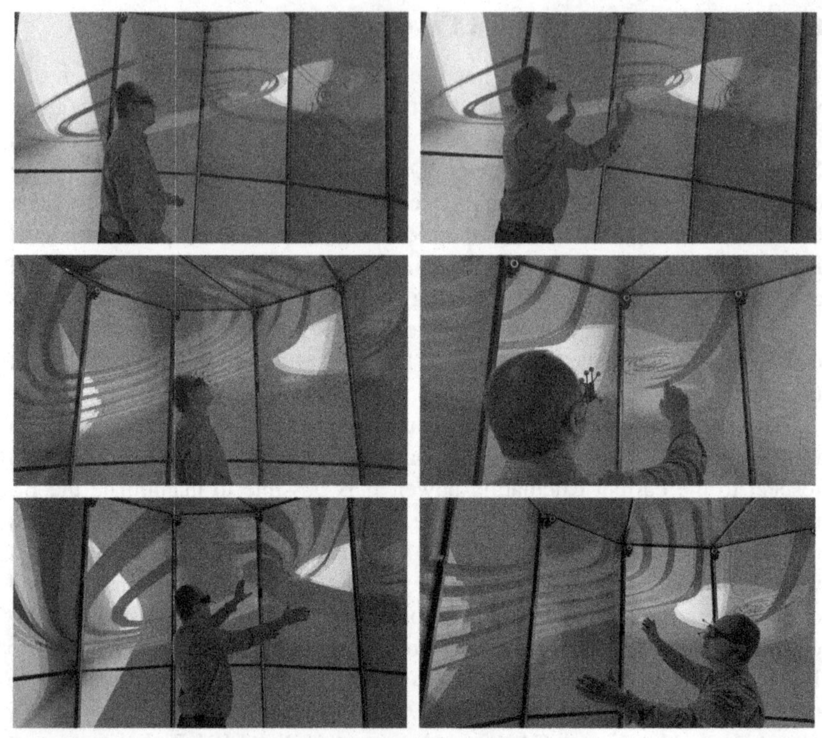

FIG. 19 Examples of CHA observer in LIRKIS CAVE environment for parameters *PC004*. Observer is located inside of BS. Parameters *PC004* are: $C_1 = 0.10443$ F, $C_2 = 1$ F, $L = 0.625$ H, $R = 1\ \Omega$, $m_0 = -1.2$ S, $m_1 = -0.714$ S, $B_P = 1$ V, $\rho = 0$.

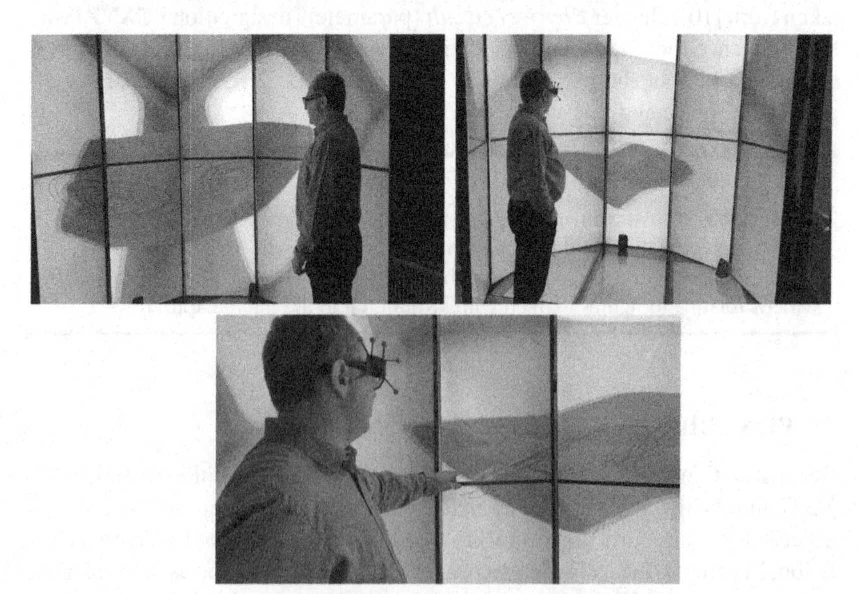

FIG. 20 Examples of a view to BS (and CHA inside transparent BS) as whole object in LIRKIS CAVE environment for parameters *PC004*.

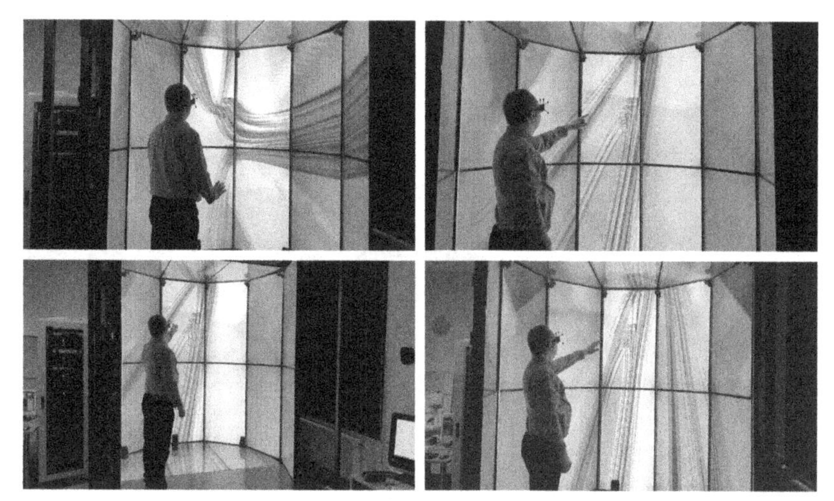

FIG. 21 Examples of a view inside BS and to CHA in LIRKIS CAVE environment for parameters *PC006*: $C_1 = -0.1333$ F, $C_2 = 11$ F, $L = 0.31$ H, $R = 1\ \Omega$, $m_0 = -0.98$ S, $m_1 = -2.4$ S, $B_P = 1$ V, $\rho = -0.1\ \Omega$.

FIG. 22 Examples of a view to BS as whole object in LIRKIS CAVE environment for parameters: *PC025* $C_1 = 1$ F, $C_2 = -1.12$ F, $L = -1.49$ H, $R = 0.03\ \Omega$, $m_0 = -0.5$ S, $m_1 = 0.0064$ S, $B_P = 1$ V, $\rho = 2.228\ \Omega$ (left) and parameters *PC001*: $C_1 = 0.10443$ F, $C_2 = 1$ F, $L = 0.625$ H, $R = 0.989119\ \Omega$, $m_0 = -1.143$ S, $m_1 = -0.714$ S, $B_P = 1$ V, $\rho = 0$ (right).

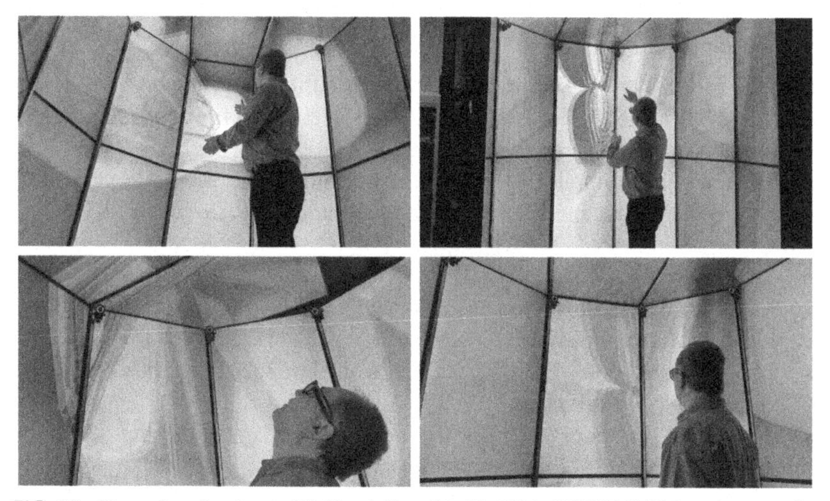

FIG. 23 Examples of a view to BS (first left) and inside BS in LIRKIS CAVE environment for parameters *PC006*.

FIG. 24 Examples of virtual environment with BS and CHA shared by multiple users in LIRKIS CAVE (parameters *PC001*).

- a PC with one or a multicore processor,
- GRID or high—performance computer clusters, and
- GPGPU technology.

7.1 Calculation of C-SBS by a PC With Single or Multicore Processor

The author of the work [64] has been dealing with solutions of differential equations of autonomous circuits since the early 1990s. It was possible, compared to a "manual" graphic calculation, to calculate thousands of particular solutions in a relatively short time with the arrival of PCs. This made it possible to calculate also the C-SBS. The system was solved by Runge-Kutta method of 4th degree. At the beginning, the C-SBS were calculated with $M \times N = 20 \times 20$ or 50×50 points (ICs for the differential equation system—Fig. 14B). As it has been found, the computing speed was not affected only by the CPU's frequency but also by the C language the compiler used to create the executable file (*.*exe*) and the CPU architecture. The graph in Fig. 25 illustrates a comparison of the one trajectory calculation speed using different CPU types, compilers and CPU architectures [64]. The data have been collected since 1994. The shortcuts in the graph refer to the used compilers, where:

- VC—is the Visual C compiler, ver. 5.0 (used in MS Windows),
- Djgpp—is a MS DOS 32-bit compiler. Its executable file could run on MS Windows, and
- Borland—is a compiler (version 3.1) working in the MS DOS operating system. Its executable file could run on Windows 98, 2000, and XP.

The Fig. 25 implies among others that:

- AMD 486 Series Processor (AMD 486–133 MHz) was faster than its Intel 486-120 MHz competitor,
- Intel Pentium—100 MHz CPU is faster from the 586 Series CPUs compared to AMD (K5-90 MHz),
- IBM Cyrix processors or WinChip performance lagged despite higher or equal CPU frequencies compared to Intel Pentium,

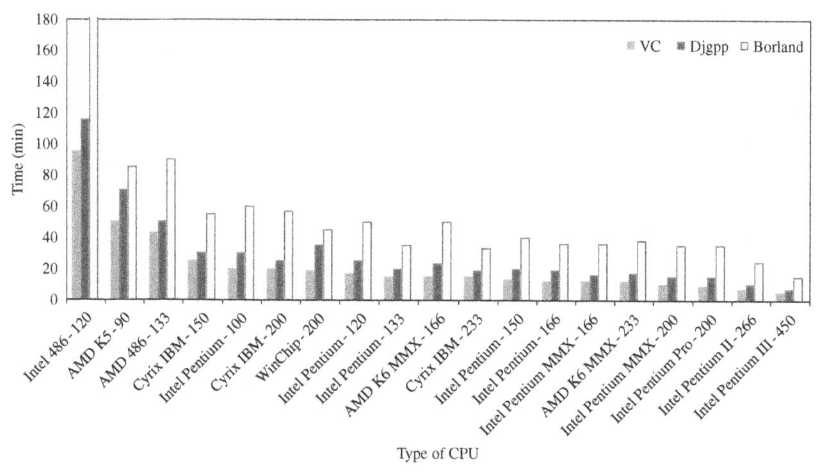

FIG. 25 The comparison of single-trajectory calculation speed for different CPUs and compilers. The number behind the dash indicates the frequency of the given CPU in MHz.

- Using the Intel Pentium 120 $\div$ 450 MHz CPU, the calculation time is reduced with the increasing CPU frequency, and
- The fastest compiler is VC. Its **.exe* file calculates the same trajectory on the same CPU 2-3 $\times$ faster than the Borland compiler.

Currently, it is possible to create an executable file **.exe* with Bloodshed Dev-C++ software on OS Windows. A good alternative for OS Linux is the *gcc* compiler with the ability to set different optimizations that can accelerate the C-SBS calculations as well.

Windows PCs have been used to test more powerful and multicore CPUs since 2004. The program was created by a fast Bloodshed Dev-C++ compiler that calculated one and the same C-SBS. One C-SBS in the calculation consisted of 440 $\times$ 440 dots—ICs. This is a total of 193,600 calculated trajectories. A large increase in computing power compared to, for example, CPU Intel Celeron 500 MHz or Pentium III 600 MHz at the start of the graph in Fig. 26 is evident from the results. Even a PC with a Pentium 133 MHz CPU (also used in Fig. 25) was tested to show the performance of an "old" PC. This calculated one C-SBS up to 3960 min (66 h) and is therefore not shown in the chart.

Fig. 26 implies among others that:

- With today's volume of computations, the CPUs with a time $t > 300$ min are inapplicable, even though they cover relatively new all-in-one PCs or 10″ netbook. The results show that the CPUs contained therein are not at all suitable for the calculation of 2D C-SBS.
- Older CPUs produced by AMD (Sempron, Duron, Athlon) at equivalent CPU frequencies compared to the Celeron—2400 MHz CPU, were much more suitable for C-SBS calculations not only in terms of CPU speed but

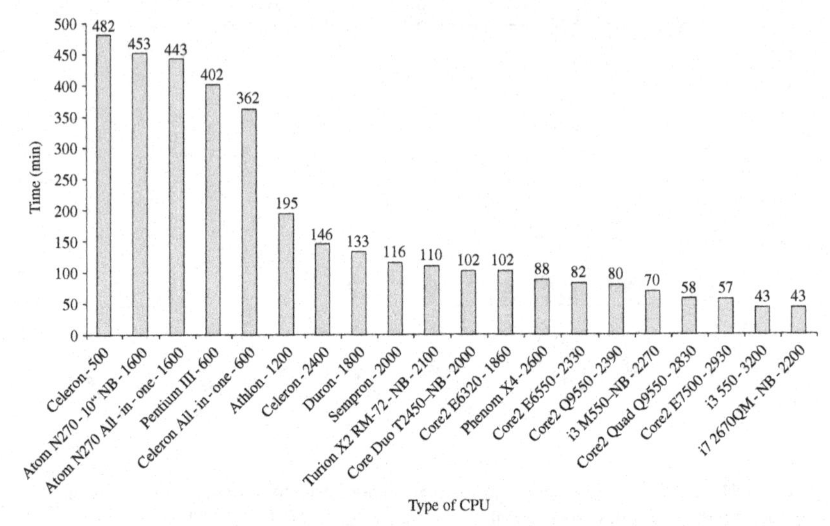

FIG. 26 The comparison of single-thread C-SBS calculation for different CPUs using the Bloodshed Dev-C++ ver. 4.9.9.2. The computer with Pentium 133 MHz CPU computed one C-SBS up to 3960 min.

also in terms of CPU or PC prices. Similar results showing significantly higher AMD CPU performance compared to Intel were found in the calculation of other C-SBS when the same C-SBS was calculated on PC with a Duron 650 MHz CPU for the same calculation time as a PC containing an Intel Pentium IV 1700 MHz CPU. CPU architecture change has proven up to 1.75 times better improvement when comparing the Celeron Pentium II 266 MHz CPU with the older Pentium CPU 233 MHz (at minimum CPU frequency change).

- Intel's new multicore CPU architecture (CPU Core2 and above) suggests that AMD CPUs are not very useful for C-SBS calculations (single thread use). This is also demonstrated by the comparison of CPU Phenom ×4 and Turion ×2 RM—72 with the CPUs of the same age or newer (the last 8 CPUs in Fig. 26).
- CPUs with recent generation architecture—Intel i7 2670QM—2200 MHz and i3 550–3200 MHz CPUs are in comparison to the previous CPU Core2 or Core Duo T2450 the best to calculate the C-SBS. At the lowest 2200 MHz CPU frequency, the i7 2670QM processor is better than any Core2 CPU even with a CPU frequency approaching 3000 MHz. This is confirmed also by the very fast modern CPU i7 7700. The question is how much power offers one CPU/thread of current AMD CPUs (Ryzen). This option has not been tested yet.

Because the last two CPUs in Fig. 26 (i7 2670QM—2200 MHz or i3 550–3200 MHz) have 4 cores/8 threads or 2 cores/4 threads and Core2 Quad CPU contains 4 cores, a single filament performance test was performed while

TABLE 1 The Comparison of Calculation Speed Running 1, 2, 4, or 8 Programs for C-SBS Calculation on Processors Intel i7 2670QM, i3 550 and Core2 Quad at the Same Time

Number of "CPU"	CPU & Computation Time (min)		
	i7 2670QM	i3 550	Core2 Quad
1	43	43	58
2	44	46	58
4	58	75	58
8	96	–	–

running 1, 2, 4, or 8 of the same C-SBS calculation programs with affinity settings—1 program/1 CPU (if the affinity has not been set, C-SBS calculation was only extended only by about 2–3 min). The results, including data for 1 CPU, are summarized in Table 1. It follows that:

- The CPU Core2 Quad calculates of C-SBS still the same time, regardless of the number of CPU kernels used.
- CPU i7 2670QM and i3 550 are fast by using $1 \div 2$ cores—calculation time is up to 46 min.
- For CPU i7 2670QM, the calculation time using the four cores is the same as for Core2 Quad CPU.
- Even though the slower C-SBS computation time when using i7 and i3 was achieved by utilization of all i7 or i3 processor threads (96 and 75 min), in case of i7 the time 96 min is acceptable, because it is possible to calculate up to 8C-SBSs at the same time! As it is shown in Table 1, the concurrent use of all 8 threads in the Intel i7 CPU is most effective, because if we wanted to compute eight C-SBS using only 4CPU cores, the time needed to calculate 4C-SBS (58 min) would actually double (116 min). So, this required time is 20 min longer than using all 8 Intel i7 CPUs. Similarly, it would be possible to comment on the calculation time of 4 C-SBS by using the Intel i3 processor.

Using CPU parallelization is an even more efficient way to compute C-SBS. The work [65] has shown a great advantage of parallelizing the C-SBS computation. Each one point of $M \times N$ represents one IC for system solution (1) (Fig. 14B). Each trajectory can be computed independently of the previous one. The designed application was written in C++. Programming language C ++ is one the fastest among modern programming languages [66] that supports multiple platforms, and there is a wide range of free libraries available for that language. Based on these facts, it was chosen for the development of the application. In addition to that, the Qt framework was chosen to power the GUI and Intel TBB for compute-intensive operations. As a compiler, Visual Studio 2013 was used, and O2 optimization was used as it is optimal for compiling. The computation of one C-SBS (only 200 × 200 points was tested, we normally use resolution 400 × 400, 500 × 500, or up to 1000 × 1000 points) was 2–3 × faster compared with O1 or Od optimization. Therefore, in Table 2, we present the efficiency of multicore CPUs usage compared to only 1 CPU usage (using O2 optimization) being standard in previous years. It is also clear from Table 2 that the fastest calculation is performed if we use 64-bit operating system and 64-bit application.

TABLE 2 The Comparison of Computation Times on Multicore Computers, a Calculation of 1 BS Cut 200 × 200 Points (ICs)

Computers						32/64-Bit Application	
PC	*CPU Model*	*Operating System*	*Bits*	*Cores*	*Threads*	*O2 t (s)*	*1 CPU, O2 t (s)*
1	Intel i7-4790 K, 4 GHz	LIN Mint 18	64	4	8	–/14	–/66
1b	Intel i7-4790 K, 4 GHz	Win10 (virt.)		4	8	15/15	78/77
2	Intel i5-5200 U, 2.2 GHz, NB	Win 8.1		2	4	30/29	83/75
3	Intel Q9550, 2.83 GHz	Win 7		4	4	39/38	204/144
4	Intel i7-2670QM, 2.2 GHz, NB	Win 7		4	8	43/44	144/149
5	Intel E5200 2.5 GHz	Win 7	32	2	2	89/–	235/–
6	AMD Athlon II 215, 2.7 GHz	Win 7		2	2	87/–	174/–
7	Intel E6600, 2.4 GHz	Win 7		2	2	130/–	319/–
8	Intel i7-3770, 3.4 GHz	Win 7		4	8	17/18	82/80
9	Intel G3250 T, 2.8 GHz	Win 7		2	2	59/57	114/112

Shortcut NB (PC2 and PC4) means: CPU was installed in notebook (mobile version of CPU).

7.2 GRID Technology or High-Performance Computer Clusters

With the need for complex RA mapping for individual attractors in the analyzed dynamic systems, the requirements for volume of C-SBS calculations was also increased. The PC with Sempron CPU calculated one C-SBS shown in Fig. 14A in 5 h. If we needed 300C-SBS, the calculation time with single-processor computer would increase up to 1500 h! Now, there was a big task that could be subdivided into a set of subtasks because the individual subtasks are independent of each other. The subtasks can therefore be calculated in a parallel way (SIMD). Grid technologies or high-performance computer clusters are appropriate for such calculations. Because, theoretically, 300 parallel subtasks can be calculated simultaneously, then the total calculation of one big task (job) could take only 5 h. Subtasks are distributed to different clusters and, depending on the number of clusters, the calculation of one task may be longer than 5 h, but significantly shorter than the 1500 h mentioned above. The results of the calculations using Grid or high-performance computer clusters have been published, for example, in the works [55, 67]. Because the solution is very extensive and time-demanding, currently it is possible to use also cloud-solutions.

7.3 GPGPU Technology Usage

In recent years, with increasing graphics cards performance, due to the constantly increasing number of cores on graphics cards, parallelization of scientific calculations is applied increasingly. After 2003, it was possible to implement matrix computations through the GPU and to implement parallel computations using GPGPU with the later expansion of the CUDA GPU architecture from NVIDIA. The later OpenCL standard provided support for parallel computing not only for the GPU but also for the CPUs.

The work [67] compares the C-SBS ternary memory computing times on PC with a single CPU (AMD Turion II P520 Dual-Core 2.3 GHz) and the card NVIDIA Tesla C1060 (240 cores). It shows that the acceleration of the calculation of 1 cut BS is 41 times or 15 times faster when a float or double type of variable is used. But even if the calculations were "only" five times faster, the use of GPGPU technology would still be interesting and perspective.

8 CONCLUSION

We presented in this chapter a visualization of RP motion, CHA, and BS of Chua's circuit in 3D state space. Although a visualization of computed CHA in 2D or 3D space is standard nowadays (e.g., using Matlab, MathCAD, Octave, or software created in one of the programming languages), it is still interesting to have an idea about the dynamics of RP movement in state space or a possibility of viewing CHA from all directions not only from the scientific but also from the pedagogical point of view. It is also standard these days to calculate and display C-SBS in 2D space. Many works are satisfied with 1 or 2C-SBS going through the start of the coordinate system. However, it is necessary to know the morphology of BS as a whole, if it is necessary to control the analyzed system or to use it for confidential communication by using chaos. If you need to calculate a few tens C-SBS in 2D space, the C-SBS calculation time is no longer negligible, and the optimization of C-SBS calculation has to be considered—these options were also highlighted in our chapter. The process of creating BS reconstruction in 3D is not easy and some of the 195 BS-s in Ref. [10] cannot be reconstructed in 3D for their very complex morphology. The result of the reconstructed BS can be fascinating not only by proper illumination of the scene or displaying along with the attractor and IV characteristics on the monitor, but also mainly with the use of virtual reality technologies. Entering into the virtual 3D space allows the observer to actually be transferred directly to the chip of integrated circuit and track, for example, BS, CHA, or changes that occur due to control pulse or parasitic elements on the chip. The presented CAVE LIRKIS system does not allow direct calculation (differential equations calculation) or CHA or BS editing in real time yet. However, it allows its stereoscopic real-time visualization as a whole and a detailed "inside" view into any analyzed autonomous differential equation system, not only into the Chua's system. The use of VR technologies is, therefore, certainly a significant step in exploring such a beautiful phenomenon such as the Chua's circuit and its state space.

REFERENCES

[1] Matsumoto T. A chaotic attractor from Chua's circuit. IEEE Trans Circ Syst 1984;31 (12):1055–8.

[2] Li C, Sprott JC. An infinite 3-D quasiperiodic lattice of chaotic attractors. Phys Lett A 2018;382(8):581–7.

[3] Gotthans T, Petržela J. New class of chaotic systems with circular equilibrium. Nonlinear Dyn 2015;81(3):1143–9.

[4] Pham V, Jafari S, Volos C, Gotthans T, Wang X, Hoang D. A chaotic system with rounded square equilibrium and with no-equilibrium. Optik 2017;130:365–71.

[5] Jafari S, Sprott J. Simple chaotic flows with a line equilibrium. Chaos, Solitons Fractals 2013;57:79–84.

[6] Lassoued A, Boubaker O. On new chaotic and hyperchaotic systems: a literature survey. Nonlinear Anal: Model Control 2016;21(6):770–89.

[7] Petržela J, Gotthans T. New chaotic dynamical system with a conic-shaped equilibrium located on the plane structure. Appl Sci 2017;7(10):976.

[8] Mobayen S, Volos CK, Kaçar S, Çavuşoglu Ü. New class of chaotic systems with equilibrium points like a three-leaved clover. Nonlinear Dyn 2017;91(2):939–56.

[9] Pham V-T, Wang X, Jafari S, Volos C, Kapitaniak T. From Wang-Chen system with only one stable equilibrium to a new chaotic system without equilibrium. Int J Bifurcat Chaos 2017;27(06).

[10] Bilotta E, Pantano P. A gallery of Chua attractors. Singapore: World Scientific; 2008.

[11] Fortuna L, Frasca M, Xibilia M. Chua's circuit implementations yesterday, today and tomorrow. Singapore: World Scientific; 2009.
[12] Kılıç R. A practical guide for studying Chua's circuits. Singapore: World Scientific; 2010.
[13] Chen G, Adamatzky A, Chua L, Chaos CNN. Memristors and beyond: a festschrift for Leon Chua. World Scientific; 2013.
[14] Mkaouar H, Boubaker O. Chaos synchronization for master slave piecewise linear systems: application to Chua's circuit. Commun Nonlinear Sci Numer Simul 2012;17(3):1292–302.
[15] Šimšík D, Galajdová A, Drutarovský M, Galajda P, Pavlov P. Wearable non-invasive computer controlled system for improving of seniors gait. Int J Rehabil Res 2009;32:S35.
[16] Boubaker O, Dhifaoui R. Robust chaos synchronization for Chua's circuits via active sliding mode control. In: Chaos, complexity and leadership. vol. 2012. Dordrecht: Springer; 2014. p. 141–51.
[17] Özkaynak F, Özer A. Cryptanalysis of a new image encryption algorithm based on chaos. Optik 2016;127(13):5190–2.
[18] El-Gohary A, Alwasel I. The chaos and optimal control of cancer model with complete unknown parameters. Chaos, Solitons Fractals 2009;42(5):2865–74.
[19] Koyuncu İ, Turan Özcerit A. The design and realization of a new high speed FPGA-based chaotic true random number generator. Comput Electr Eng 2017;58:203–14.
[20] Xu G, Xu J, Xiu C, Liu F, Zang Y. Secure communication based on the synchronous control of hysteretic chaotic neuron. Neurocomputing 2017;227:108–12.
[21] Mkaouar H, Boubaker O. Robust control of a class of chaotic and hyperchaotic driven systems. Pramana 2016;88(1).
[22] Mkaouar H, Boubaker O. In: On electronic design of the piecewise linear characteristic of the Chua's diode: application to chaos synchronization. 16th IEEE mediterranean electrotechnical conference, Yasmine Hammamet; 2012. p. 197–200.
[23] Zhang W. Differential equations, bifurcations, and chaos in economics. Singapore: World Scientific; 2005.
[24] Špány V. The analysis of a one-tunnel diode binary. Proc IEEE 1967;55(6):1089–90.
[25] Špány V. Graphical solution of a non-linear circuit by the method of a m-dimensional phase state. Elektrotech Čas 1969;20(4):233–48 [in slovak].
[26] Špány V, Galajda P, Guzan M, Pivka L, Olejár M. Chua's singularities: great miracles in circuit theory. Int J Bifurcat Chaos 2010;20(10):2993–3006.
[27] Matsumoto T, Chua L, Komuro M. The double scroll. IEEE Trans Circ Syst 1985;32 (8):797–818.
[28] Špány V, Pivka L. Boundary surfaces in sequential circuits. Int J Circ Theory Appl 1990;18 (4):349–60.
[29] Špány V. Special surfaces and trajectories of the multidimensional state space. Proceedings of scientific papers of the Technical university in Košice, part 1; 1978. p. 123–51[in slovak].
[30] Špány V. A circuits model and a complex analysis of a memory cell by means of the boundary surface. *Elektrotech Čas* 1985;36(5):355–72 [in slovak].
[31] Špány V. Generate initial conditions for the trajectory of the boundary area of the sequence circuit. Proceedings of scientific papers of the Technical university in Košice, part 3; 1978. p. 95–9 [in slovak].
[32] Tang S, Chen H, Hwang S, Liu J. Message encoding and decoding through chaos modulation in chaotic optical communications. IEEE Trans Circ Syst: Fund Theory Appl 2002;49(2):163–9.
[33] Galajda P, Guzan M, Špány V. The control of a memory cell with the multiple stable states. Radioelektronika. Brno: Brno University of Technology; 2011. p. 211–4.

[34] Špány V, Pivka L. Dynamic properties of flip-flop sensors. Electr Eng 1996;47(7–8):169–78.

[35] Sobota B, Korečko Š. Virtual reality technologies in handicapped persons education. Adv Inform Sci Appl 2014;1:134–8.

[36] Hrozek F, Korečko Š, Sobota B. Solutions for time estimation of tactile 3D models creation process. Advanced machine learning technologies and applications: second international conference AMLTA 2014. Cairo: Springer International Publishing; 2014. p. 498–505.

[37] Sobota B, Korečko Š, Jacko M, Jacho L, Szabó C. Applied research in the field of virtual-reality technologies for 3D stereoscopic display techniques and their implementation using advanced components of human-computer interface. Potential and services of USP Technicom for efficient development of entrepreneurship and research collaboration with industry. Košice: Elfa; 2015. p. 21–5.

[38] Bačíková M, Porubän J, Lakatoš D. Defining domain language of graphical user interfaces. Slate 2013: 2nd symposium on languages, applications and technologies, Porto; 2013. p. 187–202.

[39] O'Neill S. 3D print a home on Mars. New Scient 2015;226(3023):27.

[40] Guzanová A, Ižaríková G, Brezinová J, Živčák J, Draganovská D, Hudák R. Influence of build orientation, heat treatment, and laser power on the hardness of Ti6Al4V manufactured using the DMLS process. Metals 2017;7(8):318.

[41] Brezinová J, Hudák R, Guzanová A, Draganovská D, Ižaríková G, Koncz J. Direct metal laser sintering of Ti6Al4V for biomedical applications: microstructure, corrosion properties, and mechanical treatment of implants. Metals 2016;6(7):171.

[42] Varga M, Sobota B, Hrozek F, Korečko Š. Augmented reality with interactive interfaces. Studia Univ Babes-Bolyai: Ser Inform 2014;59(2):21–33.

[43] Hudák M, Korečko Š, Sobota B. On architecture and performance of LIRKIS CAVE system. 8th IEEE international conference on cognitive infocommunications Debrecen 2017;295–300.

[44] Guzan M, Galajda P, Pivka L, Špány V. Element of singularity is a key to laws of chaos. Radioelektronika 2005: 15th international Czech-Slovak scientific conference. Brno: Brno University of Technology; 2005. p. 33–6.

[45] Špány V, Pivka L, Galajda P, Guzan M, Kalakaj P, Šak A. The calculation of symmetrizing voltage in flip flop sensors. Radioelectron Koš 1994;22–7.

[46] Kennedy M. ABC (adventures in bifurcations and chaos): a program for studying chaos. J Franklin Inst 1994;331(6):631–58.

[47] Guzan M, Sobota B. 3D visualisation dynamics (Pembroke, ON) of Chua's circuit in state space. SEKEL 2009. Brno: VUT Brno; 200991–8 [in slovak].

[48] Guzan M, Sobota B. Visualization of chaos. J Electr Electron Eng 2009;2(1):48–51.

[49] Sobota B, Guzan M. Visualization of chaotic attractor of modified Chua's circuit. SEKEL 2011. Ostrava: VŠB-TU; 2011. p. 115–8 [in slovak].

[50] Sobota B, Guzan M. In: Visualization of non-linear dynamics systems using web technologies. Nové trendy v prevádzke technických systémov '09, Prešov; 2009. p. 89–92 [in slovak].

[51] Siderskiy V. Chua's circuit simulator, Available from:http://www.chuacircuits.com/sim.php. Accessed 29 January 2018.

[52] Sobota B, Guzan M. Utilization of virtual reality for gallery of chaotic attractors. SEKEL 2012. Bratislava: STU; 2012. p. 172–8 [in slovak].

[53] Galajda P, Špány V, Guzan M. The state space mystery with negative load in multiple-valued logic. Radioengineering 1999;8(2):2–7.

[54] Špány V, Galajda P, Guzan M. The state space mystery in multiple-valued logic circuit with load plane—part I. Acta Electrotech Inform 2001;1(1):17–22.

[55] Guzan M. Boundary surface of a ternary memory in the absence of limit cycles. Radioelektronika 2012. Brno; 2012. p. 1–4.
[56] Sobota B. 3D modelling of Chua's circuit boundary surface. Acta Electrotech Inform 2011;11(1).
[57] Špány V, Galajda P, Guzan M. In: Boundary surfaces of one-port memories. Tesla III Milenium. 5th Int conference Beograd; 1996. p. 131–7.
[58] Keppel E. Approximating complex surfaces by triangulation of contour lines. IBM J Res Dev 1975;19(1):2–11.
[59] Sobota B, Guzan M. Recontructing of boundary surface Chua's circuit. SEKEL 2010. Liberec: Technická univerzita v Liberci; 2010. p. 151–6 [in slovak].
[60] Bajaj C, Coyle E, Lin K. Arbitrary topology shape reconstruction from planar cross sections. Graph Models Image Process 1996;58(6):524–43.
[61] Ekoule A, Peyrin F, Odet C. A triangulation algorithm from arbitrary shaped multiple planar contours. ACM Trans Graph 1991;10(2):182–99.
[62] Lorensen W, Cline H. Marching cubes: a high resolution 3D surface construction algorithm. ACM SIGGRAPH Comput Graph 1987;21(4):163–9.
[63] Korečko Š, Sorád J. Using simulation games in teaching formal methods for software development. Innovative teaching strategies and new learning paradigms in computer programming: Advances in Higher Education and Professional Development (AHEPD). Hershey: IGI Global; 2015. p. 106–30.
[64] Guzan M. Calculate boundary surface using a PC. Posterus 2012;5(9):1–9 [in slovak].
[65] Racz Z, Sobota B, Guzan M. Parallelizing boundary surface computation of Chua's circuit. Radioelektronika 2017. Brno: IEEE; 2017. p. 1–4.
[66] Fourment M, Gillings M. A comparison of common programming languages used in bioinformatics. BMC Bioinform 2008;9(1):82.
[67] Guzan M, Sobota B, Astaloš J. Calculate boundary surface using grid and GPGPU technologies. Posterus 2012;5(10):1–9 [in slovak].

Chapter 9

Some New Chaotic Maps With Application in Stochastic

Ezzedine Mliki*, Navid Hasanzadeh[†], Fahimeh Nazarimehr[†], Akif Akgul[‡], Olfa Boubaker[§] and Sajad Jafari[†]

[*]Department of Mathematics, College of Science, Imam Abdulrahman Bin Faisal University, Dammam, Saudi Arabia, [†]Biomedical Engineering Department, Amirkabir University of Technology, Tehran, Iran, [‡]Department of Electrical and Electronics Engineering, Sakarya University, Adapazarı, Turkey, [§]National Institute of Applied Sciences and Technology, Tunis, Tunisia

1 INTRODUCTION

Nonlinear dynamical systems can be categorized in two groups, continuous (flow) and discrete (map) systems. Such systems can show different attractors such as stable equilibria, limit cycle, quasiperiodic and chaos [1]. Chaos is a rare type of dynamic in nonlinear systems. In fact, a narrow pass in the nonlinear dynamical system's world has the ability of showing chaos. Chaotic systems and their dynamical properties have been a hot topic in recent years [2]. Many studies have been done on those systems. For many years, researchers have believed that the existence of chaotic attractors is related to saddle equilibria [3]. Recently, some papers have proposed chaotic systems without any equilibria [4–11], with a line or curve of equilibria [12–17], with stable equilibria [18–21], and with many other interesting features [22–33]. Those counterexamples show that the existence of a saddle point is not necessary for having a chaotic attractor [34].

Leonov and Kuznetsov have proposed two categories of attractors, hidden and self-excited attractors [35–37]. An attractor is called hidden if its basin of attraction does not intersect with a small neighborhood of any equilibrium point of the system [38–45]. It is called self-excited if the basin of attraction intersects with an unstable equilibrium. Hidden attractors are very important because they can cause problems in real-world systems [34, 42, 46–49]. A dynamical system with hidden attractors can change its dynamic to an unwanted attractor with a small perturbation. In this situation, perturbation causes changing in the initial conditions. So, the system can be placed in the basin of attraction of hidden attractor and the dynamic varies to an unwanted hidden attractor. The most

Recent Advances in Chaotic Systems and Synchronization. https://doi.org/10.1016/B978-0-12-815838-8.00009-1

studies on chaotic dynamics and hidden attractors have been done on flows. In 2017, a paper [44] investigated the presence of hidden attractors in chaotic maps. Multistability has been a hot topic in the study of nonlinear dynamics. Systems with multistability have some attractors which coexist with each other. In other words, in systems with multistability, the attractor can be changed by varying initial conditions [50–52]. Chaotic modeling of biological systems has been a hot topic recently [53].

White noise is a type of random signal in which the power density function is distributed uniformly in all frequencies. Autocorrelation of white noise is a Dirac delta function. If this noise follows a Gaussian distribution, then it is called white Gaussian noise [54]. Modeling of white Gaussian noise has many applications in telecommunications [55], electronics [56], imaging [57], and especially, generating random numbers [58]. Increasing accuracy in modeling white Gaussian noise can lead to better results in different applications. Montiero and his coauthors have introduced a method for generating pseudo-Gaussian two-dimensional distribution [59]. In this method, two uniform distributions were constructed using a Tent map. Then, with the help of a quasi-polarized transform, the obtained uniform distributions were converted to a two-dimensional pseudo-Gaussian distribution.

Attention deficit disorder (ADD) is a neurological disorder in which patients have difficulty paying attention. In 2014, a paper [60] proposed a simple non-linear neural network to model the dynamics of ADD. The model represents the interactions of inhibitory and excitatory parts of brain actions. It contained different dynamics such as periodic and chaotic attractors.

In this chapter, we investigate two chaotic maps. Different dynamical properties of these two maps are studied. The first map is discussed in Section 2. It is a modified method for generating quasi-Gaussian distribution in one dimension. It increases the similarity with normal distributions generated by algorithms such as ziggurat [61]. The second map is studied in Section 3. This map is an ADD model. In this chapter, we investigate some dynamical properties of ADD model which no one has mentioned before. Also, some applications of those systems are discussed. Finally, the paper is concluded in Section 4.

2 ONE-DIMENSIONAL WHITE GAUSSIAN NOISE GENERATOR

Tent map $\Lambda : (0,1) \rightarrow (0,1)$ is defined as,

$$z(n+1) = \Lambda(z(n)) = \begin{cases} Az(n) & 0 < z(n) < 0.5 \\ A(1-z(n)) & 0.5 \leq z(n) < 1 \end{cases} \tag{1}$$

If we assume $A = 2$, map plots of the first to forth iterates of tent map will be as Fig. 1.

The tent map with $A = 2$ has two unstable equilibrium points. Due to the constant gradient in each $z(n)$, the output of this map is uniform. So, it can be used to generate random numbers uniformly in the interval $[0, 1]$.

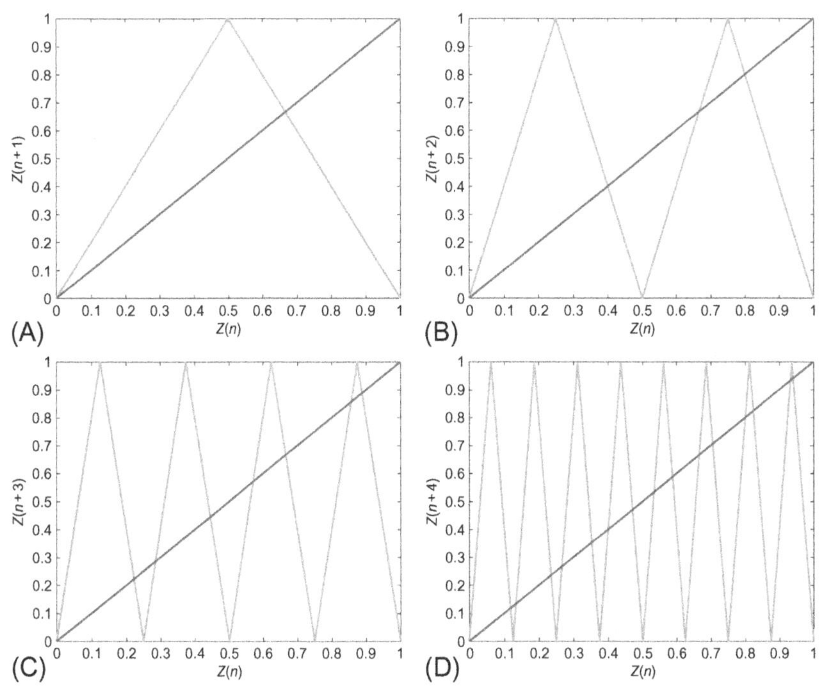

FIG. 1 (A) Map plot of first iterate of tent map. (B) Map plot of second iterate of tent map. (C) Map plot of third iterate of tent map. (D) Map plot of fourth iterate of tent map.

Figs. 2–4 show 2-D numbers generated by the Tent map in $A=2$, a histogram of its dynamic in $A=2$, and its bifurcation diagram with respect to changes in parameter A.

In order to convert the one-dimensional map with uniform distribution to a Gaussian one, a function can be defined as Eq. (2). It transforms values in the interval [0.5,1] to $[-\infty,0]$ and values in the interval [0,0.5] to $[0,\infty]$. Thus, a pseudo-Gaussian distribution is created.

$$Y = H(X) = (U_1 - U_2)\sigma^2 \ln\left(\frac{0.5}{|0.5 - X|}\right) \tag{2}$$

By assuming $\sigma^2 = 1$, we have,

$$H(X) = (U_1 - U_2) \ln\left(\frac{0.5}{|0.5 - X|}\right) \tag{3}$$

where U_1 and U_2 are step functions which are shown in Figs. 5 and 6, respectively.

Then, we change these two functions to be more realizable. An exponential function with a small rise time can be used as follows,

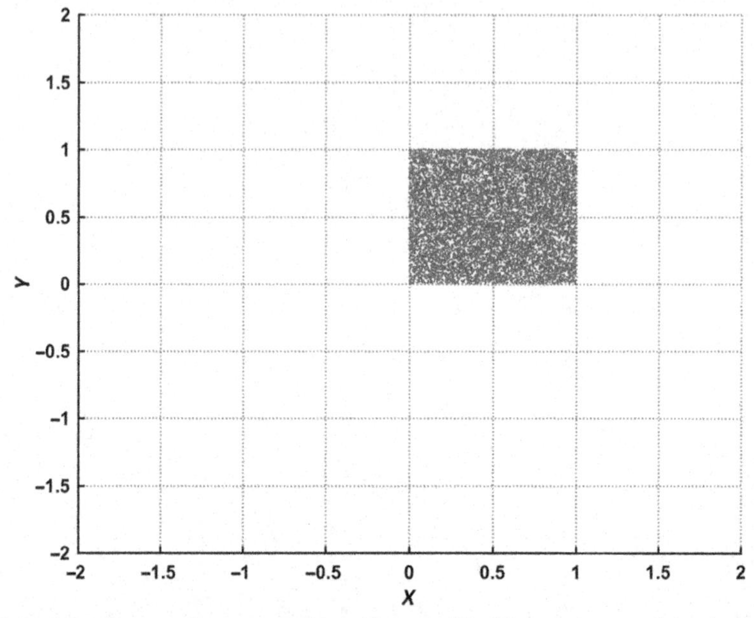

FIG. 2 2-D uniform numbers generated by the Tent map. *X* and *Y* are two generated chaotic signals with different initial conditions.

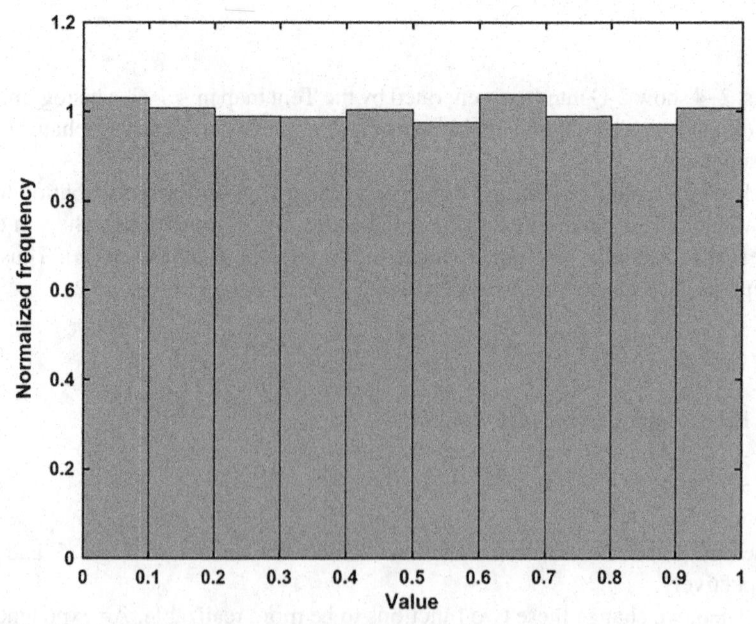

FIG. 3 Histogram of the uniform distribution formed by the Tent map.

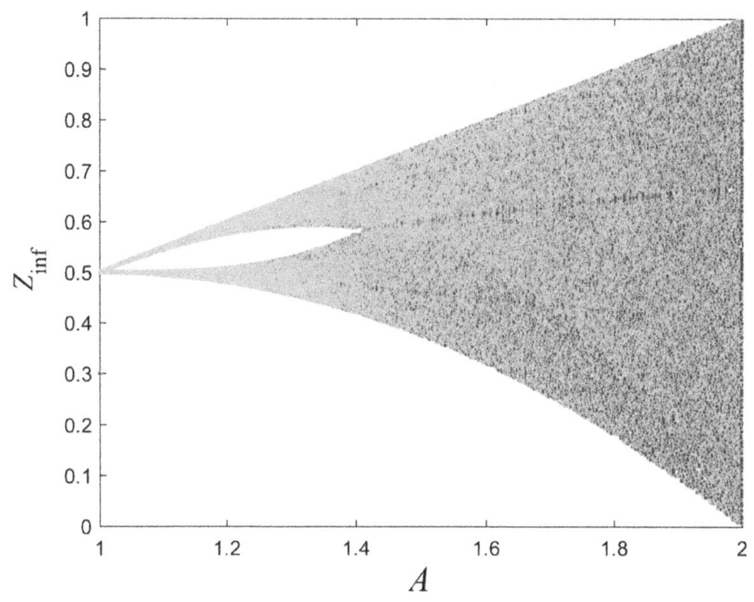

FIG. 4 Bifurcation diagram of the Tent map with respect to changing parameter A and constant initial condition $z(0) = 0.3$.

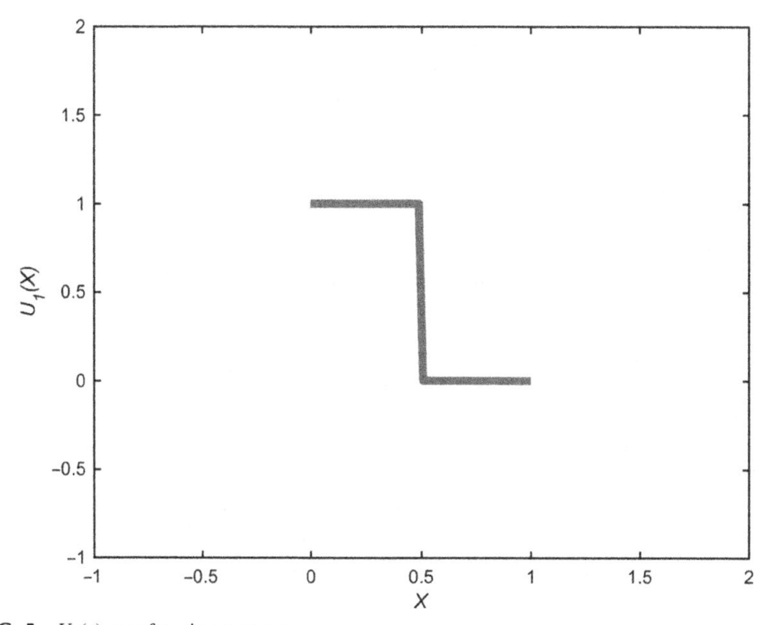

FIG. 5 $U_1(x)$ step function versus x.

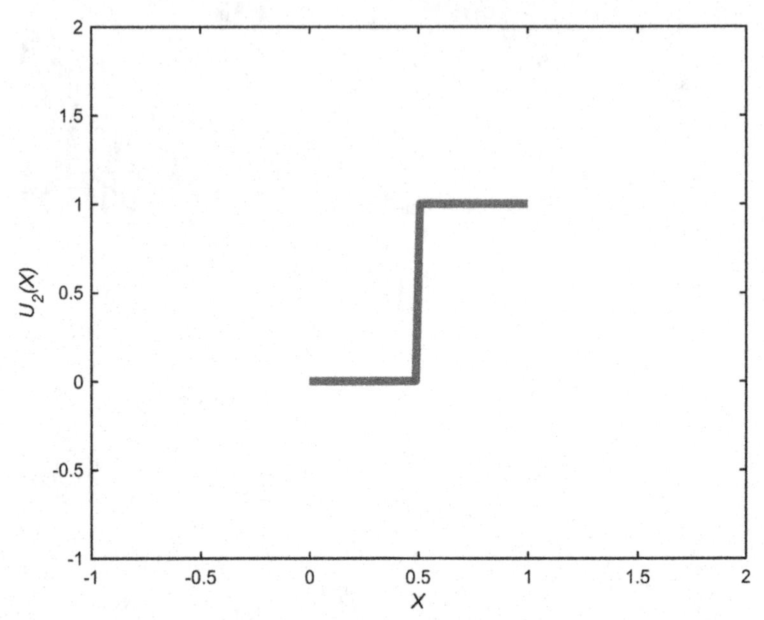

FIG. 6 $U_2(x)$ step function versus x.

$$U_1 = \frac{1}{1 + \exp\left(-K(0.5 - X_1)\right)}$$
$$U_2 = \frac{1}{1 + \exp\left(-K(X_1 - 0.5)\right)} \tag{4}$$

where K is a very large number (e.g., 50,000,000).

The output histogram which is resulted by applying H transform to the uniform distribution generated by the Tent map is shown in Fig. 7.

Although the resulting histogram is similar to a Gaussian distribution, there are some differences. So, we need to improve our work. To do this, the following function is applied,

$$f(X) = A\sigma^2 \log\left(1 + \left(\frac{X}{B}\right)\right) \operatorname{sgn}(X) \tag{5}$$

Assuming parameters $A = 1$ and $B = 1.8$, and $\sigma = 1$, the function f is shown in Fig. 8.

As can be seen in Fig. 8, this function is approximately one-to-one for inputs which are close to zero. When the absolute value of input increases, f represents smaller values. In a nutshell, by applying this function to H outputs, the density of numbers around zero increases and makes the histogram in Fig. 7 more similar to the histogram of an expected Gaussian distribution. The result is shown in Fig. 9.

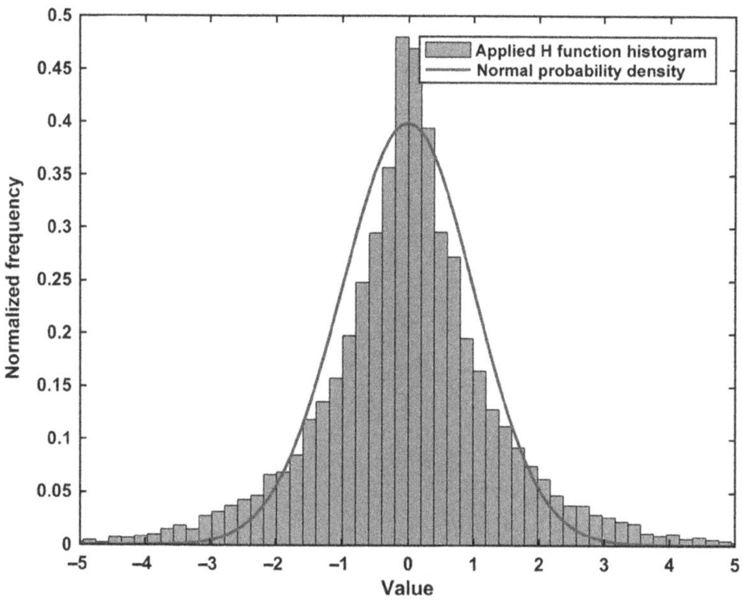

FIG. 7 Histogram which is resulted by applying H transform to the uniform distribution generated by the Tent map with $\mu = 0$ and $\sigma^2 = 1$.

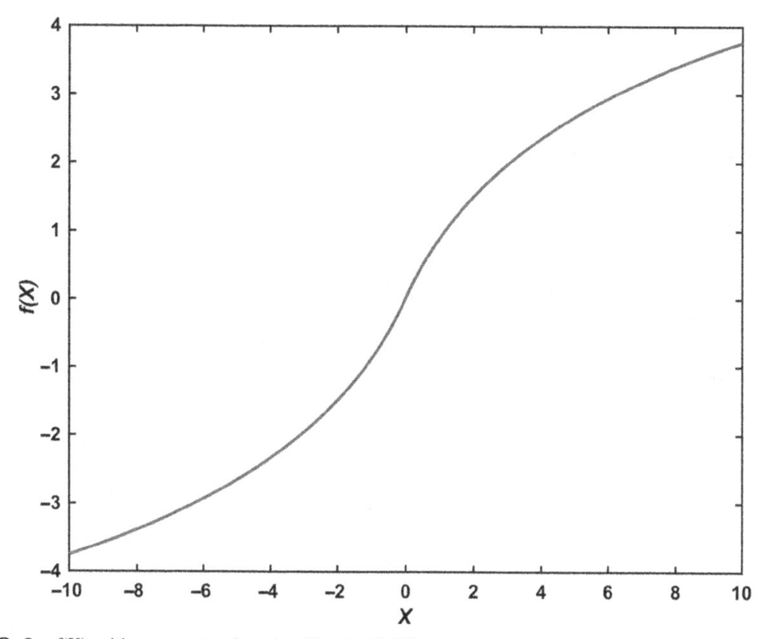

FIG. 8 $f(X)$ with respect to changing $X \in [-10, 10]$.

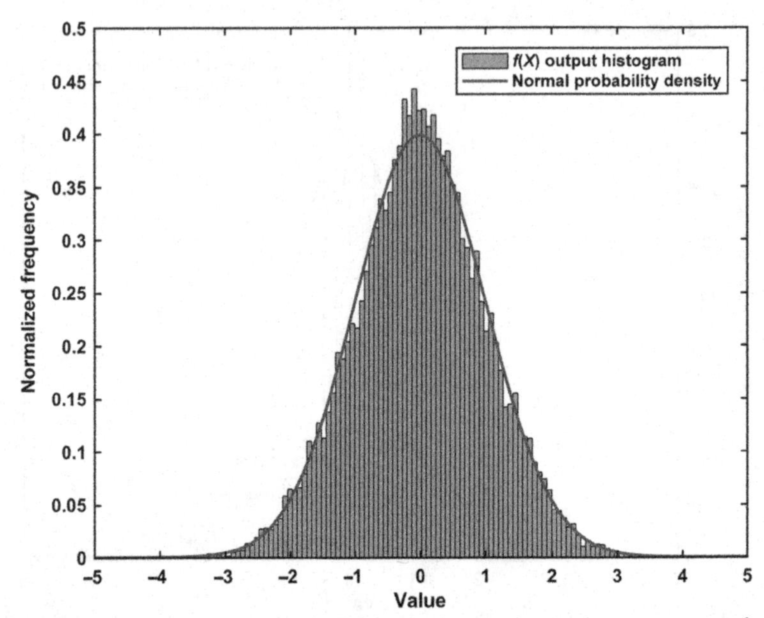

FIG. 9 Histogram resulted by applying function f to outputs of function H with $\mu = 0$ and $\sigma^2 = 1$.

Function f is one-to-one. So, its inverse function exists and is obtained as follows,

$$f^{-1}(X) = \frac{B}{\sigma^2}\operatorname{sgn}(X)\left|\exp\left(\left|\frac{X}{A}\right|\right) - 1\right| \tag{6}$$

Finally, we define map N, which can generate Gaussian distribution from a one-dimensional uniform distribution in the Tent map (Eq. 7).

$$N = f \circ H \circ T \circ H^{-1} \circ f^{-1} \tag{7}$$

So, the system of Eq. (8) has the ability to generate time series with a Gaussian distribution. In other words, Eq. (8) is an iterative map. Each initial condition evolves with iterations through the system. Until now, we have generated our desired distribution using one iteration and different initial conditions. In order to convert it to an iterative equation, we need to remove effects of the previous iteration in the new one (Eq. 7). So, the chaotic dynamic of the Tent map produces various initial conditions, and $f \circ H \circ T$ applied to each of them. Therefore, if $x(0)$ is a chaotic initial condition for N, Gaussian distribution numbers will be generated using the chaotic map of Eq. (8).

$$x(n+1) = N(x(n)) \tag{8}$$

As shown in Fig. 9, the resultant histogram is very similar to the histogram of a Gaussian distribution. Fig. 10 shows bifurcation diagram of map N with

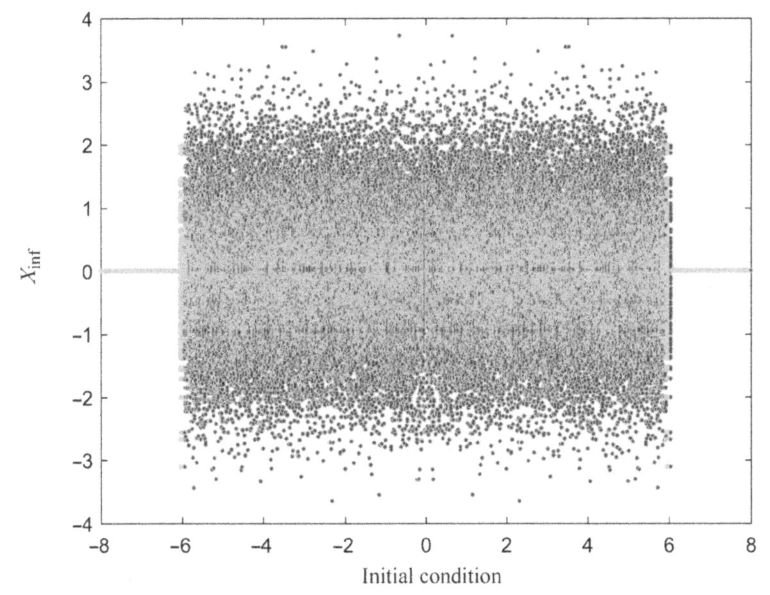

FIG. 10 Bifurcation diagram of map N with respect to changing initial condition.

respect to changing initial condition. Fig. 11 indicates an example of the distribution obtained from this map in two dimensions.

Autocorrelation coefficients are useful for showing how our generated distribution is accurate. Fig. 12 shows the obtained autocorrelation coefficient of map N with different lags. Because autocorrelation coefficients of the proposed map with $lag \neq 0$ are approximately zero, it can be claimed that the distribution is applicable for white noise modeling.

We used different normal distribution tests to check our generated distribution similarity with a normal distribution. The null hypothesis of these tests is the normality of distribution. Lilliefors [62], Z-test [63], Shapiro-Wilk [64], and Jarque-Bera [62] tests are used with a confidence level of 0.05. We used these tests 10,000 times on the distributions with 10,000 samples. Table 1 shows how many of the 10,000 tests on 10,000 distributions with different initial conditions have not been ruled out of normality.

In order to have a better visualization and compare our results with a normal distribution, a quantile-quantile plot is used [65] (Fig. 13). In this plot, if the two distributions are the same, the curve will be placed approximately on the identity line. The obtained quantile-quantile curve depicts that our generated distribution is very similar to a normal distribution. The results are obtained by ziggurat algorithm [61].

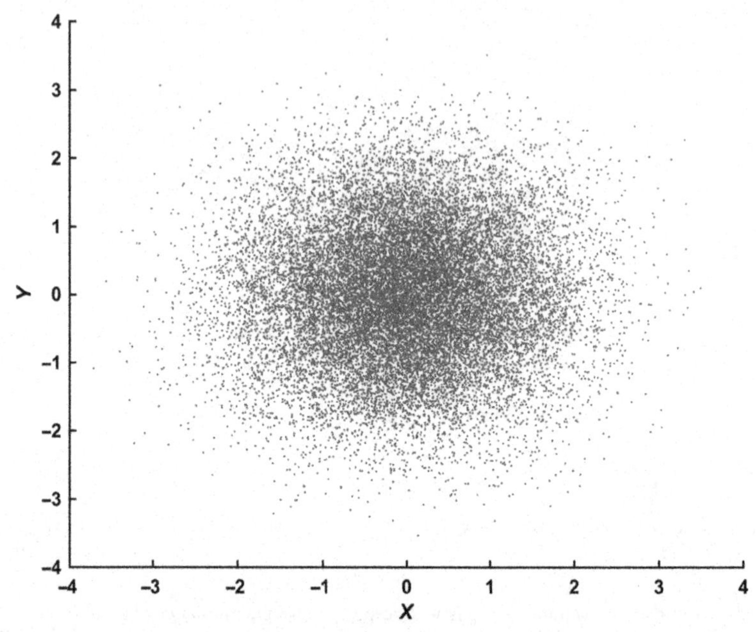

FIG. 11 2-D Gaussian numbers generated by map N. X and Y are two generated chaotic signals with different initial conditions.

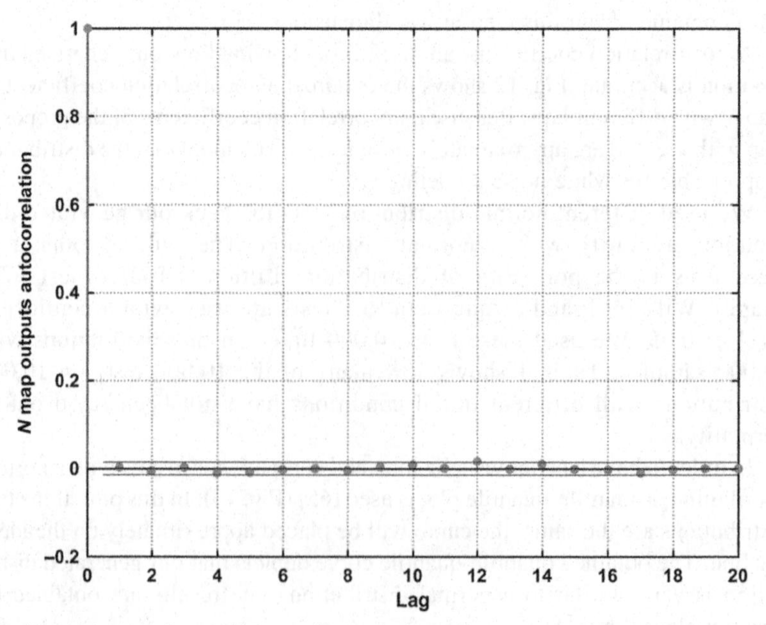

FIG. 12 Autocorrelation coefficients of outputs of map N.

TABLE 1 Results of Normality Tests on 10,000 Distributions With Different Initial Conditions

Test	Result
Lilliefors	3058
Z-test	9530
Shapiro-Wilk	3939
Jarque-Bera	9442

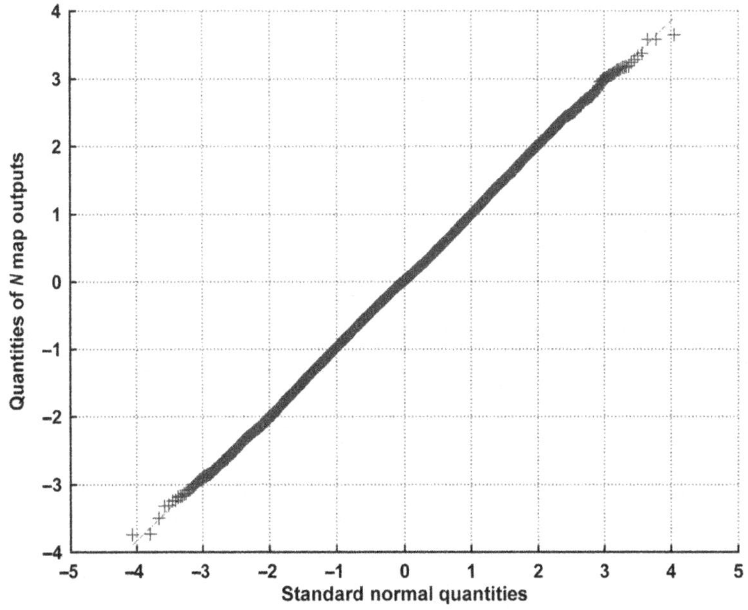

FIG. 13 Quantile-quantile plot of outputs of map *N*.

3 ATTENTION DEFICIT DISORDER MODEL

ADD model is a one-dimensional discrete system [66]. It has two tanh nonlinearities. The system is as follows,

$$x_{k+1} = f(x_k) = B\tanh(w_1 x_k) - A\tanh(w_2 x_k) \tag{9}$$

where $B = 5.821$, $w_1 = 1.487$, $w_2 = 0.2223$ are constant parameters and A is the bifurcation parameter. Discrete systems in one-dimension have the ability of generating chaotic attractors. System (9) can show different dynamics.

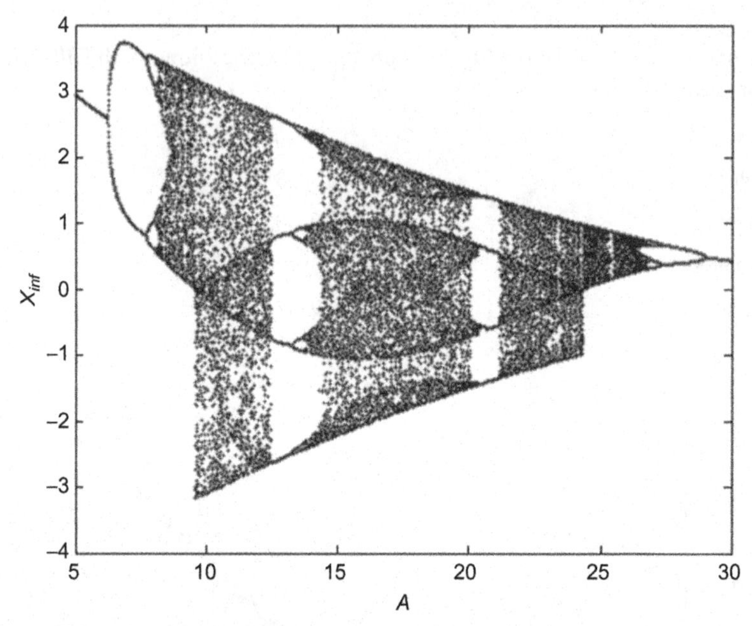

FIG. 14 Bifurcation diagram of system (9) with respect to changing parameter *A*.

Fig. 14 presents bifurcation diagram of system (9) with respect to changing parameter A in the interval [5,30]. As the figure shows, the system has different dynamics such as stable fixed point (period-one); period-two and its dynamic goes to chaos with a period-doubling route. Then, an inverse route of period doubling happens until it reaches a period-one dynamic. Also, some periodic windows can be seen among chaotic behaviors of the system.

In order to investigate dynamical properties of the system, consider the map plot of system (9) in four different values of parameter A (Fig. 15). The figure shows that the system has two horizontal asymptotes. tanh(x) has asymptotes as,

$$\begin{aligned} &\text{if}\, x \rightarrow +\infty \,\text{then}\, \tanh(x) \rightarrow +1 \\ &\text{if}\, x \rightarrow +\infty \,\text{then}\, \tanh(x) \rightarrow -1 \end{aligned} \tag{10}$$

So we can calculate two horizontal asymptotes of system (9) as follows,

$$\begin{aligned} &\text{if}\, x \rightarrow +\infty \,\text{then}\, f(x) \rightarrow B - A \\ &\text{if}\, x \rightarrow -\infty \,\text{then} f(x) \rightarrow A - B \end{aligned} \tag{11}$$

The result is matched with Fig. 15.

In order to analyze fixed points of system (9), consider the map plots of Fig. 16 with $A = 5$. It is shown that the map plot intersects with the identity line in three points. So, the system has three fixed points in this situation. Origin is an unstable fixed point because its gradient is higher than one. The two other fixed points are stable because their absolute value of gradient is lower than one. So,

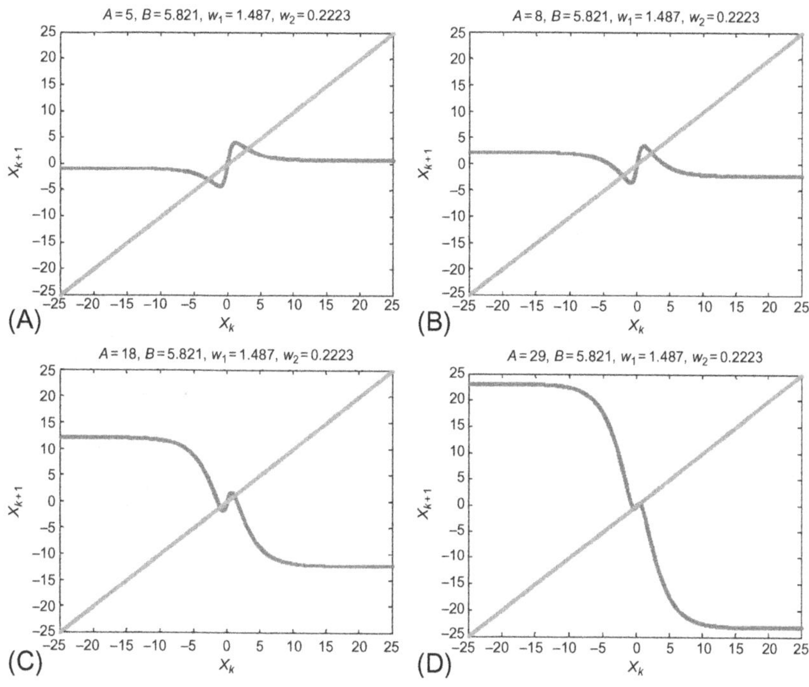

FIG. 15 (A) Map plot of system (9) in parameter $A = 5$. (B) Map plot of system (9) in parameter $A = 8$. (C) Map plot of system (9) in parameter $A = 18$. (D) Map plot of system (9) in parameter $A = 29$.

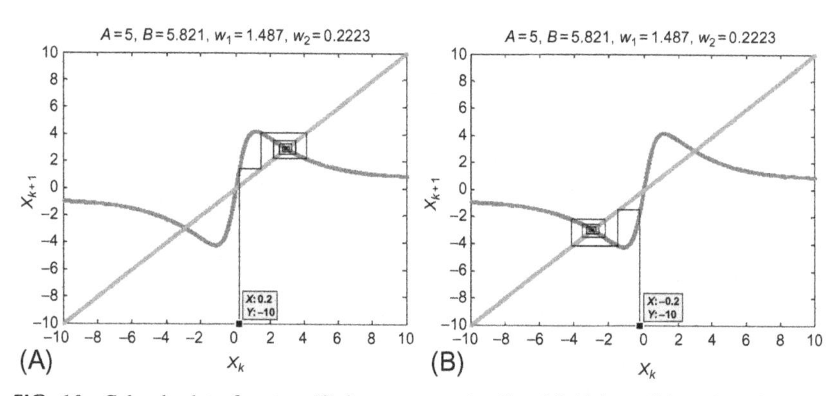

FIG. 16 Cobweb plot of system (9) in parameter $A = 5$ and initial condition (A) $x(0) = 0.2$, (B) $x(0) = -0.2$.

the system has two stable fixed points simultaneously which causes a kind of multistability. The coexistence of these fixed points causes different initial conditions to go to different attractors. A cobweb plot of the system with two different initial conditions (+0.2 and −0.2) is shown on the map plots of Fig. 16. In other words, the system has two separate parts in its map plot. So, depending on the initial condition, the trajectory traps in one of the lobes and remains in the lobe till the end. Fig. 17 shows bifurcation diagram of system (9) for the two different initial conditions $x(0) = 0.2$ and $x(0) = -0.2$. The figure shows that in approximately $A = 10$ two bifurcations join together until $A = 24$. In order to explain this behavior, consider the cobweb plot of Fig. 18. As the figure shows, in such a situation, the range of one lobe intersects with the domain of another lobe and so the attractor can cross through these two parts. However, while it is a necessary condition, it is not sufficient, because for joining two lobes, we should have the condition $f(f(x)) < 0$ for $x > 0$ and $f(f(x)) > 0$ for $x < 0$ where x is the possible value in system (9) range for each parameter A. After approximately $A = 24$, conditions are not satisfied anymore, and two lobes are separated. In other words, the system creates two separate attractors in two lobes. So, the system becomes multistable again, as can be seen in Fig. 17. A cobweb plot of the system with $A = 29$ is shown in Fig. 19.

Another interesting dynamic of system (9) is the multistability of a period-two attractor with other studied attractors in high values of parameter A. This behavior has never been studied before. In order to have a better visualization, consider the cobweb plot of Fig. 20 with parameter $A = 29$ (as in Fig. 19 but

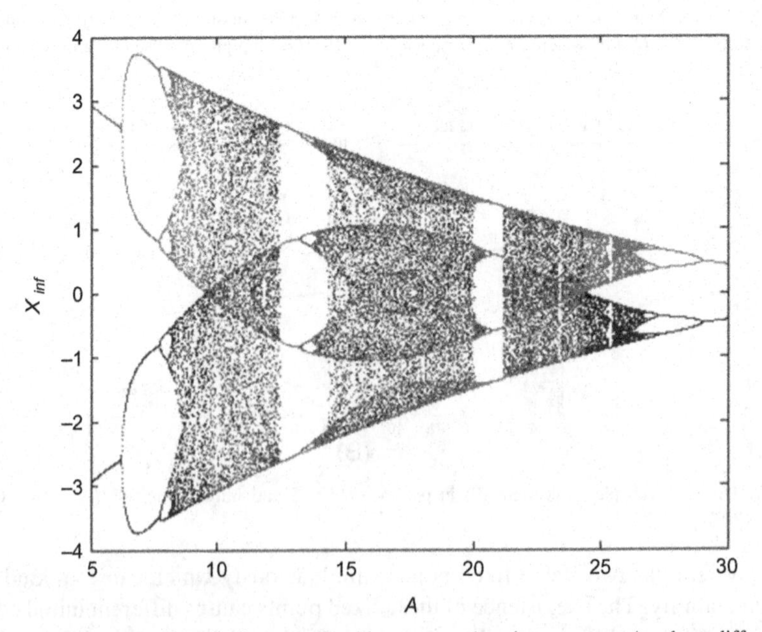

FIG. 17 Bifurcation diagram of system (9) with respect to changing parameter A and two different initial conditions ($x(0) = 0.2$ and $x(0) = -0.2$).

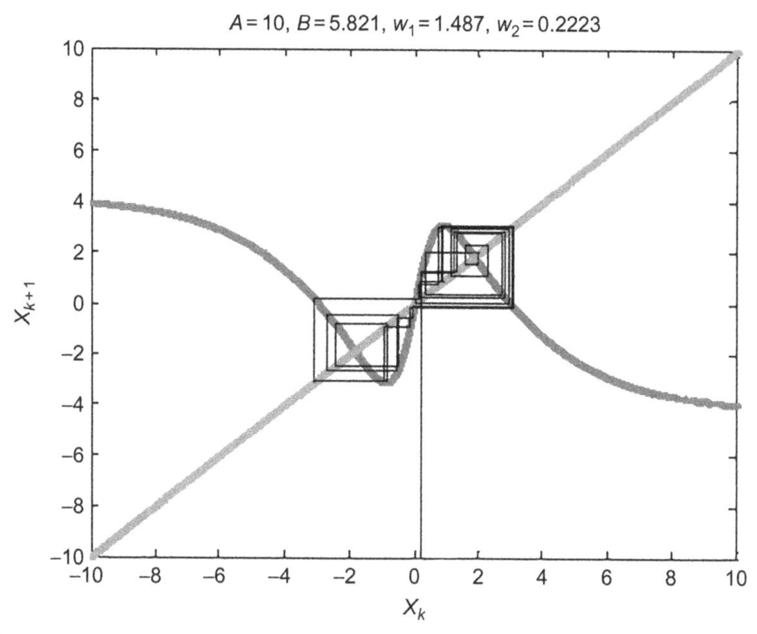

FIG. 18 Cobweb plot of system (9) in parameter $A = 10$ and initial condition $x(0) = 0.2$.

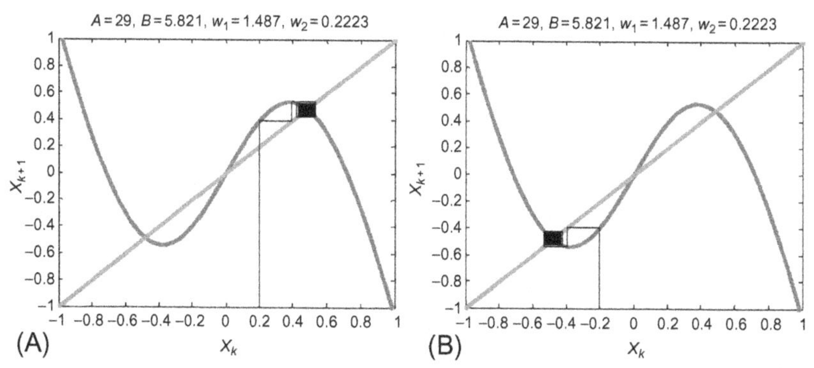

FIG. 19 Cobweb plot of system (9) with parameter $A = 29$ and initial condition (A) $x(0) = 0.2$, (B) $x(0) = -0.2$.

with different initial condition $x(0) = 1.2$ and a larger view of map plot). By comparing the two figures, we find that small initial conditions lead to one of the two stable fixed points in small lobes, and large initial conditions force the system to trap to a large period-two attractor.

In order to investigate this behavior, consider map plots of the second iterate of system (9) as shown in Fig. 21. The figure shows map plot of system (9) in

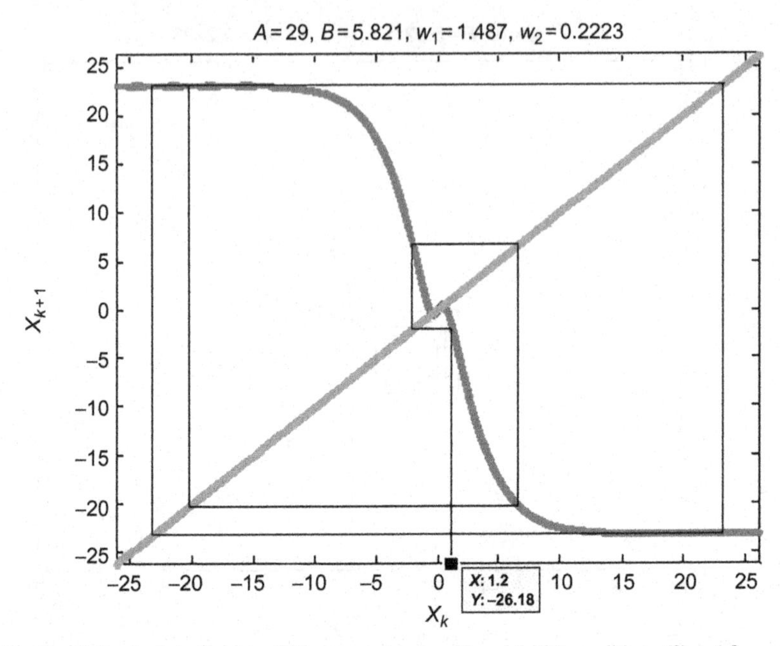

FIG. 20 Cobweb plot of system (9) in parameter $A = 29$ and initial condition $x(0) = 1.2$.

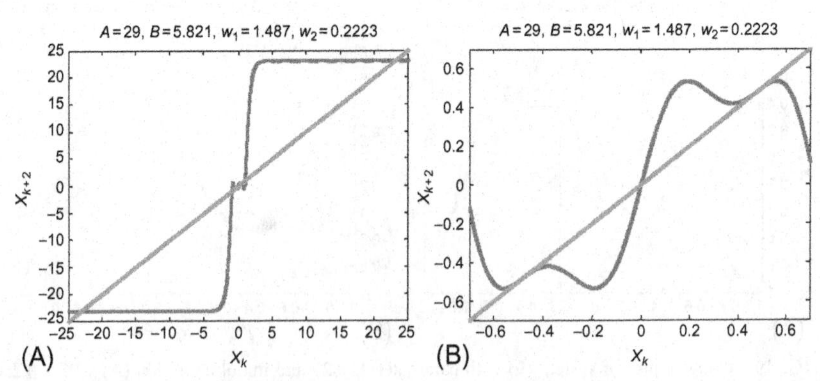

FIG. 21 Map plot of second iterate of system (9) with parameter $A = 29$ (A) in a large view, (B) in a zoomed view.

two large and zoomed views. The second iterate has four intersects with the identity line, so it has four fixed points. Two of those which can be seen in part b of Fig. 21 are equilibria of the first iterate of system (9). So, they show two period-one attractors of the system, which is discussed in Fig. 19. The two other fixed points cannot be seen in the first iterate, and so, they generate a period-two trajectory. Furthermore, in parameter $A = 29$ there are three attractors which are coexisting together, two period-one attractors and one period-two attractor.

In order to find the interval of parameter A in which the period-two dynamic is coexisted with other attractors, we investigate bifurcation diagram of system (9) with a large initial condition $x(0) = 5.5$ (Fig. 22).

The figure shows that the period-two attractor has existed in the interval $A \in [13.46, \infty]$.

4 DISCUSSION AND CONCLUSION

In this chapter, we have investigated dynamical properties of two interesting discrete systems. The first one was a map which was proposed as a new method for constructing a one-dimensional chaotic Gaussian distribution. Results have been tested using normal distribution and visual tests. Also, by plotting the autocorrelation coefficient diagram, we have shown that the distribution could have the characteristics of a white noise distribution. So, it can be used as a model of white Gaussian noise in telecommunication systems and electronics. The second studied system was an ADD model. The system has different dynamics. The most interesting behavior of this system was the coexistence of a period-two attractor with different attractors such as two periodic dynamics, two chaotic dynamics or one larger attractor. The system gives us the idea that

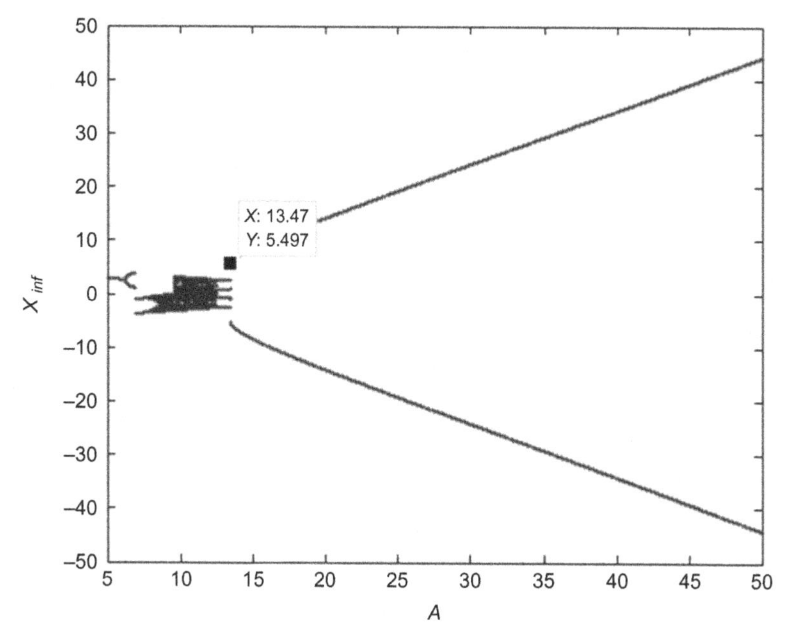

FIG. 22 Bifurcation diagram of system (9) with respect to changing parameter $A \in [5, 50]$ and initial condition $x(0) = 5.5$.

existence of horizontal asymptotes can help a map with two symmetric lobes show coexistence of periodic behaviors with different periods or coexistence of periodic and chaotic attractors. ADD is one of the most common disorders, and its modeling can help in its diagnosis and treatment. So, investigating dynamical properties of the ADD model can be helpful for future research work on this disorder.

As to some suggestions for future work, we can mention the following:

(a) We suggest generalizing the results replacing the map $x_{k+1} = f(x_k)$ by continuous time dynamical system, $x_{t+h} = f_h(x_t)$, t, $h \geq 0$, in which $f_{s+t} = f_s$ of t; s; $t \geq 0$.
(b) We suggest generalizing the results by replacing the map $x_{k+1} = f(x_k)$ by a dynamical system with randomness (Markov process), $X_{t+h} = f_h(X_t; w_t^{t+h})$; t; $h \geq 0$ in which X_t is Markov process.
(c) We suggest investigating stochastic-chaotic systems to study some new chaotic applications [67–71].

REFERENCES

[1] Sprott JC, Sprott JC. Chaos and time-series analysis. vol. 69. Citeseer; 2003.
[2] Buscarino A, Fortuna L, Frasca M, Sciuto G. Design of time-delay chaotic electronic circuits. IEEE Trans Circ Syst I: Regul Pap 2011;58:1888–96.
[3] Špány V, Galajda P, Guzan M, Pivka L, Olejár M. Chua's singularities: great miracle in circuit theory. Int J Bifurcat Chaos 2010;20:2993–3006.
[4] Pham V-T, Volos C, Jafari S, Wang X. Generating a novel hyperchaotic system out of equilibrium. Optoelectron Adv Mater—Rapid Commun 2014;8:535–9.
[5] Pham V-T, Volos C, Jafari S, Wei Z, Wang X. Constructing a novel no-equilibrium chaotic system. Int J Bifurcat Chaos 2014;24:1450073.
[6] Jafari S, Pham V-T, Kapitaniak T. Multiscroll chaotic sea obtained from a simple 3D system without equilibrium. Int J Bifurcat Chaos 2016;26:1650031.
[7] Pham V-T, Vaidyanathan S, Volos C, Jafari S, Kingni ST. A no-equilibrium hyperchaotic system with a cubic nonlinear term. Optik 2016;127:3259–65.
[8] Pham V-T, Vaidyanathan S, Volos C, Jafari S, Kuznetsov NV, Hoang T-M. A novel memristive time-delay chaotic system without equilibrium points. Eur Phys J Spec Top 2016;225:127–36.
[9] Kengne J, Jafari S, Njitacke Z, Khanian MYA, Cheukem A. Dynamic analysis and electronic circuit implementation of a novel 3D autonomous system without linear terms. Commun Nonlinear Sci Numer Simul 2017;52:62–76.
[10] Pham V-T, Akgul A, Volos C, Jafari S, Kapitaniak T. Dynamics and circuit realization of a no-equilibrium chaotic system with a boostable variable. AEU Int J Electron Commun 2017;78:134–40.
[11] Pham V-T, Volos C, Jafari S, Kapitaniak T. Coexistence of hidden chaotic attractors in a novel no-equilibrium system. Nonlinear Dyn 2017;87:2001–10.
[12] Barati K, Jafari S, Sprott JC, Pham V-T. Simple chaotic flows with a curve of equilibria. Int J Bifurcat Chaos 2016;26:1630034.

[13] Kingni ST, Pham V-T, Jafari S, Kol GR, Woafo P. Three-dimensional chaotic autonomous system with a circular equilibrium: Analysis, circuit implementation and its fractional-order form. Circ Syst Signal Process 2016;35:1933–48.
[14] Pham V-T, Jafari S, Volos C, Giakoumis A, Vaidyanathan S, Kapitaniak T. A chaotic system with equilibria located on the rounded square loop and its circuit implementation. IEEE Trans Circ Systems II: Express Briefs 2016;63:878–82.
[15] Pham V-T, Jafari S, Wang X, Ma J. A chaotic system with different shapes of equilibria. Int J Bifurcat Chaos 2016;26:1650069.
[16] Pham VT, Volos C, Kapitaniak T, Jafari S, Wang X. Dynamics and circuit of a chaotic system with a curve of equilibrium points. Int J Electron 2017;1–13.
[17] Pham V-T, Jafari S, Volos C. A novel chaotic system with heart-shaped equilibrium and its circuital implementation. Optik 2017;131:343–9.
[18] Kingni ST, Jafari S, Simo H, Woafo P. Three-dimensional chaotic autonomous system with only one stable equilibrium: analysis, circuit design, parameter estimation, control, synchronization and its fractional-order form. Eur Phys J Plus 2014;129:1–16.
[19] Lao S-K, Shekofteh Y, Jafari S, Sprott JC. Cost function based on Gaussian mixture model for parameter estimation of a chaotic circuit with a hidden attractor. Int J Bifurcat Chaos 2014;24:1450010.
[20] Pham V-T, Jafari S, Kapitaniak T, Volos C, Kingni ST. Generating a chaotic system with one stable equilibrium. Int J Bifurcat Chaos 2017;27:1750053.
[21] Pham V-T, Wang X, Jafari S, Volos C, Kapitaniak T. From Wang-Chen system with only one stable equilibrium to a new chaotic system without equilibrium. Int J Bifurcat Chaos 2017;27:1750097.
[22] Jafari S, Nazarimehr F, Sprott JC, Hashemi Golpayegani SMR. Limitation of perpetual points for confirming conservation in dynamical systems. Int J Bifurcat Chaos 2015;25:1550182.
[23] Lassoued A, Boubaker O. On new chaotic and hyperchaotic systems: a literature survey. Nonlinear Anal: Model Control 2016;21:770–89.
[24] Pham V-T, Jafari S, Kapitaniak T. Constructing a chaotic system with an infinite number of equilibrium points. Int J Bifurcat Chaos 2016;26:1650225.
[25] Pham V-T, Volos C, Jafari S, Vaidyanathan S, Kapitaniak T, Wang X. A chaotic system with different families of hidden attractors. Int J Bifurcat Chaos 2016;26:1650139.
[26] Jafari MA, Mliki E, Akgul A, Pham V-T, Kingni ST, Wang X, et al. Chameleon: the most hidden chaotic flow. Nonlinear Dyn 2017;1–15.
[27] Kingni ST, Jafari S, Pham V-T, Woafo P. Constructing and analyzing of a unique three-dimensional chaotic autonomous system exhibiting three families of hidden attractors. Math Comput Simul 2017;132:172–82.
[28] Mobayen S, Kingni ST, Pham V-T, Nazarimehr F, Jafari S. Analysis, synchronisation and circuit design of a new highly nonlinear chaotic system. Int J Syst Sci 2017;1–14.
[29] Nazarimehr F, Jafari S, Golpayegani SMRH, Sprott J. Categorizing chaotic flows from the viewpoint of fixed points and perpetual points. Int J Bifurcat Chaos 2017;27:1750023.
[30] Pham V-T, Jafari S, Volos C, Kapitaniak T. Different families of hidden attractors in a new chaotic system with variable equilibrium. Int J Bifurcat Chaos 2017;27:1750138.
[31] Rajagopal K, Akgul A, Jafari S, Karthikeyan A, Koyuncu I. Chaotic chameleon: dynamic analyses, circuit implementation, FPGA design and fractional-order form with basic analyses. Chaos Solitons Fractals 2017;103:476–87.
[32] Tlelo-Cuautle E, de la Fraga LG, Pham V-T, Volos C, Jafari S, de Jesus Quintas-Valles A. Dynamics, FPGA realization and application of a chaotic system with an infinite number of equilibrium points. Nonlinear Dyn 2017;1–11.

[33] Wang Z, Xu Z, Mliki E, Akgul A, Pham V-T, Jafari S. A new chaotic attractor around a pre-located ring. Int J Bifurcat Chaos 2017;27:1750152.

[34] Jafari S, Sprott J, Nazarimehr F. Recent new examples of hidden attractors. Eur Phys J Spec Top 2015;224:1469–76.

[35] Kuznetsov N, Leonov G, Seledzhi S. Hidden oscillations in nonlinear control systems. In: IFAC proceedings volumes (IFAC-Papers Online). vol. 18. 2011. p. 2506–10.

[36] Leonov G. Hidden oscillation in dynamical systems. In: From physics to control through an emergent view. vol. 15. 2010. p. 3.

[37] Stankevich NV, Kuznetsov NV, Leonov GA, Chua LO. Scenario of the birth of hidden attractors in the Chua circuit. Int J Bifurcat Chaos 2017;27:1730038.

[38] Pham V-T, Volos C, Jafari S, Wang X, Vaidyanathan S. Hidden hyperchaotic attractor in a novel simple memristive neural network. Optoelectron Adv Mater—Rapid Commun 2014;8:1157–63.

[39] Jafari S, Sprott JC, Nazarimehr F. Recent new examples of hidden attractors. Eur Phys J Spec Top 2015;224:1469–76.

[40] Pham V-T, Vaidyanathan S, Volos C, Jafari S. Hidden attractors in a chaotic system with an exponential nonlinear term. Eur Phys J Spec Top 2015;224:1507–17.

[41] Shahzad M, Pham V-T, Ahmad MA, Jafari S, Hadaeghi F. Synchronization and circuit design of a chaotic system with coexisting hidden attractors. Eur Phys J Spec Top 2015;224:1637–52.

[42] Dudkowski D, Jafari S, Kapitaniak T, Kuznetsov NV, Leonov GA, Prasad A. Hidden attractors in dynamical systems. Phys Rep 2016;637:1–50.

[43] Goudarzi S, Jafari S, Moradi MH, Sprott JC. NARX prediction of some rare chaotic flows: Recurrent fuzzy functions approach. Phys Lett A 2016;380:696–706.

[44] Jafari S, Pham V-T, Golpayegani SMRH, Moghtadaei M, Kingni ST. The relationship between chaotic maps and some chaotic systems with hidden attractors. Int J Bifurcat Chaos 2016;26:1650211.

[45] Pham V-T, Jafari S, Vaidyanathan S, Volos C, Wang X. A novel memristive neural network with hidden attractors and its circuitry implementation. Sci China Technol Sci 2016;59:358–63.

[46] Nazarimehr F, Saedi B, Jafari S, Sprott JC. Are perpetual points sufficient for locating hidden attractors? Int J Bifurcat Chaos 2017;27:1750037.

[47] Brezetskyi S, Dudkowski D, Kapitaniak T. Rare and hidden attractors in Van der Pol-Duffing oscillators. Eur Phys J Spec Top 2015;224:1459–67.

[48] Kapitaniak T, Leonov GA. Multistability: uncovering hidden attractors. Eur Phys J Spec Top 2015;224:1405–8.

[49] Dudkowski D, Prasad A, Kapitaniak T. Perpetual points and hidden attractors in dynamical systems. Phys Lett A 2015;.

[50] Maistrenko Y, Kapitaniak T, Szuminski P. Locally and globally riddled basins in two coupled piecewise-linear maps. Phys Rev E 1997;56:6393.

[51] Sprott JC, Jafari S, Khalaf AJM, Kapitaniak T. Megastability: Coexistence of a countable infinity of nested attractors in a periodically-forced oscillator with spatially-periodic damping. Eur Phys J Spec Top 2017;226:1979–85.

[52] Tanga Y, Khalaf AJM, Rajagopald K, Pham V-T, Jafari S, Tian Y. Whirlpool: A new nonlinear oscillator with infinite number of coexisting hidden and self-excited attractors. Chin Phys B 2018;.

[53] Zhang G, Wang C, Alzahrani F, Wu F, An X. Investigation of dynamical behaviors of neurons driven by memristive synapse. Chaos Solitons Fractals 2018;108:15–24.

[54] Haykin SS, Moher M. Communication systems. Wiley; 2010.

[55] Chitode J. Communication theory. Technical Publications; 2010.
[56] Suárez A. Analysis and design of autonomous microwave circuits. vol. 190. John Wiley & Sons; 2009.
[57] Al-Ghaib H, Adhami R. On the digital image additive white Gaussian noise estimation. 2014 international conference on industrial automation, information and communications technology (IAICT); 2014. p. 90–6.
[58] Toral R, Chakrabarti A. Generation of Gaussian distributed random numbers by using a numerical inversion method. Comput Phys Commun 1993;74:327–34.
[59] Eisencraft M, Monteiro LH, Soriano DC. White Gaussian chaos. IEEE Commun Lett 2017;21:1719–22.
[60] Baghdadi G, Jafari S, Sprott JC, Towhidkhah F, Hashemi Golpayegani SMR. A chaotic model of sustaining attention problem in attention deficit disorder. Commun Nonlinear Sci Numer Simul 2015;20:174–85.
[61] Marsaglia G, Tsang WW. The ziggurat method for generating random variables. J Stat Softw 2000;5:1–7.
[62] Öztuna D, Elhan AH, Tüccar E. Investigation of four different normality tests in terms of type 1 error rate and power under different distributions. Turk J Med Sci 2006;36:171–6.
[63] Sprinthall RC. Basic statistical analysis. Pearson Allyn & Bacon; 2012.
[64] Peat J, Barton B. Medical statistics: a guide to data analysis and critical appraisal. John Wiley & Sons; 2008.
[65] Razali NM, Wah YB. Power comparisons of Shapiro-Wilk, Kolmogorov-Smirnov, Lilliefors and Anderson-Darling tests. J Stat Model Anal 2011;2:21–33.
[66] Baghdadi G, Jafari S, Sprott J, Towhidkhah F, Golpayegani MH. A chaotic model of sustaining attention problem in attention deficit disorder. Commun Nonlinear Sci Numer Simul 2015;20:174–85.
[67] Faranda D, Sato Y, Saint-Michel B, Wiertel C, Padilla V, Dubrulle B, et al. Stochastic chaos in a turbulent swirling flow. Phys Rev Lett 2017;119:014502.
[68] Mejri H, Mliki E. On the abstract subordinated exit equation. In: Abstract and applied analysis. 2010.
[69] Billings L, Bollt EM, Schwartz IB. Phase-space transport of stochastic chaos in population dynamics of virus spread. Phys Rev Lett 2002;88:234101.
[70] Kyrtsou C, Terraza M. Stochastic chaos or arch effects in stock series?: A comparative study. Int Rev Financ Anal 2002;11:407–31.
[71] Evans LC. An introduction to stochastic differential equations. vol. 82. American Mathematical Soc; 2012.

Chapter 10

Chaotic Solutions in a Forced Two-Dimensional Hindmarsh-Rose Neuron

Zahra Rostami*, Mohsen Mousavi*, Karthikeyan Rajagopal†, Olfa Boubaker‡ and Sajad Jafari*

**Biomedical Engineering Department, Amirkabir University of Technology, Tehran, Iran, †Center for Nonlinear Dynamics, Defence University, Bishoftu, Ethiopia, ‡National Institute of Applied Sciences and Technology, Tunis, Tunisia*

1 INTRODUCTION

Chaos theory application ranges from pattern formation in physical, chemical, and even biological systems [1,2], space programs [3], and the discovered chaotic superhighways between the planets [1], weather forecasting having to do with chaotic demonstrations first uncovered by Edward Norton Lorenz with his famous paper on deterministic nonperiodic flow [4], etc. In chaos theory, we can study how highly complex demonstrations which may seem to be random arise from determinism [5]. Determinism means the changes in the state of the system is determined by a specific rule [6]. Especially in biological systems, it is very important that chaotic demonstrations play an essential role. For example, it is confirmed that chaotic heartbeat rate or brain signals guarantee the heart and brain respectively being in their healthy state [1,7]. Therefore, it is important to develop our knowledge in this field of science and be able to distinguish chaotic behavior from other behaviors in such dynamical systems [8]. In fact, we need appropriate tools to recognize chaotic behavior both qualitatively and quantitatively [9–11]. One of the main characteristics of chaotic behavior is sensitive dependence on initial conditions [6,12]. Additionally, one positive Lyapunov exponent can denote chaotic demonstrations [6,13]. Therefore, these features provide us an acceptable range of estimation for chaos measurement [14,15].

It is believed that dynamical models are the best tools to explain the real world [6]. As a simple definition, a system is dynamical if its state change in

Recent Advances in Chaotic Systems and Synchronization. https://doi.org/10.1016/B978-0-12-815838-8.00010-8

time [6]. In other words, a dynamical system consists of some possible states and their evolution over the time, which is governed by a specific law [16], that may not be known to us, necessarily. Considering the studies in the biological and neuronal fields of science, there have been some qualitative dynamical neuronal models proposed for the purpose of getting closer to both a neuron's activities and its physiological conditions [17–25]. Actually, these two factors have to do with neuromodulator concentrations and applied currents or synaptic inputs. These qualitative models focus on the qualitative features of oscillatory dynamics or excitable fluctuations of the neuron's membrane [26]. Fundamentally, the concepts from qualitative theory of nonlinear differential equations accompanied by basics of bifurcation theory are employed in these models.

A neuron is the basic nonlinear unit of the neuronal system that is responsible for all the biological behaviors in the body [27]. An excitable tissue consisting of a large number of neurons is a good example for complex dynamical systems [28]. Any tissue that can have electrical activities can be considered an excitable tissue, like the heart and the brain tissue. The brain's great ability in information processing by the use of meaningful interactions between the sensory neurons, motor neurons, and interneurons [27] through electrical propagations have made it a perfect organizing station. This processing is accomplished through information transference between the neurons under a dynamical law. It is also a good example to mention heartbeat variation and the required rhythmic contraction of the heart muscle. In fact, the electrical signal holding the specified information results in mechanical contraction of the cardiac muscle [29]. Therefore, generally, the neuronal system is a dynamical system with serious complexities. Accordingly, it is not only because of our tremendous lack of knowledge about an individual neuron as the basic block of an excitable tissue, but also the connections and interactions that govern the large arrangement of the neurons forming a greatly intricate network. That is why we need to focus on the fundamental dynamical representations while we accept a specific range of reductions in order to release from these complexities. In fact, complexity is the undeniable part of the natural phenomena including in various physical, chemical, and biological systems [30–38]. An interesting characterization of complex systems is their representation of organized behaviors with no central organizing unit [39]. This phenomenon is basically possible due to the capability of representing collective dynamical behaviors. Therefore, there are several studies on complex systems using some useful concepts like synchronization [40], cooperative and synergic evolutions [39], nonlinear dynamics [41], chaotic behavior [40], self-organization [39], self-adaptation, and so on. Thus, on the way to modeling such systems, understanding the related concepts and using appropriate tools and methods close to the desired natural system's behavior are highly needed.

Considering the required perception of complex dynamical systems, many studies focus on wave propagation and pattern recognition in a dynamical system [42,43]. These patterns contain the intrinsic dynamical law of the system

[44]. That is why this field of study has attracted so much attention, especially in recent decades. Spatiotemporal patterns are in the branch of dynamic patterns governed by the laws that can represent their spatial properties only during time. In other words, despite the static patterns, which have only some characteristics in space domain, the actual complexity of the spatiotemporal patterns is recognizable only by the passage of time. The emergence of possible spatiotemporal patterns in a complex excitable tissue is investigated frequently [29,45]. There are several studies from different aspects focusing on possible mechanisms of emergence or maintenance of different spiral waves [28,46,47], target waves [48], or any other particular spatiotemporal pattern. For example, spatiotemporal patterns resulted from defects in an excitable tissue [46,49], CO oxidation on a Pt(110) surface [50–52], etc. In fact, this point of view can help researchers obtain a higher perception of the dynamical systems with their complex characteristics and properties. As an example, it is worth mentioning the regulating role of some specific spatiotemporal patterns like spiral pattern affecting the ongoing cortical activities [53]. Alternatively, some of the pathological forms of behavior like epileptic seizures are found to be the outcome of defects in wave propagation among the neurons in the brain [54]. Moreover, studying the spatiotemporal pattern can help predict some future results of such systems, too. Regarding the body as a harmonious biological system, the importance of these predictions is more understood when it comes to abnormalities [18] of different parts of the body system. For example, it is believed that some heart disorders like tachycardia and ventricular fibrillations can be predicted by detecting the initiation of reentry in the propagating wave that can trigger spiral seed in the cardiac tissue [5,29,55].

Fundamentally, wave propagation is based on the collective behaviors of the units making up a network [56] or become suppressed due to some properties of the network [57]. It can be initiated from an excitation in a local part of the network and start to travel to the other parts. This means that the local activity of a particular region can influence the activity of other parts. For example, it is confirmed that target waves can be initiated from a local activity caused by any internal or external reason [48,58]. Generally, the theory of local activity benefits the determination of emergent behaviors in a regular network of dynamical systems [39]. Although it arises from circuit theory [39], it can have an extremely general concept which is employed in different fields of biological, physical, chemical, thermal, and optical studies [59]. Here, the resulting behaviors can have an evolutionary pattern in the network. This means that time plays a determinative role in the resultant dynamic pattern. Some studies also have focused on variation of these patterns over the time [48,60].

The rest of the chapter is organized as follows:

In the next section, the neuronal model equations with respective variables and parameters are represented. In addition, for more illustration some explanations are also given. In Section 3, we explain our method for the numerical study focusing on bifurcation diagrams and the calculated Lyapunov exponent.

Section 4 is dedicated to the numerical analysis on spatiotemporal patterns and the interesting properties of the designed network as an excitable tissue. Additionally, the results of our simulations are depicted and the observations are described in details. Finally, the conclusions are debated in Section 5.

2 MODEL AND DESCRIPTION

In neuronal studies, the four variable Hodgkin-Huxley (H-H) model [61] with its nonlinear differential equations is often used to describe the basic dynamical behaviors of the neuron. Furthermore, some simplified versions are proposed because it is generally difficult to deal with a four-variable system in both analytical and numerical studies. Thus, regardless of the reductions and simplifications, which is the inseparable part of any modeling process, it is greatly suitable to use such neuronal models for neuronal studies and investigations. Among these reduced models, the Hindmarsh-Rose (H-R) neuronal model [62] derived from H-H model is one of the successful ones in reproducing the basic states of neuronal dynamics. In this study, the two-dimensional H-R neuronal model is proposed. This model contains a quadratic function for the variations of its second variable (see Eq. 1) having a parabolic nullcline with $g(x \cdot y) = 0$. Noticing Eq. (1), for $z = 0$, there can be found some bifurcation analysis and the bifurcation diagram plus some phase portraits within the nullclines in Ref. [63]. The differential equations of this model are as follows:

$$\begin{aligned} \frac{dx}{dt} &= f(x \cdot y; z) = c\left(x - \frac{x^3}{3} - y + z\right) \\ \frac{dy}{dt} &= g(x \cdot y) = \frac{x^2 + dx - by + a}{c} \end{aligned} \tag{1}$$

where x and y represent the cell membrane potential and a recovery variable, respectively. z is an external stimulus. As mentioned earlier, z is considered zero in Ref. [63]. Here we study the forced two-dimensional H-R model by allocating nonzero values to z. Thus, the potential possibility of different behaviors including chaotic behaviors can be investigated.

For an individual neuron, it sounds more reliable to regard a time varying behavior for z as time varying external effect from the neuron's external environment. Hence, we simplify the case to a periodical behavior like sinusoid function. As a result, we assume $z = A \sin(\omega t)$ for the desired time varying external stimulus. The nominal values of the parameters are set as $a = 0.2$, $b = 1$, $c = 3$, $d = 2.2$, and the parameters of periodical external stimulus are $A = 1.25$, $\omega = 3$.

As explained earlier, it is not just about an individual neuron's activity when we talk about the neuronal system as a complex dynamical system, but also the connections and interactions between the neurons. In fact, it is insufficient to limit our investigation to one individual H-R neuron. Therefore, we develop the case to a large number of H-R neurons in the form of a network. For this purpose, we rewrite the equations of Eq. (1) in a network formation (see Eq. 2). The related noticeable explanations will be discussed in Section 4 in more details. The considered two-dimensional network system is as follows:

$$\frac{dx_{ij}}{dt} = f\left(x_{ij}\cdot y_{ij}; z\right) = c\left(x_{ij} - \frac{x_{ij}^3}{3} - y_{ij} + z\delta_{i\alpha}\delta_{j\beta}\right.$$
$$\left. + D\left(x_{i+1j} + x_{i-1j} + x_{ij+1} + x_{ij-1} - 4x_{ij}\right)\right) \frac{dy_{ij}}{dt} = g\left(x_{ij}\cdot y_{ij}\right) = \frac{x_{ij}^2 + dx_{ij} - by_{ij} + a}{c} \tag{2}$$

$$z = A\sin(\omega t) \tag{3}$$

Where the subscript ij shows position of each neuron in the two-dimensional network and parameter D denotes coupling intensity between the neurons. We choose $D = 1$ as a constant coupling intensity. $\delta_{i\alpha} = 1$ for $i = \alpha$, $\delta_{i\alpha} = 0$ for $i \neq \alpha$; $\delta_{j\beta} = 1$ for $j = \beta$, $\delta_{j\beta} = 0$ for $j \neq \beta$. We impose the external stimulation in the center of the network by $\alpha = \beta = 50$.

3 NUMERICAL RESULTS OF THE BIFURCATION ANALYSIS

In this section, the results of numerical analysis are represented. These numerical results are calculated by using Runge-Kutta 4th-order method. The initial states are selected as $(x_0, y_0) = (5\ 5)$. Actually, the dynamic behavior of the system affected by the periodic external stimulus can be revealed through the bifurcation diagram and Lyapunov exponent spectra. The bifurcation diagrams indicate the qualitative change of solutions as function of parameters and initial conditions. Note that to plot a proper bifurcation diagram some points should be considered, especially checking the existence of multistability and hysteresis [64]. Here, we calculate the Lyapunov exponent and plot the bifurcation diagram for all the parameters of the system, which are introduced in Section 2. In this way, it is possible to track the changes of the original system's dynamics influenced by the mentioned external stimulus, under the changes of each parameter. In each case, the bifurcation diagram is plotted, and the corresponding Lyapunov exponents are calculated for changes of each parameter, while the other parameters are fixed in their nominal values.

First, we examine the changes of parameter A in the range of 0–3. The resulting dynamics are shown in Fig. 1. For more completeness, some sample solutions of time series and state space of the system are plotted for this case. These results for $A = 0.7$, $A = 1.25$, $A = 1.5$ are represented in Figs. 2, 3, and 4, respectively, where other parameters are fixed in their nominal values. As is clear, the resulting plots of time series and state space are in accordance with Fig. 1.

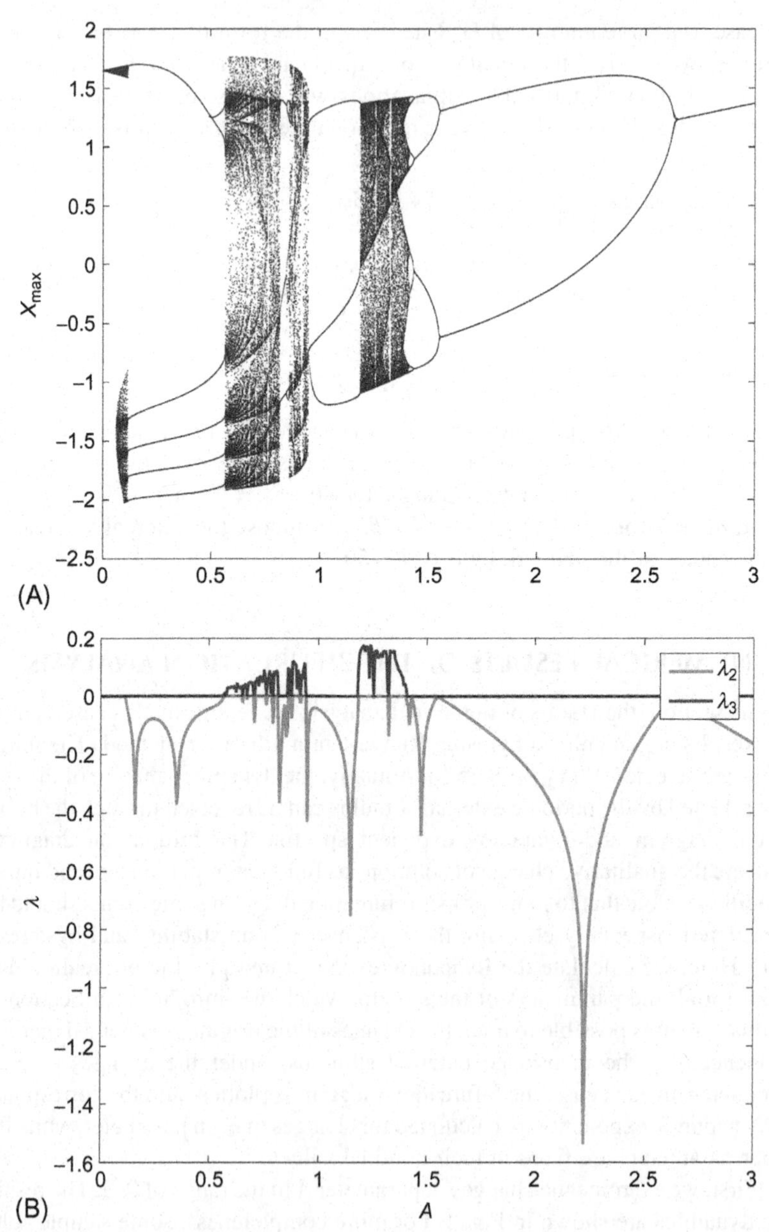

FIG. 1 (A) Bifurcation diagram of an individual forced H-R neuron with variation of parameter A. (B) The second and the third Lyapunov exponents (LEs) in the same range of parameter A (the first LE is out of scale). For both (A) and (B) the initial conditions are $(x \cdot y) = (5\ 5)$ and the initial conditions for every iteration are reinitialized with the ending value of the states from the previous iteration.

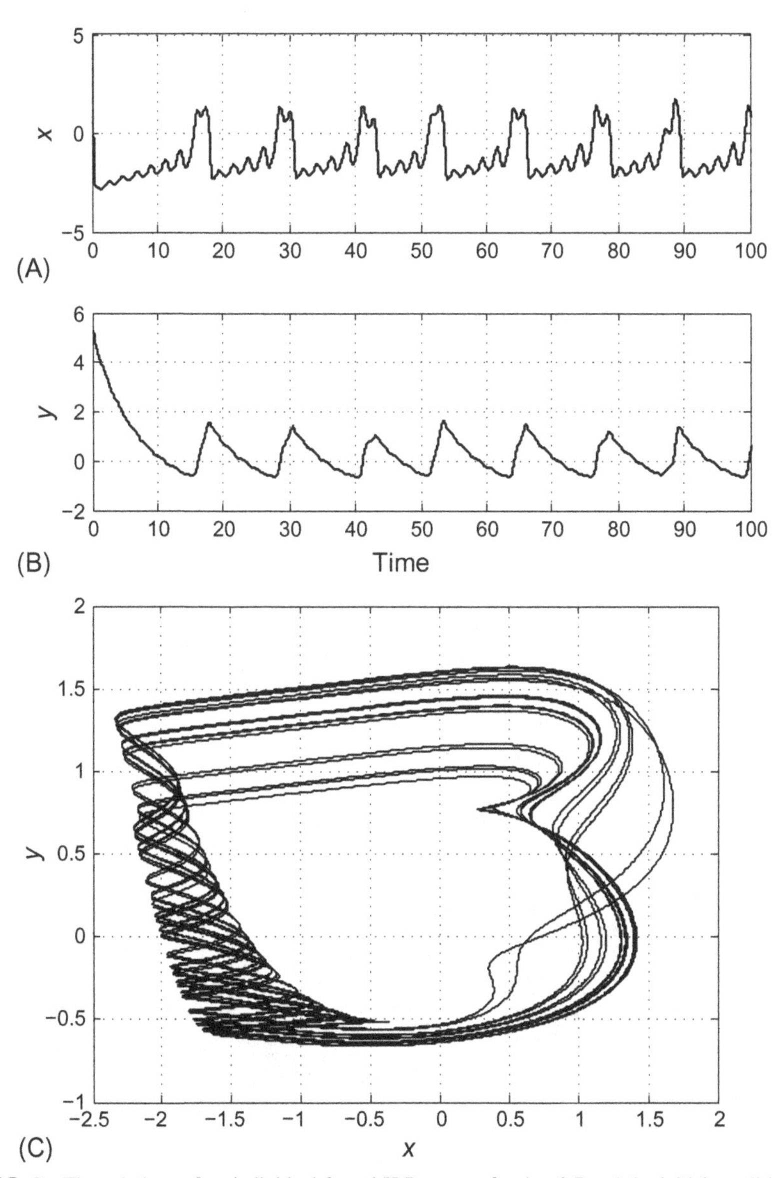

FIG. 2 The solutions of an individual forced H-R neuron for $A = 0.7$ and the initial conditions $(x \cdot y) = (5\ \ 5)$. Other parameters are fixed in their nominal values. (A,B) The time series. (C) The state space.

Bifurcation diagrams and evolution of second and third Lyapunov exponents of an individual forced H-R neuron are also shown for different ranges of variations of parameters ω, a, b, and d in Figs. 5–11, respectively.

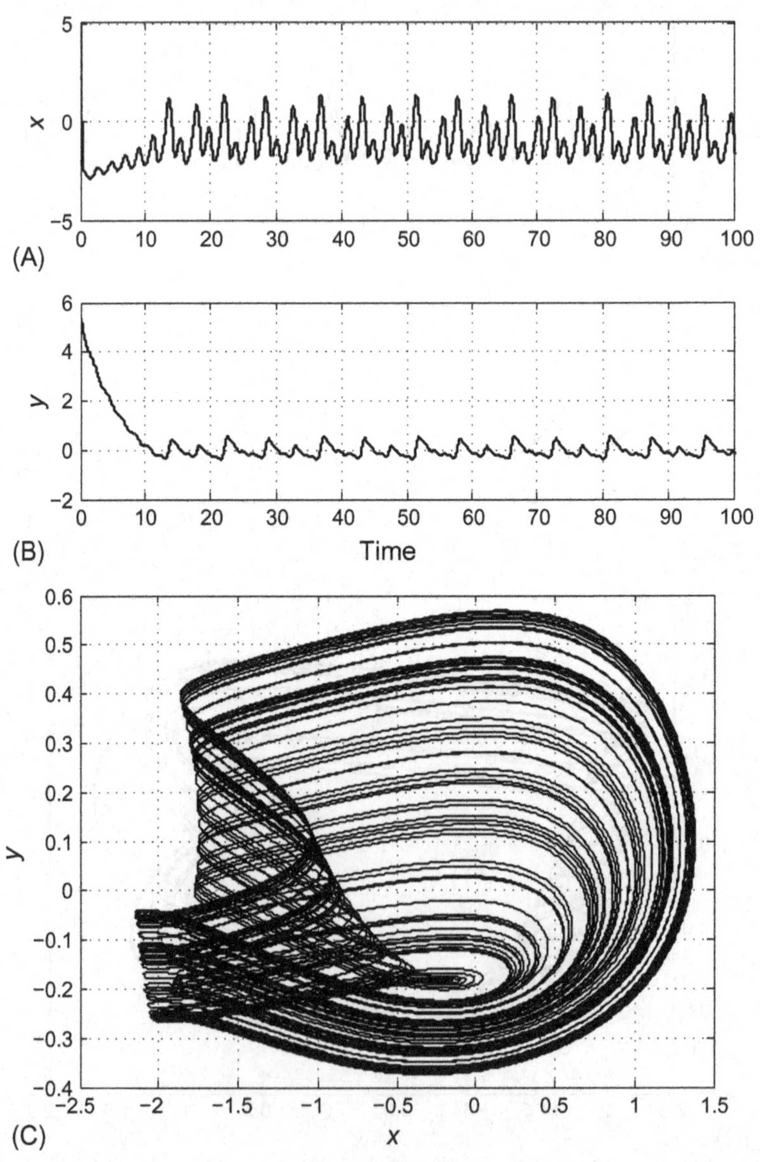

FIG. 3 The solutions of an individual forced H-R neuron for $A = 1.25$ and the initial conditions $(x \cdot y) = (5\ 5)$. Other parameters are fixed in their nominal values. (A,B) The time series. (C) The state space.

4 NUMERICAL RESULTS OF WAVE PROPAGATION IN THE DESIGNED NETWORK

As mentioned earlier, the best tool to study a large number of nonlinear oscillatory systems interrelated to each other is to focus on the spatiotemporal

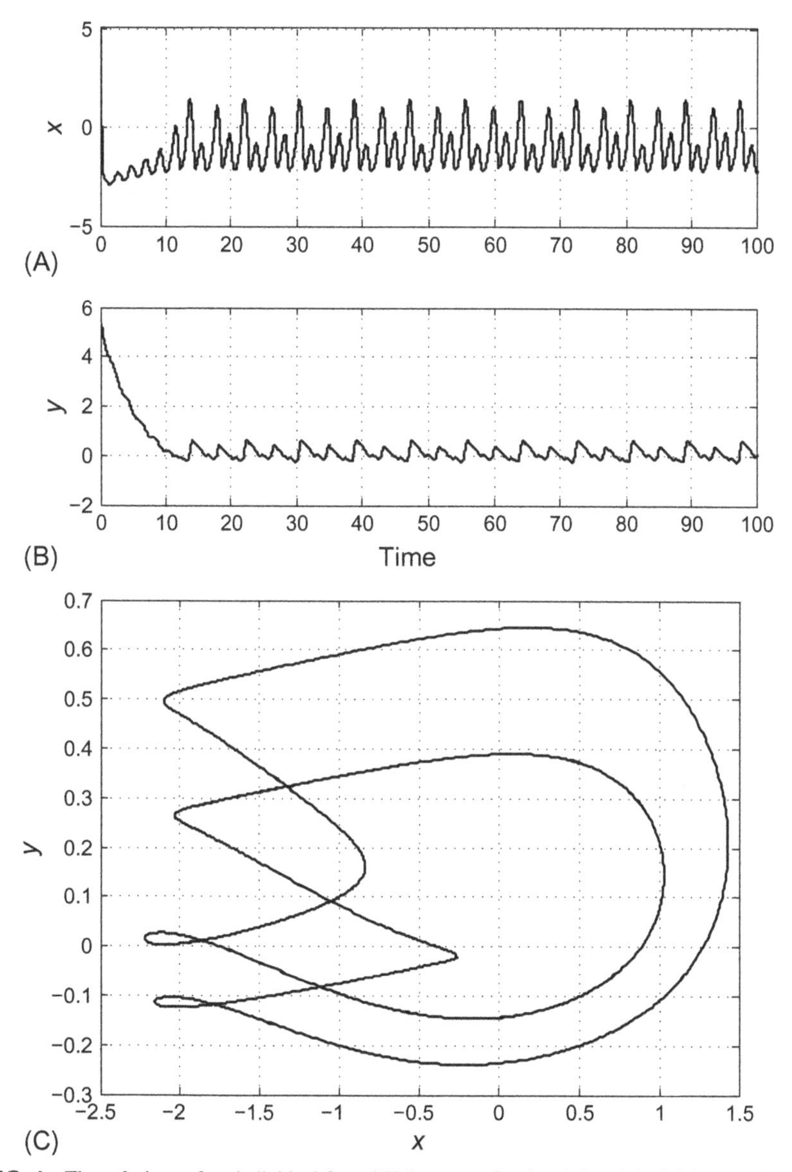

FIG. 4 The solutions of an individual forced H-R neuron for $A = 1.5$ and the initial conditions $(x \cdot y) = (5\ \ 5)$. Other parameters are fixed in their nominal values. (A,B) The time series. (C) The state space.

dynamics arising from the whole network. Following this perspective, in this section, some numerical analysis focusing on the network of a large number of neurons (as a model of an excitable tissue) is represented. Some snapshots of the resulting wave patterns over the time are shown, and the related

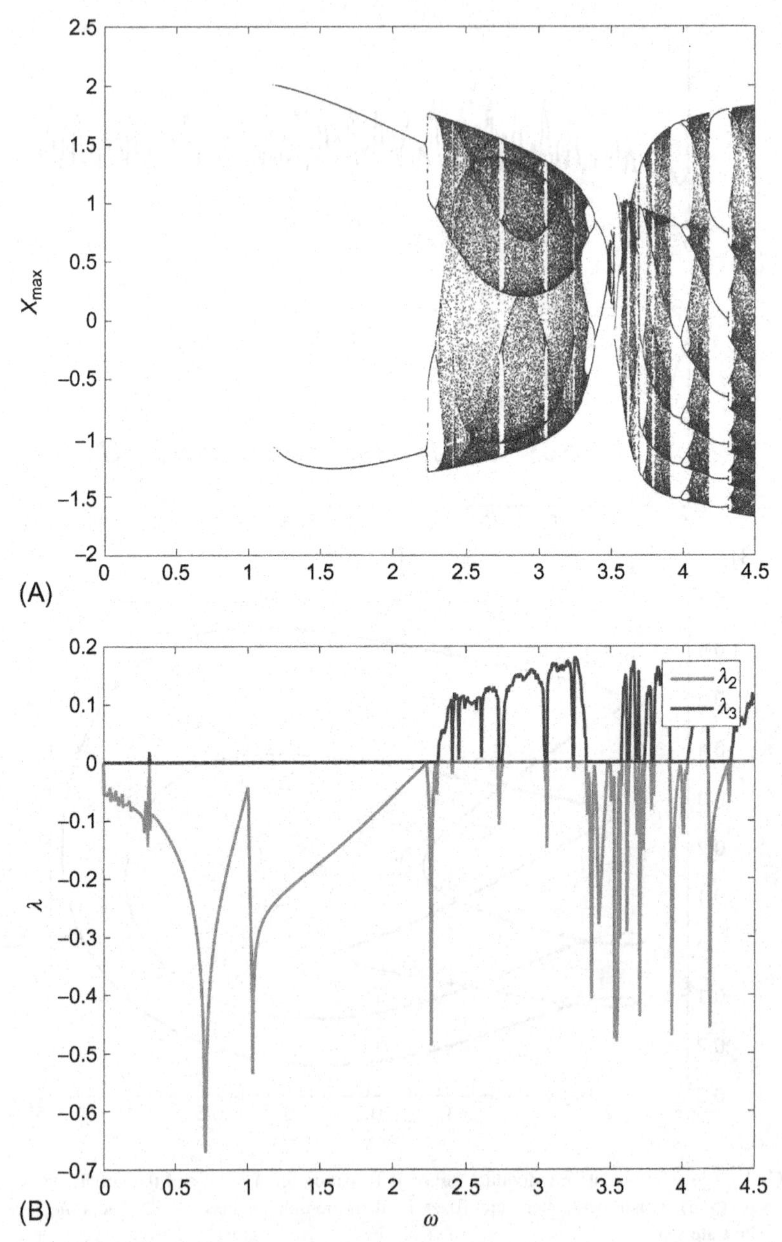

FIG. 5 (A) Bifurcation diagram of an individual forced H-R neuron with variation of parameter ω. (B) The second and the third Lyapunov exponents (LEs) in the same range of parameter ω (the first LE is out of scale). For both (A) and (B) the initial conditions are $(x \cdot y) = (5\ 5)$ and the initial conditions for every iteration are reinitialized with the ending value of the states from the previous iteration.

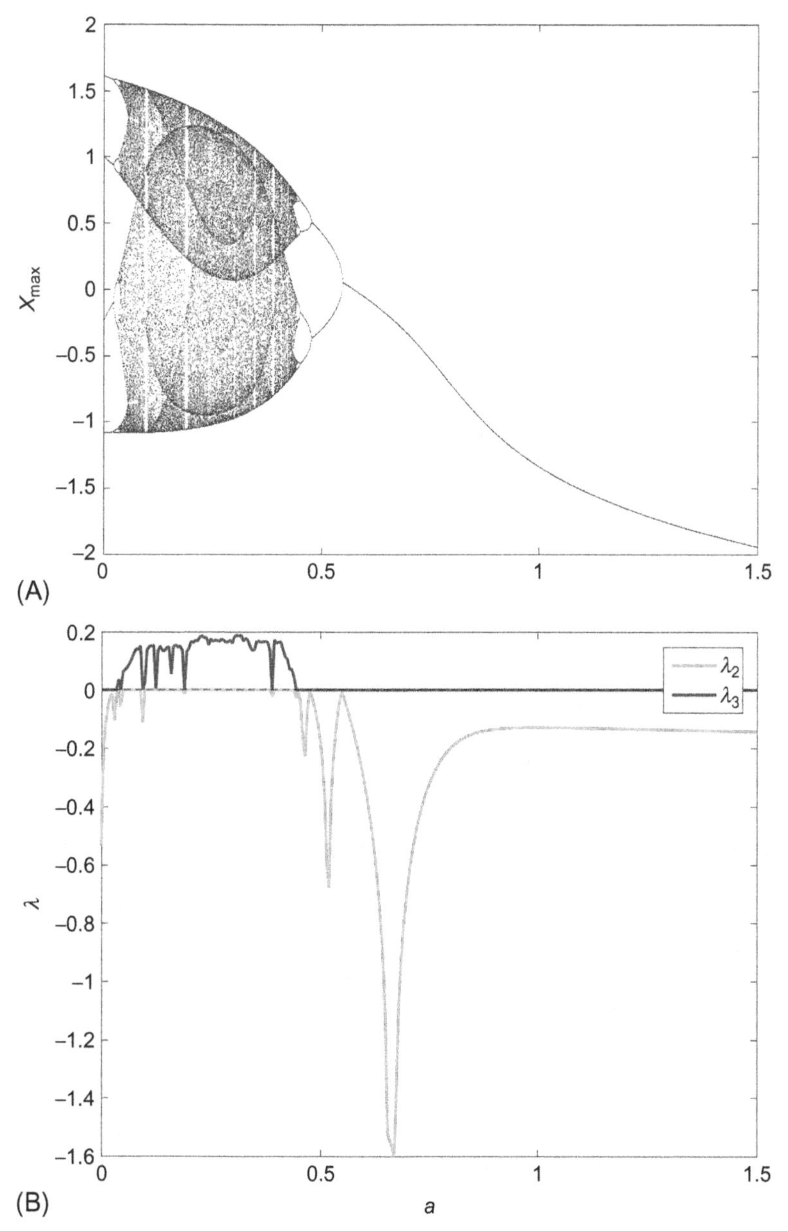

FIG. 6 (A) Bifurcation diagram of an individual forced H-R neuron with variation of parameter a. (B) The second and the third Lyapunov exponents (LEs) in the same range of parameter a (the first LE is out of scale). For both (A) and (B) the initial conditions are $(x \cdot y) = (5\ 5)$ and the initial conditions for every iteration are reinitialized with the ending value of the states from the previous iteration.

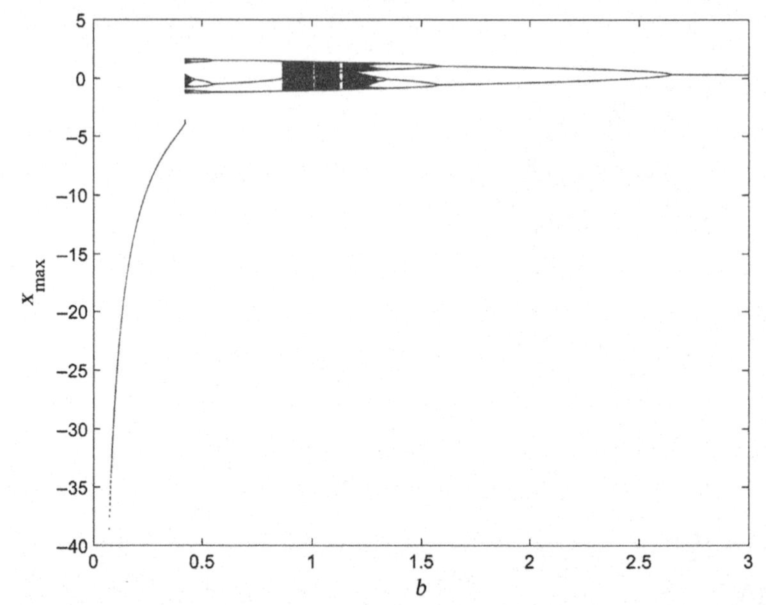

FIG. 7 Bifurcation diagram of an individual forced H-R neuron with variation of parameter b from 0 to 3. The initial conditions are $(x \cdot y) = (5\ 5)$ and the initial conditions for every iteration are reinitialized with the ending value of the states from the previous iteration.

explanations are given, as well. Here, calculation of the numerical study is performed by using Runge-Kutta 4th-order method under Neumann (no flux) boundary condition. We consider a square array network of 100×100 neurons with nearest neighbor connection in which the local dynamics of each individual neuron are governed by the two-dimensional H-R model (recall the described network model illustrated in Eq. 2). The initial condition of the variables is considered $(x \cdot y) = (5\ 5)$.

In fact, the main idea beyond the designing such a network is to discover how an individual neuron, which is investigated in Section 3, behaves in connection with some other neurons. This network can be a model of an excitable tissue capable of transmitting a local stimulation initiated for any reason from one region to another. We show the collective electrical activity of all the neurons. The arrangement of the patterns belong to each neuron's fluctuation during the time brings the tissue a specific spatiotemporal pattern. Therefore, it is interesting to see the effect of a time-varying external stimulus, not only for one H-R neuron, but a large array of neurons interrelated with each other. For this purpose, as indicated in Eq. (3), again we choose $z = A \sin(\omega t)$ as an external stimulus applied to center of the network, which is regarded as a model of an excitable tissue. Actually, this stimulus can be caused by the variation of any factor in the external environment or maybe such an artificial stimulation for a special goal. We examine the considered sinusoidal stimulus with different

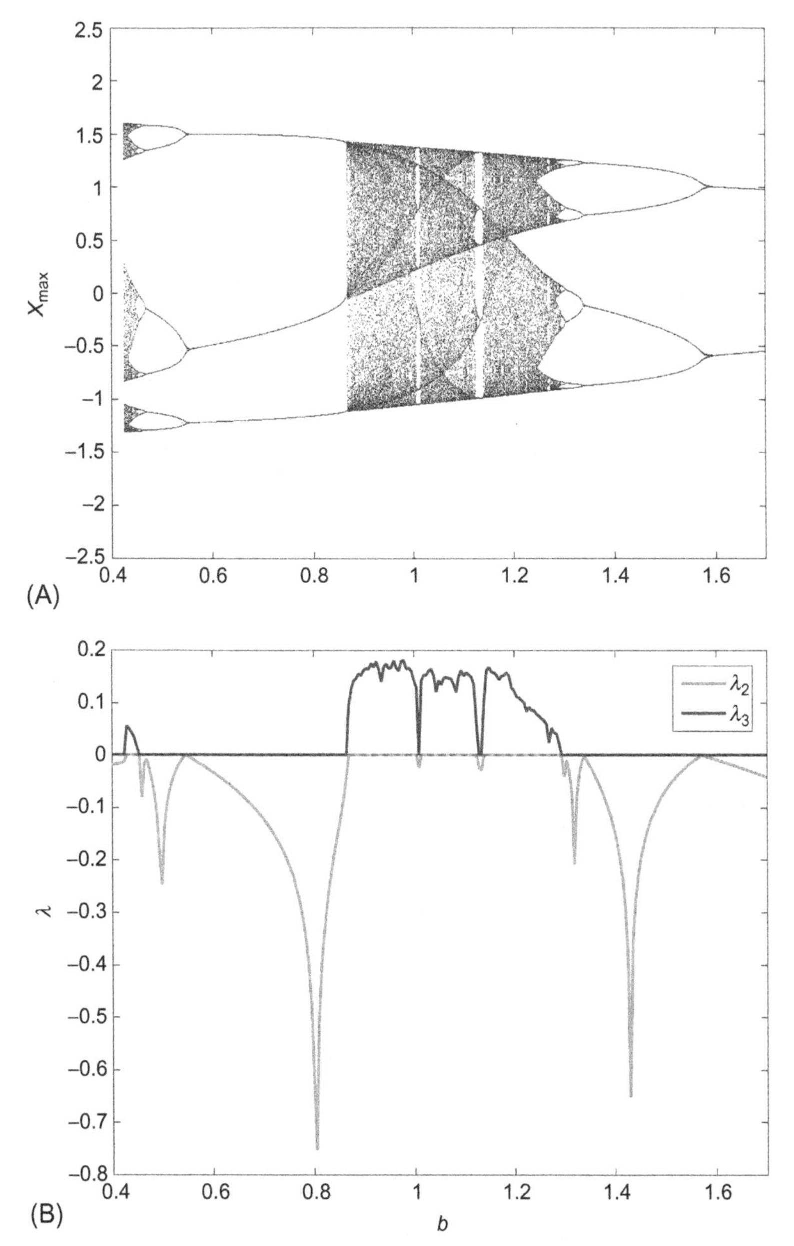

FIG. 8 (A) Bifurcation diagram of an individual forced H-R neuron with variation of parameter b from 0.4 to 1.7. (B) The second and the third Lyapunov exponents (LEs) in the same range of parameter b (the first LE is out of scale). For both (A) and (B) the initial conditions are $(x \cdot y) = (5\ 5)$ and the initial conditions for every iteration are reinitialized with the ending value of the states from the previous iteration.

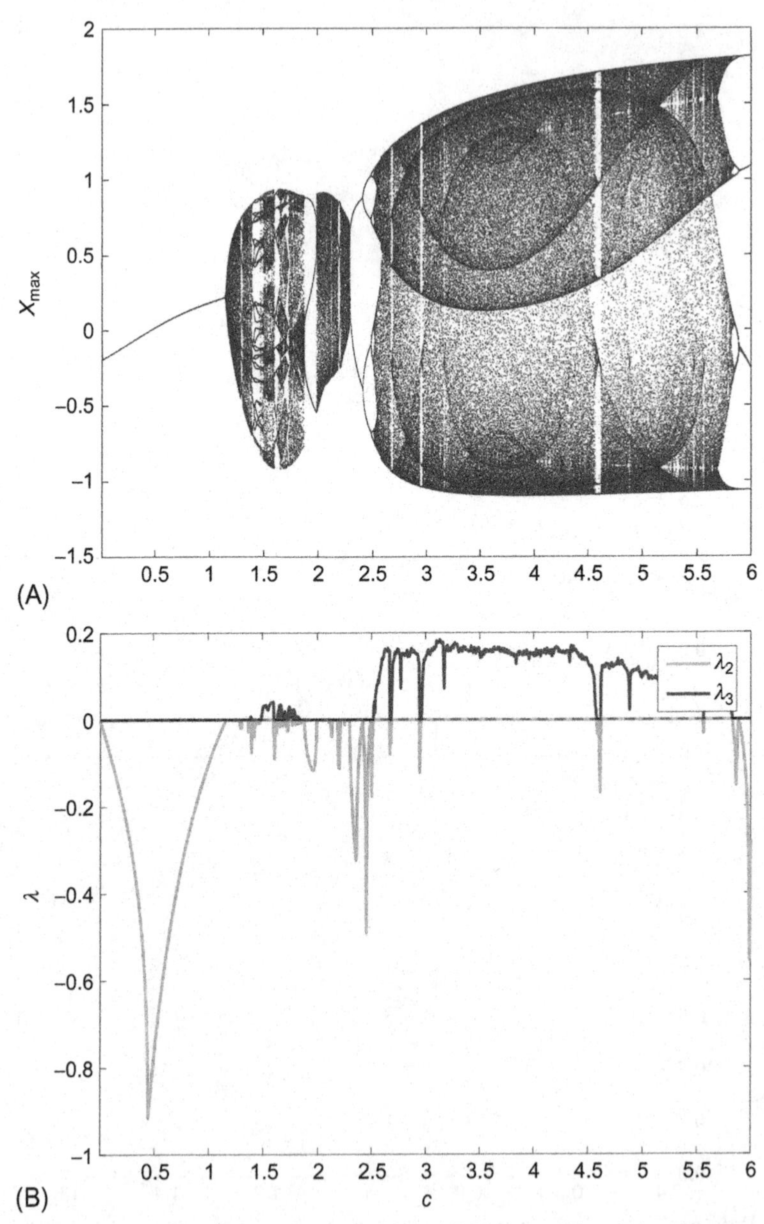

FIG. 9 (A) Bifurcation diagram of an individual forced H-R neuron with variation of parameter c. (B) The second and the third Lyapunov exponents (LEs) in the same range of parameter c (the first LE is out of scale). For both (A) and (B) the initial conditions are $(x \cdot y) = (5\ 5)$ and the initial conditions for every iteration are reinitialized with the ending value of the states from the previous iteration.

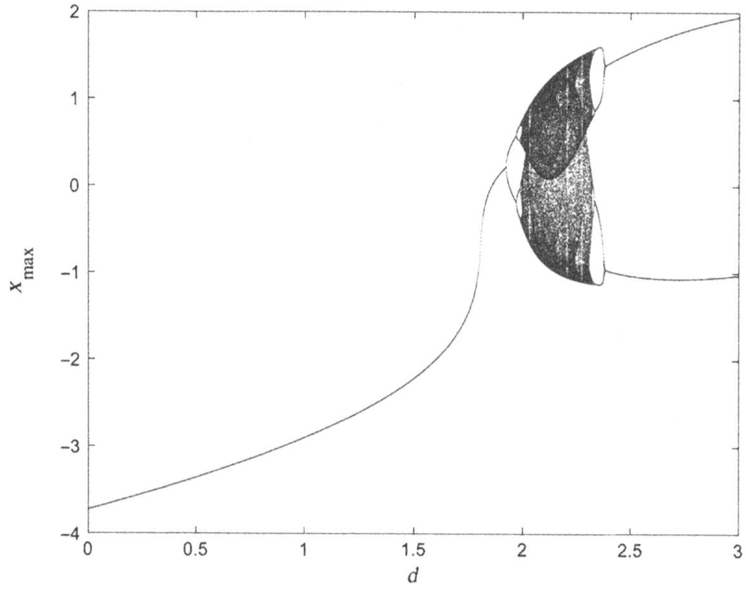

FIG. 10 Bifurcation diagram of an individual forced H-R neuron with variation of parameter d from 0 to 3. The initial conditions are $(x \cdot y) = (5\ 5)$ and the initial conditions, for every iteration are reinitialized with the ending value of the states from the previous iteration.

levels of amplitude (A) and frequency (ω). How do each of these two parameters affect the final pattern of propagated waves in the network? Which one is more determinative? In the following, we set some different amplitudes and frequencies to see how the level of these two parameters specifies the tissue response for holding transmission of a wave front.

First, we choose $\omega = 0.0001$. We examine the case for $A = 0.3$ as a lower level and $A = 3$ as higher-level amplitude for the induced external stimulus. Some snapshots of the stimulation results over the considered time span are displayed in Fig. 12. When we set the lower level for the amplitude, the stimulation point represents a low rate flashing behavior during the whole time span without any propagation, as is observable in Fig. 12A–C. It seems that the excitable tissue does not receive enough power to support the wave transmission. That is why the propagation takes place, and some target waves start to grow after a short period of twinkling demonstrations when we increase the amplitude by $A = 3$. In fact, as soon as the target waves arise, they start to struggle with the primary integrated manner of the tissue. After a short while, they can overcome this homogeneous discipline and regulate the fluctuations based on their own dynamics. As the result, the traveling target waves are visible by this set of parameters, as is observable in Fig. 12D–F.

Second, we set $\omega = 0.001$ by having two levels of amplitude $A = 0.3$ and $A = 3$. The related resulting wave fronts are depicted in Fig. 13. In fact, it seems

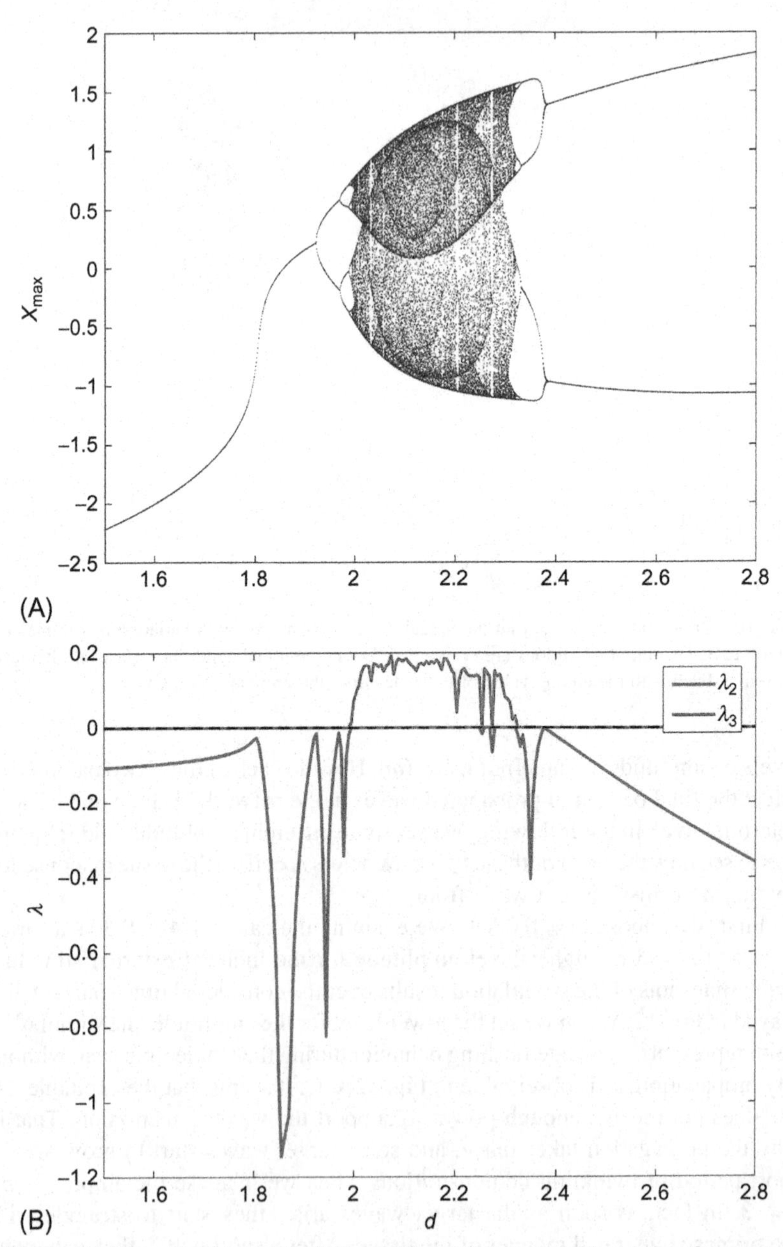

FIG. 11 (A) Bifurcation diagram of an individual forced H-R neuron with variation of parameter d from 1.5 to 2.8. (B) The second and the third Lyapunov exponents (LEs) in the same range of parameter d (the first LE is out of scale). For both (A) and (B) the initial conditions are $(x \cdot y) = (5\ 5)$, and the initial conditions for every iteration are reinitialized with the ending value of the states from the previous iteration.

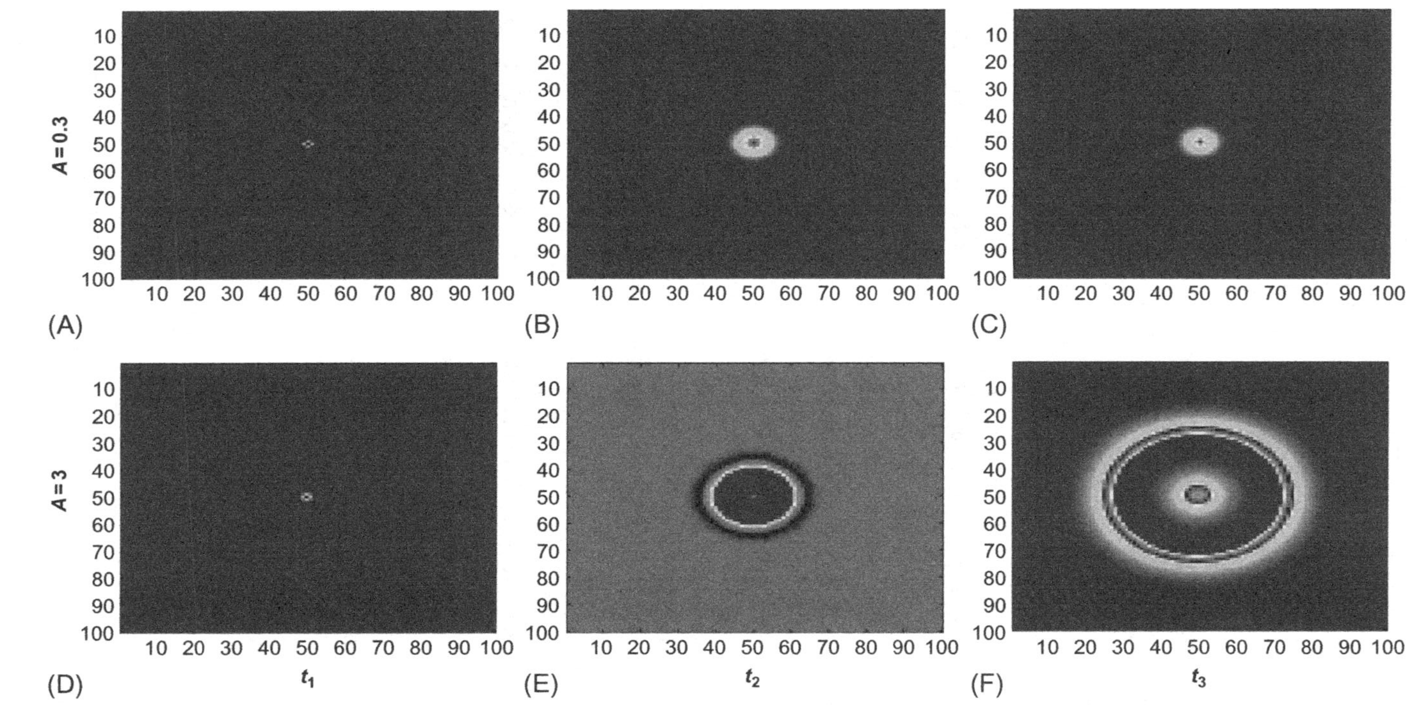

FIG. 12 The snapshots of spatial distribution of membrane potential for neurons in the square array network for $\omega = 0.0001$, $A = 0.3$, and $A = 3$ at (A), (D) $t_1 = 200$ time units (B), (E) $t_2 = 50{,}000$ time units (C), (F) $t_3 = 100{,}000$ time units.

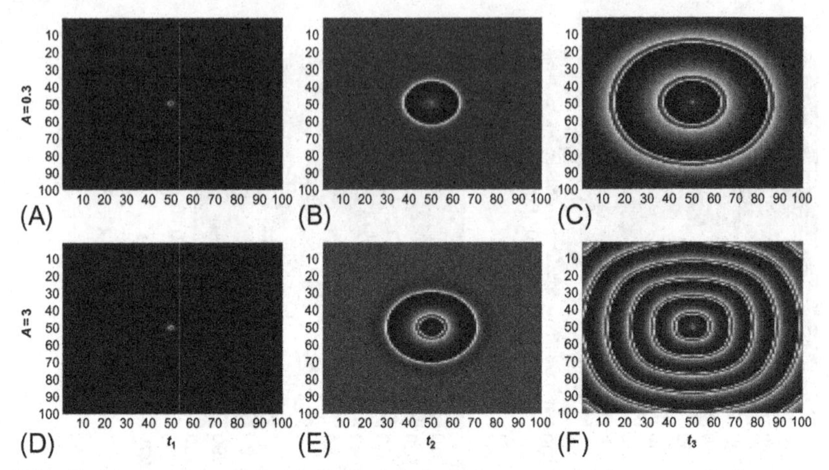

FIG. 13 The snapshots of spatial distribution of membrane potential for neurons in the square array network for $\omega = 0.001$, $A = 0.3$, and $A = 3$ at (A), (D) $t_1 = 200$ time units (B), (E) $t_2 = 50{,}000$ time units (C), (F) $t_3 = 100{,}000$ time units.

that higher amplitude provides the tissue more power to support the propagation process, while the low level of frequency lets the tissue prepare itself to achieve a rhythmical fluctuation in the formation of a wave front. The results show that, each time, the generated signals face a kind of opposition against their propagation in the beginning of stimulation. Therefore, a kind of continuous twinkling with no propagation is observable for a limited time span. However, this manner is just a transition. In fact, the tissue needs a short period to adapt itself and become capable of transmitting the generated signals. After this transition period, the stimulated fluctuations of the whole tissue achieve a rhythmical order and cause the tissue to go far from such a primary twinkling behavior. Moreover, it changes the spatiotemporal pattern in the network when the seed of the target wave in the central part comes to power supported by adaptation of the tissue. As a result, the new spatiotemporal pattern governs the dynamics of the entire tissue (Fig. 13D and F).

For the next step, we pick a higher level of frequency to see the changes in the behavior of the excitable tissue. In this case, we set $\omega = 0.01$. Fig. 14 shows the results of this case for two amplitudes of $A = 0.3$ and $A = 3$. Here, it is interesting that the transition period (mentioned in the previous case), which is needed for preparation of the tissue gets shorter. This means that the external stimulus subjugates the primary integrated discipline of the tissue and starts to generate some target waves, traveling the whole surface in all directions. Furthermore, by providing a higher amount of A ($A = 3$), some spiral seeds find existence but not remain too much and break down after a few seconds. This occurrence causes some additional circular waves, which are observable in Fig. 14F due to breakdown of the spiral seeds near the boundaries.

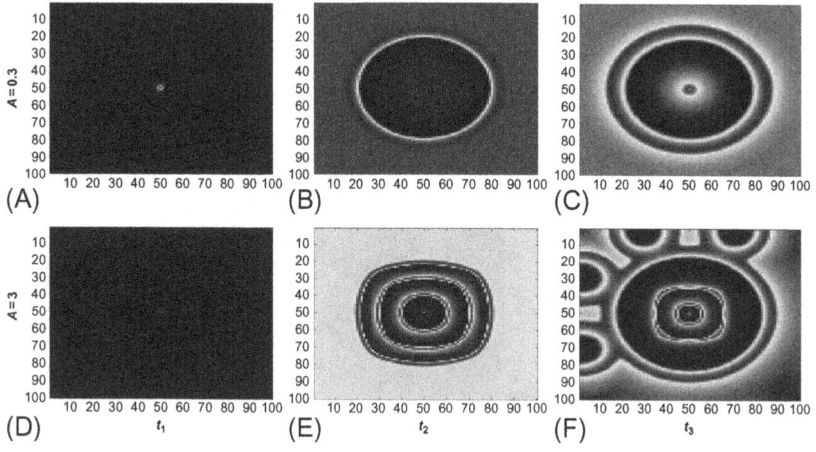

FIG. 14 The snapshots of spatial distribution of membrane potential for neurons in the square array network for $\omega = 0.01$, $A = 0.3$, and $A = 3$ at (A), (D) $t_1 = 200$ time units (B), (E) $t_2 = 50{,}000$ time units (C), (F) $t_3 = 100{,}000$ time units.

For $\omega = 0.1$, it also takes the tissue some time to adapt its intrinsic dynamic with the rhythm of the induced vibration. After this time span, the rhythm of the fluctuations reaches a recognizable order in the form of a wave front. It seems that, by passing this period, the tissue finds the opportunity to become released from the primary disturbed behavior caused by initiation of the stimulation. Therefore, it becomes a supportive tool for transmission of the information in the form of rhythmic signal propagation. In this case, by this level of frequency, the generated target waves start to move and expand to cover the entire tissue, however, they subside before reaching the boundaries. In other words, this higher level of frequency impedes the slippery movement of the wave fronts and causes them not to reach the boundaries for a relatively long period. This time span becomes shorter by assigning greater amplitude. This delicate difference is recognizable by comparing the results in Fig. 15C and F. Clearly, the traveling circular waves can be found, as displayed in Fig. 15.

Moreover, by further increase in the frequency of the external stimulus by $\omega = 3$, it seems that the frequency of the stimulation is higher than what it needs to be, so that no propagation take place in the tissue (Fig. 16). Actually, the influence of the stimulation shows up only in such a twinkling behavior, and the generated signals do not move from the central area of the tissue. It can be said that the tissue does not find enough time for its adaptation performance, which is necessary for propagation progress, when we apply a higher frequency beyond an approximate threshold. Even though greater amplitude provides the tissue more potential power, the transmission of the signals cannot take place in this case because it is not just about the potential power of the induced stimulation, but also the required accompaniment of the tissue, which emphasizes the importance of adaptation period.

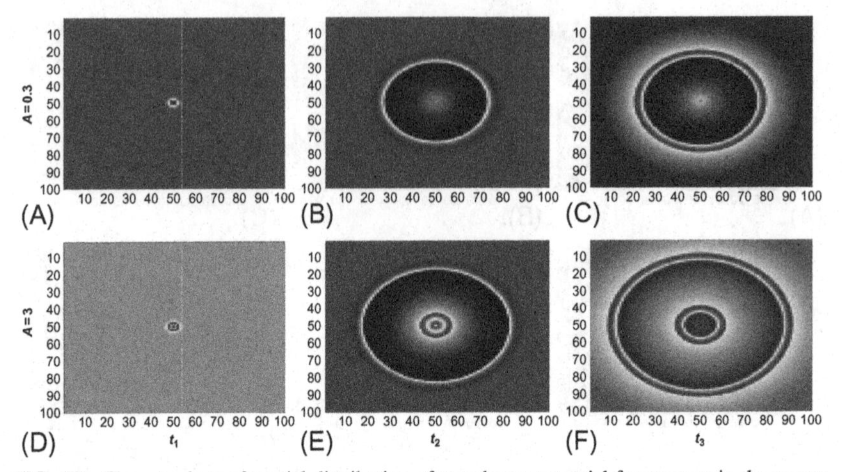

FIG. 15 The snapshots of spatial distribution of membrane potential for neurons in the square array network for $\omega = 0.1$, $A = 0.3$, and $A = 3$ at (A), (D) $t_1 = 200$ time units (B), (E) $t_2 = 50{,}000$ time units (C), (F) $t_3 = 100{,}000$ time units.

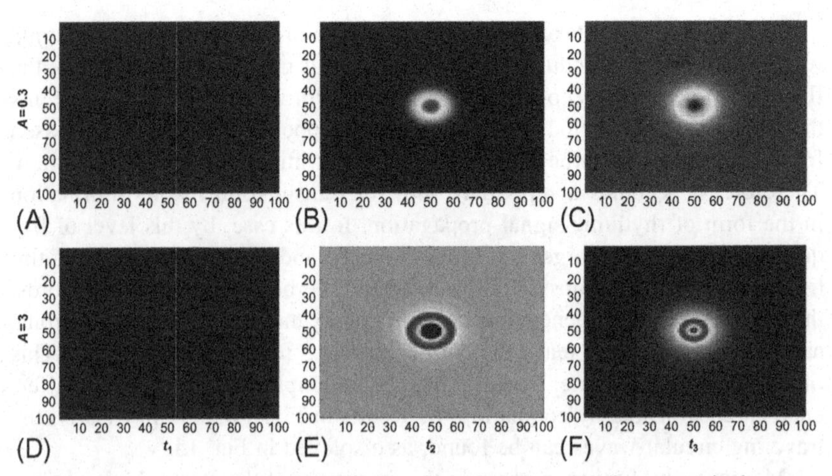

FIG. 16 The snapshots of spatial distribution of membrane potential for neurons in the square array network for $\omega = 3$, $A = 0.3$, and $A = 3$ at (A), (D) $t_1 = 200$ time units (B), (E) $t_2 = 50{,}000$ time units (C), (F) $t_3 = 100{,}000$ time units.

5 CONCLUSION

In this study, an individual forced Hindmarsh-Rose neuron is investigated. The dynamics of the system are studied via bifurcation demonstrations. The bifurcation diagrams of the forced system are displayed for variation of each of the system's parameters. Additionally, calculation of the Lyapunov exponent corresponding to each of the bifurcation diagrams is plotted, as well. The bifurcation diagrams show the route of period doubling, as well as chaotic demonstrations in some ranges of the parameters.

For further investigation, a square array network of the two-dimensional H-R neurons representing a model of an excitable tissue is designed. Consequently, the emergence of possible spatiotemporal patterns under the considered circumstances is scrutinized. In this way, the effect of amplitude and frequency of the forced current regarded as an external stimulus is examined. As the result, the frequency of the external stimulus is the main determinant of adaptation capability of the tissue. However, the sufficient potential power is also needed for the propagation process. To some extent, this required power is provided by the amplitude of the stimulation.

REFERENCES

[1] Thompson JMT. Chaos, fractals and their applications. Int J Bifurcation Chaos 2016;26 (13):1630035.

[2] El Naschie M, Kapitaniak T. Soliton chaos models for mechanical and biological elastic chains. Phys Lett A 1990;147(5–6):275–81.

[3] Doroshin AV. Chaos as the hub of systems dynamics. The part I—the attitude control of spacecraft by involving in the heteroclinic chaos. Commun Nonlinear Sci Numer Simul 2018;59:47–66.

[4] Lorenz EN. Deterministic nonperiodic flow. J Atmos Sci 1963;20(2):130–41.

[5] Weiss JN, et al. Chaos and the transition to ventricular fibrillation. Circulation 1999;99 (21):2819–26.

[6] Sprott JC. Elegant chaos: algebraically simple chaotic flows. London, UK: World Scientific; 2010. https://doi.org/10.1142/7183.

[7] Chialvo DR, Michaels DC, Jalife J. Supernormal excitability as a mechanism of chaotic dynamics of activation in cardiac Purkinje fibers. Circ Res 1990;66(2):525–45.

[8] Nazarimehr F, et al. Fuzzy predictive controller for chaotic flows based on continuous signals. Chaos, Solitons Fractals 2018;106:349–54.

[9] Pham V-T, Jafari S, Volos C. A novel chaotic system with heart-shaped equilibrium and its circuital implementation. Optik 2017;131:343–9.

[10] Jafari S, Smith LS. Can Lionel Messi's brain slow down time passing? Chronobiol Int 2016;33 (4):462–3.

[11] Jafari S, Sprott JC, Golpayegani SMRH. Layla and Majnun: a complex love story. Nonlinear Dynam 2016;83(1–2):615–22.

[12] Bao H, et al. Initial condition-dependent dynamics and transient period in memristor-based hypogenetic jerk system with four line equilibria. Commun Nonlinear Sci Numer Simul 2018;57:264–75.

[13] Nazarimehr F, et al. Can Lyapunov exponent predict critical transitions in biological systems? Nonlinear Dynam 2017;88(2):1493–500.

[14] Shekofteh Y, et al. A gaussian mixture model based cost function for parameter estimation of chaotic biological systems. Commun Nonlinear Sci Numer Simul 2015;20(2):469–81.

[15] Jafari S, Golpayegani SH, Jafari A. A novel noise reduction method based on geometrical properties of continuous chaotic signals. Sci Iran 2012;19(6):1837–42.

[16] Kuznetsov YA. Elements of applied bifurcation theory. vol. 112. Berlin, Heidelberg: Springer Science & Business Media; 2013.

[17] Jafari S, et al. A novel viewpoint on parameter estimation in a chaotic neuron model. J Neuropsychiatry Clin Neurosci 2013;25(1):E19.
[18] Rostami Z, Jafari S. Defects formation and spiral waves in a network of neurons in presence of electromagnetic induction. Cogn Neurodyn 2018;12(2):235–54.
[19] Aram Z, et al. Using chaotic artificial neural networks to model memory in the brain. Commun Nonlinear Sci Numer Simul 2017;44:449–59.
[20] Baghdadi G, et al. A chaotic model of sustaining attention problem in attention deficit disorder. Commun Nonlinear Sci Numer Simul 2015;20(1):174–85.
[21] Falahian R, et al. Artificial neural network-based modeling of brain response to flicker light. Nonlinear Dynam 2015;81(4):1951–67.
[22] Hadaeghi F, et al. Toward a complex system understanding of bipolar disorder: a chaotic model of abnormal circadian activity rhythms in euthymic bipolar disorder. Aust N Z J Psychiatry 2016;50(8):783–92.
[23] Jafari S, et al. Is attention a "period window" in the chaotic brain? J Neuropsychiatry Clin Neurosci 2013;25(1):E05.
[24] Jafari S, et al. Is attention deficit hyperactivity disorder a kind of intermittent chaos? J Neuropsychiatry Clin Neurosci 2013;25(2):E02.
[25] Molaie M, et al. Artificial neural networks: powerful tools for modeling chaotic behavior in the nervous system. Front Comput Neurosci 2014;8.
[26] Jafari S, Golpayegani SMH, Gharibzadeh S. Is there any geometrical information in the nervous system? Front Comput Neurosci 2013;7:1–2. https://doi.org/10.3389/fncom.2013.00121.
[27] Kandel ER, et al. Principles of neural science. 4. New York: McGraw-hill; 2000.
[28] Zhang J, et al. The dynamics of spiral tip adjacent to inhomogeneity in cardiac tissue. Physica A 2018;491:340–6.
[29] Cherry EM, Fenton FH. Visualization of spiral and scroll waves in simulated and experimental cardiac tissue. New J Phys 2008;10(12).
[30] Tabatabaei SS, et al. Extensions in dynamic models of happiness: effect of memory. Int J Happiness Dev 2014;1(4):344–56.
[31] Molaei SR, et al. Optimization of a nonlinear electrical–thermal model of the skin. Biomed Eng Appl Basis Commun 2013;25(03):1350039.
[32] Hosseini SS, Nazarimehr F, Jafari S. Investigation of seasonal and latitudinal effects on the expression of clock genes in drosophila. Int J Bifurcation Chaos 2017;27(10):1750153.
[33] Panahi S, et al. Modeling of epilepsy based on chaotic artificial neural network. Chaos, Solitons Fractals 2017;105:150–6.
[34] Molaie M, et al. A chaotic viewpoint on noise reduction from respiratory sounds. Biomed Signal Process Control 2014;10:245–9.
[35] Pham V-T, et al. Hidden hyperchaotic attractor in a novel simple memristive neural network. Optoelectron Adv Mater Rapid Commun 2014;8(11 – 12):1157–63.
[36] Pham VT, et al. A novel memristive neural network with hidden attractors and its circuitry implementation. SCIENCE CHINA Technol Sci 2016;59(3):358–63.
[37] Shabestari, P.S., et al., A novel approach to numerical modeling of metabolic system: Investigation of chaotic behavior in diabetes mellitus. Complexity; in press. https://www.hindawi.com/journals/complexity/aip/6815190/.
[38] Abdolmohammadi SJH, Rajati M, Gharibzadeh SSS. A modified artificial immune system for data classification, In: 16th Iranian conference on electrical engineering; 2008. p. 173–8.
[39] Arena P, et al. Locally active hindmarsh–rose neurons. Chaos, Solitons Fractals 2006;27(2):405–12.

[40] Gu H, Pan B, Xu J. Experimental observation of spike, burst and chaos synchronization of calcium concentration oscillations. Europhys Lett 2014;106(5).

[41] Zhang C, et al. Robust outer synchronization between two nonlinear complex networks with parametric disturbances and mixed time-varying delays. Physica A 2018;494:251–64.

[42] Fenton FH, et al. Modeling wave propagation in realistic heart geometries using the phase-field method. Chaos 2005;15(1):013502.

[43] Clayton R, et al. Models of cardiac tissue electrophysiology: progress, challenges and open questions. Prog Biophys Mol Biol 2011;104(1):22–48.

[44] Wang C, et al. Synchronization stability and pattern selection in a memristive neuronal network. Chaos 2017;27(11):113108.

[45] Fenton FH, et al. Multiple mechanisms of spiral wave breakup in a model of cardiac electrical activity. Chaos 2002;12(3):852–92.

[46] Ma J, et al. Defects formation and wave emitting from defects in excitable media. Commun Nonlinear Sci Numer Simul 2016;34:55–65.

[47] Chen J-X, Guo M-M, Ma J. Termination of pinned spirals by local stimuli. Europhy Lett 2016;113(3):38004.

[48] Ma J, et al. Electromagnetic induction and radiation-induced abnormality of wave propagation in excitable media. Physica A 2017;.

[49] Hildebrand M, Bär M, Eiswirth M. Statistics of topological defects and spatiotemporal chaos in a reaction-diffusion system. Phys Rev Lett 1995;75(8):1503.

[50] Bertram M, et al. Pattern formation on the edge of chaos: experiments with CO oxidation on a Pt (110) surface under global delayed feedback. Phys Rev E 2003;67(3):036208.

[51] Beta C, et al. Excitable CO oxidation on Pt (110) under nonuniform coupling. Phys Rev Lett 2004;93(18):188302.

[52] Beta C, et al. Controlling turbulence in a surface chemical reaction by time-delay autosynchronization. Phys Rev E 2003;67(4):046224.

[53] Huang X, et al. Spiral wave dynamics in neocortex. Neuron 2010;68(5):978–90.

[54] Folias SE, Bressloff PC. Breathing pulses in an excitatory neural network. SIAM J Appl Dynam Sys 2004;3(3):378–407.

[55] Weiss JN, et al. Ventricular fibrillation. Circ Res 2000;87(12):1103–7.

[56] Ma J, et al. Pattern selection in neuronal network driven by electric autapses with diversity in time delays. Int J Modern Phys B 2015;29(01):1450239.

[57] Guo S, et al. Collective response, synapse coupling and field coupling in neuronal network. Chaos, Solitons Fractals 2017;105:120–7.

[58] Ma J, et al. Simulating the formation of spiral wave in the neuronal system. Nonlinear Dynam 2013;73(1–2):73–83.

[59] Chua LO. CNN: a paradigm for complexity. vol. 31. London, UK: World Scientific; 1998. https://doi.org/10.1142/3801.

[60] Ma J, et al. Prediction for breakup of spiral wave in a regular neuronal network. Nonlinear Dynam 2016;84(2):497–509.

[61] Hodgkin AL, Huxley AF. A quantitative description of membrane current and its application to conduction and excitation in nerve. J Phys 1952;117(4):500–44.

[62] Hindmarsh J, Rose R. A model of the nerve impulse using two first-order differential equations. Nature 1982;296(5853):162–4.

[63] Tsuji S, et al. Bifurcations in two-dimensional Hindmarsh–Rose type model. Int J Bifurcation Chaos 2007;17(03):985–98.

[64] Kengne J, Njikam S, Signing VF. A plethora of coexisting strange attractors in a simple jerk system with hyperbolic tangent nonlinearity. Chaos, Solitons Fractals 2018;106:201–13.

Chapter 11

PD Bifurcation and Chaos Behavior in a Predator-Prey Model With Allee Effect and Seasonal Perturbation

Afef Ben Saad*, Olfa Boubaker* and Zeraoulia Elhadj†
**National Institute of Applied Sciences and Technology, Tunis, Tunisia, †Department of Mathematics and Computing, University of Tebessa, Tebessa, Algeria*

1 INTRODUCTION

Recently, dynamical relationship between species has been intensively studied and will continue to be one of the most important themes of ecology [1–5]. Such systems are generally depicted by nonlinear polynomial models which are based on nonlinear Lotka-Volterra model [6] and have the following basic form [7].

$$\begin{cases} \dfrac{dx_1}{dt} = ax_1 - f(x_1)x_2 \\ \dfrac{dx_2}{dt} = cx_1x_2 - cmx_2 \end{cases} \tag{1}$$

where x_1 and x_2 are the size of the prey population and the predator population, respectively. a is the prey's growth rate in absence of the predator. $f(x_1)$ is the functional response of the predator to prey density. c represents the predator's conversion efficiency, and m is the mortality rate of the predator depending on the predator's efficiency.

On the other hand, during the last few years, bifurcation analysis of predator-prey models which incorporate one or more extra factors has been considered as a challenging research topic. The most famous extra factors are time delay effect [8, 9], impulsive effect [10], seasonal effect [11], and Allee effects [12, 13].

Recent Advances in Chaotic Systems and Synchronization. https://doi.org/10.1016/B978-0-12-815838-8.00011-X

In [14], a bifurcation analysis of the predator-prey Bazykin and Berezovskaya (BB) model affected by Allee effect considered in [15], and introduced noninteger polynomials, is carried out. To extend the set of information related to this model, the authors propose modeling the prey's growth rate and the predator's growth rate using polynomials with noninteger elements such as:

$$a = x_1^{q_1 - 1}(x_1 - l)(1 - x_1), \; f(x_1) = x_1^{q_2} \tag{2}$$

where l is the Allee effect threshold , q_1 and q_2 are two noninteger elements. Using Eq. (2), system (1) is then given by

$$\begin{cases} \dfrac{dx_1}{dt} = x_1^{q_1}(x_1 - l)(1 - x_1) - x_1^{q_2} x_2 \\ \dfrac{dx_2}{dt} = c(x_1 - m)x_2 \end{cases} \tag{3}$$

For the system described by Eq. (3), several bifurcation points are detected according to the Allee effect's type and the noninteger elements but with standard bifurcation types such as Hopf bifurcation (H) and branch point bifurcation (BP).

In this chapter, a modified structure of prey's growth rate and predator's efficiency based on noninteger elements is proposed. Detailed analyses of the nonperturbed and the perturbed system are investigated by considering the two cases study of Strong and Weak Allee effects.

This chapter is arranged as follows: Section 2 presents the predator-prey BB model with noninteger elements in the prey's growth rate and predator's efficiency structure. Preliminaries are given in Section 3. Stability and Hopf bifurcation analysis are investigated in Section 4. In Section 5, numerical simulations are depicted. Finally, the research work undertaken is discussed in Section 6.

2 MATHEMATICAL MODELING

For the system modeled by Eq. (1):

$$a = (x_1^{q_1} - l)(1 - x_1), \; f(x_1) = x_1, \; c = x_2^{q_2 - 1}$$

We have proposed a modified algebraic structure-based noninteger polynomials described by

$$\begin{cases} \dfrac{dx_1}{dt} = x_1(x_1^{q_1} - l)(1 - x_1) - x_1 x_2 \\ \dfrac{dx_2}{dt} = (x_1 - m)x_2^{q_2} \end{cases} \tag{4}$$

where q_1 and q_2 are two noninteger elements.

For dynamical analysis of system (4), two case studies for the Allee effect can be considered

- The strong Allee effect when $l \in [0\ 1]$,
- The weak Allee effect when $l \in [-1\ 0]$.

3 PRELIMINARIES

Definition 1 Consider a system with a single noninteger element whose polynomial equation is given by [16]

$$x^q = b = |b|e^{j\theta_b} \tag{5}$$

where x is a variable and q being a rational or irrational number. b is a complex parameter and θ_b is the phase of x^q. The roots of Eq. (5) are given by

$$x = |b|^{\frac{1}{q}} e^{j\left(\frac{\theta_b \pm 2n\pi}{q}\right)}, \quad n \in \mathbb{N} \tag{6}$$

When b is a pure real, $\theta_b = 0$ if $b > 0$ and $\theta_b = \pi$ if $b < 0$.

Definition 2 Let us consider the two-dimensional system defined by

$$\begin{cases} \dfrac{dx}{dt} = f(x,y,\alpha) \\ \dfrac{dy}{dt} = g(x,y,\alpha) \end{cases} \tag{7}$$

with α is a system parameter. Suppose that $(x(\alpha), y(\alpha))$ are the equilibrium point and $\lambda_{1,\,2}$ are the eigenvalues evaluated at this equilibrium and described by

$$\lambda_{1,2} = \mu(\alpha) \pm i\omega(\alpha)$$

with $\mu(\alpha)$ and $\omega(\alpha)$ are the real and imaginary parts of the eigenvalues $\lambda_{1,2}$, respectively.

A Hopf or Poincare-Andronov-Hopf bifurcation occurs when the stability properties of the equilibrium involves a change to a periodic solution. This latter is obtained when the Jacobian matrix of system described by Eq. (7) has a pair of purely imaginary eigenvalues at a critical value α_0 of α. In other words, when

$$\lambda_{1,2} = \pm i\omega(\alpha_0) \quad \text{and} \quad \mu(\alpha_0) = 0$$

4 DYNAMICAL ANALYSIS OF SYSTEM FOR BOTH TYPES OF ALLEE EFFECT

Proposition 1 *System (4) with strong Allee effect admits a varied equilibrium number according to the value of* m*:*

- *Three equilibrium points* $E_1(0,0)$, $E_2(l^{\frac{1}{q_1}},0)$, $E_3(1,0)$, *when* $m \in [0, l^{\frac{1}{q_1}}]$.
- *Four equilibrium points* E_1, E_2, E_3 *and* $E_4(m,\ (m^{q_1} - l)(1-m))$, *when* $m \in [l^{\frac{1}{q_1}}, 1]$.

System (4) with weak Allee effect admits a fixed equilibrium number. It has three equilibrium points $E_1(0, 0)$, $E_3(1, 0)$, *and* $E_4(m, (m^{q_1} - l)(1 - m))$ $\forall m \in [0\ 1]$.

Proof Let resolving the following system:

$$\begin{cases} \frac{dx_1}{dt} = 0 \\ \frac{dx_2}{dt} = 0 \end{cases} \tag{8}$$

For both cases of strong and weak Allee effect, system (4) admits an extinction trivial equilibrium $E_1(0,0)$ and a nonisolated trivial equilibrium $E_3(1,0)$ $\forall m \in [0\ 1]$. In addition, system (4) admits an equilibrium E_2, $\forall m \in [0\ 1]$, such as

$$x_1^{q_1} = l \text{ and } x_2 = 0$$

- For the case of strong Allee effect, l is a pure positive real. Thus, according to Definition 1, $\theta_1 = 0$ and the coordinates of the equilibrium E_2 are given by:

$$x_1 = l^{1/q_1} \text{ and } x_2 = 0$$

- For the case of weak Allee effect, l is a pure negative real. Thus, according to Definition 1, $\theta_1 = \pi$ and the coordinates of the equilibrium E_2 are given by:

$$x_1 = -|l|^{1/q_1} \text{ and } x_2 = 0$$

The last solution cannot be acceptable considering the biological meaning conditions of $x_1$and x_2 which should be positive. Therefore, system (4) admits the equilibrium E_2 only in the case of strong Allee effect. Furthermore, system (4) admits a fourth equilibrium E_4 with the following coordinates:

$$x_1 = m \text{ and } x_2 = (m^{q_1} - l)(1 - m)$$

As the biological meaning requires that x_2 should be positive, the equilibrium E_4 exists if and only if $l^{q_1} \leq m \leq 1$ for the case of strong Allee effect. Whereas, for the case of weak Allee effect, x_2 is positive $\forall m$. Thus, system (4) with weak Allee effect admits an equilibrium $E_4(m, (m^{q_1} - l)(1 - m))$ $\forall m \in [0\ 1]$. □

Proposition 2 *For the strong Allee effect, system (4) has a stable node at* E_1 *and saddle points at* E_2 *and* E_3. *However, at the equilibrium* E_4, *the singularity of the equilibrium splits into a center, a stable, and an unstable focus according to the values of the parameters* m *and* q_1.

For the weak Allee effect, system (4) has saddle points at E_1 and E_3. Whereas, at E_4, the singularity of the equilibrium splits too into a center, a stable and an unstable focus, like the case of the strong Allee effect but according to other values of m and q_1.

Proof The Jacobian matrix of system (4) at each equilibrium E_i ($i \in [1, 4]$) is given by:

$$J_i = \begin{pmatrix} a_{11} & a_{12} \\ a_{21} & a_{22} \end{pmatrix} \tag{9}$$

where

$$\begin{aligned} a_{11} &= -x_1^{q_1+1}(2+q_1) + (1+q_1)x_1^{q_1} + 2lx_1 - (l+x_2) \\ a_{12} &= -x_1 \\ a_{21} &= x_2^{q_2} \\ a_{22} &= q_2 x_2^{q_2-1}(x_1 - m) \end{aligned}$$

and its characteristic equation is given by:

$$\lambda^2 - tr(J)\lambda + det(J) = 0 \tag{10}$$

where

$$\begin{aligned} tr(J) &= a_{11} + a_{22} \\ det(J) &= a_{11}a_{22} - a_{12}a_{21} \end{aligned}$$

The solutions of Eq. (10) at each equilibrium are shown in Tables 1 and 2. As it is demonstrated, one of each equilibrium eigenvalues is zero indicating that the phase portrait can degenerate in some sense. Thus,

- The equilibrium E_1 can be a stable node or a saddle point for the strong Allee effect case study. However, for the weak Allee effect case study, the equilibrium point becomes unstable since it admits a positive eigenvalue. It becomes a saddle point or an unstable node.
- The equilibrium $E_2(l^{1/q_1}, 0)$ can be a saddle point or an unstable node.
- The equilibrium E_3 can be a saddle point or an unstable node for the strong Allee effect case study. However, for the weak Allee effect case study, it can be either a saddle point or a stable node.
- The stability of E_4 depends on the value of $tr(J_4)$ which in turn depends on the parameters q_1 and m for the two cases of Allee effect.
  * If $tr(J_4) > 0$ the system (4) admits an unstable focus point.
  * If $tr(J_4) = 0$ the system (4) admits a center point.
  * If $tr(J_4) < 0$ the system (4) admits a stable focus point. □

Remark 1 The description of prey's growth rate and predator's efficiency with noninteger polynomials is proposed in order to describe further the system behavior by providing more bifurcations. These latter are detected by a numerical continuation from the bifurcation points obtained by the integer-order system analysis. In the following, the existence of a Hopf bifurcation for system (4) will be then verified by setting $q_1 = 1$ and proceeding to a continuation.

TABLE 1 Eigenvalues of System (4) With Strong Allee Effect

Parameter m	Equilibriums	Eigenvalues
$m \in [0, l^{1/q_1}]$	$E_1(0, 0)$	$\lambda_1 = 0$
		$\lambda_2 = -l$
	$E_2(l^{1/q_1}, 0)$	$\lambda_1 = 0$
		$\lambda_2 = q_1 l(1 - l^{1/q_1})$
	$E_3(1, 0)$	$\lambda_1 = (l - 1)$
		$\lambda_2 = 0$
$m \in [l^{1/q_1}, 1]$	$E_1(0, 0)$	$\lambda_1 = 0$
		$\lambda_2 = -l$
	$E_2(l^{1/q_1}, 0)$	$\lambda_1 = 0$
		$\lambda_2 = q_1 l(1 - l^{1/q_1})$
	$E_3(1, 0)$	$\lambda_1 = (l - 1)$
		$\lambda_2 = 0$
	$E_4(m, (m^{q_1} - l)(1 - m))$	$\lambda_1 = \frac{tr(J_4) - i\sqrt{\delta}}{2}$
		$\lambda_2 = \frac{tr(J_4) + i\sqrt{\delta}}{2}$

TABLE 2 Eigenvalues of System (4) With Weak Allee Effect

Parameter m	Equilibriums	Eigenvalues
$m \in [0, 1]$	$E_1(0, 0)$	$\lambda_1 = 0$
		$\lambda_2 = -l$
	$E_3(1, 0)$	$\lambda_1 = (l - 1)$
		$\lambda_2 = 0$
	$E_4(m, (m^{q_1} - l)(1 - m))$	$\lambda_1 = \frac{tr(J_4) - i\sqrt{\delta}}{2}$
		$\lambda_2 = \frac{tr(J_4) + i\sqrt{\delta}}{2}$

Proposition 3 *For the two cases of strong and weak Allee effects, the dynamical behavior of system (4) is sensitive to the parameter* q_l *and undergoes a supercritical Andronov Hopf bifurcation when* $q_1=1$.

Proof For both cases of Allee effect, the Jacobian matrix J_4 of system (4) at the equilibrium E_4 has a pair of complex conjugate eigenvalues as shown in Tables 1 and 2, where

$$\begin{aligned} tr(J_4) &= -m^{q_1+1}(1+q_1)+q_1 m^{q_1}+lm \\ \delta &= tr(J_4)^2 - 4det(J_4) \\ det(J_4) &= m[(m^{q_1}-l)(1-m)]^{q_2} \end{aligned}$$

When the parameter q_1 crosses the critical value $q_{10}=1$, the equilibrium E_4 has a purely imaginary eigenvalues $\lambda_{1,\,2} = \pm i\omega(q_{10})$ for the strong and weak Allee effects. For such cases, $tr(J_4)= 0$ and then $m=\frac{1+l}{2}$. Thus, according to Definition 2, system (4) with both cases of Allee effects admits an Andronov Hopf bifurcation when $q_1=1$. Due to the complexity of the analytical calculus of noninteger polynomials, the criticality of Hopf bifurcation is detected numerically in the next section. □

5 NUMERICAL ANALYSIS

5.1 Predator-Prey System With Strong Allee Effect

Analytic stability analysis is strengthened numerically by the different phase portraits given in Figs. 1–3 using the Matlab package MATCONT [17]. For the numerical simulation of system (4) with strong Allee effect case, we take $q_1 = 0.5$. In this case, when one of the equilibrium's eigenvalue is zero, we cannot conclude analytically the equilibrium's singularity. It is proven also that this latter depends on the value of either the parameter q_1 or q_2.

Furthermore, a detailed bifurcation analysis of system (4) is presented in Figs. 4 and 5. The first bifurcation continuation shown in Fig. 4 shows the existence of only one Hopf bifurcation point which is denoted by H. It appears when a limit cycle periodic solution occurs in a small neighborhood of the equilibrium E_4. This Hopf point corresponds to the same Hopf point of system (4) and it admits a negative First Lyapunov Coefficient (FLC) equal to − 0.8393214 indicating the existence of a supercritical Hopf bifurcation.

However, by contrast with system (3), system (4) admits a numerical continuation of the bifurcation analysis according to the parameter q_1 as it is shown in Fig. 5. Computing the solution forward and backward in time from the Hopf codimension 1, bifurcation point shows that the system (4) exhibits the following bifurcation points:

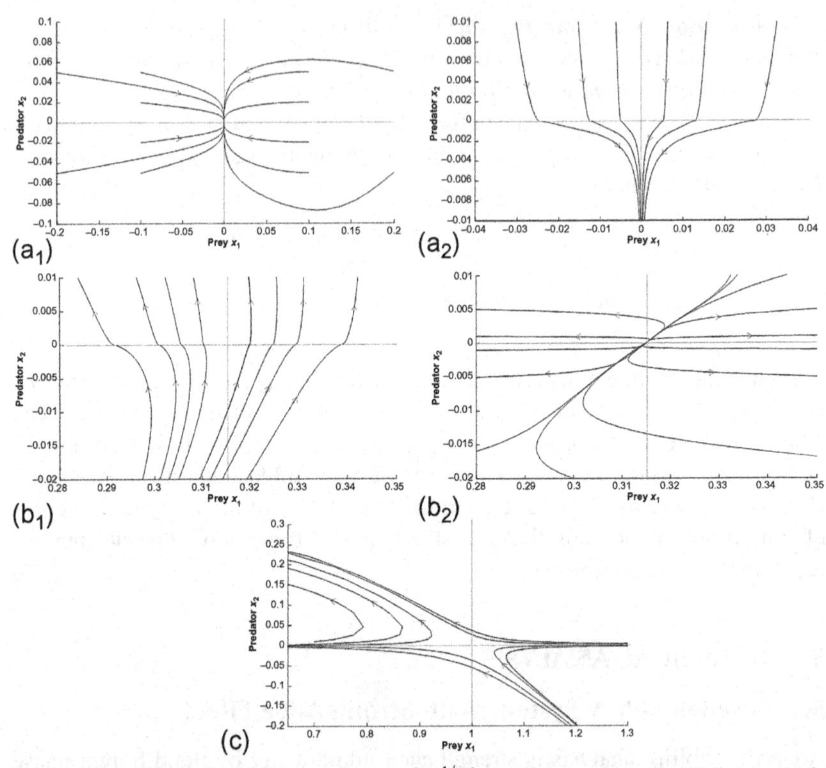

FIG. 1 Singularities of equilibriums when $m \in [0, l^{1/q_1}]$: (a_1) E_1 is a stable node if $0.5 < q_2 \leq 1.9$, (a_2) E_1 is a saddle point if $0 < q_2 \leq 0.5$, (b_1) E_2 is a saddle point if $0 < q_2 \leq 0.5$, (b_2) E_2 is an unstable node if $0.5 < q_2 \leq 1.1$, and (c) E_3 is a saddle point: strong Allee effect case study.

- A period doubling codimension 1 bifurcation, labeled as PD. This latter occurs when the parameter q_1 attends the critical value $q_1 = 1.036593$ only when $q_2 = 0.5$. In other word, modeling the predator's efficiency c by $1/\sqrt{x_2}$ undergoes a periodic solution. The period of the obtained PD bifurcation is equal to 16.7535 and the normal form is positive and equal to $7.762788e^3$. These results indicate that the bifurcation is supercritical and the obtained PD cycle is stable.
- Two limit point cycle bifurcations, labeled as LPC. This bifurcation type is created due to the occurrence of saddle point bifurcation and called fold bifurcation. In this point, there is a collision and disappear of two limit cycles. The first one occurs, by integration forward in time, near the PD cycle when the parameter q_1 attends the critical value $q_1 = 1.035979$. However, the second one occurs by integration backward in time when the parameter q_1 attends the critical value $q_1 = 1$.

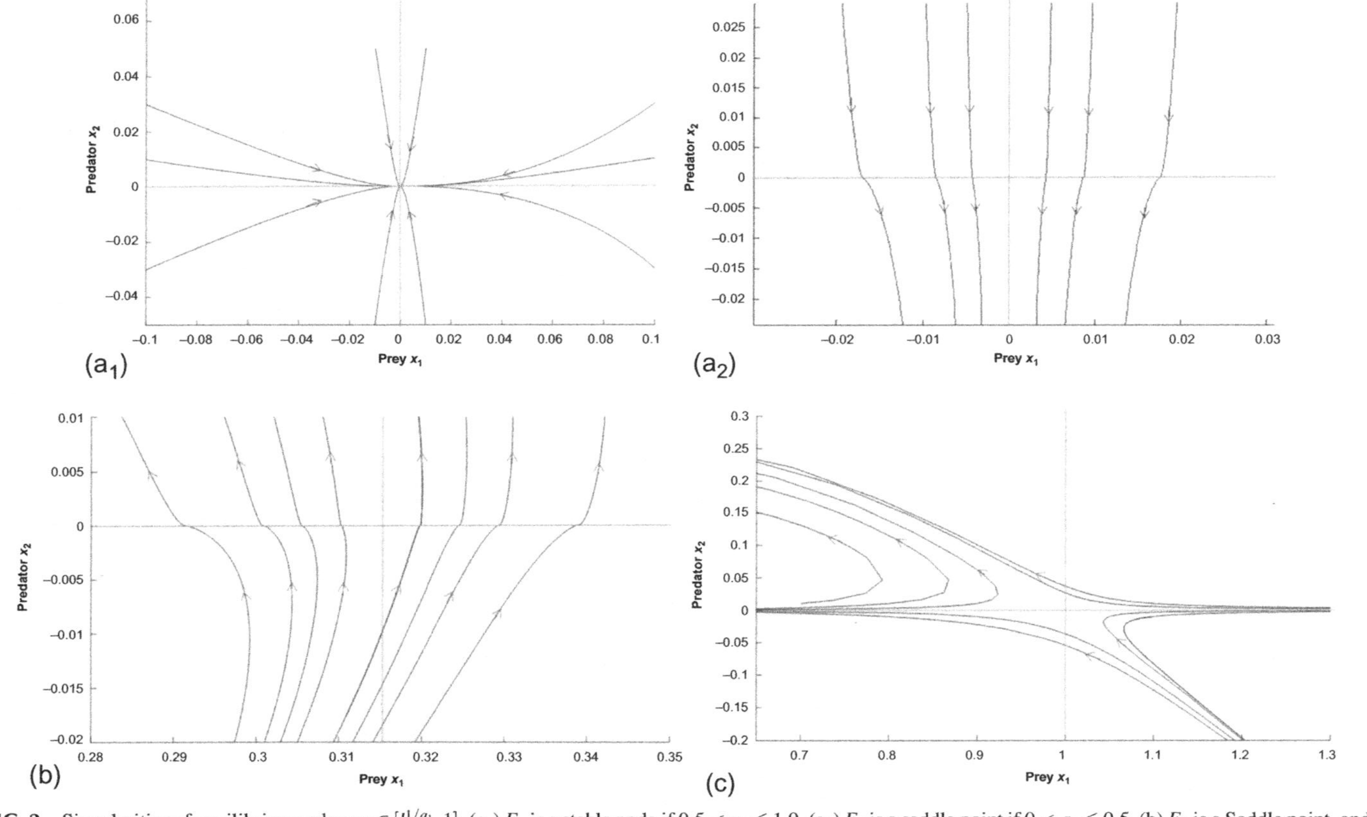

FIG. 2 Singularities of equilibriums when $m \in [l^{1/q_1}, 1]$: (a_1) E_1 is a stable node if $0.5 < q_2 \le 1.9$, (a_2) E_1 is a saddle point if $0 < q_2 \le 0.5$, (b) E_2 is a Saddle point, and (c) E_3 is a saddle point: strong Allee effect case study.

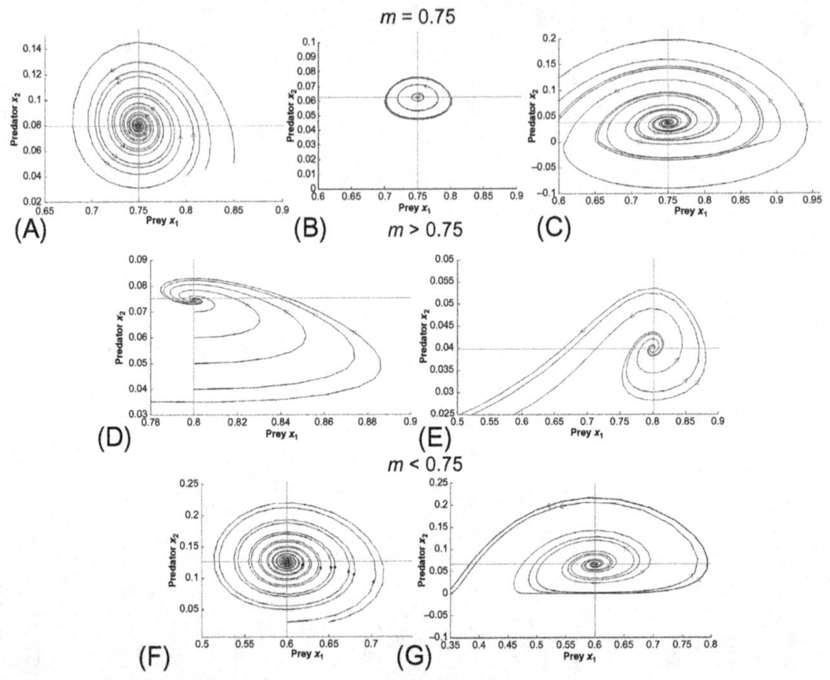

FIG. 3 Singularities of the equilibrium E_4 for $m \in [l^{1/q_1}, 1]$: (A), (D), and (F) stable focus if $q_1 < 1$; (B) center if $q_1 = 1$; (C), (E), and (G) unstable focus if $q_1 > 1$: strong Allee effect case study.

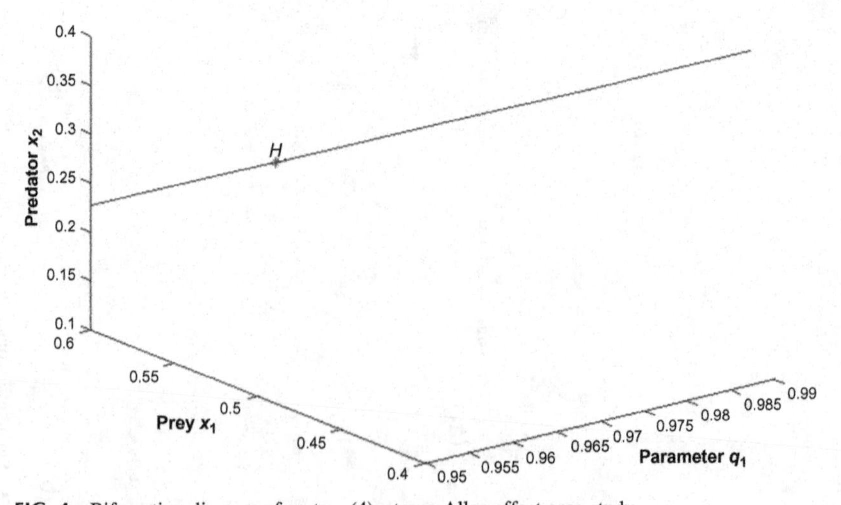

FIG. 4 Bifurcation diagram of system (4): strong Allee effect case study.

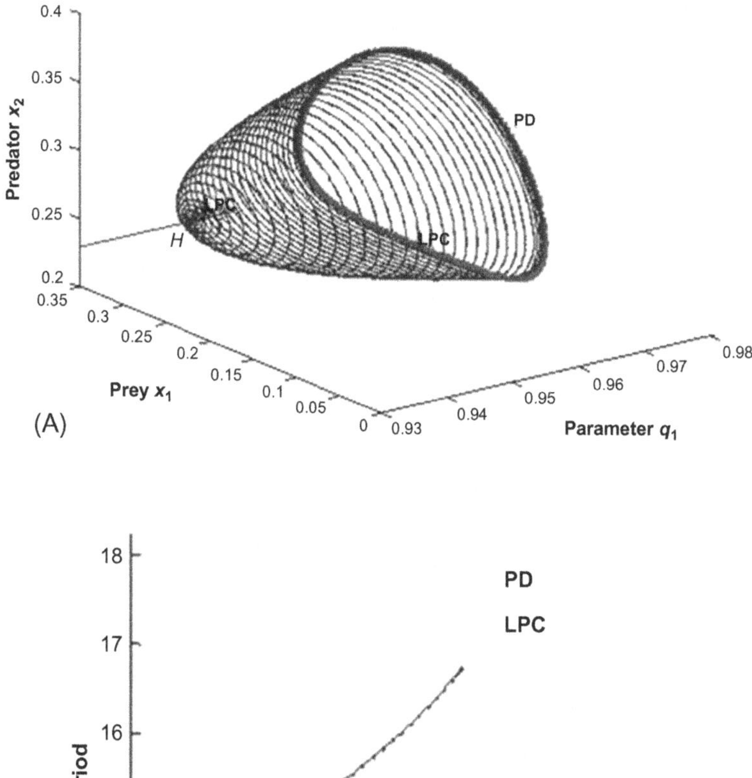

FIG. 5 Numerical continuation of system (4) for $q_2 = 0.5$: strong Allee effect case study.

5.2 Predator-Prey System With Weak Allee Effect Case Study

For the case of the weak Allee effect, the analytic stability and bifurcation analysis of the system (4) are proved numerically by drawing the phase portrait and the bifurcation diagrams presented in Figs. 6–10, respectively.

For the numerical simulation of system (4) with weak Allee effect case, we take $q_1 = -0.2$. As shown in Fig. 6, the nonisolated equilibrium E_4 admits more

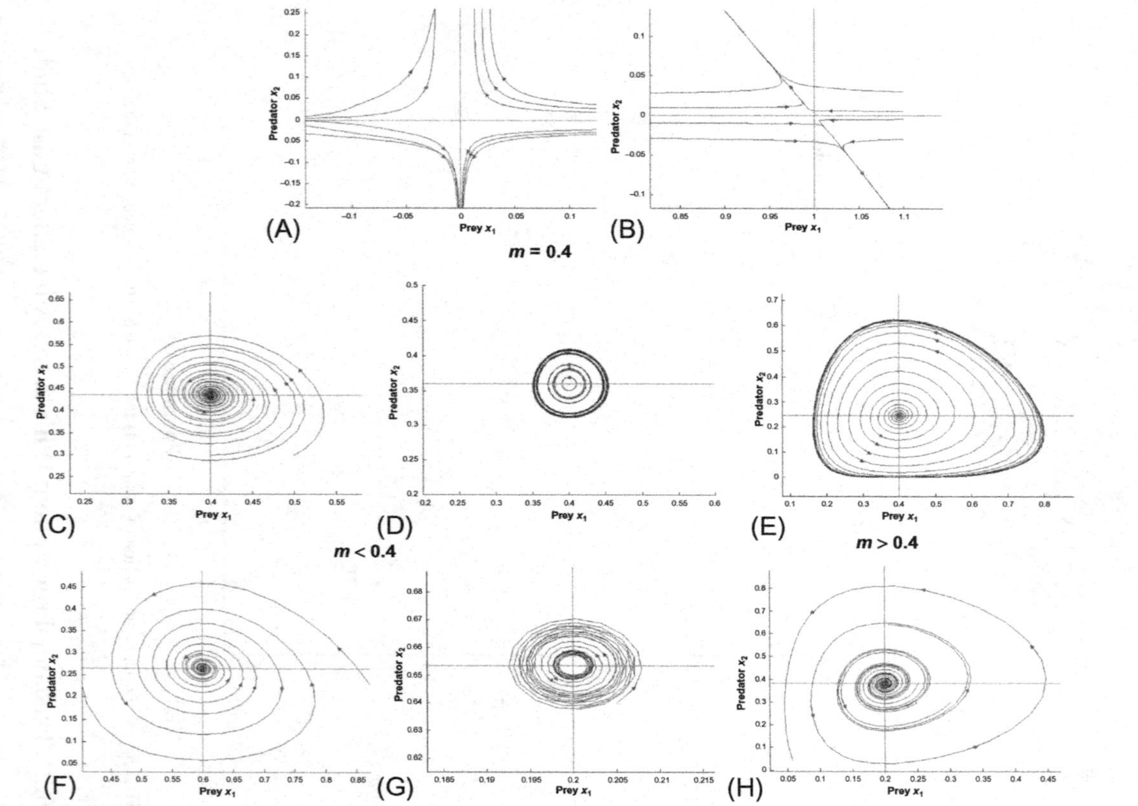

FIG. 6 Singularities of equilibriums when $m \in [0, 1]$: (A) E_1 is a saddle point, (B) E_3 is a saddle point, (C) and (F) E_4 is stable focus if $q_1 < 1$ and $q_1 < 0.4$, respectively, (D) and (G) E_4 is center if $q_1 = 1$, (E) and (H) E_4 is unstable focus if $q_1 > 1$ and $q_1 \geq 0.4$, respectively: weak Allee effect case study.

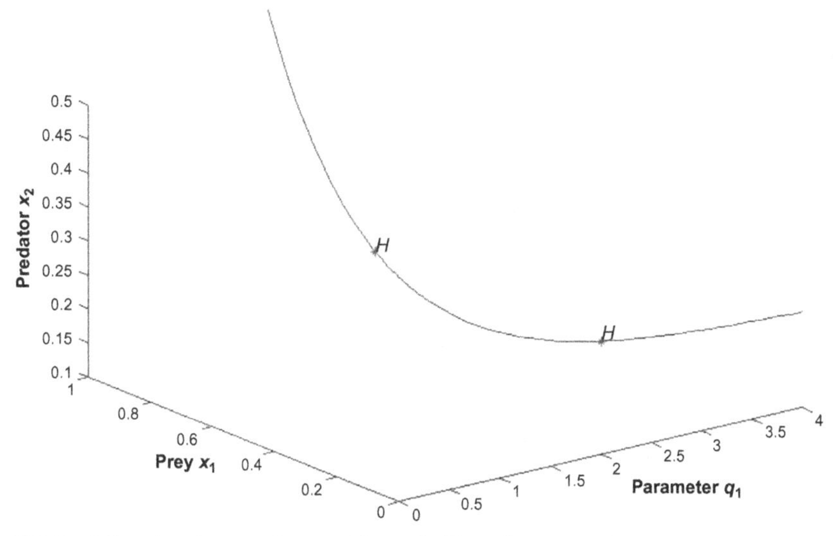

FIG. 7 Bifurcation diagram of system (4): weak Allee effect case study.

types of singularities according to the value of the parameter q_1 for each value of the parameter m.

Therefore, a numerical bifurcation analysis according to the parameter q_1 is investigated. Fig. 7 shows that system (4) admits two Hopf bifurcation points which are denoted by H_1 and H_2 and described as follows:

- The first point, H_1, occurs when $q_1 = 0.999967$ and admits a supercritical Hopf point since the FLC is negative and equal to $-$ 0.4000004.
- The second point, H_2, occurs when $q_1 = 3.223516$ and admits a subcritical Hopf point since the FLC is equal to 0.004301. This point is only detected numerically, and it is verified analytically according to Definition 2.

In addition, the numerical continuation of the bifurcation from each of these two Hopf points, H_1 and H_2, exhibits new types of bifurcation.

- If $q_2 = 0.5$
  * The continuation from the first Hopf bifurcation H_1 exhibits a PD and two LPC codimension 1 bifurcations as shown in Fig. 8. The PD bifurcation occurs by integration forward in time when the parameter q_1 attends the critical value $q_1 = 1.773590$. It admits a stable period cycle because it has a positive normal form and a period equal to 16.33313. Moreover, the two LPC bifurcations are obtained by integration backward in time. The first one is detected when $q_1 = 1$ and the second is detected when $q_1 = 1.773618$.

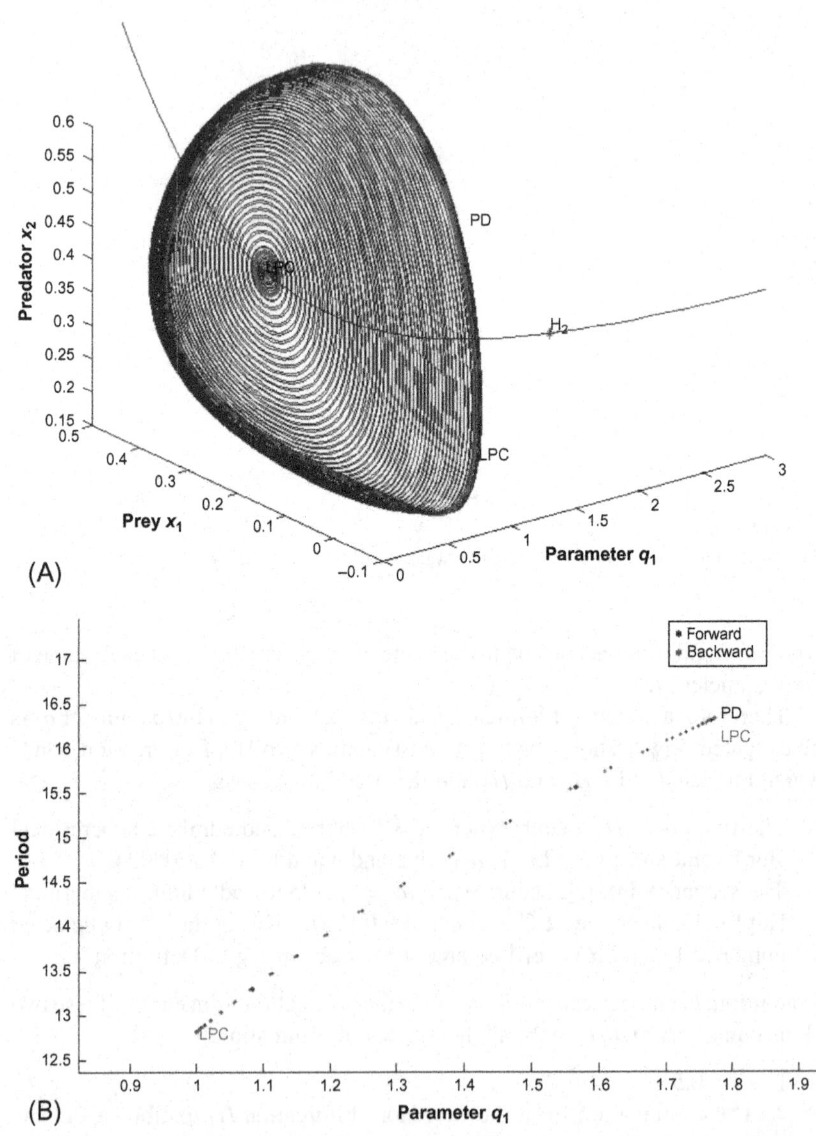

FIG. 8 Numerical continuation of system (4) from the first Hopf bifurcation H_1 for $q_2 = 0.5$: weak Allee effect case study.

* The continuation, by integration only forward in time, from the second Hopf point H_2 exhibits the same type of bifurcation points as those detected from the first Hopf point, but with new critical values of the parameter q_1 as shown in Fig. 9. The first detected bifurcation is an LPC bifurcation which occurs when $q_1 = 3.223605$. Subsequently, the

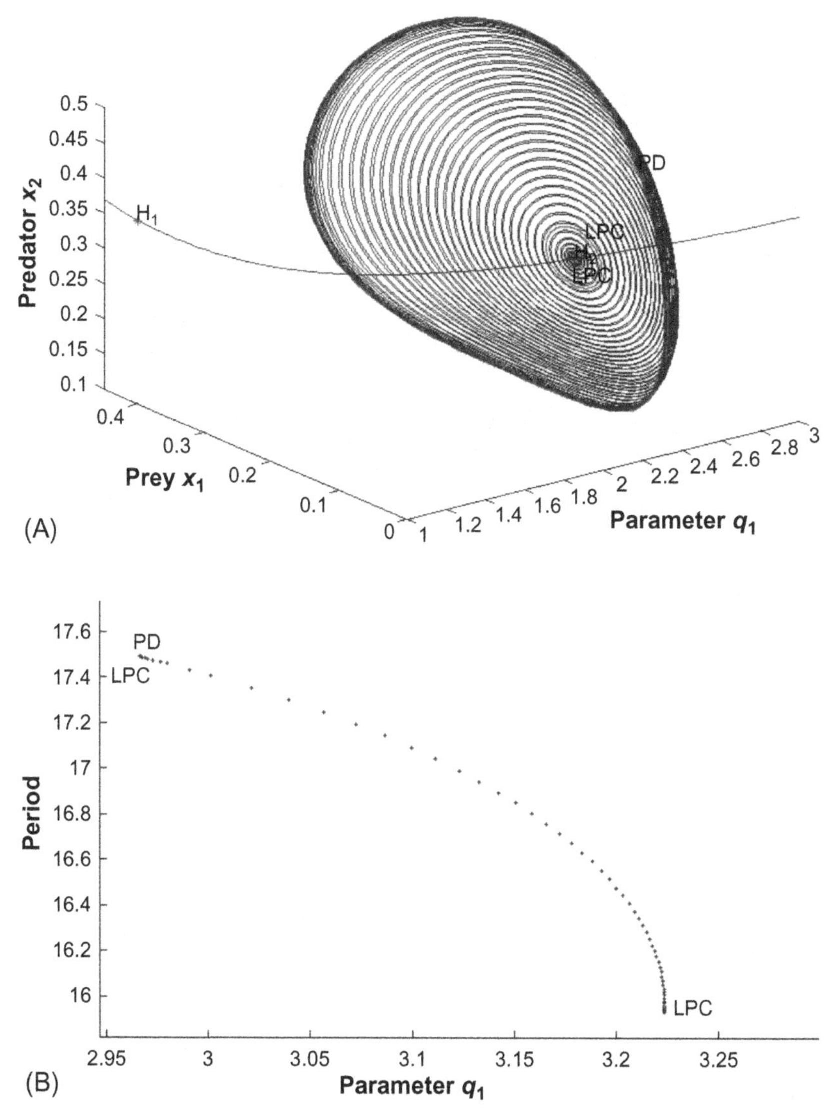

FIG. 9 Numerical continuation of system (4) from the second Hopf bifurcation H_2 for $q_2 = 0.5$: weak Allee effect case study.

second LPC bifurcation appears when the parameter q_1 attends the critical value 2.966445. Finally, the PD bifurcation occurs when $q_1 = 2.966560$. Its period is equal to 17.49135, and it admits a positive normal form which indicates that the obtained PD bifurcation is stable.

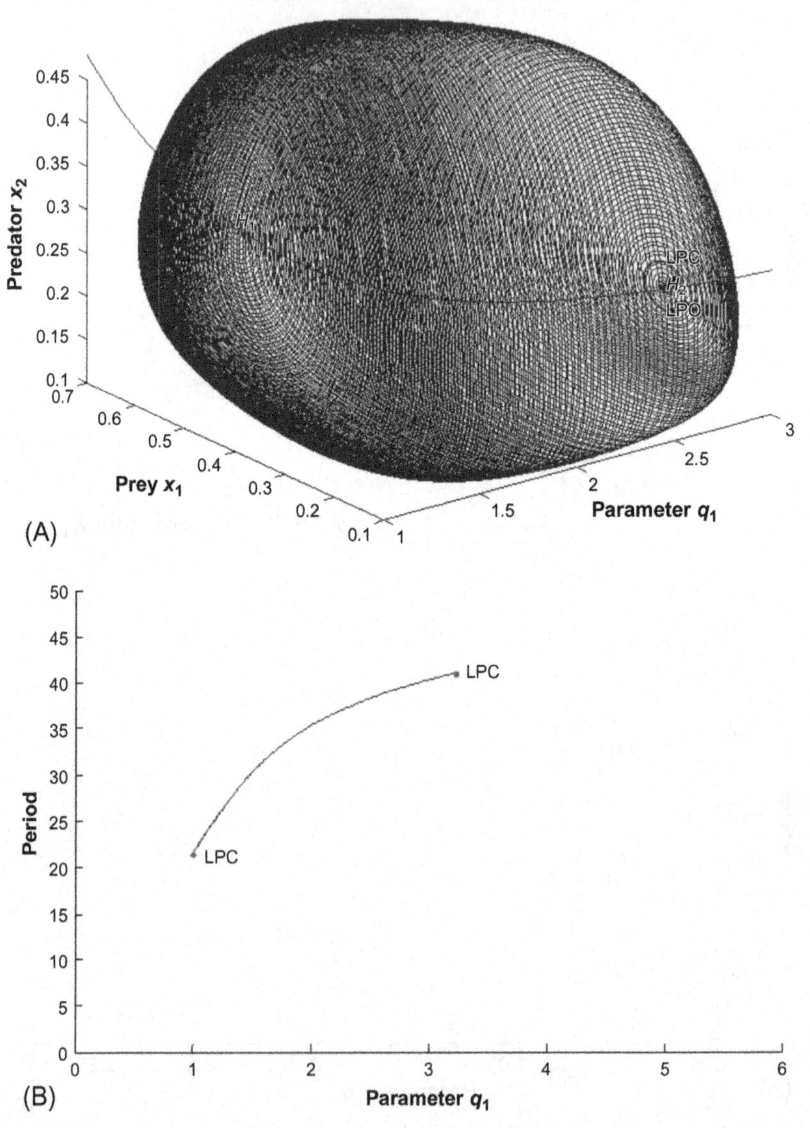

FIG. 10 Numerical continuation of system (4) for $q_2 \neq 0.5$: weak Allee effect case study.

– For all $q_2 \neq 0.5$, the continuation exhibits a new periodic behavior which is descried by a limit cycle curve and bounded between the two detected Hopf points as shown in Fig. 10. An LPC bifurcates from the Hopf point H_2 when the parameter q_1 attends the critical value 3.223605.

5.3 Seasonally Perturbed System

To identify the role that seasons play on the dynamic behavior of system (4), a seasonally external perturbation is applied for which the system becomes:

$$\begin{cases} \dfrac{dx_1}{dt} = x_1(x_1^{q_1} - l)(1 - x_1) - x_1x_2 + \lambda x_1(1 - \epsilon\cos(\omega t)) \\ \dfrac{dx_2}{dt} = (x_1 - m)x_2^{q_2} \end{cases} \tag{11}$$

- *Strong Allee effect case study (l > 0)*

For numerical simulation, the most elegant set of noninteger elements giving a strange attractor are $q_1 = 1$ and $q_2 = 1.8$. The three-dimensional strange attractor and its projections are presented in Figs. 11 and 12, respectively.

To calculate the Lyapunov exponents, the famous Wolf's method [18] has been applied in our work. It is important to note that the Lyapunov exponents depend on the computation time and the initial point of trajectory [19–21]. Therefore, we fixed the initial point as $(x_1(0), x_2(0), x_3(0)) = (0.7, 0.06, 2\pi)$, where $x_3 = \omega t$, and the time of computation is 600. Fig. 13 shows that system (11) admits three Lyapunov exponents $LE_1 = 0.3270$, $LE_2 = 0$, $LE_3 = -1.72$, for $q_1 = 1$ and $q_2 = 1.8$, implying that there is a strong evidence that we have a chaotic attractor [22–24].

Moreover, Lyapunov dimension D_L [25] which is commonly called "Kaplan-Yorke dimension" is computed by applying the following formula:

$$D_L = k + \frac{LE_1 + LE_2 + \cdots + LE_k}{|LE_{k+1}|} \tag{12}$$

where k is the maximum number of LE_i such that $LE_i \geq 0$.

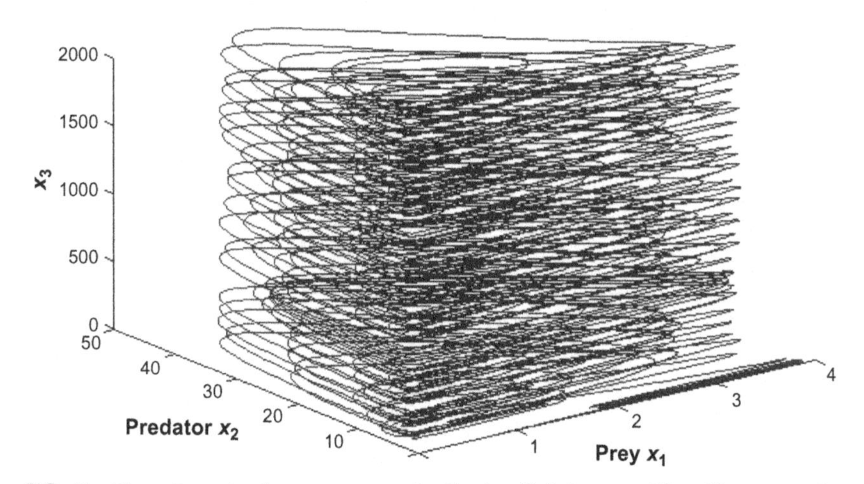

FIG. 11 Three-dimensional strange attractor for $(\lambda, \omega) = (5, 2\pi)$: strong Allee effect case study.

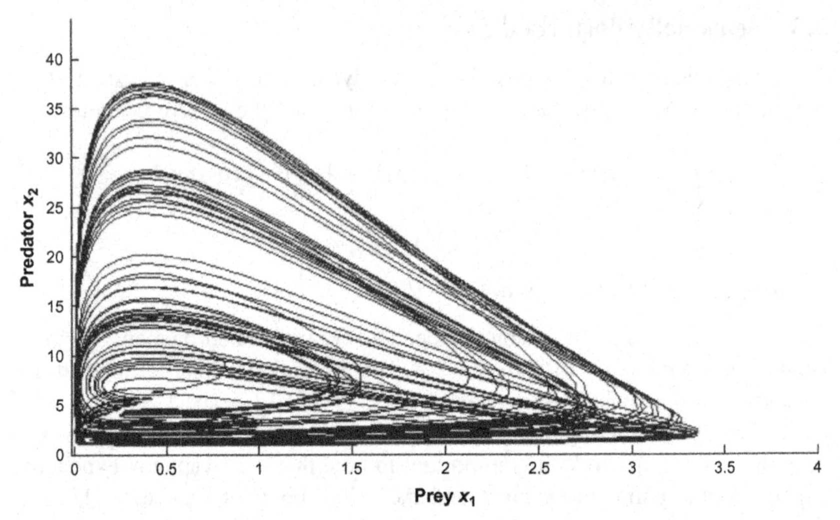

FIG. 12 Projection of chaotic attractor in x_1–x_2 plane for $(\lambda, \omega) = (5, 2\pi)$: strong Allee effect case study.

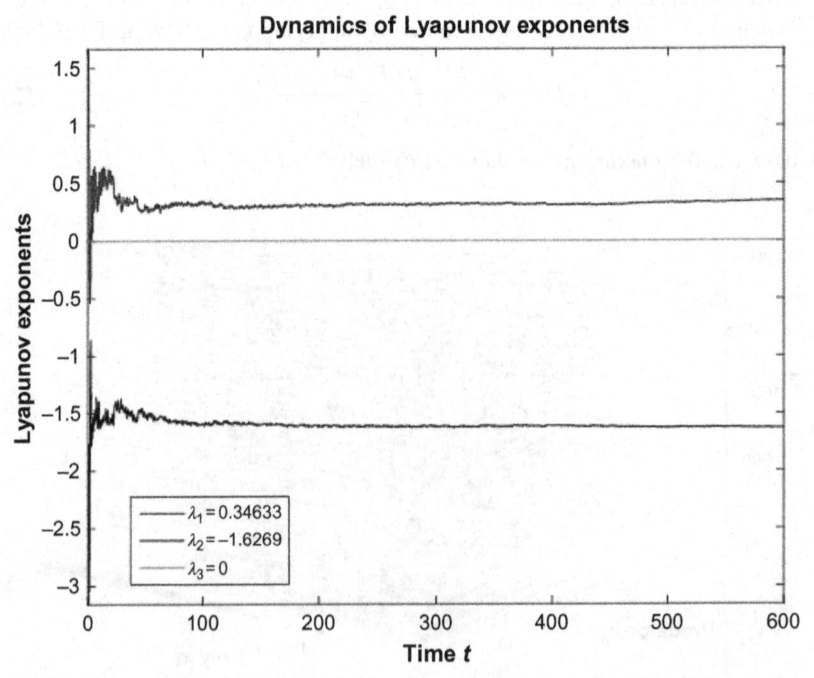

FIG. 13 Lyapunov exponent spectrum of the chaotic system (11): strong Allee effect case study.

The obtained $D_L = 2.19$ belongs to the interval]2, 3[. Thus, we can conclude that the seasonally perturbed system (11) admits a chaotic behavior when it's prey's growth rate is affected by a strong Allee effect and when the predator's efficiency is modeled by $c = x_2^{0.8}$.

– *Weak Allee effect case study ($l < 0$)*

The most elegant set of noninteger elements giving a strange attractor are $q_2 = 0.5$ and $q_1 = 1$. The three-dimensional strange attractor and its projections are presented in Figs. 14 and 15.

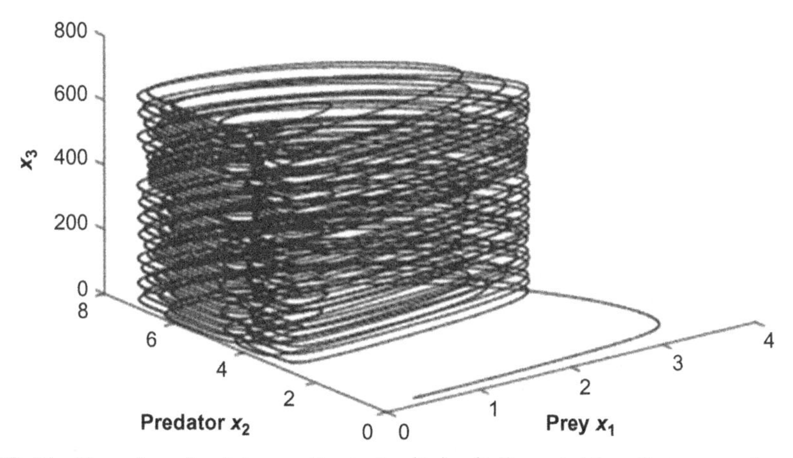

FIG. 14 Three-dimensional strange attractor for $(\lambda, \epsilon) = (5, \frac{\pi}{2})$: weak Allee effect case study.

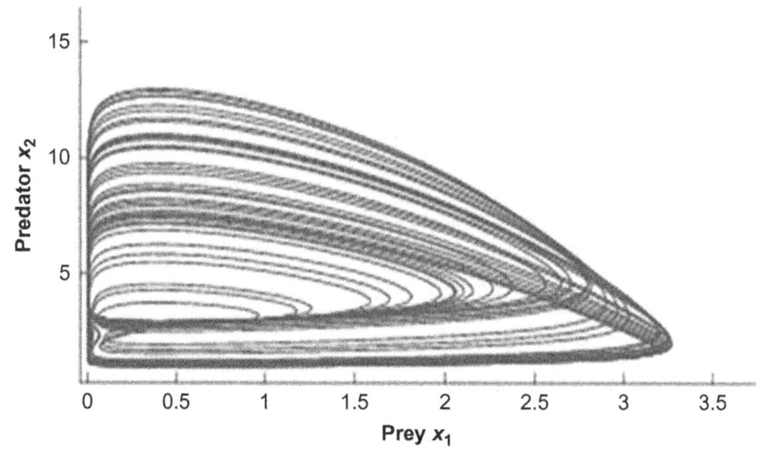

FIG. 15 Projection of strange attractor in x_1–x_2 plane for $(\lambda, \omega) = (5, \frac{\pi}{2})$: weak Allee effect case study.

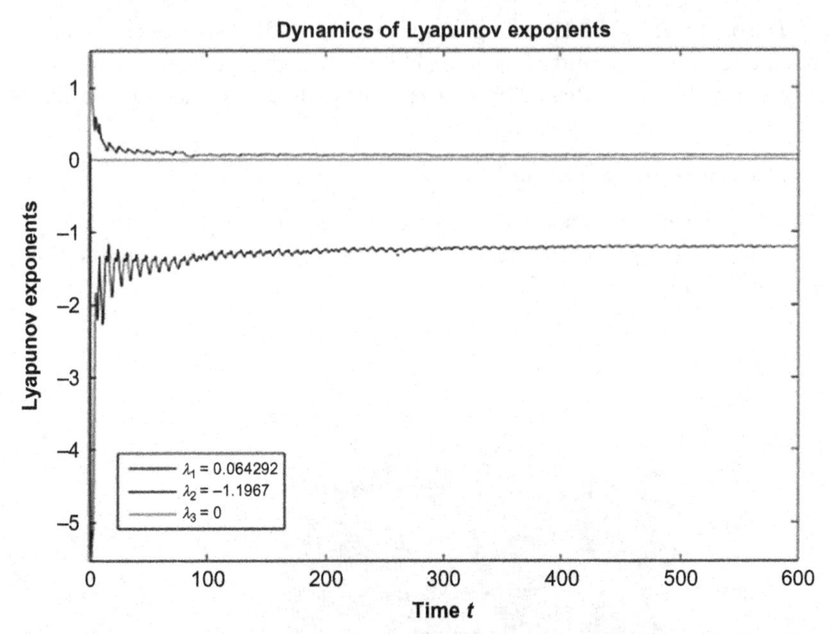

FIG. 16 Lyapunov exponent spectrum of the strange attractor of system (11): weak Allee effect case study.

Applying Wolf's method, system (11) admits three Lyapunov exponents $LE_1 = 0.048$, $LE_2 = 0$, $LE_3 = -1.175$, as it is shown in Fig. 16. In this case, we fixed the initial point as $(x_1(0), x_2(0), x_3(0)) = (0.4, 0.3, \frac{\pi}{2})$, the time of computation is 600, $q_2 = 0.5$, and $q1 = 1$. The obtained results indicate that the system has a strange attractor.

To detect the chaotic attractor, we compute the Kaplan-Yorke dimension given by the formula (12):

$$D_L = 2 + \frac{LE_1 + LE_2}{|LE_3|} = 2.041$$

Thus, we can conclude that the perturbed system (11) admits a chaotic behavior when its prey's growth rate is affected by a weak Allee effect and when the predator's efficiency is modeled by $c = x_2^{-0.5}$, with $x_2 \neq 0$.

6 CONCLUSION

In this chapter, a modified algebraic structure of the predator-prey BB model with strong and weak Allee effects is proposed. Bifurcation analysis gives the opportunity to divide the parameter space into more regions and depict more complex dynamic behaviors. For the two case studies, the dynamical behavior analysis of the nonperturbed proposed system proves that the noninteger element modeling the prey's growth rate q_1 destabilizes the system and creates more singularities for each equilibrium point. For that, q_1 is chosen as a bifurcation parameter during the system's analysis. In addition, the noninteger element modeling the predator's efficiency q_2 influences mainly the dynamic behavior of the system and permits us to bring out new bifurcation points. When the predator's efficiency is modeled by $1/\sqrt{x_2}$, the proposed system bifurcates to a stable periodic orbits PD bifurcations for the two cases study of Allee effect. However, under the effect of seasonal external force, the system loses its stability and becomes chaotic for certain structures of the predator's efficiency. Thus, the perturbed system admits a chaotic attractor when the predator's efficiency is modeled by $x_2^{0.8}$ or by $1/\sqrt{x_2}$, for the strong or weak Allee effect case study, respectively. In fact, we can conclude that the predator's efficiency algebraic structure and the periodic external perturbation have an important effect on the topology and complexity of populations' interactions dynamics.

REFERENCES

[1] Jiang J, Yu P. Multistable phenomena involving equilibria and periodic motions in predator-prey systems. Int J Bifurcation Chaos 2017;27(3):28.

[2] Zhang L, Liu J, Banerjee M. Hopf and steady state bifurcation analysis in a ratio-dependent predator-prey model. Commun Nonlinear Sci Numer Simul 2017;44:S52–73.

[3] Hu D, Cao H. Stability and bifurcation analysis in a predator-prey system with Michaelis-Menten type predator harvesting. Nonlinear Anal Real World Appl 2017;33:S58–82.

[4] Xie T, Yang X, Li X, Wang H. Complete global and bifurcation analysis of a stoichiometric predator-prey model. J Dyn Diff Equat 2016;30(2)S1–26.

[5] Shyam PB, Alakes M, Guruprasad S. Stochastic analysis of a prey-predator model with herd behaviour of prey. Nonlinear Anal Model Control 2016;21(3)S345–61.

[6] Volterra V. Variations and fluctuations of the number of individuals in animal species living together. ICES J Mar Sci 1928;3(1)S3–51.

[7] Berryman AA. The origins and evolution of predator-prey theory. Ecology 1992;73(5) S1530–5.

[8] Hulang C, Cao J, Xiao M, Alsaedi A, Alsaadi FE. Controlling bifurcation in a delayed fractional predator prey system with incommensurate orders. Appl Math Comput 2017;293: S293–310.

[9] Jana D, Gopal R, Lakshmanan M. Complex dynamics generated by negative and positive feedback delays of a prey-predator system with prey refuge: Hopf bifurcation to Chaos. Int J Dyn Control 2016;5:S1–15.

[10] Li S, Liu W. A delayed Holling type III functional response predator-prey system with impulsive perturbation on the prey. Adv Differ Equat 2016;42:42.

[11] Banrjee C, Das P. Impulsive effect on tri-trophic food chain model with mixed functional responses under seasonal perturbations. Differ Equat Dyn Syst 2016;157–76.

[12] Li Y, Wang J. Spatiotemporal patterns of a predator-prey system with an Allee effect and Holling type III functional response. Int J Bifurcation Chaos 2016;26(5)S20.

[13] Olivares EG, Palma AR, Yanez BG. Multiple limit cycles in a Leslie-Gower-type predator-prey model considering weak Allee effect on prey. Nonlinear Anal Modell Control 2017;22 (3)S347–65.

[14] Ben Saad A, Boubaker O. A new fractional-order predator-prey system with Allee effect, In: Azar A, Vaidyanathan S, Ouannas A, editors. Stud Comput IntellFractional order control and synchronization of chaotic systems688:; 2017857–77.

[15] Van Voorn AKG, Hemerik L, Boer MP, Kooi BW. Heteroclinic orbits indicate overexploitation in predator-prey systems with a strong Allee effect. Math Biosci 2007;209:S451–69.

[16] Lassoued A, Boubaker O. Dynamic analysis and circuit design of a novel hyperchaotic system with fractional-order terms. Complexity 2017;2017:S10.

[17] Dhooge A, Govaerts W, Yu AK, Meijer HGE, Sautois B. New features of the software MatCont for bifurcation analysis of dynamical systems. J Math Comput Model Dyn Syst 2008;14 (2)S147–75.

[18] Wolf A, Swift JB, Swinney HL, Vastano JA. Determining Lyapunov exponents from a time series. Physica D 1985;16(3)S285–317.

[19] Leonov GA, Kuznetsov NV, Korzhemanova NA, Kusakin DV. Lyapunov dimension formula for the global attractor of the Lorenz system. Commun Nonlinear Sci Numer Simul 2016;41: S84–103.

[20] Kuznetsov NV. The Lyapunov dimension and its estimation via the Leonov method. Phys Lett 2016;380:S2142–9.

[21] Kuznetsov NV, Alexeeva TA, Leonov GA. Invariance of Lyapunov exponents and Lyapunov dimension for regular and irregular linearizations. Nonlinear Dyn 2016;85:S195–201S195–201.

[22] Wang Z, Xu Z, Mliki E, Akgul A, Pham VT, Jafari S. A new chaotic attractor around a prelocated ring. Int J Bifurcation Chaos 2017;27(10)S10.

[23] Lassoued A, Boubaker O. On new chaotic and hyperchaotic systems: a literature survey. Nonlinear Anal Modell Control 2016;21(6)S770–89.

[24] Jafari S, Reza SM, Golpayegani H, Jafari AH, Gharibzadeh S. Letter to the editor some remarks on chaotic systems. Int J Gen Syst 2012;41(3)S329–30.

[25] Pham VT, Volos C, Kapitaniak T, Jafari S, Wang X. Dynamics and circuit of a chaotic system with a curve of equilibrium points. Int J Electron 2017;105(3)S385–97.

Chapter 12

Chaotic Path Planning for a Two-Link Flexible Robot Manipulator Using a Composite Control Technique

Kshetrimayum Lochan*,†, Jay Prakash Singh*, Binoy Krishna Roy* and Bidyadhar Subudhi‡

**Department of Electrical Engineering, National Institute of Technology Silchar, Silchar, India, †Department of Mechatronics Engineering, Manipal Institute of Technology, Manipal Academy of Higher Education, Manipal, India, ‡Department of Electrical Engineering, National Institute of Technology Rourkela, Rourkela, India*

1 Introduction

Chaotic systems have aperiodic long time behavior [1]. In the last decade, many chaotic/hyperchaotic systems have been reported in the literature based on their various behaviors and characteristics [2–4]. Chaotic/hyperchaotic systems are used in fields such as secure communication [5], information theory [6], image processing [5], structural engineering [7], security [8], economics [9], biomedical [9], robotics [10–12], etc. But the use of chaotic systems in many other areas remains little explored. This chapter seeks to explore the use of chaotic signals of a chaotic system as the desired trajectory for a flexible manipulator (FM).

Many chaotic systems are reported based on the characteristics of their equilibrium point. Reported chaotic systems can be classified into two groups: chaotic systems having self-excited attractors and chaotic systems having hidden attractors [13–18]. Chaotic systems having (a) a stable equilibrium point [19] or (b) no equilibrium point [20] are considered to be under hidden attractors. Recently, chaotic systems with infinitely many equilibrium points [21–25] and systems with curve [26], plane [27], or surface of equilibria [25,28] have also become considered to be under hidden attractors. Chaotic systems like the Lorenz [29], Chen [30], Lu [31], and Sprott [32] systems in [33–38] are categorized under the self-excited attractors chaotic system. Hidden attractors are

Recent Advances in Chaotic Systems and Synchronization. https://doi.org/10.1016/B978-0-12-815838-8.00012-1

also seen in many electromechanical systems like induction motors [39], drilling systems [17], and many others.

FMs are being used for many applications in industry [40], aerospace [41], medical science [42], home use [43], education [40,44], etc. These applications are increasing gradually because of FMs' many inherent advantages [40]. The literature considers various types of control problems for FMs. The most commonly used problem is trajectory tracking for the joint angle and tip position [40]. Various types of the desired trajectory for FMs are considered in the literature and are listed in Table 1. It should be noted from Table 1 that the use of a chaotic signal as the desired trajectory in an FM is still less explored in the literature.

The precise operation of FMs depends on the modeling method used for the design of the controller. The commonly used modeling methods for FMs are the assumed modes method (AMM) [51], the lumped parameter method (LPM) [11,52], and the finite element method (FEM) [40,53]. Among these, the most widely used method is AMM [40]. Another interesting modeling method used in coordination with the AMM is the singular perturbation (SP) technique [40]. In this technique, a two-time scale separation principle is used in which the dynamics are divided into two parts: slow and fast subsystems [51,54]. The slow system consists of the rigid dynamics, and the fast subsystem consists of the flexible dynamics [54]. Using this method, it is easy to design separate control inputs to achieve desired performances.

Suppression of link deflection of a flexible robot manipulator is also an interesting control problem. The literature reports various control techniques, along with modeling methods, for quick suppression of link deflection of a FM [55–57]. SP is also used for suppression of link deflection. The SP modeling method is more appropriate for suppression of link deflections because separate control can be designed using the fast subsystem for this purpose.

Many types of control techniques are reported in the literature for control of a TLFM (two-link flexible manipulator), including some classical control techniques like state feedback control [58] and observer-based control [59], and

TABLE 1 Types of the Desired Trajectory Used for a Flexible Manipulator

Sl. No.	Desired Trajectory	Paper References
1.	Bang-bang	[45]
2.	Circular	[46]
3.	Exponentially varying	[47,48]
4.	Straight link	[49]
5.	Chaotic signal	[11,50]

some robust control techniques like backstepping control [60], extended state observer [61], sliding mode control (SMC) [62–65], adaptive control [45], adaptive SMC [62,66,67], second-order SMC [68,69], hybrid control technique [70], etc. Some soft computing techniques (intelligent control techniques) are also used to control FMs, including fuzzy logic control [71], artificial neural network [71], genetic algorithm [71], etc. Most of the reported control techniques for a TLFM use a particular modeling method like AMM, LPM, or FEM. But few use the SP modeling approach along with AMM for designing a controller for a two-link FM. The available literature on use of the SP modeling approach with AMM for a TLFM is classified in Table 2. It is apparent from Table 2 that the use of the SP for designing the controller for a TLFM uses various controllers and demands exploration of other potential controllers.

TABLE 2 Categorization of Composite Controllers Applied on the Singular Perturbation Model of a TLFM

References of Paper	Slow Subsystem	Fast Subsystem
	Types of Controller	*Types of Controller*
[72]	PID	PID
[73]	Feedback linearization	Linear quadratic regulator (LQR)
[51]	Computed torque control	LQR
[74]	PD	State feedback control
[48]	VSC	Virtual force control
[49,75]	PID + ANN	H_∞
[76]	Adaptive normal SMC with H_∞	LQR
[77]	PID feedback control	PID
[78]	Fuzzy TSMC	Observer-based LQR
[79]	PD	Lyapunov-based
[80]	SMC	H_∞
[81]	VSC	Lyapunov-based
[82]	NN	LQR
[83]	Nonsingular TMC	Observer-based LQR
[84]	LMI-SMC	LMI-based state feedback control
This work	Dynamic surface control	Backstepping

Motivated by the above discussion, this chapter attempts to design a composite control using the SP technique for the chaotic trajectory tracking of a TLFM.

In this chapter, we design a composite controller for chaotic trajectory tracking control of a two-link FM. The composite controller consists of a dynamic surface control for the slow subsystem and a backstepping control for the fast subsystem. The dynamic surface control is designed for chaotic trajectory tracking and backstepping controller is designed for quick suppression of link deflection.

Following are the contributions of the chapter:

(i) A composite control is proposed in this chapter for the chaotic trajectory tracking/path planning and quick link deflection suppression for a TLFM.
(ii) The composite controller is designed with dynamic surface control for the slow subsystem and a backstepping control for the fast subsystem.
(iii) The dynamic surface control is designed on rigid body dynamics for chaotic trajectory tracking, and backstepping control is designed on the flexible body dynamics for quick link deflection suppression.

The organization of the chapter is as follows. Section 2 discusses modeling a two-link FM. Modeling using the SP is shown in Section 3. Section 4 presents the design of a composite control for chaotic trajectory tracking and quick link deflection suppression of a two-link FM. The dynamics of a chaotic system used for the generation of the desired trajectory are presented in Section 5. Results and discussion are given in Section 6. This chapter is concluded in Section 7.

2 Modeling of the Two-Link Flexible Manipulator

A view of a TLFM is shown in Fig. 1. The variables are defined in Table 3.

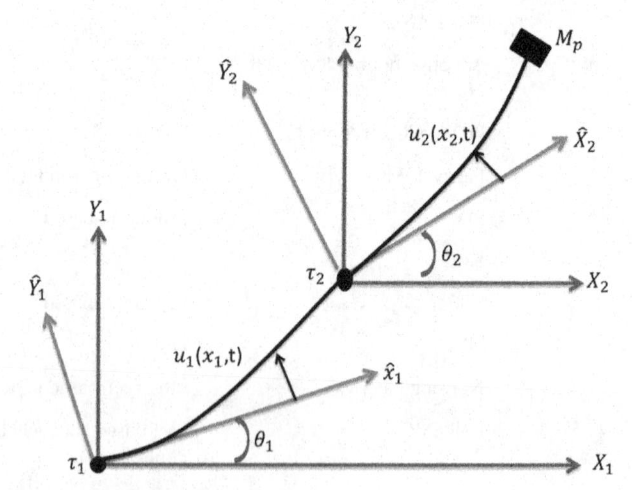

FIG. 1 Schematic representation of a TLFM.

TABLE 3 Variables and Their Description for TLFM Are Shown in Fig. 1

Symbol/Variable	Description
τ_i	Actuator torque applied at *i*th link
θ_i	Joint angle of *i*th link
$u_i(l_i, t)$	Link deflection of *i*th link
M_p	Payload mass attached with tip of link-2
(X_i, Y_i)	Rigid body coordinate frame with *i*th link
$(\hat{X}_i, \hat{Y}_i)$	Flexible coordinate frame

The rigid body dynamics are presented using θ_i and the flexible body motion is presented using $u_i(l_i,t)$. The dynamic model of a TLFM is obtained by using the Euler-Bernoulli beam theory. The dynamic model is shown in the form of a partial differential equation (PDE) along with the boundary conditions describing the links' motion. Dynamics of the system energy are obtained using the Lagrangian formation approach with the help of assumed modes method (AMM) [45,51]. The Lagrangian dynamics of a flexible motion are described in Eq. (1).

$$\frac{d}{dt}\frac{\partial\left((E_K)_i-(E_P)_i\right)}{\partial q_i}-\frac{\partial\left((E_K)_i-(E_P)_i\right)}{\partial q_i}=\tau_i \tag{1}$$

where $(E_K)_i$ and $(E_P)_i$ are the total kinetic energy and potential energy, respectively, of the ith link, and q_i is the generalized coordinates consisting of joint angles, joint velocities, and modal coordinates [51]. The total kinetic energy is obtained as $(E_K)_i=$ (total KE due to ith joint) + (total KE due to ith link) + (total KE due M_p) in the absence of gravity. The PDE of the link deflections is described as

$$(EI)_i\frac{\partial^4 u_i(l_i,t)}{\partial l_i^4}+\rho_i\frac{\partial^2 u_i(l_i,t)}{\partial l_i^2}=0 \tag{2}$$

where

i	*i*th link
$(EI)_i$	Flexural rigidity
l_i	Length
ρ_i	Density
t	Time
$u_i(l_i, t)$	Deflection

A solution of Eq. (2) can be obtained by applying proper boundary conditions.

Suppose that the mass of the links is negligible compared with the mass of the payload; we can write it as [85–87];

$$\begin{cases} (EI)_i \dfrac{\partial^4 u_i(l_i,t)}{\partial l_i^2} = -J_{eq_i} \dfrac{d^2}{dt^2}\left(\dfrac{\partial u_i(l_i,t)}{\partial l_i}\right) \\ (EI)_i \dfrac{\partial^3 u_i(l_i,t)}{\partial l_i^3} = M_{eq_i} \dfrac{d^2}{dt^2}(u_i(l_i,t)) \end{cases} \quad (3)$$

where J_{eq_i} and M_{eq_i} are the mass inertia and moment of inertias at the end of the ith link. Finite dimensional equation of the links flexibility $u_i(l_i,t)$ can be expressed using AMM [85] as

$$u_i(l_i,t) = \sum_{j=1}^{n} \varphi_{ij}(l_i)\delta_{ij}(t) \quad (4)$$

where

φ_{ij} jth mode shapes (special coordinates)
δ_{ij} jth modal coordinates (time coordinates)
n Number of assumed modes

The solution of Eq. (1) can be obtained using Eq. (4) in the form of time-harmonic function and space eigenfunction as in Eq. (5) [85].

$$\begin{cases} \delta_{ij}(t) = e^{j\varnothing_{ij}t} \\ \varphi_{ij} = C_{1,i}\sin(\alpha_i, l_i) + C_{2,i}\cos(\alpha_i, l_i) + C_{3,i}\sinh(\alpha_i, l_i) + C_{4,i}\cosh(\alpha_i, l_i) \end{cases} \quad (5)$$

where $\varnothing_{ij}$ is the natural frequency and $\alpha_i^4 = \varnothing_i^4 \rho_i/(EI)_i$. Now, using the clamed free boundary condition for AMM [45,51], the constants in Eq. (5) are obtained as:

$$\begin{cases} C_{3,i} = -C_{1,i}, C_{4,i} = -C_{2,i} \\ [f(\alpha_i, l_i)] \begin{bmatrix} C_{1,i} \\ C_{2,i} \end{bmatrix} = 0 \end{cases} \quad (6)$$

The values of α_i can be obtained by solving Eqs. (5), (6), and also the finite solution of the link deformation using Eq. (2) can be obtained. Finally, using the Lagrangian expression (1), the dynamic model of motion equation of a TLFM obtained using AMM is described as [51,85].

$$B(\theta_i,\delta_i)\begin{bmatrix} \ddot{\theta}_i \\ \ddot{\delta}_i \end{bmatrix} + \begin{bmatrix} H_1(\theta_i,\delta_i,\dot{\theta}_i,\dot{\delta}_i) \\ H_2(\theta_i,\delta_i,\dot{\theta}_i,\dot{\delta}_i) \end{bmatrix} + K\begin{bmatrix} 0 \\ \delta_i \end{bmatrix} + D\begin{bmatrix} \dot{\theta}_i \\ \dot{\delta}_i \end{bmatrix} = \begin{bmatrix} \tau_i \\ 0 \end{bmatrix} \quad (7)$$

where

τ_i	Actuated torques
$\delta_i, \dot{\delta}_i$	Modal displacements and velocities
$\theta_i, \dot{\theta}_i$	Joint angle and velocity
B	Positive definite mass inertia matrix
H_1, H_2	Vectors of centrifugal and Coriolis forces
K	Positive definite stiffness matrix
D	Positive definite damping matrix

3 Singular Perturbation Modeling of A TLFM

In this section, the dynamics of a TLFM are divided into slow and fast subsystems. This is achieved by using the SP technique. The slow subsystem consists of the rigid body dynamics of the manipulator, and the fast subsystem consists of the flexible mode dynamics of the manipulator. Now, with the help of these subsystems, two separate controllers can be designed to obtain the desired performances.

The dynamics model in Eq. (7) of a TLFM can be rewritten as:

$$B\begin{bmatrix}\ddot{\theta}\\ \ddot{\delta}\end{bmatrix}+\begin{bmatrix}H_r+D_r\dot{\theta}\\ H_f+D_f\dot{\delta}+K\delta\end{bmatrix}=\begin{bmatrix}\tau_i\\ 0\end{bmatrix} \tag{8}$$

or in a simplified form as

$$\ddot{\theta}=-B_{11}\left(H_r+D_r\dot{\theta}\right)-B_{12}\left(H_f+D_f\dot{\delta}+K\delta\right)+B_{11}\tau_i \tag{9}$$

$$\ddot{\delta}=-B_{21}\left(H_r+D_r\dot{\theta}\right)-B_{22}\left(H_f+D_f\dot{\delta}+K\delta\right)+B_{21}\tau_i \tag{10}$$

where

$\theta=[\theta_1,\theta_2]^T\in R^2$	Vector of joint angle
$\delta=[\delta_{i1},\delta_{i2}]^T\in R^4$	Vector of flexible modes
$D_r\in R^{2\times 2}$, $D_f\in R^{4\times 4}$	Damping matrices
$K\in R^{4\times 4}$	Stiffness matrix
$H_r\in R^2$, $H_f\in R^4$	Matrices contacting gravitational, Coriolis, & centripetal forces
$B\in R^{6\times 6}$	Inertia matrix

The inertial matrix B can be represented as

$$B=\begin{bmatrix}B_{11} & B_{12}\\ B_{21} & B_{22}\end{bmatrix}=\begin{bmatrix}B_r & B_{rf}\\ \left(B_{rf}\right)^T & B_f\end{bmatrix}^{-1} \tag{11}$$

where $B_{11}\in R^{2\times 2}$, $B_{12}\in R^{2\times 4}$, $B_{21}\in R^{4\times 2}$, $B_{22}\in R^{4\times 4}$ and

$$B_r=\left[B_{11}-B_{12}(B_{22})^{-1}B_{21}\right]^{-1} \tag{12}$$

Considering new state variables $\delta = \varepsilon q$ and $K_s = \varepsilon K$, where ε is a new parameter which is defined as $\varepsilon = \frac{1}{K_m}$, K_m is the value of smallest stiffness. Using the new state variable, the singularly perturbed model of the flexible robot manipulator dynamics (8) can be presented as:

$$\ddot{\theta} = -B_{11}\left(H_r + D_r\dot{\theta}\right) - B_{12}\left(H_f + D_f\varepsilon\dot{q} + K_s q\right) + B_{11}\tau_i \tag{13}$$

$$\varepsilon\ddot{q} = -B_{21}\left(H_r + D_r\dot{\theta}\right) - B_{22}\left(H_f + D_f\varepsilon\dot{q} + K\delta\right) + B_{21}\tau_i \tag{14}$$

A composite control τ_i is described as:

$$\tau_i = \tau_s + \tau_f \tag{15}$$

where τ_s and τ_f are the slow and fast control inputs, respectively.

3.1 Dynamic Model of the Slow Subsystem

The slow subsystem dynamics of the FM in Eq. (7) are obtained by choosing $\varepsilon = 0$ in Eq. (14) and solving for q

$$\overline{q} = K_s^{-1}\left(\overline{B}_{22}\right)^{-1}\left(\overline{B}_{21}\overline{D}_r\dot{\overline{\theta}} + \overline{B}_{21}\overline{H}_r + \overline{B}_{22}\overline{H}_f - \overline{B}_{21}\tau_s\right) \tag{16}$$

where over-bar in the terms is obtained with $\varepsilon = 0$.

Substituting Eq. (16) in Eq. (13), we get

$$\ddot{\overline{\theta}} = \left(\overline{B}_{11} - \overline{B}_{12}\left(\overline{B}_{22}\right)^{-1}\overline{B}_{21}\right)\left(-\overline{H}_r - \overline{D}_r\dot{\overline{\theta}} + \tau_s\right) \tag{17}$$

which represents the rigid body dynamics of the flexible robot manipulator. Using Eq. (12), the dynamics of the slow subsystem are written as:

$$\ddot{\overline{\theta}} = \left(\overline{B}_r\right)^{-1}\left(-\overline{H}_r - \overline{D}_r\dot{\overline{\theta}} + \tau_s\right) \tag{18}$$

Now, in order to obtain the dynamics of the fast subsystem, a two-time scale method is used. Considering a fast time scale $t = \tau\sqrt{\varepsilon}$ and boundary correction terms $z_1 = q - \overline{q}$ and $z_2 = \sqrt{\varepsilon}\dot{q}$. Thus, using Eq. (14) the boundary layer system is written as:

$$\begin{cases} \dfrac{dz_1}{d\tau} = z_2 \\ \dfrac{dz_2}{d\tau} = -B_{21}\left(H_r + D_r\dot{\theta}\right) - B_{22}\left(H_f + D_f\varepsilon\dot{q} + K\delta\right) + B_{21}\tau_i \end{cases} \tag{19}$$

3.2 Dynamic Model of the Fast Subsystem

Using the SP method, the slow dynamics variables can be treated as negligible [54], thus, $\frac{d\overline{q}}{d\tau} = \sqrt{\varepsilon}\dot{\overline{q}} = 0$. Using Eq. (16) into Eq. (19) with $\varepsilon = 0$,we can write as

$$\frac{dz_2}{d\tau} = -\overline{B}_{22}K_s y_1 + B_{21}\tau_f \tag{20}$$

Or the dynamics of the fast subsystem are written in Eq. (21).

$$\dot{y} = A_f z + B_f \tau_f \tag{21}$$

where $z = [z_1, z_2]^T \in R^8$ and

$$A_f = \begin{bmatrix} 0 & 1 \\ -\overline{B}_{22}K_s & 0 \end{bmatrix}, B_f = \begin{bmatrix} 0 \\ \overline{B}_{21} \end{bmatrix} \tag{22}$$

The state space dynamics of the fast subsystem (21) represent the linear system with $\overline{\theta}$ as the parameter.

4 Design of A Composite Control

This section presents the design of a composite control input $\tau_i = \tau_s + \tau_f$ for the TLFM dynamics in Eq. (11). This is achieved by designing separate controllers τ_s and τ_f for the slow and fast subsystems, respectively.

4.1 Dynamic Surface Control of the Slow Subsystem

The dynamics surface control is designed to track the desired trajectory of the TLFM dynamics in Eq. (11). The controller is designed using the slow subsystem dynamics of TLFM (18).

Dynamics of the slow subsystem of TLFM are described as.

$$\ddot{\overline{\theta}} = \left(\overline{B}_r\right)^{-1}\left(-\overline{H}_r - \overline{D}_r\dot{\overline{\theta}} + \tau_s\right) \tag{23}$$

Consider, $x_1 = \overline{\theta}, x_2 = \dot{\overline{\theta}}$, the dynamics Eq. (23) can be written as:

$$\begin{cases} \dot{x}_1 = x_2 = \dot{\overline{\theta}} \\ \dot{x}_2 = \left(\overline{B}_r\right)^{-1}\left(-\overline{H}_r - \overline{D}_r x_2 + \tau_s\right) \end{cases} \tag{24}$$

The design of the dynamic surface control for the tracking control of the slow system is achieved in two stages.

Stage 1: In this stage, the goal is to design a surface to track the chaotic desired trajectory θ_d. Consider a surface (S_1) in Eq. (25).

$$S_1 = x_1 - x_{1d} \tag{25}$$

where x_{1d} is a twice differentiable desired trajectory and $\overline{\theta}$ is the states of the slow subsystem. Here x_{1d} is considered a signal of chaotic system, given in results and discussion section. The time derivative of Eq. (25) is given as:

$$\dot{S}_1 = x_2 - \dot{x}_{1d} \tag{26}$$

Now, using the idea of the design procedure of the dynamic surface control and introducing a low pass filter, we can write this as:

$$\overline{x}_2 = \dot{\theta}_d - k_1 s_1 \tag{27}$$

$$\tau_1 \dot{x}_{2d} + x_{2d} = \overline{x}_2, x_{2d}(0) = \overline{x}_2(0) \tag{28}$$

Stage 2: In this stage, the goal is to find the control input after designing of the surface. Consider a surface as:

$$S_2 = x_2 - x_{2d} \tag{29}$$

The time derivative of (29) results:

$$\dot{S}_2 = \dot{x}_2 - \dot{x}_{2d} \tag{30}$$

$$\dot{S}_2 = \left(\overline{B}_r\right)^{-1}\left(-\overline{H}_r - \overline{D}_r x_2 + \tau_s\right) - \dot{x}_{2d}$$

Now control torque input for slow subsystem can be obtained as:

$$\tau_s = \left(\overline{B}_r\right)^{-1}\left(\dot{x}_{2d} - k_2 S_2 + \left(\overline{B}_r\right)^{-1}\overline{H}_r\right) \tag{31}$$

where k_1, k_2 are the controller gains and τ_1 is the filter coefficient.

Next, we need to assign the controller gain (k_1, k_2) and filter coefficient (τ_1) to guarantee stability and error boundedness.

4.2 Backstepping Control for the Fast Subsystem

In this subsection, a backstepping control technique is designed for the fast subsystem.

Dynamics of the fast subsystem can be written as:

$$\begin{cases} \dot{z}_1 = z_2 \\ \dot{z}_2 = -A_{f3} z_1 + B_{f2} \tau_f \end{cases} \tag{32}$$

where $A_{f3} = \overline{B}_{22} K_s$ and $B_{f2} = \overline{B}_{21}$.

Suppose, z_d is a twice differentiable desired link deflection and v_d is a virtual control term. Link deflection errors are defined as:

$$e_{1ft} = z_d - z_1 \tag{33}$$

$$e_{2ft} = v_d - z_2 \tag{34}$$

The errors dynamics can be obtained as:

$$\dot{e}_{1ft} = \dot{z}_d - z_2 \tag{35}$$

$$\dot{e}_{2ft} = \dot{v}_d + A_{f3} z_1 - B_{f2} \tau_f \tag{36}$$

The control law designed for control of the fast subsystem is obtained using Theorem 1.

Theorem 1 *Suppose the backstepping control law defined in Eq. (37) using the error variable Eqs. (35), (36); then, the fast subsystem of manipulator dynamics Eq. (32) follow the desired trajectory z_d, that is, the link deflection of the manipulator is suppressed to zero properly.*

$$\tau_f = (B_{f2})^{-1}\left(\dot{v_d} + A_{f3}z_1 + k_{2b}e_{2ft}\right) \tag{37}$$

Proof. The design of a backstepping controller for the fast subsystem of TLFM (11) is achieved using the following steps:

Step 1: Consider a Lyapunov function candidate as:

$$v_{1f} = \frac{1}{2}e_{1ft}^2 \tag{38}$$

Time derivative of (38) results by using Eqs. (35), (36) as:

$$\dot{v}_{1ft} = e_{1f}\left(\dot{z}_d + e_{2ft} - v_d\right) = e_{1ft}\dot{z}_d - e_{1ft}v_d + e_{1ft}e_{2ft} \tag{39}$$

Now considering the virtual control variable v_d as:

$$v_d = \dot{z}_d + k_{1b}e_{1ft} + e_{2ft} \tag{40}$$

where $k_{1b} > 0$ is a positive definite matrix. Using Eq. (36), the derivative (40) can be written as:

$$\dot{v}_{1f} = -k_{1b}e_{1ft}^2 \tag{41}$$

It is seen from Eq. (41) that the time derivative of the Lyapunov function candidate $\dot{v}_{1f}$ is a negative definite function. Thus, the first state variable of the fast subsystem in Eq. (32) is stabilized. Next step is to show the stability of second state variable and to obtain the control input τ_f gap for the fast subsystem.

Step 2: Considering another Lyapunov function candidate as:

$$v_{2f} = v_{1f} + \frac{1}{2}e_{2ft}^2 \tag{42}$$

Using Eqs. (36), (41), the time derivative of Eq. (42) can be written as:

$$\dot{v}_{2f} = -k_{1b}e_{1ft}^2 + e_{2ft}\left(\dot{v_d} + A_{f3}z_1 - B_{f2}\tau_f\right) \tag{43}$$

Now, we can obtain the actual torque input as:

$$\tau_f = (B_{f2})^{-1}\left(\dot{v_d} + A_{f3}z_1 + k_{2b}e_{2ft}\right) \tag{44}$$

Using Eq. (44), $\dot{v}_{2f}$ in Eq. (43) is given as:

$$\dot{v}_{2f} = -\left(k_{1b}e_{1ft}^2 + k_{2b}e_{2ft}^2\right) \tag{45}$$

Because k_{1b}, k_{2b} are positive constant matrices, then, using the Lyapunov stability theory, we can say that Eq. (45) is a negative definite function.

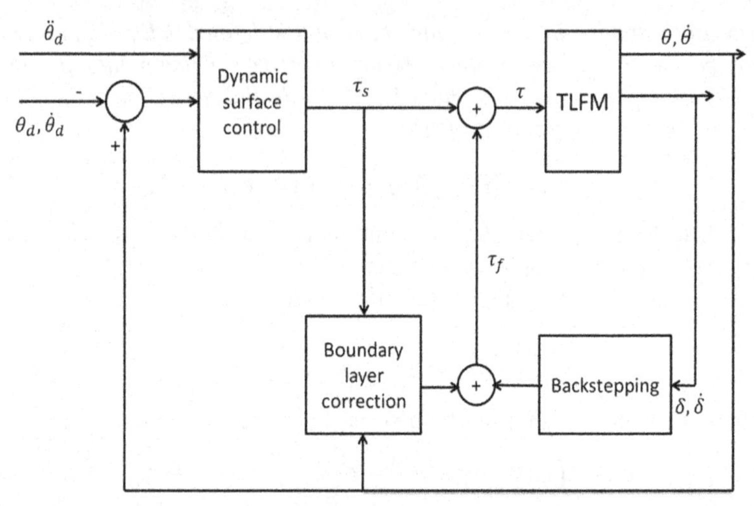

FIG. 2 Block diagram of the designed composite control technique.

Thus, the error variables e_{1ft} and e_{2ft} asymptotically converge to origin with suitable choice of constant matrices k_{1b}, k_{2b}. Therefore, the links' deflection of the fast subsystem of the TLFM is supressed to their desired values, that is, at zero.

The structure of the composite control technique designed for the chaotic trajectory tracking is presented in Fig. 2.

5 Chaotic Signal as the Desired Trajectory

The dynamics of a chaotic system whose signal is used as the desired trajectory for the TLFM are described as [88]:

$$\begin{cases} \dot{y}_1 = y_2 \\ \dot{y}_2 = y_3 \\ \dot{y}_3 = -0.44y_3 - 2y_2 + y_1^2 - 1 \end{cases} \tag{46}$$

The system in Eq. (46) is chaotic with initial conditions $y(0) = (0,0,0)^T$ where Lyapunov exponents are $L_i = (0.105, 0, -0.545)$. The chaotic attractors and chaotic signals of system (46) with initial conditions $y(0) = (0,0,0)^T$ are shown in Figs. 3 and 4, respectively.

In this paper, the signals y_1, y_2, y_3 are used as the desired trajectories for the TLFM as $\theta_{1d} = \theta_{2d} = y_1$, $\dot{\theta}_{1d} = \dot{\theta}_{2d} = y_2$ and $\ddot{\theta}_{1d} = \ddot{\theta}_{2d} = y_3$.

6 Results and Discussion for the Composite Control

This section discusses the results and discussion on the chaotic trajectory tracking control of the TLFM. The parameters of the TLFM used for system (7) are given in Table 4.

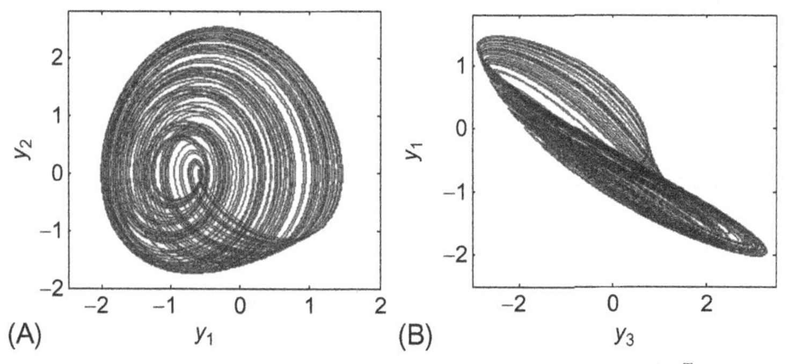

FIG. 3 Chaotic attractors in different planes of system (46) with $y(0) = (0,0,0)^T$.

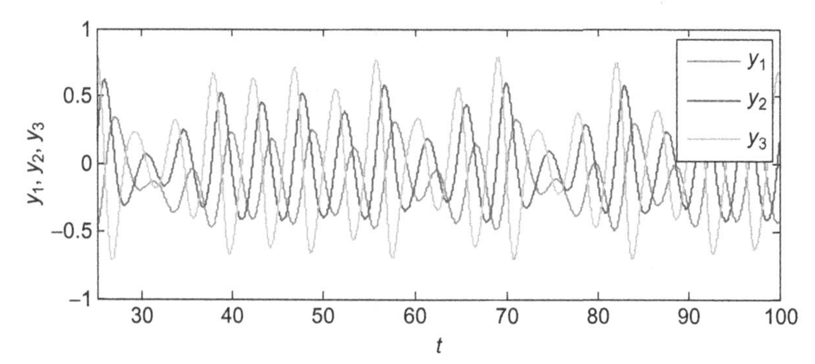

FIG. 4 Signals of chaotic system (46).

The initial condition considered for simulating the TLFM dynamics in Eq. (7) and adaptation laws (35) are $(\theta(0), \delta(0)) = (0.1, 0.1, 0.0, 0.0, 0.0, 0.0)^T$, $(\dot{\theta}(0), \dot{\delta}(0)) = (0, 0, 0, 0, 0, 0)^T$. The value of the gains used for the dynamic surface control and the backstepping controller is $k_1 = \begin{bmatrix} 7 & 0 \\ 0 & 7 \end{bmatrix}$, $k_2 = \begin{bmatrix} 7 & 0 \\ 0 & 7 \end{bmatrix}$, $k_{1b} = \begin{bmatrix} 10 & 0 & 0 & 0 \\ 0 & 10 & 0 & 0 \\ 0 & 0 & 10 & 0 \\ 0 & 0 & 0 & 10 \end{bmatrix}$, $k_{2b} = \begin{bmatrix} 5 & 0 & 0 & 0 \\ 0 & 5 & 0 & 0 \\ 0 & 0 & 5 & 0 \\ 0 & 0 & 0 & 5 \end{bmatrix}$. These gains are considered in a manner to achieve better tracking performances but using less control effort.

6.1 Simulation Results With the Nominal Payload (0.145 kg)

Chaotic trajectory tracking with the nominal payload (0.145 kg) for the two-link FM (7) is discussed here.

Fig. 5 shows the chaotic trajectory tracking for both the links of the TLFM. It is observed from Fig. 6 that the chaotic trajectory tracking is achieved within 0.5 s. The modes of links with the nominal payload of 0.145 kg are shown in Figs. 6

TABLE 4 Parameters of a Physical Two-Link Flexible Manipulator [89]

Mass of link 1, $m_1 = 0.15268$ kg	**Coefficients of viscous damping, $B_{eq1} = 4$ Nms/rad, $B_{eq2} = 1.5$ Nms/rad**
Mass of link 2, $m_2 = 0.0535$ kg	Efficiency of gear boxes, $\eta_{g1} = 0.85$, $\eta_{g2} = 0.9$
Length of link 1, $L_1 = 0.201$ m	Efficiency of motors, $\eta_{m1} = 0.85$, $\eta_{m2} = 0.85$
Length of link 2, $L_2 = 0.201$ m	Constants of back emf, $K_{m1} = 0.119$ v/rad, $K_{m2} = 0.0234$ v/rad
Resistance of armatures, $R_{m1} = 11.5$ Ω, $R_{m2} = 2.32$ Ω	Gear ratio, $K_{g1} = 100$, $K_{g2} = 50$
Equivalent MI at load, $J_{eq1} = 0.17043$ kgm^2	Motor torque constants $K_{t1} = 0.119$ Nm/A, $K_{t2} = 0.0234$ Nm/A
Equivalent MI at load, $J_{eq2} = 0.0064387$ kgm^2	Stiffness of the links, $K_{s1} = 22$ Nm/rad, $K_{s2} = 2.5$ Nm/rad
Link-1 MI, $J_{arm1} = 0.002035$ kgm^2	Link-2 MI, $J_{arm2} = 0.0007204$ kgm^2

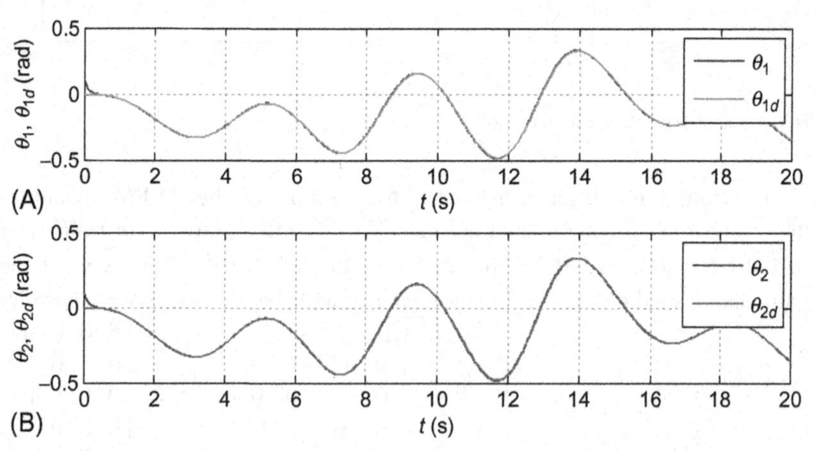

FIG. 5 Chaotic trajectory tracking of the joint angles with the nominal payload for both the links.

and 7. It is noted from Figs. 6 and 7 that the flexible modes are supressed quickly and are of low value. The nature of the tip deflections for both the links are given in Fig. 8. It is observed from Fig. 8 that the tip deflections of both the links are supressed within 10^{-2} and 10^{-3} mm. The required control torque inputs in the slow subsystem are shown in Fig. 9. The control inputs required in the fast subsystem are shown in Fig. 10. The nature of the composite control inputs consisting of dynamic surface control and backstepping control are given in

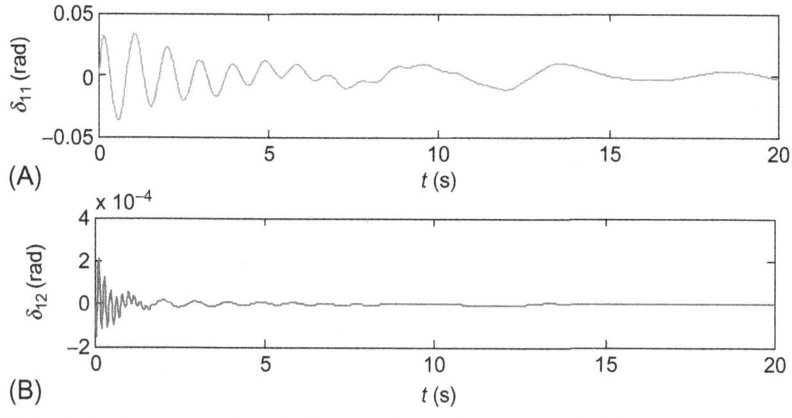

FIG. 6 Behavior of the modes of link-1 with the nominal payload of 0.145 kg.

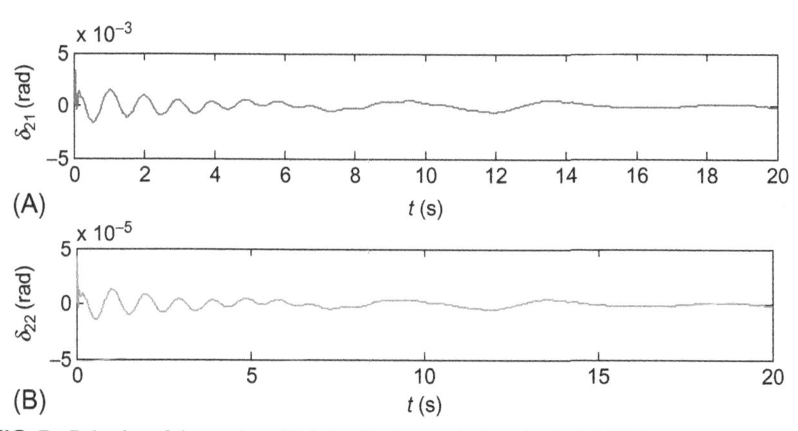

FIG. 7 Behavior of the modes of link-2 with the nominal payload of 0.145 kg.

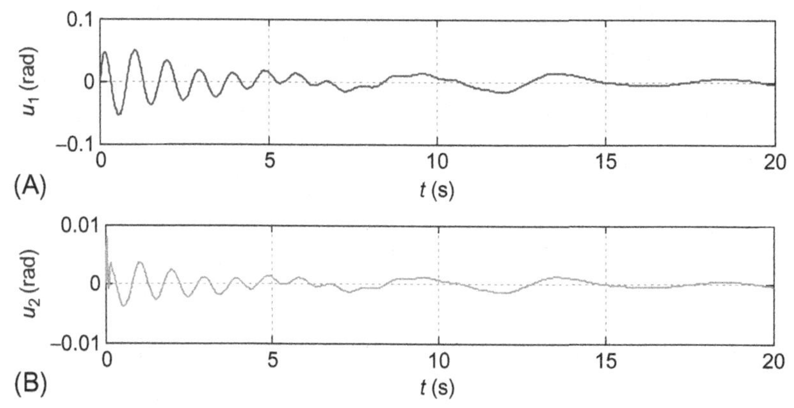

FIG. 8 Behavior of tip deflections of both the link for the chaotic trajectory tracking with the nominal payload of 0.145 kg.

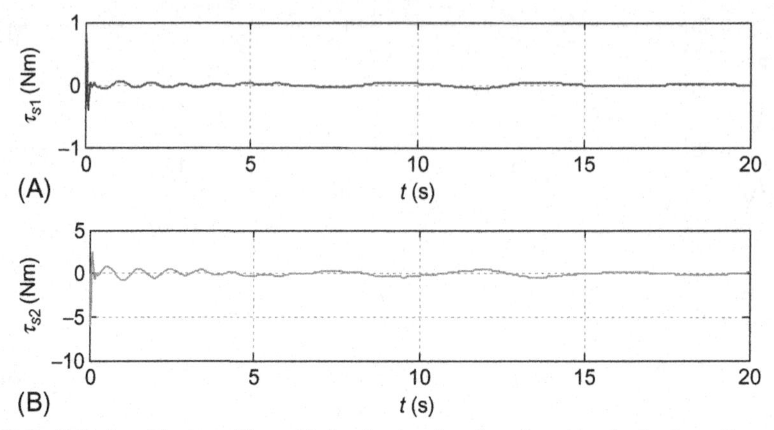

FIG. 9 Behavior of the control inputs during the chaotic trajectory tracking for the slow subsystem with the nominal payload of 0.145 kg.

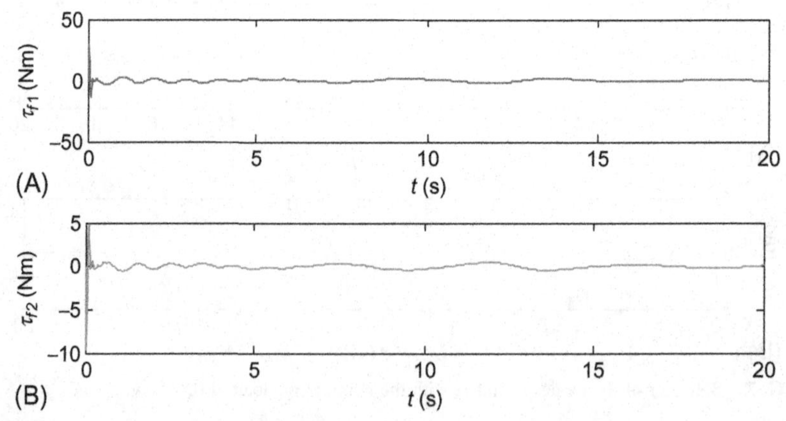

FIG. 10 Behavior of the control inputs during the chaotic trajectory tracking for the fast subsystem with the nominal payload of 0.145 kg.

Fig. 11. It is noted from the Fig. 11 that the required control inputs using the composite control inputs for both the links are initially high, but after some time, decayed within small values.

6.2 Chaotic Trajectory Tracking With a 0.3 kg Payload

This subsection describes the robustness of the composite controllers designed for chaotic tracking of the two-link FM in Eq. (7) with a payload of 0.3 kg. The chaotic signal tracking of both the links with a 0.3 kg payload is shown in Fig. 12. The behaviors of modes of both the links with a 0.3 kg payload are

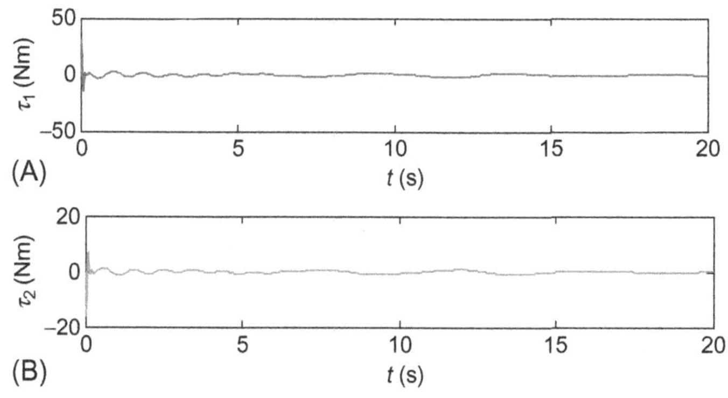

FIG. 11 Behavior of the composite control inputs during the chaotic trajectory tracking with the nominal payload of 0.145 kg.

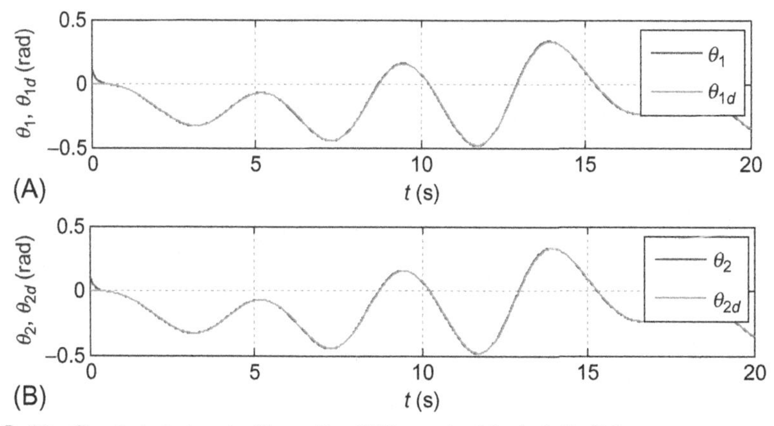

FIG. 12 Chaotic trajectory tracking with a 0.3 kg payload for both the links.

shown in Figs. 13 and 14. The tip deflections of both the links are shown in Fig. 15. It is noted from Fig. 15 that with an increase of a payload, there is a small increase in the magnitude of deflection for both the links. Behavior of the control inputs used for the slow and fast subsystem with the increase in payload 0.3 kg payload are shown in Figs. 16 and 17. Behavior of the composite control inputs with a 0.3 kg payload is shown in Fig. 18. It is noted from Fig. 18 that with an increase in the payload, the required control inputs are more comparable with the nominal payload.

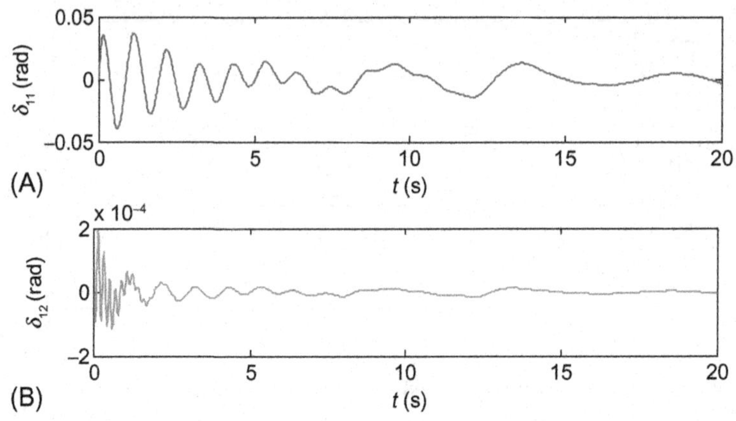

FIG. 13 Behavior of the modes of the link-1 with a 0.3 kg payload.

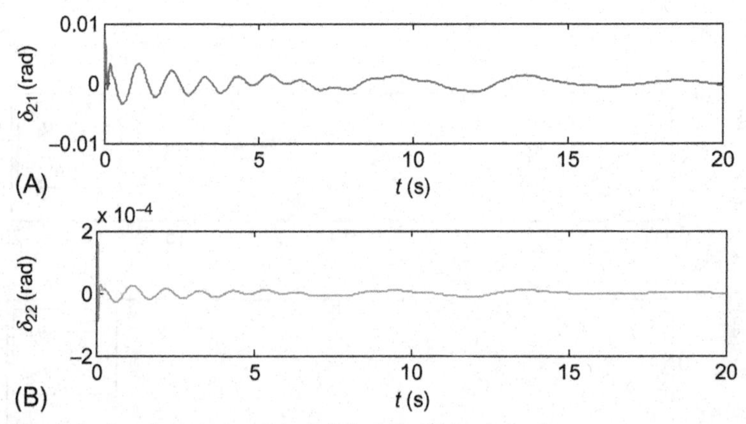

FIG. 14 Behavior of the modes of the link-2 with a 0.3 kg payload.

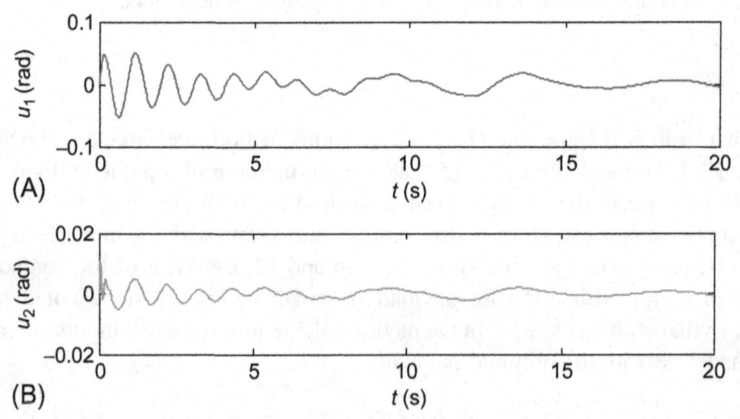

FIG. 15 Behavior of tip deflections of both the links for the chaotic trajectory tracking with a 0.3 kg payload.

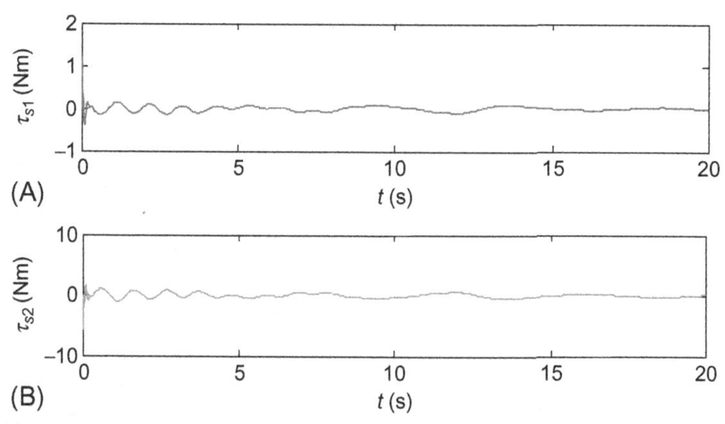

FIG. 16 Behavior of the control inputs during the chaotic trajectory tracking for the slow subsystem with payload 0.3 kg.

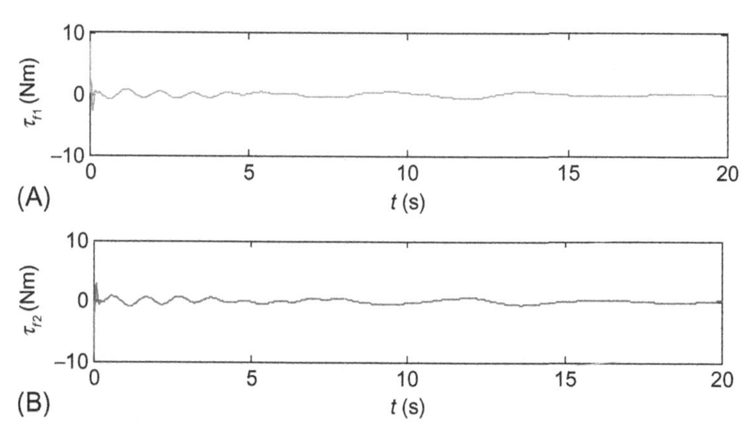

FIG. 17 Behavior of the control inputs during the chaotic trajectory tracking for the fast subsystem with payload 0.3 kg.

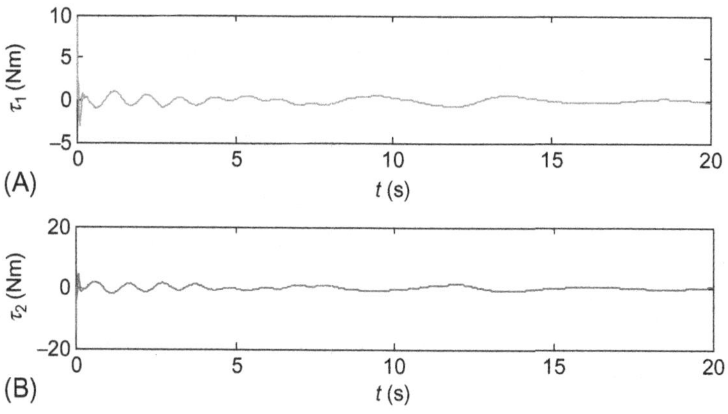

FIG. 18 Behavior of the composite control inputs during the chaotic trajectory tracking with payload 0.3 kg.

7 Conclusions

In this chapter, the chaotic path planning (trajectory tracking) problem for a two-link FM is reported. We know that the dynamics of a two-link FM are nonlinear and complex, making the chaotic trajectory tracking (path planning) for such a system challenging. Therefore, the dynamics of a two-link FM are divided into two subsystems consisting of slow and fast subsystems. The slow subsystem deals with rigid body dynamics of the manipulator, and the fast subsystem deals with the flexible mode dynamics of the manipulator. Here, this is achieved with the help of a two-time scale separation principle using the SP method. Then, a composite controller is designed using the controller of each subsystem. The composite controller consists of a dynamic surface control in the slow subsystem for the chaotic trajectory tracking and a backstepping controller in the fast subsystem for quick suppression of link deflection. The MATLAB simulation results reveal that the proposed composite controller has better performance in terms of smaller steady-state errors, quick and smaller tip deflection, and less control effort.

REFERENCES

[1] Lassoued A, Boubaker O. Dynamic analysis and circuit design of a novel hyperchaotic system with fractional-order terms. Complexity 2017;2017:1–11. https://doi.org/10.1155/2017/3273408.

[2] Pham V, Jafari S, Volos C, Kapitaniak T. A gallery of chaotic systems with an infinite number of equilibrium points. Chaos, Solitons Fractals 2016;93:58–63.

[3] Lassoued A, Boubaker O. On new chaotic and hyperchaotic systems: a literature survey. Nonlinear Anal Modell Control 2016;21(6):770–89.

[4] Dudkowski D, Jafari S, Kapitaniak T, Kuznetsov NV, Leonov GA, Prasad A. Hidden attractors in dynamical systems. Phys Rep 2016;637:1–50.

[5] Tlelo-Cuautle E, Carbajal-Gomez VH, Obeso-Rodelo PJ, Rangel-Magdaleno JJ, Núñez-Pérez JC. FPGA realization of a chaotic communication system applied to image processing. Nonlinear Dynamics 2015;82:1879–92. https://doi.org/10.1007/s11071-015-2284-x.

[6] Esteban T-C, Rangel MJdJ, de la Fraga LG. Engineering applications of FPGAs: chaotic systems, artificial neural networks, random number generators, and secure communication systems. Switzerland: Springer; 2016.

[7] Nichols JM, Todd MD, Wait JR. Using state space predictive modeling with chaotic interrogation in detecting joint preload loss in a frame structure. Smart Mater Struct 2003;12:580–601.

[8] Pham V-T, Volos C, Jafari S, Wei Z, Wang X. Constructing a novel non-equilibrium chaotic system. Int J Bifurc Chaos 2014;24. https://doi.org/10.1142/S0218127414500734.

[9] Andrievskii BR, Fradkov AL. Control of Chaos: methods and applications. II. Applications. Autom Remote Control 2004;65:505–33.

[10] Tlelo-Cuautle E, Ramos-López HC, Sánchez-Sánchez M, Pano-Azucena AD, Sánchez-Gaspariano LA, Núñez-Pérez JC, Camas-Anzueto JL. Application of a chaotic oscillator in an autonomous mobile robot. J Electr Eng 2014;65:157–62. https://doi.org/10.2478/jee-2014-0024.

[11] Lochan K, Roy BK, Subudhi B. SMC controlled chaotic trajectory tracking of two-link flexible manipulator with PID sliding surface. IFAC-Papers OnLine 2016;49:219–24. https://doi.org/10.1016/j.ifacol.2016.03.056.
[12] Singh JP, Lochan K, Kuznetsov NV, Roy BK. Coexistence of single- and multi-scroll chaotic orbits in a single-link flexible joint robot manipulator with stable spiral and index-4 spiral repellor types of equilibria. Nonlinear Dynamics 2017;90:1277–99. https://doi.org/10.1007/s11071-017-3726-4.
[13] Leonov GA, Kuznetsov NV, Vagaitsev VI. Localization of hidden Chuas attractors. Phys Lett A 2011;375:2230–3.
[14] Leonov GA, Kuznetsov NV, Vagaitsev VI. Hidden attractor in smooth Chua systems. Physica D 2012;241:1482–6.
[15] Leonov GA, Kuznetsov NV. Hidden attractors in dynamical systems: from hidden oscillations in Hilbert-Kolmogorov, Aizerman, and Kalman problems to hidden chaotic attractor in Chua circuits. Int J Bifurc Chaos 2013;23:1330002.
[16] Leonov GA, Kuznetsov NV, Kuznestova OA, Seledzhi SM, Vagaitsev VI. Hidden oscillations in dynamical systems system. Trans Syst Control 2011;6:1–14.
[17] Leonov GA, Kuznetsov NV, Kiseleva MA, Solovyeva EP, Zaretskiy AM. Hidden oscillations in mathematical model of drilling system actuated by induction motor with a wound rotor. Nonlinear Dynamics 2014;77:277–88.
[18] Leonov GA, Kuznetsov NV, Mokaev TN. Hidden attractor and homoclinic orbit in Lorenz-like system describing convective fluid motion in rotating cavity. Commun Nonlinear Sci Numer Simul 2015;28:166–74. https://doi.org/10.1016/j.cnsns.2015.04.007.
[19] Singh JP, Roy BK. A new four-dimensional chaotic system with first Lyapunov exponent $\approx$ 22, hyperbolic curve and circular paraboloid types of equilibria and its switching synchronization by an adaptive global integral sliding mode control. Chinese Phys B 2018;27:040503–17. https://doi.org/10.1088/1674-1056/27/4/040503.
[20] Singh JP, Roy BK. Multistability and hidden chaotic attractors in a new simple 4-D chaotic system with chaotic 2-torus behaviour. Int J Dyn Control 2017. https://doi.org/10.1007/s40435-017-0332-8.
[21] Pham VT, Jafari S, Volos C, Giakoumis A, Vaidyanathan S, Kapitaniak T. A chaotic system with equilibria located on the rounded square loop and its circuit implementation. IEEE Trans Circuits Syst II Express Briefs 2016;63:878–82.
[22] Jafari S, Sprott JC. Simple chaotic flows with a line equilibrium. Chaos, Solitons Fractals 2013;57:79–84. https://doi.org/10.1016/j.chaos.2013.08.018.
[23] Singh JP, Roy BK. The simplest 4-D chaotic system with line of equilibria, chaotic 2-torus and 3-torus behaviour. Nonlinear Dynamics 2017;89:1845–62. https://doi.org/10.1007/s11071-017-3556-4.
[24] Singh JP, Roy BK. Coexistence of asymmetric hidden chaotic attractors in a new simple 4-D chaotic system with curve of equilibria. Optik (Stuttg) 2017;145:209–17. https://doi.org/10.1007/s40435-017-0332-8.
[25] Singh JP, Roy BK, Jafari S. New family of 4-D hyperchaotic and chaotic systems with quadric surfaces of equilibria. Chaos, Solitons Fractals 2017;106:243–57.
[26] Barati K, Jafari S, Sprott JC, Pham V. Simple chaotic flows with a curve of equilibria. Int J Bifurc Chaos 2016;26:1630034–40.
[27] Jafari S, Sprott JC, Molaie M. A simple chaotic flow with a plane of equilibria. Int J Bifurc Chaos 2016;26:1650098–104.
[28] Jafari S, Sprott JC, Pham V, Volos C, Li C. Simple chaotic 3D flows with surfaces of equilibria. Nonlinear Dynamics 2016;86:1349–58.

[29] Lorenz EN. Deterministic nonperiodic flow. J Atmos Sci 1963;20:130–41.

[30] Chen G, Ueta T. Yet another chaotic attractor. Int J Bifurc Chaos 1999;9:14651999. https://doi.org/10.1142/S0218127499001024.

[31] Lü J, Chen G, Cheng D, Celikovsky S. Bridge the gap between the Lorenz system and the Chen system. Int J Bifurc Chaos 2002;12:2917–26.

[32] Sprott JC. Some simple chaotic flow. Phys Rev E 1994;50:647–50.

[33] Singh JP, Roy BK. The nature of Lyapunov exponents is (+ , + , − , −). Is it a hyperchaotic system? Chaos, Solitons Fractals 2016;92:73–85.

[34] Singh JP, Roy BK. Crisis and inverse crisis route to chaos in a new 3-D chaotic system with saddle, saddle foci and stable node foci nature of equilibria. Optik (Stuttg) 2016;127:11982–2002.

[35] Singh JP, Roy BK. Comment on "theoretical analysis and circuit verification for fractional-order chaotic behavior in a new hyperchaotic system" Math Probl Eng 2016;2014:1–4. https://doi.org/10.1155/2014/682408.

[36] Singh JP, Roy BK. Analysis of an one equilibrium novel hyperchaotic system and its circuit validation. Int J Control Theory Appl 2015;8:1015–23.

[37] Singh JP, Roy BK. A novel asymmetric hyperchaotic system and its circuit validation. Int J Control Theory Appl 2015;8:1005–13.

[38] Singh PP, Singh JP, Roy BK. Synchronization and anti-synchronization of Lu and Bhalekar-Gejji chaotic systems using nonlinear active control. Chaos, Solitons Fractals 2014;69:31–9.

[39] Kiseleva M, Kondratyeva N, Kuznetsov N, Leonov G. Hidden oscillations in electromechanical systems. In: Dyn control adv struct mach. Berlin Heidelberg: Springer-Verlag; 2017. p. 119–24. https://doi.org/10.1007/978-3-319-43080-5.

[40] Lochan K, Roy BK, Subudhi B. A review on two-link flexible manipulators. Annu Rev Control 2016;42:346–67. https://doi.org/10.1016/j.arcontrol.2016.09.019.

[41] Sabatini M, Gasbarri P, Monti R, Palmerini GB. Vibration control of a flexible space manipulator during on orbit operations. Acta Astronaut 2012;73:109–21. https://doi.org/10.1016/j.actaastro.2011.11.012.

[42] Arora A, Ambe Y, Kim TH, Ariizumi R, Matsuno F. Development of a maneuverable flexible manipulator for minimally invasive surgery with varied stiffness. Artif Life Robot 2014;19:340–6. https://doi.org/10.1007/s10015-014-0184-7.

[43] Nakamura T, Saga N, Nakazawa M, Kawamura T. Development of a soft manipulator using a smart flexible joint for safe contact with humans. In: IEEE/ASME int conf adv intell mechatronics. AIM; 2003. p. 441–6. https://doi.org/10.1109/AIM.2003.1225136.

[44] Kiang CT, Spowage A, Yoong CK. Review of control and sensor system of flexible manipulator. J Intell Robot Syst Theory Appl 2014;77:187–213. https://doi.org/10.1007/s10846-014-0071-4.

[45] Pradhan SK, Subudhi B. Nonlinear adaptive model predictive controller for a flexible manipulator: an experimental study. IEEE Trans Control Syst Technol 2014;22:1–15. https://doi.org/10.1109/TCST.2013.2294545.

[46] Zhang L, Liu J. Observer-based partial differential equation boundary control for a flexible two-link manipulator in task space. IET Control Theory Appl 2012;6:2120–33. https://doi.org/10.1049/iet-cta.2011.0545.

[47] Pradhan SK, Subudhi B. Real-time adaptive control of a flexible manipulator using reinforcement learning. IEEE Trans Autom Sci Eng 2012;9:237–49. https://doi.org/10.1109/TASE.2012.2189004.

[48] Lee SH, Lee CW. Hybrid control scheme for robust tracking of two-link flexible manipulator. J Intell Robot Syst 2002;34:431–52. https://doi.org/10.1023/A:1014286327294.

[49] Li Y, Liu G, Hong T, Liu K. Robust control of a two-link flexible manipulator with quasi-static deflection compensation using neural networks. J Intell Robot Syst 2005;44:263–76. https://doi.org/10.1007/s10846-005-9019-z.

[50] Lochan K, Roy BK, Subudhi B. Generalized projective synchronization between controlled master and multiple slave TLFMs with modified adaptive SMC. Trans Inst Meas Control 2016;https://doi.org/10.1177/0142331216674067.

[51] Subudhi B, Morris AS. Dynamic modelling, simulation and control of a manipulator with flexible links and joints. Robot Auton Syst 2002;41:257–70. https://doi.org/10.1016/S0921-8890(02)00295-6.

[52] Lochan K, Roy BK. Position control of two-link flexible manipulator using low chattering SMC techniques. Int J Control Theory Appl 2015;8:1137–46.

[53] Korayem M, Haghpanahi M. Finite element method and optimal control theory for path planning of elastic manipulators. New Adv Intell Decis Technol 2009;199:117–26. https://doi.org/10.1007/978-3-642-00909-9_12.

[54] Siciliano B, Book WJ. A singular perturbation approach to control of lightweight flexible manipulators.

[55] Özer A, Semercigil SE. Effective vibration supression of maneuvering two-link flexible arm with an event-based stiff. In: Proc IMAC-XXVIII. 2010. p. 323–30. https://doi.org/10.1007/978-1-4419-9834-7.

[56] Chu Z, Cui J. Experiment on vibration control of a two-link flexible manipulator using an input shaper and adaptive positive position feedback. Adv Mech Eng 2015;7:1–13. https://doi.org/10.1177/1687814015610466.

[57] Karagulle H, Malgaca L, Dirilmis M, Akdag M, Yavuz S. Vibration control of a two-link flexible manipulator. J Vib Control 2015;https://doi.org/10.1177/1077546315607694.

[58] Lochan K, Roy BK, Subudhi B. A review on two-link flexible manipulators. Annu Rev Control 2016;42:346–67. https://doi.org/10.1016/j.arcontrol.2016.09.019.

[59] Zhang L, Liu J. Observer-based partial differential equation boundary control for a flexible two-link manipulator in task space. IET Control Theory Appl 2012;6:2120–33. https://doi.org/10.1049/iet-cta.2011.0545.

[60] Lochan K, Roy BK. Trajectory tracking control of an AMM modelled TLFM using backstepping method. Int J Control Theory Appl 2016;9:239–46.

[61] Yu Y, Yuan Y, Fan X, Yang H. Back-stepping control of two-link flexible manipulator based on extended state observer. Adv Sp Res 2015;56:2312–22.

[62] Lochan K, Suklabaidya S, Roy BK. Sliding mode and adaptive sliding mode control approaches of two link flexible manipulator. 2nd conf adv robot, BIts, Goa; 2015https://doi.org/10.1145/2783449.2783508.

[63] Lochan K, Suklabaidya S, Roy BK. Comparison of chattering in single link flexible manipulator with sliding mode controllers. 2015 int conf energy, power environ towar sustain growth, 2015; 2015. https://doi.org/10.1109/EPETSG.2015.7510106.

[64] Lochan K, Roy BK. Position control of two-link flexible manipulator using low chattering SMC techniques position control of two-link flexible manipulator using low chattering SMC techniques. Int J Control Theory Appl 2015;8:1137–45.

[65] Lochan K, Roy BK, Subudhi B. SMC controlled chaotic trajectory tracking of two-link flexible manipulator with PID sliding. IFAC-Papers OnLine 2016;49:219–24. https://doi.org/10.1016/j.ifacol.2016.03.056.

[66] Lochan K, Roy BK, Subudhi B. Generalized projective synchronization between controlled master and multiple slave TLFMs with modified adaptive SMC. Trans Inst Meas Control 2016;1–23. https://doi.org/10.1177/0142331216674067.

[67] Singh JP, Roy BK. Second order adaptive time varying sliding mode control for synchronization of hidden chaotic orbits in a new uncertain 4-D conservative chaotic system. Trans Inst Meas Control 2017;1–14. https://doi.org/10.1177/0142331217727580.

[68] Lochan K, Roy BK, Subudhi B. Robust tip trajectory synchronisation between assumed modes modelled two-link flexible manipulators using second-order PID terminal SMC. Robot Auton Syst 2017;97:108–24.

[69] Lochan K, Roy BK. Second-order SMC for tip trajectory tracking and tip deflection suppression of an AMM modelled nonlinear TLFM. Int J Dyn Control 2017;1–25. https://doi.org/10.1007/s40435-017-0371-1.

[70] Pradhan SK, Subudhi B. Real-time adaptive control of a flexible manipulator using reinforcement learning. IEEE Trans Autom Sci Eng 2012;9:237–49. https://doi.org/10.1109/TASE.2012.2189004.

[71] Subudhi B, Morris AS. Soft computing methods applied to the control of a flexible robot manipulator. Appl Soft Comput 2009;9:149–58. https://doi.org/10.1016/j.asoc.2008.02.004.

[72] Khorrami F, Jain S, Tzes A. Experiments on rigid body-based controllers with input preshaping for a two-link flexible manipulator. IEEE Trans Robot Autom 1994;10:55–65. https://doi.org/978 0 7340 3893 7.

[73] Khorrami F, Jain S. Non-linear control with end-point acceleration feedback for a two-link flexible manipulator: Experimental results. J Robot Syst 1993;10:505–30. https://doi.org/10.1002/rob.4620100407.

[74] Bo XU, Bakakawa Y. Control two-link flexible manipulators using controlled Lagrangian method. SICE annu conf Sapporo; 2004. p. 289–94.

[75] Li Y, Liu G, Hong T, Liu K. Robust control of a two-link flexible manipulator with quasi-static deflection compensation using neural networks. J Intell Robot Syst 2005;44:263–76. https://doi.org/10.1007/s10846-005-9019-z.

[76] Zhang Y, Mi Y, Zhu M, Lu F. Adaptive sliding mode control for two-link flexible manipulator with H infinity tracking. Proc fourth int conf mach learn cybern, Guangzhou; 2005. p. 702–7.

[77] Matsuno F, Yamamoto K. Dynamic hybrid force/position control of two degree of freedom flexible manipulator. J Robot Syst 1994;11:355–66.

[78] Wang Y, Feng Y, Yu X. In: Fuzzy terminal sliding mode control of two-link flexible manipulators. 34th IEEE annu conf ind electron; 2008. p. 1620–5. https://doi.org/10.1109/IECON.2008.4758196.

[79] Ashayeri A, Farid M. In: Trajectory tracking for two-link flexible arm via two-time scale and boundary control methods. Proc IMECE2008 2008 ASME Int Mech Eng Congr Expo; 2008. p. 1–9.

[80] Li YC, Tang BJ, Shi ZX, Lu YF. Experimental study for trajectory tracking of a two-link flexible manipulator. Int J Syst Sci 2010;31:3–9. https://doi.org/10.1080/002077200291398.

[81] Mirzaee E, Eghtesad M, Fazelzadeh SA. Maneuver control and active vibration suppression of a two-link flexible arm using a hybrid variable structure/Lyapunov control design. Acta Astronaut 2010;67:1218–32. https://doi.org/10.1016/j.actaastro.2010.06.054.

[82] Yue-jiao D, Xi C, Ming Z, Jun R. Anti-windup for two-link flexible arms with actuator saturation using neural network. 2010 int conf E-product E-service E-entertainment (ICEEE); 2010. p. 6–9.

[83] Wang Y, Han F, Feng Y, Xia H. Hybrid continuous nonsingular terminal sliding mode control of uncertain flexible manipulators. 40th IEEE annu conf ind electron soc; 2014. p. 190–6. https://doi.org/10.1109/IECON.2009.5415316.

[84] Lochan K, Dey R, Roy BK, Subudhi B. Tracking control with Vibration suppression of a two-link flexible manipulator using singular perturbation with composite control design.

In: Balas VE, Jain LC, editors. Soft comput appl SOFA 2016. Adv intell syst comput. Berlin Heidelberg: Springer; 2018. p. 365–77.
[85] De Luca A, Siciliano B. Closed-form dynamic model of planar multilink lightweight robots. IEEE Trans Syst Man Cyber 1991;21:826–39. https://doi.org/10.1109/21.108300.
[86] Subudhi B, Ranasingh S, Swain AK. Evolutionary computation approaches to tip position controller design for a two-link flexible manipulator. Arch Control Sci 2011;21:269–85. https://doi.org/10.2478/v10170-010-0043-2.
[87] Subudhi B, Pradhan SK. A flexible robotic control experiment for teaching nonlinear adaptive control. Int J Electr Eng Educ 2016;https://doi.org/10.1177/0020720916631159.
[88] Molaie M, Jafari S, Sprott JC, Golpayegani SMRH. Simple chaotic flows with one stable equilibrium. Int J Bifurc Chaos 2013;23. https://doi.org/10.1142/S0218127413501885.
[89] QUANSER. Equation for the frist (second) stage of the 2DOF serial flexible link robot. QUANSER; 2006.

Part III

New Trends in Chaos Synchronization

Chapter 13

Robust Synchronization of Master Slave Chaotic Systems: A Continuous Sliding-Mode Control Approach With Experimental Study

Hafiz Ahmed*, Héctor Ríos† and Ivan Salgado‡

*School of Mechanical, Aerospace and Automotive Engineering, Coventry University, Coventry, United Kingdom, †División de Estudios de Posgrado e Investigación, CONACYT—Tecnológico Nacional de México/I.T. La Laguna, Torreón, Coahuila, México, ‡Centro de Innovación y Desarrollo Tecnológico en Cómputo, Instituto Politécnico Nacional, Mexico City, Mexico

1 INTRODUCTION

Over the centuries, oscillators have attracted the attention of researchers in various scientific disciplines. An oscillating behavior is pervasive in nature, technology, and human society. Oscillation represents repetitive or periodic processes and has several remarkable features [1–4]. Chaotic oscillators are a particular class of nonlinear oscillators. Chaotic oscillators are very sensitive to change in parameters or initial conditions. For chaotic systems, it is frequently very difficult to predict the asymptotic regime on which this system will attain asymptotically for the given set of initial conditions. Chaotic behavior exists in many natural and artificial systems, for example, weather and climate [5], road traffic [6], Ferroresonant overvoltages or undervoltages in electric power distribution system [7], randomization in network routing [8], multiphase chemical reactors [9], eye movement signals [10]. As such the study of chaotic behavior attracted lot of attention in recent decades [11, 12].

Although chaotic behavior is difficult to predict, it is very useful in several application areas. As of today, numerous successful applications have been

Recent Advances in Chaotic Systems and Synchronization. https://doi.org/10.1016/B978-0-12-815838-8.00013-3

reported in the literature related to chaos or chaos synchronization, for example, road traffic prediction [13], PID controller parameter tuning [14], rolling bearing performance monitoring [15], secure communication in OFDM systems [16].

Master-slave synchronization is a major focus in the study of chaotic system. Master-slave synchronization in general is very useful in several application areas, for example, bilateral teleoperations [17]. In this case, master system moves freely while slave system needs to follow the master system using the information available from the master. Because this problem has numerous applications, it has been rigorously studied. Numerous research results are reported in the literature for the synchronization of master-slave chaotic systems [18–20]. Various control approaches have been applied for master-slave synchronization of chaotic systems, for example, uncertainty identification-based control [21], quasicontinuous sliding-mode [22], sliding-mode with adaptation [23], proportional-derivative controller [24], immersion and invariance [25], passivity-based approach [26], adaptive backstepping [27], nonfragile fuzzy output feedback [28]. In general, the performance of these controllers is satisfactory. Some of the controllers require the information of all state variables of the master system, some use output-feedback approach to estimate the unmeasured states, some use robust approach to maintain the performance despite the presence of external disturbance.

The control approaches mentioned in the previous paragraph are mostly model based. As such, the performance depends heavily on the quality of the model. It is well known that real-life systems are difficult to model due to various practical considerations like uncertainties and external disturbances. Several results have recently been reported to overcome the problem of model-based control [29, 30]. The main idea here is to approximate the system dynamics by a local model described by an appropriate input-output relationship. In this chapter similar approach is going to be used and a continuous singular terminal sliding-mode (CSTSM) [31] controller has been selected for this purpose.

In this chapter, an approximation of the synchronization error system is first obtained. To obtain the approximation, relative degree concept has been used. Based on this approximated synchronization error model, a CSTSM controller is designed for the purpose of master-slave synchronization. The CSTSM controller uses only the output measurement while the unmeasured states are estimated using the finite-time convergent higher-order sliding mode observer (HOSM-O) [32].

The rest of the chapter is organized as follows: Some preliminaries are given in Section 2. The problem statement is given in Section 3, and the proposed HOSM-O CSTSM control strategy can be found in Section 4. A numerical simulation study using the Duffing-Holmes chaotic oscillator is given in Section 5, while an experimental study using the Van der Pol oscillator can be found in Section 4.1. Finally, Section 7 concludes this article. The notion of relative degree is given in the preliminaries.

2 PRELIMINARIES

Consider the following nonlinear system

$$\dot{x} = f(x) + g(x)u, \tag{1a}$$

$$y = h(x), \tag{1b}$$

where $x \in \mathbb{R}^n$ is the state vector, $u \in \mathbb{R}$ is the input, $y \in \mathbb{R}$ is the output variable of the system, and f and g are smooth vector fields. A vector field is said to be *complete* if all solutions to the $\dot{x} = f(x)$ are defined for all $t \geq 0$ [33].

Definition 1 (Global Uniform Relative Degree [34]). The *global uniform relative degree* r of Eq. (1) is defined as the integer such that

$$L_g L_f^i h(x) = 0, \quad \forall x \in \mathbb{R}^n, \ 0 \leq i \leq r-2,$$
$$L_g L_f^{r-1} h(x) \neq 0, \quad \forall x \in \mathbb{R}^n.$$

We say that $r = \infty$ if

$$L_g L_f^i h(x) = 0, \quad \forall x \in \mathbb{R}^n, \ \forall i \geq 0.$$

Definition 2 (Khalil [35]). The system (1a) is input-to-state stable (ISS) if there exist $\beta \in \mathcal{KL}$ and $\gamma \in \mathcal{K}$ such that for any initial state $x(t_0)$ and any bounded input $u(t)$

$$\| x(t) \| \leq \beta(\| x(t_0) \|, t - t_0) + \gamma\left(\sup_{t_0 \leq \tau \leq t} \| u(\tau) \| \right), \tag{2}$$

where $\|\cdot\|$ denotes the Euclidean norm.

3 PROBLEM STATEMENT

Let us consider a master system given by

$$\sum\nolimits_M : \begin{cases} \dot{x}_M = f_M(x_M) + g_M(x_M)u_M, \\ y_M = h_M(x_M), \end{cases} \tag{3}$$

where $x_M \in \mathbb{R}^n$ is the state vector, $u_M \in \mathbb{R}$ ($u_M : \mathbb{R}_+ \to \mathbb{R}$ is locally essentially bounded and measurable signal) is the input of the master system, $y_M \in \mathbb{R}$ is the output; $f_M : \mathbb{R}^n \to \mathbb{R}^n$, $h_M : \mathbb{R}^n \to \mathbb{R}$, and $g_M : \mathbb{R}^n \to \mathbb{R}^n$ are sufficiently smooth functions, and system (3) has global uniform relative degree $r_M = n$ with respect to the output y_M (see the preliminaries for the definition). Next, consider the slave system given by

$$\sum\nolimits_S : \begin{cases} \dot{x}_S = f_M(x_S) + g_M(x_S)u_S, \\ y_S = h_S(x_S), \end{cases} \tag{4}$$

where $x_S \in \mathbb{R}^n$ is the state vector, $u_S \in \mathbb{R}$ ($u_S : \mathbb{R}_+ \to \mathbb{R}$ is locally essentially bounded and measurable signal) is the input, $y_S \in \mathbb{R}$ is the output; $f_S : \mathbb{R}^n \to \mathbb{R}^n$, $h_S : \mathbb{R}^n \to \mathbb{R}$, and $g_S : \mathbb{R}^n \to \mathbb{R}^n$ are sufficiently smooth functions, and system (4) has global uniform relative degree $r_S = n$ with respect to the output y_S.

Consider two chaotic systems: one is working as a master and moves freely, and the other is working as a slave system and has access to limited information from the master. Then, the synchronization problem becomes finding a control law that will force the trajectory of the slave system to align with the free-moving trajectory of the master system.

The misalignment or mismatch can be considered the synchronization error defined as $\varepsilon := x_M - x_S$. Formally, the master-slave synchronization can be defined as:

Definition 3. A slave system (4) exhibits master-slave synchronization with the master system (3), if

$$\lim_{t\to\infty} \varepsilon = 0, \tag{5}$$

for all $t \geq 0$ and all initial conditions (i.e., $x_M(t_0) - x_S(t_0)$).

4 OUTPUT-FEEDBACK-BASED CONTINUOUS SINGULAR TERMINAL SLIDING-MODE (CSTSM) CONTROLLER DESIGN

This section details the CSTSM control strategy that will be used for the synchronization purpose. For this purpose, the following general nonlinear single-input single-output system (affine in control) is considered:

$$\dot{x} = f(x) + g(x)u, \tag{6a}$$

$$y = h(x), \tag{6b}$$

where $x \in \mathbb{R}^n$ is the state, $u \in \mathbb{R}$ ($u: \mathbb{R}_+ \to \mathbb{R}$ is locally essentially bounded and measurable signal) is the input, $y \in \mathbb{R}$ is the output; $f: \mathbb{R}^n \to \mathbb{R}^n$, $g: \mathbb{R}^n \to \mathbb{R}^n$, and $h: \mathbb{R}^n \to \mathbb{R}$ are sufficiently smooth functions.

Assumption 1. The system (6) has global uniform relative degree $r = n$.

Under this assumption for the system (6), there is a diffeomorphic transformation of coordinates $T: \mathbb{R}^n \to \mathbb{R}^n$ such that

$$\xi = T(x), \tag{7}$$

where $\xi \in \mathbb{R}^n$ is the states of the system in the new coordinate and the system (6) can be represented in the normal form [33, 34]:

$$\dot{\xi} = A\xi + B[\alpha(\xi) + \beta(\xi)u], \tag{8a}$$

$$y = C\xi, \tag{8b}$$

where $\alpha: \mathbb{R}^n \to \mathbb{R}^n$ and $\beta: \mathbb{R}^n \to \mathbb{R}$ are sufficiently smooth functions, β is separate from zero, and

$$A = \begin{bmatrix} 0 & 1 & 0 & \dots & 0 & 0 \\ 0 & 0 & 1 & \dots & 0 & 0 \\ \vdots & \vdots & \ddots & & \vdots & \\ 0 & 0 & 0 & \dots & 0 & 1 \\ 0 & 0 & 0 & \dots & 0 & 0 \end{bmatrix}, \quad B = \begin{bmatrix} 0 \\ 0 \\ \vdots \\ 0 \\ 1 \end{bmatrix},$$

$$C = [1 \ 0 \ \dots \ 0],$$

are in the canonical form. To simplify the presentation of the forthcoming synchronization control design, let us assume that $u + d = \alpha(\xi) + \beta(\xi)u$, where $d \in \mathbb{R}$ is a new disturbance signal in Eq. (8). Since β is not singular, such a representation always exists [1, 34].

Assumption 2. The disturbance signal $d : \mathbb{R}_+ \to \mathbb{R}$ is continuously differentiable for almost all $t \geq 0$, and there is a constant $0 < \nu^+ < \infty$ such that $\text{ess sup}_{t \geq 0} |\dot{d}(t)| \leq \nu^+$.

Assumption 3. The system (8) is ISS with respect to the input u [36].

The only property we need here is the boundedness of the variables ξ for bounded u and d. Roughly speaking, the main idea of approximate model-based control is to replace an unknown "complex" mathematical model by a simple ultra-local model without any assumption on the relative degree of the unknown "complex" mathematical model. However, in our case, we will locally approximate the system (8) by the following simple local model

$$y^{(\nu)} = F + \kappa u, \tag{9}$$

where ν is the derivative of order $\nu \geq 1$ of y, F is the compensation term, which carries the unknown and/or nonlinear dynamics of the system as well as the time-varying external disturbances and $\kappa \in \mathbb{R}$ is a "nonphysical" constant parameter for scaling. In this work ν is considered to be equal to the global relative degree of the system (i.e., n). Then, by derivating Eq. (8b) $\nu = n$ times, it is possible to write locally the system (8) as system (9) in the input-output form. For further development, model (9) will be considered instead of model (8), and for computational simplicity, we will consider $n = 2$ only. The control objective here is to track a desired reference trajectory y_d using the measurement of output only.

Remark 1. For practical systems, it is not always so easy to find the relative degree. In that case, practical relative degree (PRD) can also be used. The PRD is informally defined as the order of the output derivative, which is explicitly affected by control. The PRD does not depend on the system mathematical description. Detail theoretical analysis about PRD can be found in [37] while the application of PRD can be consulted from [38].

To design the tracking controller, the tracking errors can be defined as $\chi_1 := y - y_d$ and $\dot{\chi}_1 = \chi_2 := \dot{y} - \dot{y}_d$. Then the tracking error dynamics can be written as

$$\dot{\chi}_1 = \chi_2, \tag{10a}$$

$$\dot{\chi}_2 = F + \kappa u - \ddot{y}_d. \tag{10b}$$

The tracking problem for system (9) is essentially the stabilization of the error dynamical system (10). To stabilize the system (10), it is necessary to design a controller u under the presence of external perturbations and/or parametric uncertainties which are included in the unknown function F. In this work, an output-feedback control strategy is adopted. As such, the first step in designing the controller is to construct an HOSM-observer (HOSM-O).

4.1 Finite-Time Sliding-Mode Observer

Consider the following HOSM-O for system (10)

$$\dot{\hat{\chi}}_1 = \hat{\chi}_2 + \hat{k}_1 \lceil \tilde{\chi}_1 \rfloor^{2/3}, \tag{11a}$$

$$\dot{\hat{\chi}}_2 = \hat{F} + \kappa u - \ddot{y}_d + \hat{k}_2 \lceil \tilde{\chi}_1 \rfloor^{1/3}, \tag{11b}$$

$$\dot{\hat{F}} = \hat{k}_3 \lceil \tilde{\chi}_1 \rfloor^{0}, \tag{11c}$$

where $\tilde{\chi}_1 = \chi_1 - \hat{\chi}_1$ is the output error, the function $\lceil . \rfloor^{\gamma} := |\cdot|^{\gamma}\mathrm{sign}(\cdot)$, for any $\gamma \in \mathbb{R}_{\geq 0}$; and some design parameters $\hat{k}_i, i = 1,2,3$. Define the state estimation error as $\tilde{\chi} := (\tilde{\chi}_1, \tilde{\chi}_2)^T \in \mathbb{R}^2$. Model (10) is a local approximation of model (8). So, a similar assumption on the property of F is also applicable in this case. Let us consider that the upper bound for F is $f^+ |\frac{dF}{dt}| \leq f^+$. Then, the following result summarizes the convergence property of the HOSM-O:

Proposition 1. *Let the observer* (11) *be applied to system* (10) *and Assumption* 2 *be satisfied for F. If the observer parameters are chosen as follows:*

$$\hat{k}_1 = 3(f^+)^{1/3}, \hat{k}_2 = 1.5(f^+)^{1/2}, \hat{k}_3 = 1.1f^+, \tag{12}$$

then the estimation error $\tilde{\chi} = 0$ *is finite time stable.*

Proof. The error dynamics between system (10) and the HOSM-O (11) are given as follows:

$$\dot{\tilde{\chi}}_1 = \tilde{\chi}_2 - \hat{k}_1 \lceil \tilde{\chi}_1 \rfloor^{2/3}, \tag{13a}$$

$$\dot{\tilde{\chi}}_2 = \tilde{F} - \hat{k}_2 \lceil \tilde{\chi}_1 \rfloor^{1/3}, \tag{13b}$$

$$\dot{\tilde{F}} = -\hat{k}_3 \lceil \tilde{\chi}_1 \rfloor^{0} + (\dot{f}^+). \tag{13c}$$

It is clear that the dynamics (13) is similar to the HOSM differentiator given by Levant [32]. The error dynamics (13) is homogeneous of degree $q = -1$ and weights $r = (3, 2, 1)$. Hence, based on homogeneity and Lyapunov theory, one can show that the error dynamics (13) are finite-time stable if the observer parameters are selected as mentioned in Eq. (12). □

With the estimated states $\hat{\chi}_1$, $\hat{\chi}_2$ and unknown input/perturbation $\hat{F}$, the following CSTSM controller is designed for the stabilization of the synchronization error system (10)

$$\phi(\hat{\chi}_1, \hat{\chi}_2) = \hat{\chi}_2 + k_1 \lceil \hat{\chi}_1 \rfloor^{2/3}, \tag{14a}$$

$$u = -u_{eq} - \ddot{y}_d + z - k_2 \lceil \phi(\hat{\chi}_1, \hat{\chi}_2) \rfloor^{1/2}, \tag{14b}$$

$$u_{eq} = \hat{F} + \hat{k}_2 \lceil \hat{e}_1 \rfloor^{1/3}, \tag{14c}$$

$$\dot{z} = -k_3 \lceil \phi(\hat{\chi}_1, \hat{\chi}_2) \rfloor^{0}, \tag{14d}$$

for some positive constants k_i, $i = 1, 2, 3$. Then the following result summarizes the main contribution of this chapter.

Theorem 1. *Let the observer* (11) *and the controller* (14) *be applied to system* (10) *and Assumption 2 be satisfied for F. Let the observer parameters be chosen as in Proposition 1. Then, for some k_i, i = 1, 2, 3, the tracking error $\chi = 0$ is finite-time stable.*

Proof. Let us substitute Eq. (14) into the tracking error dynamics (10), that is

$$\dot{\chi}_1 = \chi_2, \tag{15a}$$

$$\dot{\chi}_2 = z - k_2 \lceil \phi(\hat{\chi}_1, \hat{\chi}_2) \rfloor^{1/2} + \tilde{F}, \tag{15b}$$

$$\dot{z} = -k_3 \lceil \phi(\hat{\chi}_1, \hat{\chi}_2) \rfloor^{0}. \tag{15c}$$

Let us define, $\bar{\chi}_2 = \chi_2 - \tilde{\chi}_2$. Then the closed-loop dynamics is given as,

$$\prod_{cst} \begin{cases} \dot{\chi}_1 = \bar{\chi}_2 + \tilde{\chi}_2, \\ \dot{\bar{\chi}}_2 = z - k_2 \phi(\hat{\chi}_1, \hat{\chi}_2) \rfloor^{\frac{1}{2}}, \\ \dot{z} = -k_3 \lceil \phi(\hat{\chi}_1, \hat{\chi}_2) \rfloor^{0}, \end{cases} \tag{16a}$$

$$\prod_2 : \begin{cases} \dot{\tilde{\chi}}_1 = \tilde{\chi}_2 - \hat{k}_1 \lceil \tilde{\chi}_1 \rfloor^{\frac{2}{3}}, \\ \dot{\tilde{\chi}}_2 = \tilde{F} - \hat{k}_2 \lceil \tilde{\chi}_1 \rfloor^{1/3}, \\ \dot{\tilde{F}} = -\hat{k}_3 \lceil \tilde{\chi}_1 \rfloor^{0} + (\dot{f}^{+}). \end{cases} \tag{16b}$$

System (16) can be viewed as a cascade system where $\prod_2$ is the input of $\prod_{cst}$. Then, based on homogeneity [39] and ISS properties [40], it is straightforward to provide the proof of finite-time stability of the closed-loop dynamics (16).

Remark 2. Controller (14) and HOSM-O (11) are given only for the case when $\nu = 2$. For a higher-order estimated local model, the extension can be easily done by following the ideas presented in [41].

5 NUMERICAL SIMULATION RESULTS

This section deals with the numerical simulation of the proposed CSTSM controller equipped with finite-time convergent HOSM-O. This will be done by synchronizing two identical chaotic Duffing-Holmes oscillator [42]. The master and slave Duffing-Holmes oscillators are given as

$$\ddot{y}_m + r_{1m} y_m + r_{2m} \dot{y}_m + r_{3m} y_m^3 = r_{4m} \cos(\omega t), \tag{17a}$$

$$\ddot{y}_s + r_{1s} y_s + r_{2s} \dot{y}_s + r_{3s} y_s^3 = r_{4s} \cos(\omega t) + u + d, \tag{17b}$$

where m denotes the master oscillator, s denotes the slave oscillator, u denotes the control signal, d denotes the external disturbance, and $r_{1m/s}$, $r_{2m/s}$, $r_{3m/s}$, $r_{4m/s}$,

ω are model parameters. By considering $y_{m/s} = x_{1m/s}$ and $\dot{y}_{m/s} = x_{2m/s}$ as states, the master and slave oscillators can be represented in the state-space form as:

$$\dot{x}_{1m} = x_{2m}, \tag{18a}$$

$$\dot{x}_{2m} = -r_{1m}x_{1m} - r_{2m}x_{2m} - r_{3m}x_{1m}^3 + r_{4m}\cos(\omega t), \tag{18b}$$

$$\dot{x}_{1s} = x_{2s}, \tag{19a}$$

$$\dot{x}_{2s} = -r_{1s}x_{1s} - r_{2s}x_{2s} - r_{3s}x_{1s}^3 + r_{4s}\cos(\omega t) + u + d. \tag{19b}$$

For simulation purposes, the parameters of Eqs. (18), (19) are considered to be $\omega = 1$, $r_{1m} = r_{1s} = -1$, $r_{2m} = r_{2s} = 0.4$, $r_{3m} = r_{3s} = 1$, and $r_{4m} = r_{4s} = 1.9$. The external disturbance signal is selected as $d = \sin(3t)$. In order to design the master-slave synchronization controller, let us define the errors among the master and slave system as $\chi_1 = x_{1m} - x_{1s}$ and $\dot{\chi}_1 = \chi_2 = x_{2m} - x_{2s}$, where m stands for master system and s for the slave system. Then the error dynamics can be written as

$$\dot{\chi}_1 = \chi_2, \tag{20a}$$

$$\dot{\chi}_2 = \psi_m(x_{1m}, x_{2m}, \omega) - \psi_s(x_{1s}, x_{2s}, \omega, d) - u, \tag{20b}$$

where $\psi_m(x_{1m}, x_{2m}, \omega) = -r_{1m}x_{1m} - r_{2m}x_{2m} - r_{3m}x_{1m}^3 + r_{4m}\cos(\omega t)$ and $\psi_s(x_{1s}, x_{2s}, \omega, d) = -r_{1s}x_{1s} - r_{2s}x_{2s} - r_{3s}x_{1s}^3 + r_{4s}\cos(\omega t) + d$. From the properties of ψ_m and ψ_s, it can be substantiated that the difference of the two functions is bounded. Define $F := \psi_m - \psi_s$ with $|\frac{dF}{dt}| \leq f^+$ and f^+ being a known positive constant. Then Eq. (20) can be written as

$$\dot{\chi}_1 = \chi_2, \tag{21a}$$

$$\dot{\chi}_2 = F - u. \tag{21b}$$

If we consider $y = e_1$ as the output, then model (21) can be written as

$$y^{(2)} = F + \kappa u, \tag{22}$$

where $\kappa = -1$. Then the design of the synchronizing controller follows directly from Section 4. To illustrate the performance of the HOSM-O, numerical simulations are performed using the following parameters for the HOSM-O: $\hat{k}_1 = 11.0521$, $\hat{k}_2 = 10.6066$, $\hat{k}_3 = 55$ and initial conditions $\hat{\chi}_1(0) = \hat{\chi}_2(0) = 0$. From Fig. 1, we can see that the estimation errors $\tilde{\chi}_1 = \chi_1 - \hat{\chi}_1$, and $\tilde{\chi}_2 = \chi_2 - \hat{\chi}_2$, converge to the origin in finite-time.

Parameters of the controller (14) are selected as: $k_1 = 1$, $k_2 = 3.6741$, and $k_3 = 6.6$. The result of the synchronization is given in Fig. 2. Such a figure shows that master-slave synchronization is achieved after applying the control.

To check the robustness of the proposed output-feedback control approach, parametric uncertainties are considered for the next simulation. In this case, the parameter $r_{3,m}$ is changed from 1 to 5 at $t = 50$ s. Moreover, the disturbance $d(t)$ acting on the slave system is also changed from $\sin(3t)$ to $\sin(30t)$ at the same time, that is, at $t = 50$ s. Synchronization error in this case can be found in Fig. 3.

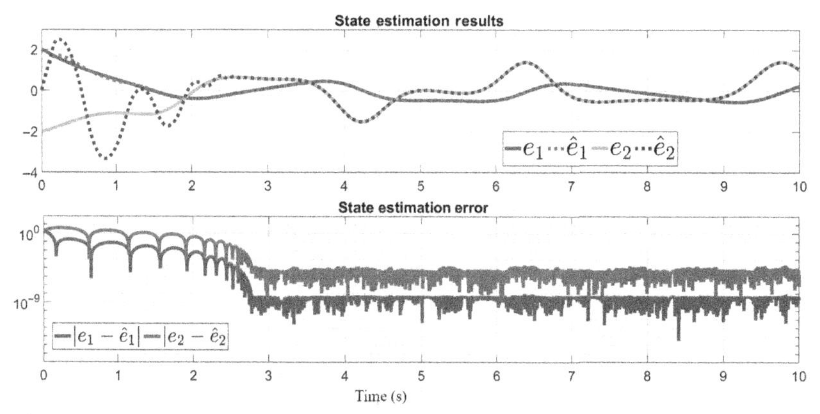

FIG. 1 State estimation results.

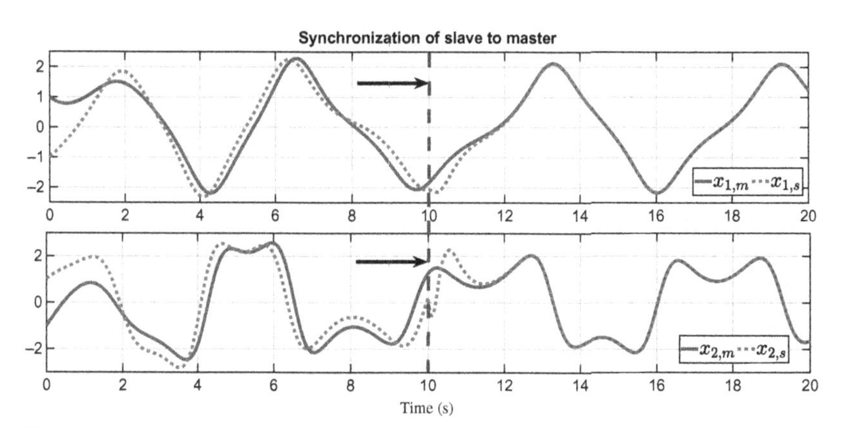

FIG. 2 Synchronization of the slave to master.

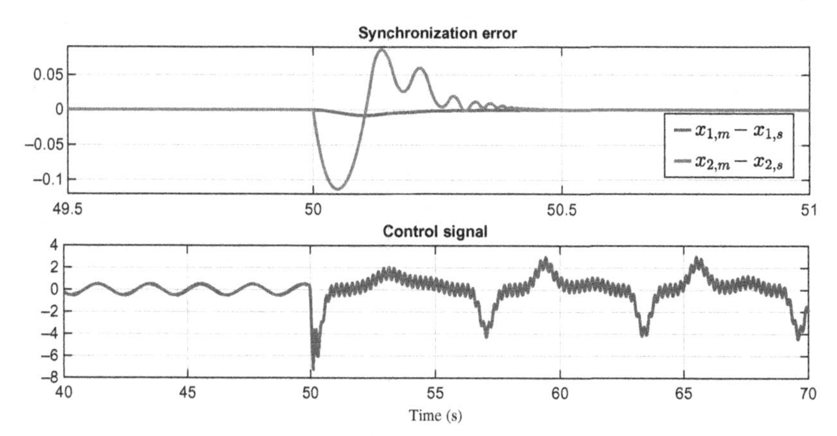

FIG. 3 Synchronization error and control signal in the presence of parametric and disturbance uncertainty.

Fig. 3 demonstrates the robustness property of the proposed control technique. The CSTSM controller immediately reacts to the parameter and disturbance change. As result, synchronization is achieved again in less than 0.5 s. These simulation results demonstrate the effectiveness of the proposed controller.

6 EXPERIMENTAL STUDY

Previous section illustrated that the proposed output-feedback controller can successfully achieve master-slave synchronization. However, numerical simulation results are not always sufficient for practical systems. In the simulation, we have considered identical chaotic oscillators. From a practical point of view, it is often very difficult to realize a family of identical oscillators. Oscillators are generally made of analog electronic components. Nonidealities and nonlinearities of the practical devices make the oscillators nonidentical in a practical settings. For example, the resistors that are used in this work have a tolerance limit of $\pm 10\%$. So, even if we take the resistors of same nominal values for individual oscillators, the final values are different due to the tolerance limit. Moreover, each oscillator was powered with a separate power supply. This also introduces some perturbation. As a result, frequently practical systems are nonidentical [43].

This section is devoted to the experimental validations of the proposed approximate model-based control. For this purpose, the Van der Pol oscillator has been selected. It is a popular benchmark second-order nonlinear model. In state-space form, the Van der Pol oscillator can be written as [1]:

$$\dot{x}_1 = x_2, \tag{23a}$$

$$\dot{x}_2 = -x_1 + \rho\left(1 - x_1^2\right)x_2 + u, \tag{23b}$$

$$y = x_1, \tag{23c}$$

where ρ is the model parameter. The Van der Pol oscillator is a limit cycle oscillator, and the value of ρ determines the shape of the limit-cycle. In this work, the parameter $\rho = 0.1$ is selected. The electronic circuit diagram can be seen in Figs. 4 and 5. The circuit parameters are: $Ri = 1\ \text{M}\Omega$, $i = \overline{1,6}$, $R7 = 130\ \Omega$,

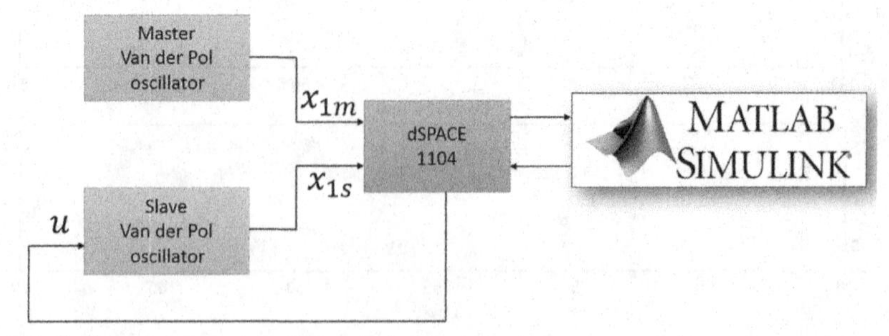

FIG. 4 Schematic overview of the practical implementation.

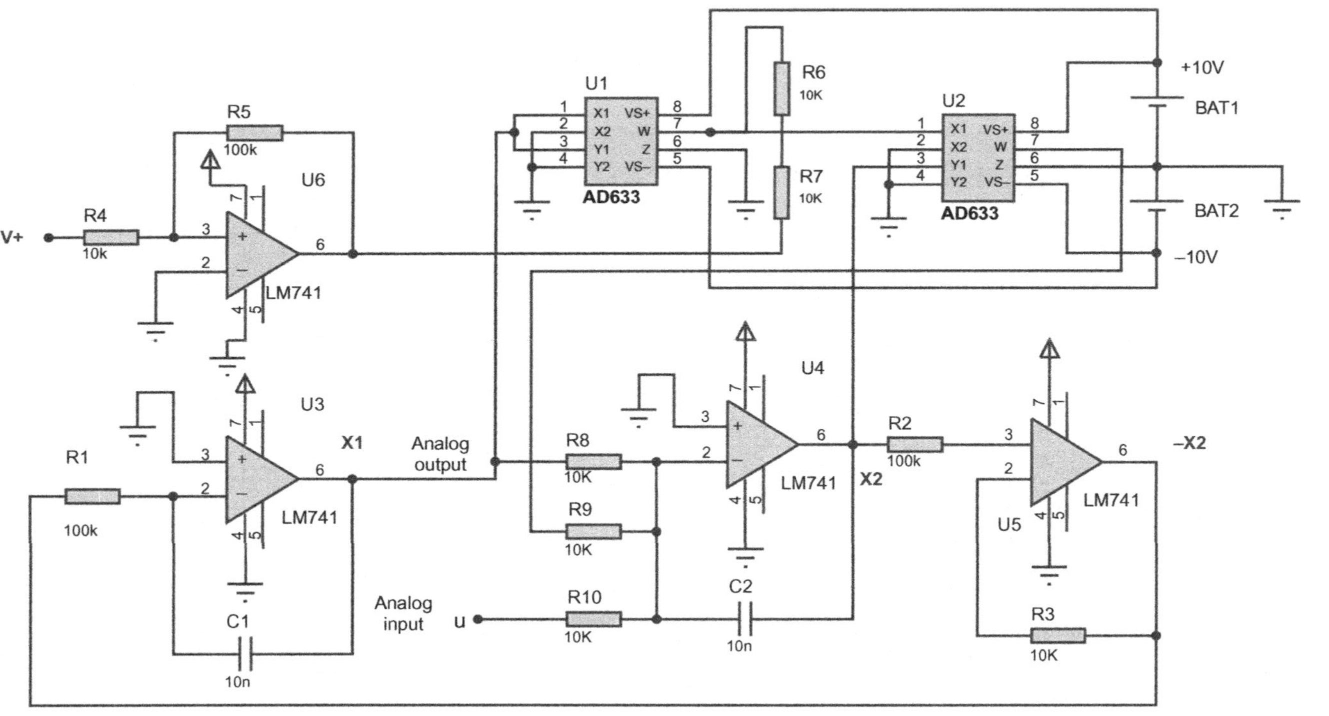

FIG. 5 Circuit diagram of the Van der Pol oscillator.

$R8 = 1.2\ \text{k}\Omega$, $R9 = 100\ \Omega$, $R10 = 1.5\ \text{k}\Omega$, $C1 = C2 = 1\ \mu\text{F}$. The parameters are adopted from Ahmed et al. [44]. dSPACE 1104 board was rapid as the rapid prototyping platform. HOSM-O observer-based CSTSM controller was implemented using Simulink. The solver was the Euler's method, and the sampling frequency was 5 kHz. According to the sliding mode control literature, Euler's discretization scheme is the preferred method. For details, Huber [45] can be consulted. dSPACE 1104 board is a complete solution for real-time prototyping. It comes with various kinds of inputs and outputs, including analog signals. dSPACE generates embedded code from Simulink model for real-time implementation. In our application, we measure the analog voltage signals of $x_{1,m}$ and $x_{1,s}$. The measured analog signals are then pass through the analog-to-digital converter of the dSPACE board. These digital signals are then used for estimating the states and generating the control signal. Once the control signal is generated, digital-to-analog converter of the dSPACE board is used to send the analog control signal to the Van der Pol oscillator circuit.

The parameters of the controller and the observer are selected the same as Section 6. Experimental synchronization results are shown in Fig. 6. The controller was activated at 10 s after the observer has converged (see Fig. 7 for estimation error). Once the controller is activated, the master-slave synchronization is achieved in a very short time. This validate the output-feedback controller developed in Section 4.

7 CONCLUSION

In this chapter, an approximate model-based master-slave synchronization of chaotic systems has been presented. The error dynamics of the master-slave systems are first approximated by a perturbed chain of integrators. The order of the model depends on the relative degree of the system. Then, synchronization

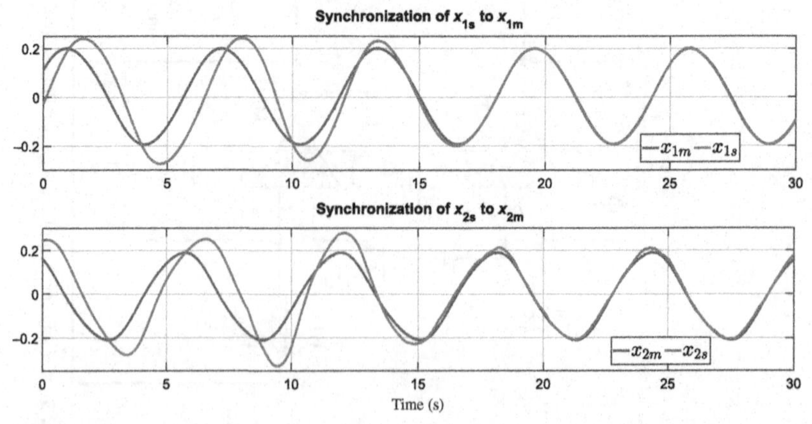

FIG. 6 Experimental synchronization of master-slave the Van der Pol oscillator systems.

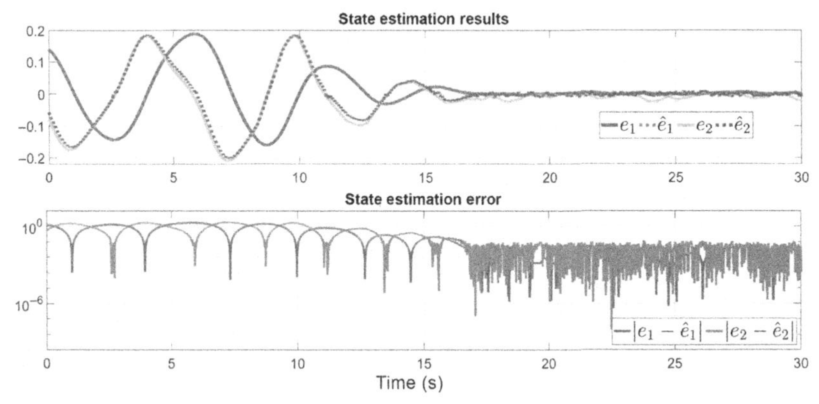

FIG. 7 Experimental state estimation results.

is achieved using control approaches based on sliding-mode. The proposed controller is output feedback based and uses a higher-order sliding mode observer to estimate the unmeasurable states. Experimental study demonstrated the effectiveness of our method using master-slave Van der Pol oscillators.

REFERENCES

[1] Ahmed H, Ushirobira R, Efimov D, Fridman L, Wang Y. Oscillatory global output synchronization of nonidentical nonlinear systems. IFAC-PapersOnLine 2017;50.

[2] Ahmed H, Ushirobira R, Efimov D. Robust global synchronization of Brockett oscillators. IEEE Trans Control Netw Syst 2018. https://doi.org/10.1109/TCNS.2018.2813927.

[3] Ahmed H, Ushirobira R, Efimov D, Perruquetti W. Robust synchronization for multistable systems. IEEE Trans Autom Control 2016;61(6):1625–30.

[4] Ahmed H, Ushirobira R, Efimov D. On robustness of phase resetting to cell division under entrainment. J Theor Biol 2015;387:206–13.

[5] Tsonis AA, Elsner JB. Chaos, strange attractors, and weather. Bull Am Meteorol Soc 1989; 70(1):14–23.

[6] Fu H, Xu J, Xu L. Traffic chaos and its prediction based on a nonlinear car-following model. J Control Theory Appl 2005;3(3):302–7.

[7] Mork BA, Stuehm DL. Application of nonlinear dynamics and chaos to ferroresonance in distribution systems. IEEE Trans Power Delivery 1994;9(2):1009–17.

[8] Konstantinidou S, Snyder L. The chaos router: a practical application of randomization in network routing. In: Proceedings of the second annual ACM symposium on parallel algorithms and architectures. ACM; 1990. p. 21–30.

[9] van den Bleek CM. Coppens MO, Schouten JC. Application of chaos analysis to multiphase reactors. Chem Eng Sci 2002;57(22–23):4763–78.

[10] Harezlak K, Kasprowski P. Searching for chaos evidence in eye movement signals. Entropy 2018;20(1):32.

[11] Lassoued A, Boubaker O. On new chaotic and hyperchaotic systems: a literature survey. Nonlinear Anal Modell Control 2016;21(6):770–89.

[12] Lassoued A, Boubaker O. Dynamic analysis and circuit design of a novel hyperchaotic system with fractional-order terms. Complexity 2017;2017. https://doi.org/10.1155/2017/3273408.
[13] Luo AQ, Xia JB, Wang HB. Application of chaos-support vector machine regression in traffic prediction. Comput Sci 2009;36(7).
[14] Davendra D, Zelinka I, Senkerik R. Chaos driven evolutionary algorithms for the task of PID control. Comput Math Appl 2010;60(4):1088–104.
[15] Xia X, Chen L. Fuzzy chaos method for evaluation of nonlinearly evolutionary process of rolling bearing performance. Measurement 2013;46(3):1349–54.
[16] Seneviratne C, Leung H. Mixing chaos modulations for secure communications in OFDM systems. Eur Phys J Spec Top 2017;226(15):3287–301.
[17] Hokayem PF, Spong MW. Bilateral teleoperation: an historical survey. Automatica 2006; 42(12):2035–57.
[18] Mkaouar H, Boubaker O. Robust control of a class of chaotic and hyperchaotic driven systems. Pramana 2017;88(1):9.
[19] Mkaouar H, Boubaker O. Chaos synchronization for master slave piecewise linear systems: application to Chua's circuit. Commun Nonlinear Sci Numer Simul 2012;17(3):1292–302.
[20] Boubaker O, Dhifaoui R. Robust chaos synchronization for Chua's circuits via active sliding mode control. In: Chaos, complexity and leadership 2012. Springer; 2014. p. 141–51.
[21] Rodriguez A, De Leon J, Fridman L. Synchronization in reduced-order of chaotic systems via control approaches based on high-order sliding-mode observer. Chaos Solitons Fractals 2009;42(5):3219–33.
[22] Rodriguez A, De Leon J, Fridman L. Quasi-continuous high-order sliding-mode controllers for reduced-order chaos synchronization. Int J Nonlinear Mech 2008;43(9):948–61.
[23] Yau HT. Design of adaptive sliding mode controller for chaos synchronization with uncertainties. Chaos Solitons Fractals 2004;22(2):341–7.
[24] Yin C, Zhong SM, Chen WF. Design PD controller for master-slave synchronization of chaotic Lure systems with sector and slope restricted nonlinearities. Commun Nonlinear Sci Numer Simul 2011;16(3):1632–9.
[25] Aguilar-Ibañez C, García-Canseco E, Martínez-Guerra R, Martínez-García JC, Suarez-Castañon MS. An I&I-based observer to solve the output-feedback synchronization problem for a class of chaotic systems. Asian J Control 2018. https://doi.org/10.1002/asjc.1650.
[26] Kocamaz UE, Çiçek S, Uyaroğlu Y. Secure communication with chaos and electronic circuit design using passivity-based synchronization. J Circuits Syst Comput 2018;27(04):1850057.
[27] Shukla MK, Sharma BB. Control and synchronization of a class of uncertain fractional order chaotic systems via adaptive backstepping control. Asian J Control 2018;20(2):707–20.
[28] Azarang A, Miri M, Kamaei S, Asemani MH. Nonfragile fuzzy output feedback synchronization of a new chaotic system: design and implementation. J Comput Nonlinear Dyn 2018;13 (1):011008.
[29] Elmetennani S, Laleg-Kirati TM. Bilinear approximate model-based robust Lyapunov control for parabolic distributed collectors. IEEE Trans Control Syst Technol 2016;25(5):1848–55.
[30] Precup RE, Radac MB, Roman RC, Petriu EM. Model-free sliding mode control of nonlinear systems: algorithms and experiments. Inf Sci 2017;381:176–92.
[31] Kamal S, Moreno JA, Chalanga A, Bandyopadhyay B, Fridman LM. Continuous terminal sliding-mode controller. Automatica 2016;69:308–14.
[32] Levant A. Higher-order sliding modes, differentiation and output-feedback control. Int J Control 2003;76(9–10):924–41.
[33] Khalil HK. Nonlinear control. Upper Saddle River, New Jersey, USA: Pearson; 2014.

[34] Marino R, Tomei P. Nonlinear control design: geometric, adaptive and robust. Upper Saddle River, New Jersey, USA: Prentice Hall International (UK) Ltd.; 1996.

[35] Khalil HK. Nonlinear systems. 2nd ed. Upper Saddle River, NJ: Prentice-Hall; 1996.

[36] Dashkovskiy SN, Efimov DV, Sontag ED. Input to state stability and allied system properties. Autom Remote Control 2011;72(8):1579.

[37] Levant A. Practical relative degree in black-box control. In: 2012 IEEE 51st annual conference on decision and control (CDC). IEEE; 2012. p. 7101–6.

[38] Hernandez AGG, Fridman L, Levant A, Shtessel Y, Leder R, Monsalve CR, et al. High-order sliding-mode control for blood glucose: practical relative degree approach. Control Eng Pract 2013;21(5):747–58.

[39] Zubov VI. Systems of ordinary differential equations with generalized-homogeneous right-hand sides. Izv Vyssh Uchebn Zaved Matematika 1958;1:80–8. [In Russian].

[40] Bernuau E, Efimov D, Perruquetti W, Polyakov A. On homogeneity and its application in sliding mode control. J Frankl Inst 2014;351(4):1866–901.

[41] Fridman L, Moreno JA, Bandyopadhyay B, Kamal S, Chalanga A. Continuous nested algorithms: the fifth generation of sliding mode controllers. In: Recent advances in sliding modes: from control to intelligent mechatronics. Springer; 2015. p. 5–435.

[42] Holmes P. A nonlinear oscillator with a strange attractor. Philos Trans R Soc Lond A 1979;292 (1394):419–48.

[43] Ahmed H, Ushirobira R, Efimov D. Experimental study of the robust global synchronization of Brockett oscillators. Eur Phys J Spec Top 2017;226(15):3199–210.

[44] Ahmed H, Salgado I, Ríos H. Robust synchronization of master-slave chaotic systems using approximate model: an experimental study. ISA Trans 2018;73(02):141–6. https://doi.org/10.1016/j.isatra.2018.01.009.

[45] Huber O. Analysis and implementation of discrete-time sliding mode control (Ph.D. thesis). Université Grenoble Alpes, 2015.

Chapter 14

A Four-Dimensional Chaotic System With One or Without Equilibrium Points: Dynamical Analysis and Its Application to Text Encryption

Victor Kamdoum Tamba*,†, Romanic Kengne†, Sifeu Takougang Kingni‡ and Hilaire Bertrand Fotsin†

*Department of Telecommunication and Network Engineering, IUT-Fotso Victor of Bandjoun, University of Dschang, Bandjoun, Cameroon, †Laboratory of Condensed Matters, Electronics and Signal Processing (LAMACETS), Department of Physics, Faculty of Science, University of Dschang, Dschang, Cameroon, ‡Department of Mechanical and Electrical Engineering, Faculty of Mines and Petroleum Industries, University of Maroua, Maroua, Cameroon

1 INTRODUCTION

Recent advancements in numerical integration methods and computer simulation algorithms have allowed us to develop chaotic systems with desired features. Examples include chaotic systems with different kinds of symmetry, with multi-scroll attractors, with multiple coexisting attractors, and with simplest equations [1–4]. These examples are about the formation and multiplication of the number of strange attractors, while some other chaotic systems have been designed based on the features of their equilibria. From the point of view of the latter consideration, the equilibrium point plays a vital role in the classification of dynamical systems. They help us to classify dynamical systems into two kinds: those with self-excited attractors and those with hidden attractors. The most chaotic systems currently studied in the literature are those with self-excited or hidden structures. A self-excited attractor has a basin of attraction that is associated with an unstable equilibrium, whereas a hidden attractor has a basin of attraction that does not intersect with small neighborhoods of any equilibrium points [5]. The well-known attractors of Lorenz, Chua, Rossler, Chen, Lu, Colpitts, and many others [6–17] belong to the family of self-excited attractors and are associated with one or more unstable equilibrium points.

Recent Advances in Chaotic Systems and Synchronization. https://doi.org/10.1016/B978-0-12-815838-8.00014-5

Chaotic behavior in such self-excited systems is examined by using Shilnikov criteria [18] and computed by standard computational methods. The family of the hidden chaotic attractors can be grouped into (i) only stable equilibrium points [19–21], (ii) no equilibria [22–24], and (iii) an infinite number of equilibria [25–29]. In contrast to self-excited chaotic attractors, the theoretical investigation, numerical localization, and computation of hidden chaotic attractors are more difficult because the equilibrium points do not help to localize them. Hidden attractors are important in engineering applications because they allow unexpected and potentially disastrous responses to perturbations in a structure like a bridge or an airplane wing [5].

Recently, Jafari and collaborators [30] have introduced a new chaotic system (namely, chameleon) with very interesting and unique properties. They proved that the proposed system displays self-excited attractors and three different families of hidden attractors depending on the values of the parameters of the system. The experimental verifications using electronic circuits as well as a text encryption application with the random numbers generated from chameleon chaotic system have been examined. In the same direction, Karthikeyan and colleagues have reported the dynamics of a chameleon hyperchaotic and chaotic systems [31,32]. The effect of fractional order on the dynamical behavior of the chameleon systems and their implementation using electronic components as well as FPGA technology have been investigated. There are few systems displaying self-excited and hidden attractors reported in the literature [30–32]. Therefore, such systems are challenging, and investigating them may be of great interest in dynamical systems.

Inspired by the above-mentioned works and by the special features of the chameleon systems, in this chapter we introduce a four-dimensional chaotic system with self-excited and hidden attractors depending on the values of the parameters of the system. It can have self-excited attractors with only one unstable equilibrium point or hidden attractors with no equilibria. The adaptive finite time synchronization of two identical four-dimensional chaotic systems with hidden attractor is exploited for a test encryption application.

The chapter layout is as follows: Section 2 deals with the description and some fundamental properties of the proposed four-dimensional system with self-excited and hidden attractors. The dynamics of the system in each configuration are investigated in Section 3. The bifurcation structures reveal that the system under study displays some interesting phenomena depending on the values of the parameters of the system. In Section 4, we design and implement an electronic circuit that can be exploited for investigation of the dynamics of a four-dimensional system with self-excited and hidden attractors. The obtained results show a very good agreement with the numerical investigations. In Section 5, the adaptive finite time synchronization of two identical four-dimensional chaotic systems with hidden attractors is discussed. The numerical simulations are performed to illustrate the effectiveness of the proposed synchronization method. We exploit in Section 6 the results for synchronization to realize a text encryption application. The simulation verifications are included to illustrate the proposed text encryption technique. Concluding remarks are finally drawn in Section 7.

2 MODEL OF PROPOSED AUTONOMOUS SYSTEM WITH ONE OR WITHOUT EQUILIBRIUM POINTS

Dynamics of a three dimensional autonomous system with golden proportion equilibria reported in Refs. [33, 34] is described as:

$$\begin{cases} \dot{x} = y - x - az \\ \dot{y} = xz - x \\ \dot{z} = b - y - xy \end{cases} \tag{1}$$

in which x, y, z are the state variables, a and b are positive control parameters. In Ref. [33], it has been demonstrated numerically and experimentally that the system (1) has two single folded chaotic attractors for specific values of parameters a and b. In Ref. [34], the authors have shown numerically and experimentally that system (1) can exhibit periodic and chaotic bursting oscillations. Using linear state feedback control in the second state equation of system (1), the dynamics of a new four-dimensional chaotic system with one or without equilibrium points is developed and described:

$$\begin{cases} \dot{x} = y - x - az \\ \dot{y} = xz - x + w \\ \dot{z} = b - y - xy \\ \dot{w} = -x \end{cases} \tag{2}$$

in which w is a state variable. It is easy to verify that system (2) is dissipative because of $\nabla V = \frac{\partial \dot{x}}{\partial x} + \frac{\partial \dot{y}}{\partial y} + \frac{\partial \dot{z}}{\partial z} = -1$. This implies that any volume element $V_0 = V(t = 0)$ will be continuously contracted by the flow (e.g., each volume element containing the trajectory shrinks to zero as time evolves to infinity). Then, all system orbits will be confined to a specific bounded subset of zero volume in state space and the asymptotic dynamics settles onto an attractor.

The equilibrium points of system (2) for $a \neq 0$ are evaluated by solving $\dot{x} = 0$, $\dot{y} = 0$, $\dot{z} = 0$ and $\dot{w} = 0$. It is simple to verify that system (2) has only one equilibrium point $E_0(0, \mathrm{b}, b/a, 0)$. The characteristic equation of the Jacobian matrix at E_0 is:

$$\lambda^4 + \lambda^3 + (1 - ab - b/a)\lambda^2 + (1 + a - b)\lambda + a = 0 \tag{3}$$

Using Routh-Hurwitz conditions, this equation has all roots with negative real parts if and only if:

$$1 + a - b > 0 \tag{4a}$$

$$b\left(a - a^2 + ab - 1/a + b/a - b\right) - 2a > 0 \tag{4b}$$

Selecting the following specific values of parameters for which the system develops self-excited chaotic oscillations (see Section 3.1): $a = 0.1$ and $b = 1$, the eigenvalues of Eq. (3) are: $\lambda_1 = -3.56$, $\lambda_2 = 2.54$, $\lambda_3 = -0.09$, and $\lambda_4 = 0.11$. According to the fact that there are eigenvalues with positive real part, the equilibrium point E_0 is an unstable saddle equilibrium point for the system (2). Thus, the system under consideration can exhibit self-excited oscillations with these values of parameters setting.

For $a = 0$ the equilibrium points of system (2) can be obtained by solving the following system equations:

$$\begin{cases} y - x = 0 \\ xz - x + w = 0 \\ b - y - xy = 0 \\ -x = 0 \end{cases} \tag{5}$$

Obviously, from the first and fourth equations of system (5), $x = y = 0$, which is inconsistent with the third equation, provided that $b > 0$. Therefore, there is no equilibrium in system (2) for $a = 0$. As a consequence, the attractors generated from the considered system are hidden, and the Shilnikov method cannot be applied to explain its chaotic behavior because it has neither homoclinic or heteroclinic orbits.

3 DYNAMICAL ANALYSIS OF PROPOSED AUTONOMOUS SYSTEM WITH ONE OR WITHOUT EQUILIBRIUM POINTS

The dynamical behaviors of proposed autonomous system with one or without equilibrium points described by system (2) can be illustrated by time series, bifurcation diagrams, Lyapunov exponents and phase portraits. Because system (2) has one or no-equilibrium points depending on the value of parameter a, it is interesting to investigate the dynamical behavior of system (2) for $a \neq 0$ and $a = 0$.

3.1 Self-Excited Attractor in Proposed Autonomous System With Only One Equilibrium Point

In this subsection, we consider the parameter $a \neq 0$. In order to know the dynamical behaviors exhibited by system (2), we plot in Fig. 1 the bifurcation diagram depicting the local maxima of the state variable $x(t)$ and the three largest Lyapunov exponents as a function of b for $a = 0.1$.

When the parameter b increases from 1.5 to 12 (see Fig. 1A), the bifurcation diagram of the output $x(t)$ shows period-1-oscillations followed by period-doubling bifurcation to chaos for $2.529 < b < 8.122$. Then period-8-oscillations are observed followed by a reverse period-doubling bifurcation. The chaotic behavior is confirmed by the largest Lyapunov exponent shown

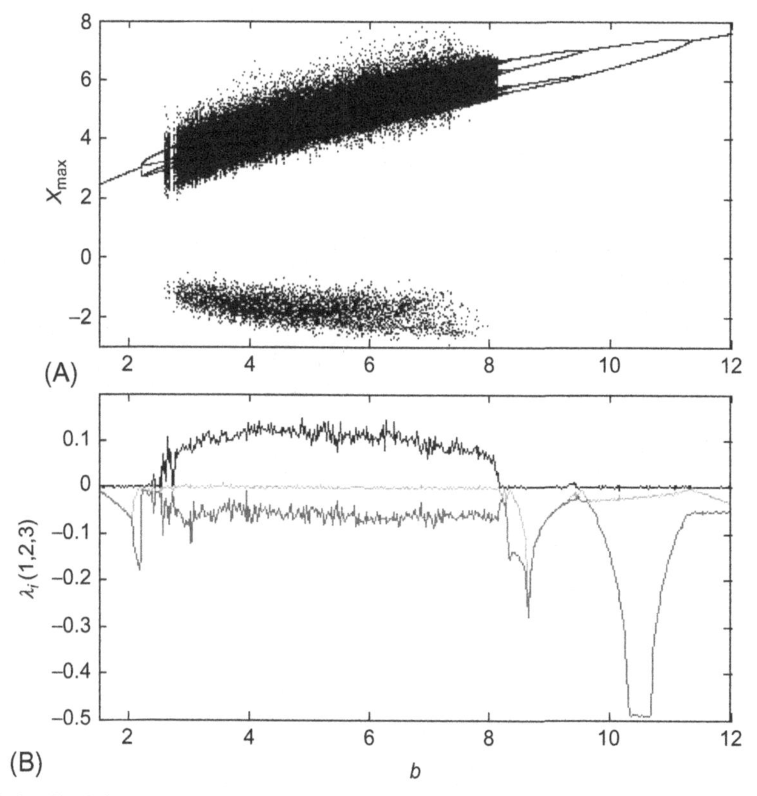

FIG. 1 The bifurcation diagrams depicting the local maxima of $x(t)$ (A) and the three largest Lyapunov exponents (B) versus the parameter b for $a = 0.1$.

in Fig. 1B. The chaotic behavior is illustrated in Fig. 2 for a specific value of parameter b.

From Fig. 2, we notice that the trajectories of chaotic attractors are swirling around the equilibrium point. This is a signature of a one-scroll chaotic attractor. It is worth noting that the chaotic attractor of Fig. 2 belongs to the family of self-excited attractors.

3.2 Hidden Attractor in Proposed Autonomous System Without Equilibrium Point

Here, we consider the parameter $a = 0$. Thus, system (2) can display hidden attractor because it has no-equilibrium points. In order to know the dynamical behaviors displayed by system (2), we plot in Fig. 3 the bifurcation diagram depicting the local maxima of $x(t)$ and the three largest Lyapunov exponents as a function of b for $a = 0$.

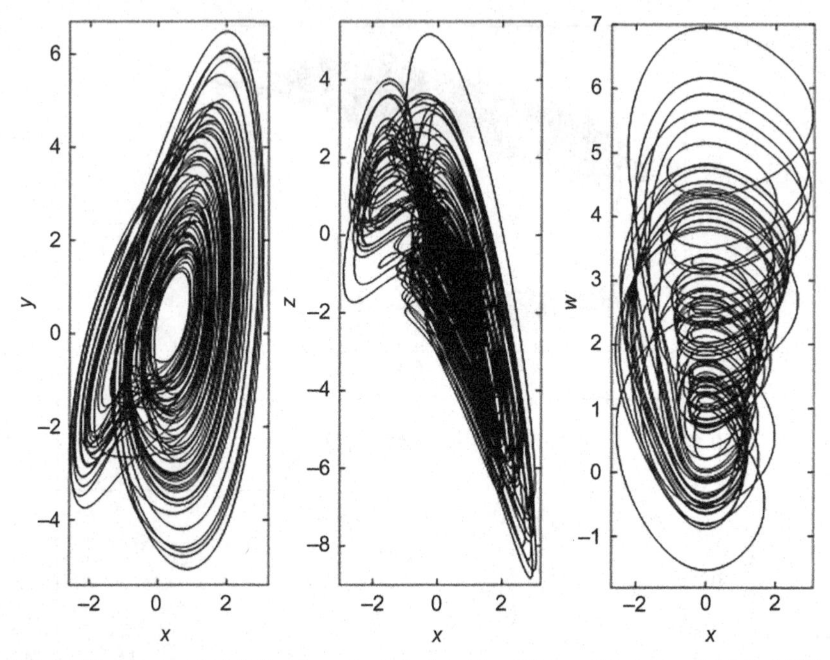

FIG. 2 The phase portrait in the planes (x, y), (x, z) and (x, w) for $a = 0.1$ and $b = 1.0$. The initial conditions are $(x(0), y(0), z(0), w(0)) = (0.1, 0.1, 0.1, 0.1)$.

When the parameter b varies from 0.1 to 3, the bifurcation diagram of the output $x(t)$ in Fig. 3A displays hidden periodic and chaotic bursting oscillations. The chaotic behavior is confirmed by the positive value of largest Lyapunov exponent shown in Fig. 3B.

The chaotic phase portraits of the system in different planes are illustrated in Figs. 4 and 5, respectively, for two specific values of parameter b.

The chaotic attractors of Figs. 4 and 5 belong to the family of hidden attractors because for $a = 0$, system (2) has no equilibrium point.

Interestingly, when $a = 0$, system (2) without equilibrium points has special feature of offset boosting control. Indeed, the state variable z appears only in the second equation of autonomous system without equilibrium points. Thus, the amplitude of the state variable z can easily change by introducing a control parameter k_z into system (2) as:

$$\begin{cases} \dot{x} = y - x \\ \dot{y} = x(z + k_z) - x + w \\ \dot{z} = b - y - xy \\ \dot{w} = -x \end{cases} \tag{6}$$

It is worth noting that this transformation has no effect on the dynamics of the original autonomous system without equilibrium points but provides a

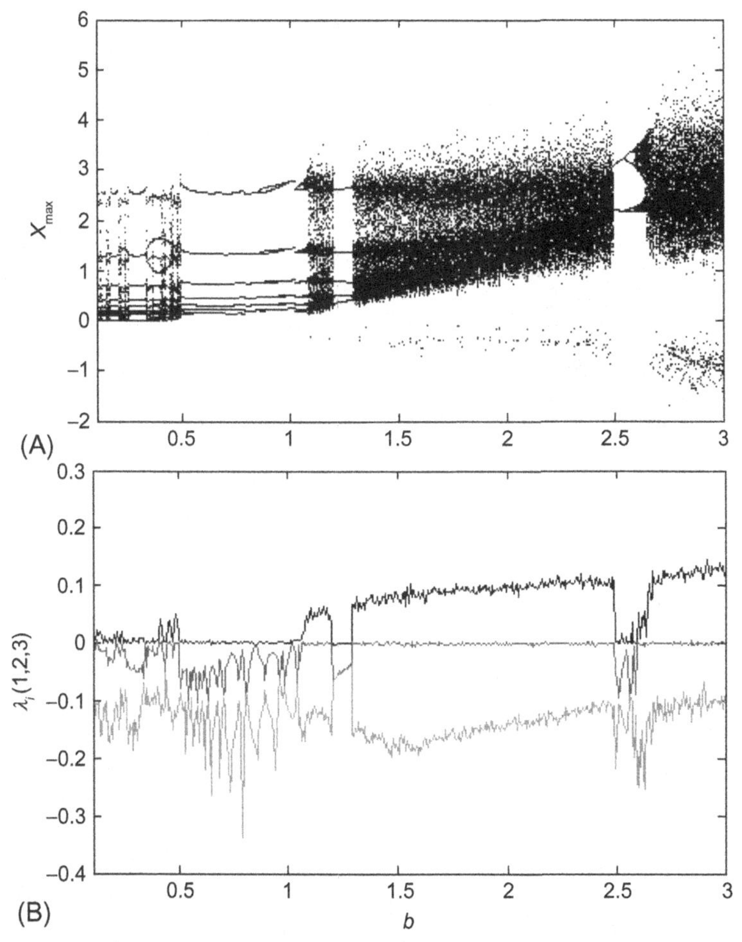

FIG. 3 The bifurcation diagrams depicting the local maxima of $x(t)$ (A) and the three largest Lyapunov exponents (B) versus the parameter b for $a = 0$.

controllable ability. Various positions of the phase portraits of system (5) with hidden attractors adjusted depending on different values of the control parameter k_z are depicted in Fig. 6.

From Fig. 6, it is observed that the positive and negative values of control parameter k_z boosts the variable z, respectively, in the negative and positive directions. The time series of the state variable z and the average values of the state variables x, y, z, and w with respect to the offset boosting parameter k_z are shown in Fig. 7A and B, respectively.

It is noted from Fig. 7A that the state variable z is effectively boosted when the offset boosting control parameter is varied. This situation is also clearly observed

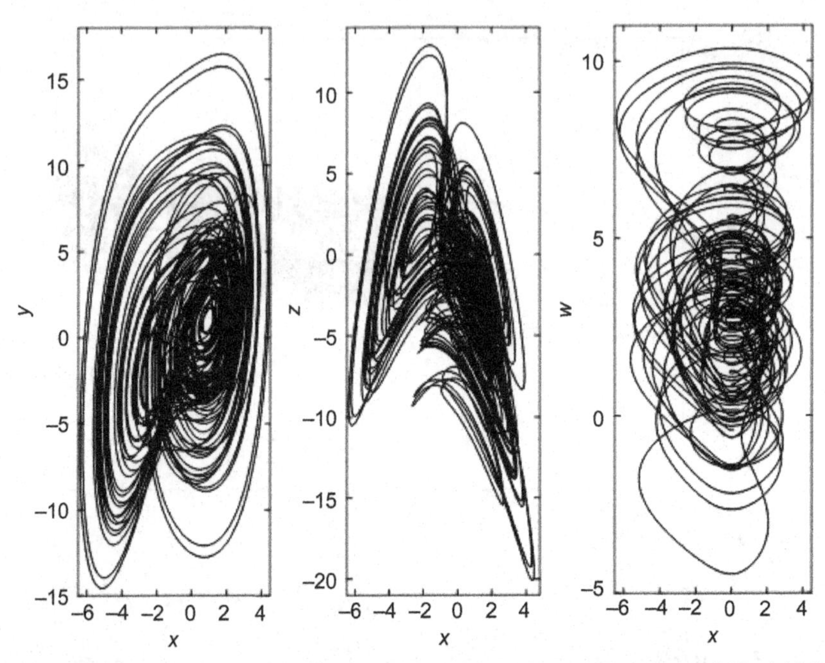

FIG. 4 The phase portrait in the plane (x, y), (x, z) and (x, w) for $a = 0.0$ and $b = 3.0$. The initial conditions are $(x(0), y(0), z(0), w(0)) = (0.1, 0.1, 0.1, 0.1)$.

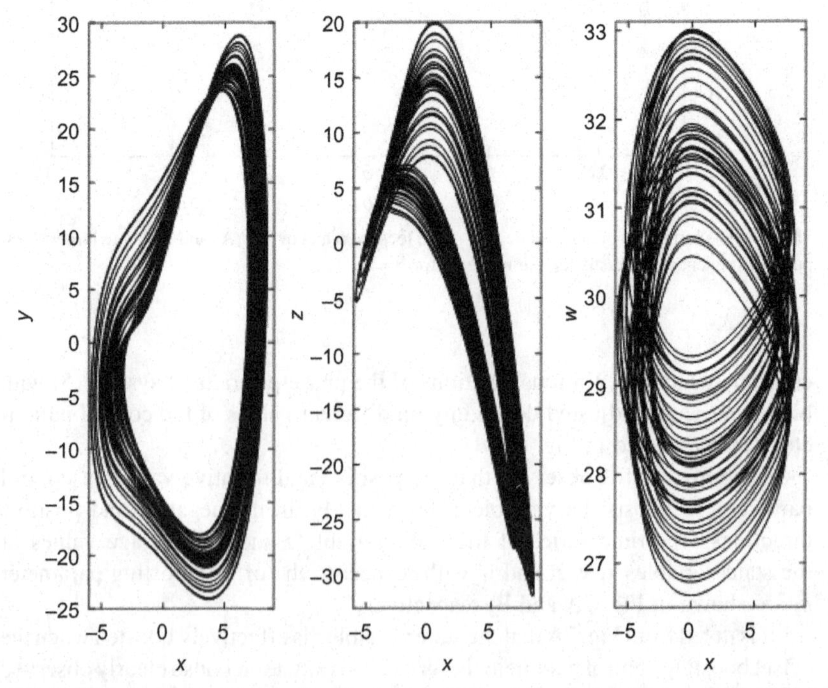

FIG. 5 The phase portrait in the plane (x, y), (x, z) and (x, w) for $a = 0.0$ and $b = 15$. The initial conditions are $(x(0), y(0), z(0), w(0)) = (0.1, 0.1, 0.1, 0.1)$.

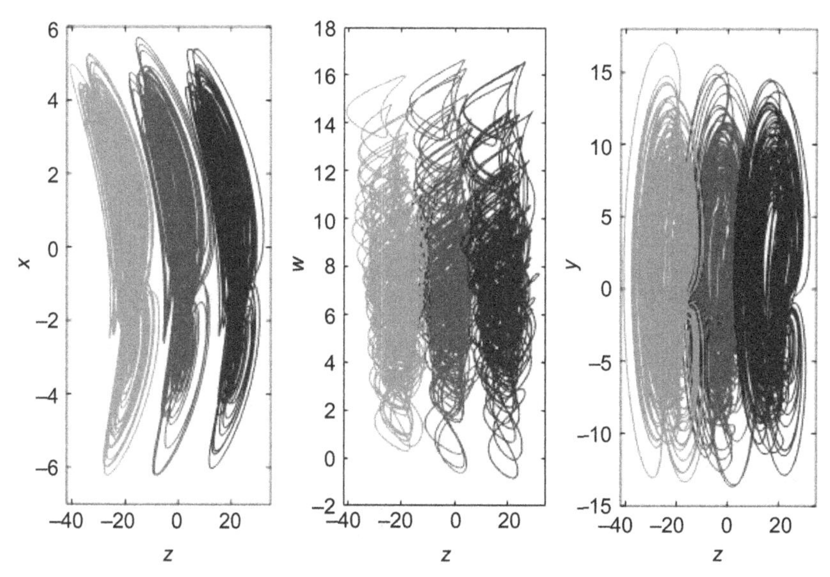

FIG. 6 The phase portrait in the planes (z,x), (z,w) and (z,y) for $a = 0.0$ and $b = 5.0$ and different values of offset boosting controller k_z. The initial conditions are $(x(0), y(0), z(0), w(0)) = (0.1, 0.1, 0.1, 0.1)$. The green, brown, and blue attractors' phase portraits are computed, respectively, for $k_z = 20$, $k_z = 0$, and $k_z = -20$. For interpretation of the references to color in this figure legend, the reader is referred to the web version of this chapter.

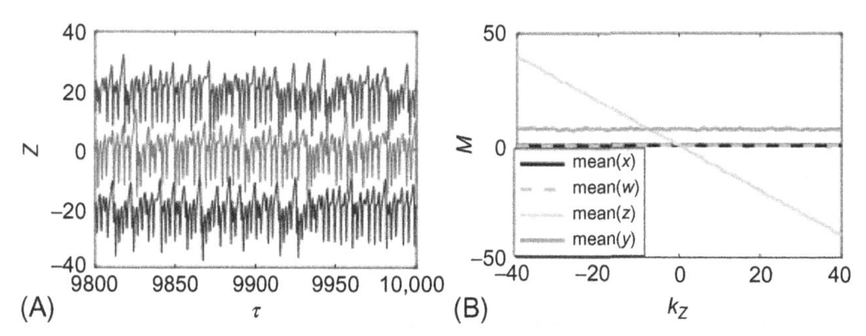

FIG. 7 Time series of the state variable z (A) and the average values of the state variables x, y, z, and w with respect to the offset boosting control k_z (B) for $b = 5$.

in Fig. 7B where one can see that the average of the state variable z decreases lineally and other three state variables (x, y, and z) unchanged when the offset boosting control parameter is increased. A system with the free control of offset boosting is of great interest for some engineering applications where the desired boosting level can be achieved by using a single constant. This fascinating feature has been previously reported in few chaotic systems [35–37].

Furthermore, it is found during the numerical investigations that the autonomous system without equilibrium points experiences periodic and chaotic bursting phenomena, as shown in Figs. 8 and 9.

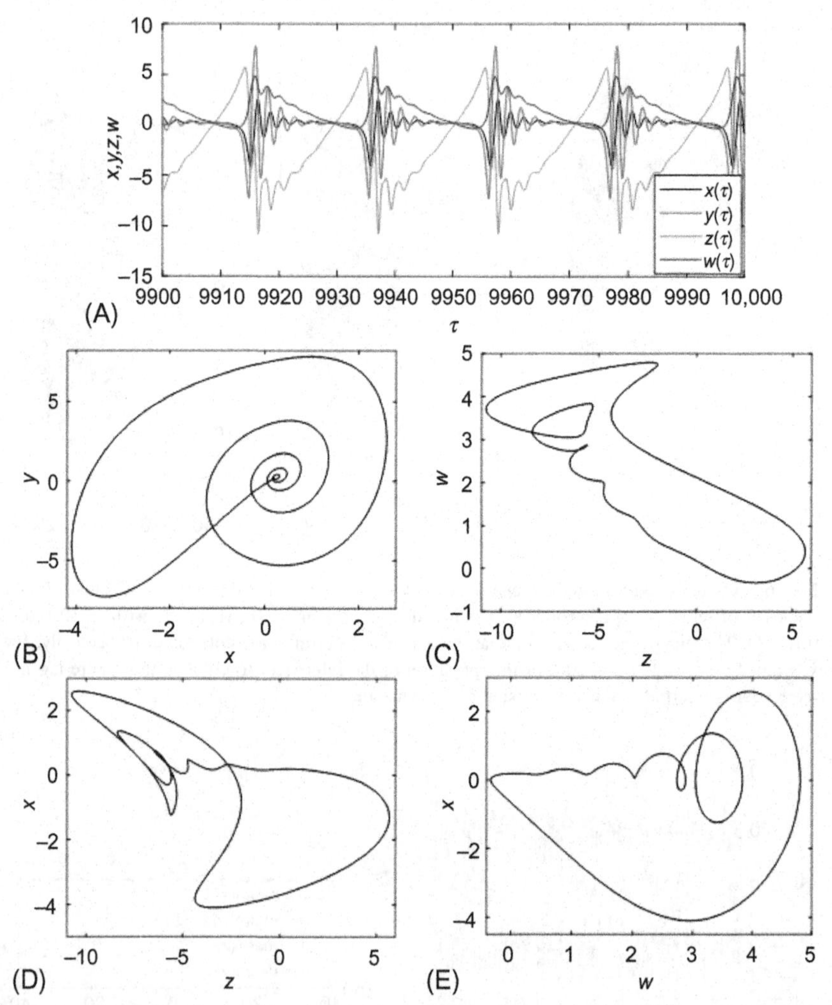

FIG. 8 Periodic bursting oscillations at $b = 1$. (A) Time series of the state variables (x, y, z, and w); phase space projections in the planes (B) (x,y), (C) (z,w), (D)(z,x) and (E) (w,x). The initial conditions are $(x(0), y(0), z(0), w(0)) = (0.1, 0.1, 0.1, 0.1)$.

It is observed from Figs. 8A and 9A that the state variables x and y describe the dynamics of relatively fast changing processes, while the state variables z and w model the relatively slowly changing quantities that modulate x and y. The occurrence of the periodic and chaotic bursting oscillations has been demonstrated in many domains such as mathematical biology [38], neuroscience [39], chemical physics [40], and so on.

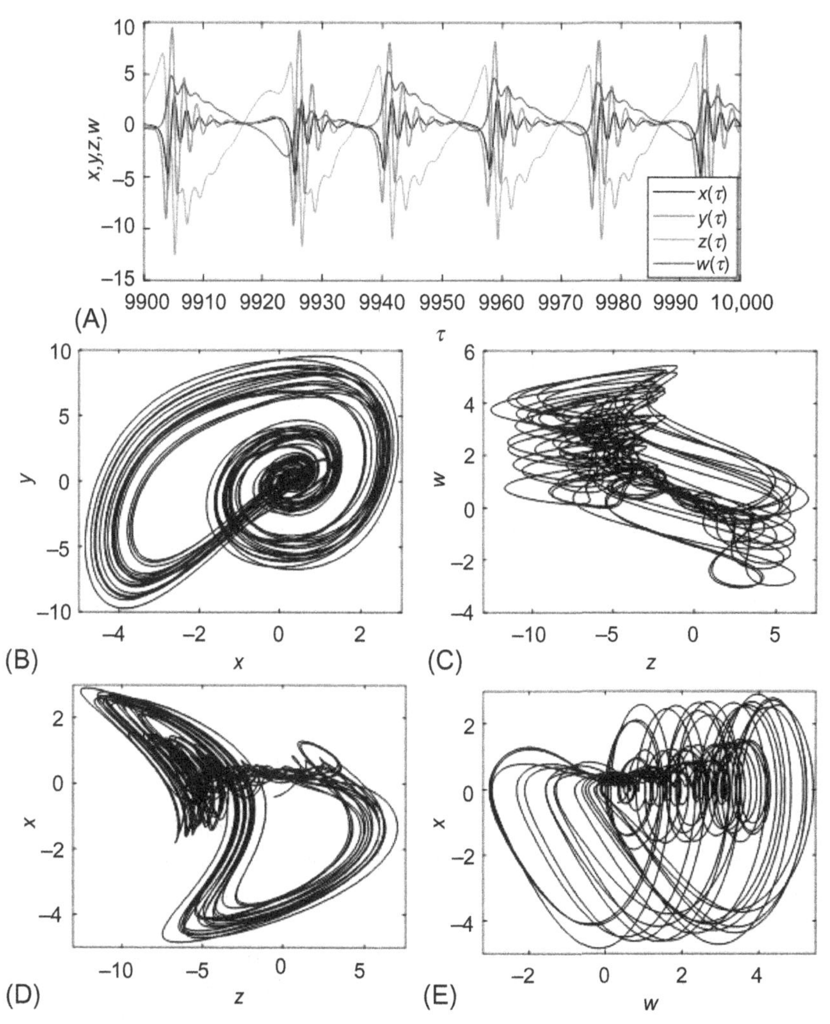

FIG. 9 Chaotic bursting oscillations at $b = 1.3$. (A) Time series of the state variables (x, y, z and w); phase space projections in the planes (B) (x, y), (C) (z, w), (D)(z, x) and (E) (w, x). The initial conditions are $(x(0), y(0), z(0), w(0)) = (0.1, 0.1, 0.1, 0.1)$.

4 ELECTRONIC CIRCUIT IMPLEMENTATION OF PROPOSED AUTONOMOUS SYSTEM WITH ONE OR WITHOUT EQUILIBRIUM POINTS

The dynamics of system (2) with one or without equilibrium points was investigated in the previous section. It has been found that the system under scrutiny exhibits self-excited and hidden attractors according to the value of parameter a. Numerical investigations revealed that the studied system experiences complex

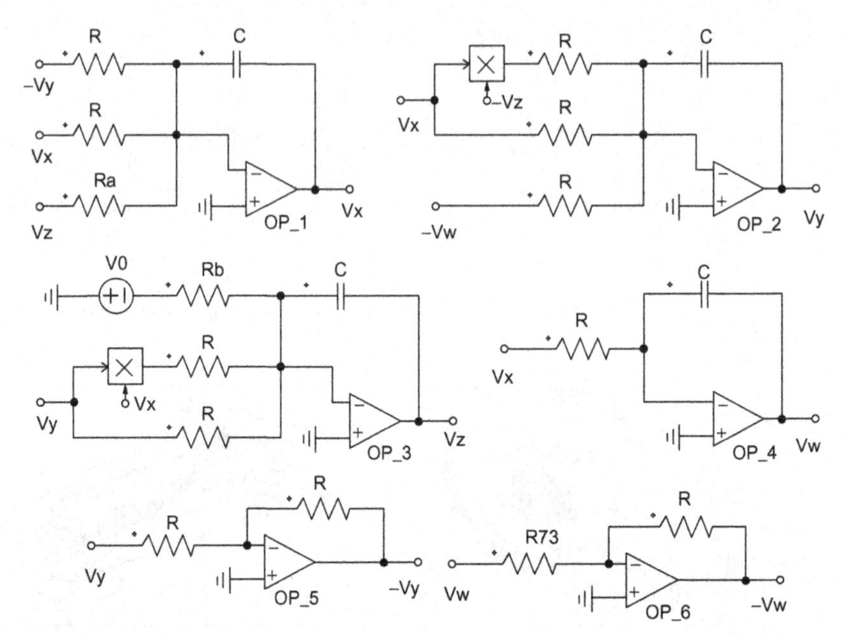

FIG. 10 Designed electronic circuit for theoretical model (2) with one or without equilibrium points.

dynamics including chaos, periodicity, quasi-periodicity, periodic and chaotic bursting oscillations, and controllable amplitude. The aim of this section is to design and implement an electronic circuit that can be exploited for investigation of the dynamics of system (2) with self-excited and hidden attractors. The electronic circuit designed for the implementation of system (2) is shown in Fig. 10.

The circuit of Fig. 10 includes two negative gain amplifiers and four channels to implement the integration, addition, and subtraction of the state variables x, y, z, and w, respectively. The nonlinear terms of the theoretical model (2) are implemented using the analogue devices AD633JN versions of the AD633 four-quadrant voltage multipliers chips. The bias is provided by a 15-V DC symmetry source. The constant term b is realized with a DC source $V_0 = 1\,\text{V}$. By applying Kirchhoff's laws into the circuit of Fig. 10, we obtain the following state equations:

$$\begin{cases} C\dfrac{dV_x}{dt} = \dfrac{V_y - V_x}{R} - \dfrac{V_z}{R_a} \\ C\dfrac{dV_y}{dt} = \dfrac{V_x V_z}{10R} - \dfrac{(V_x + V_w)}{R} \\ C\dfrac{dV_z}{dt} = \dfrac{V_0}{R_b} - \dfrac{V_y}{R} - \dfrac{V_x V_y}{10R} \\ C\dfrac{dV_w}{dt} = -\dfrac{V_x}{R} \end{cases} \tag{7}$$

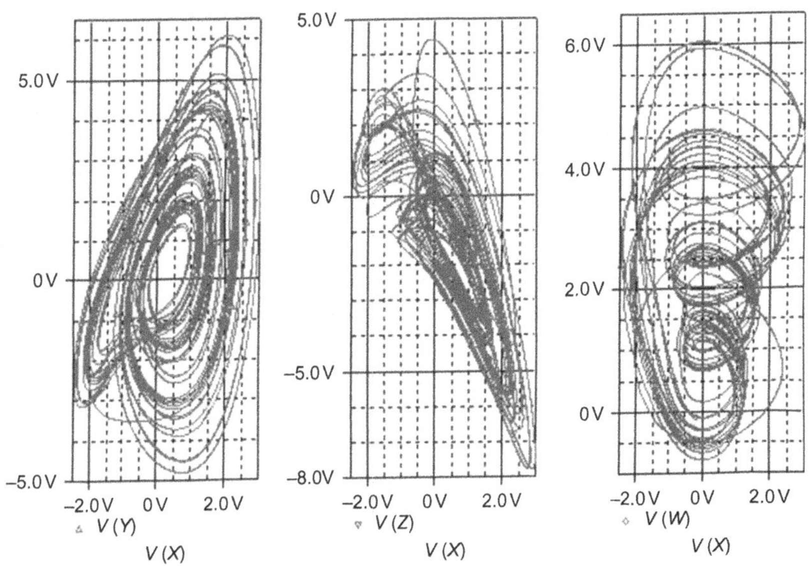

FIG. 11 Pspice simulation results depicting the self-excited phase portrait of system (2) in the (V_x, V_y), (V_x, V_z) and (V_x, V_w) planes for $R_a = 100\,\text{k}\Omega$ and $R_b = 10\,\text{k}\Omega$. The initial values of output voltages are $(V_x(0), V_y(0), V_y(0), V_w(0)) = (0.1, 0.1, 0.1, 0.1)$.

where V_x, V_y, V_z and V_w are the output voltages of the operational amplifiers OP_1, OP_2, OP_3 and OP_4, respectively. Adopting the time unit of 10^{-4} s and $C = 10\,\text{nF}$, the parameters of the theoretical model (2) can be expressed as follows: $a = R/R_a$ and $b = R/R_b$. The self-excited chaotic attractors shown in Fig. 11 in (V_x, V_y), (V_x, V_z) and (V_x, V_w) planes are obtained by using $R_a = 100\,\text{k}\Omega$ (for $a = 0.1$) and $R = R_b = 10\,\text{k}\Omega$ (for $b = 1$).

The hidden chaotic attractors of system (2) shown in Figs. 12 and 13 in (V_x, V_y), (V_x, V_z) and (V_x, V_w) planes are obtained by selecting, respectively, $R_b = 3.33\,\text{k}\Omega$ (for $b = 3$) and $R_b = 0.66\,\text{k}\Omega$ (for $b = 15$).

The periodic and chaotic bursting oscillations found numerical in the autonomous system without equilibrium points are also verified in Pspice as shown in Figs. 14 and 15.

From Figs. 11–15, very good similarity between Pspice simulation results and numerical phase portraits can be observed. These results confirm that the designed circuit emulates well the dynamics of theoretical model (2) with one or without equilibrium points (i.e., system (2) with self-excited or hidden attractors). In the next section, we synchronize two identical structures of system (2) with hidden attractors in order to exploit for text encryption application. System (2) without equilibrium points is chosen for synchronization and application for text encryption due only not for its hidden structure but also for its complicated dynamics compared to the self-excited version.

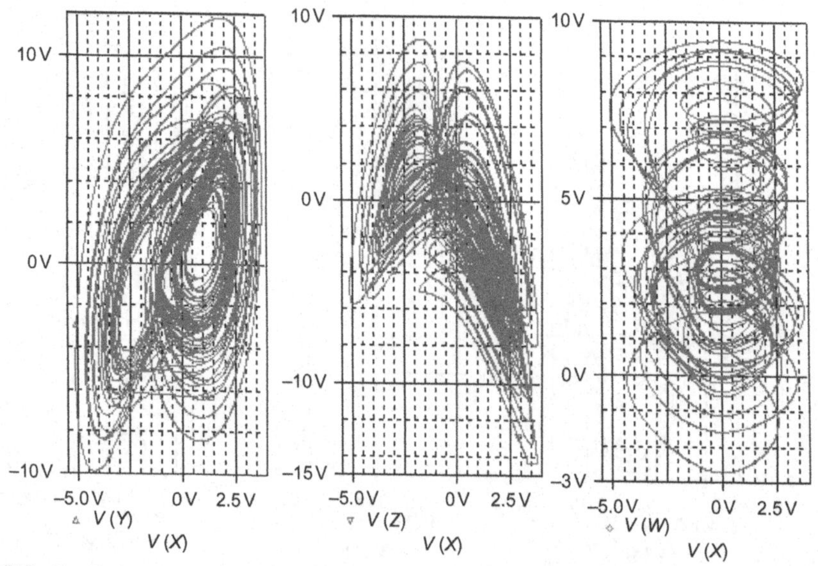

FIG. 12 Pspice simulation results depicting the hidden phase portrait of system (2) in the planes (V_x, V_y), (V_x, V_z) and (V_x, V_w) for $R_b = 3.33\,\mathrm{k\Omega}$. The initial values of output voltages are $(V_x(0), V_y(0), V_y(0), V_w(0)) = (0.1, 0.1, 0.1, 0.1)$.

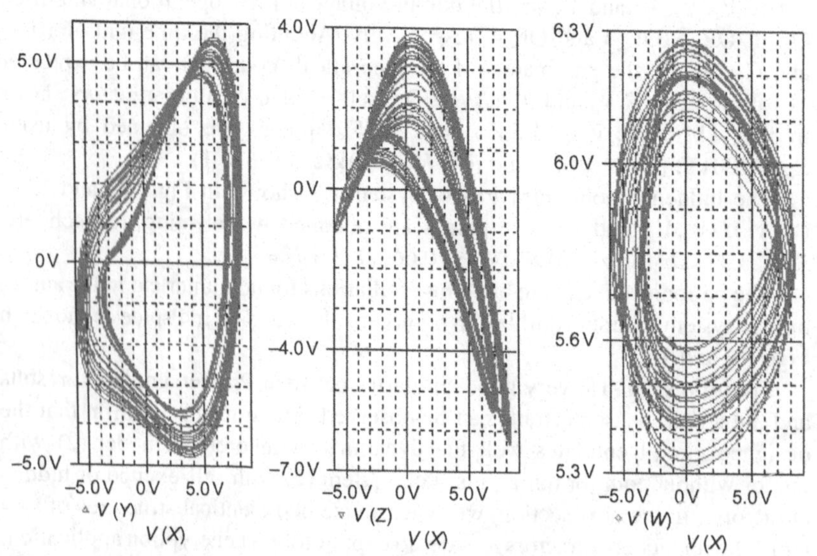

FIG. 13 Pspice simulation results depicting the self-excited phase portrait of system (2) in the (V_x, V_y), (V_x, V_z) and (V_x, V_w) planes for $R_b = 0.66\,\mathrm{k\Omega}$. The initial values of output voltages are $(V_x(0), V_y(0), V_y(0), V_w(0)) = (0.1, 0.1, 0.1, 0.1)$.

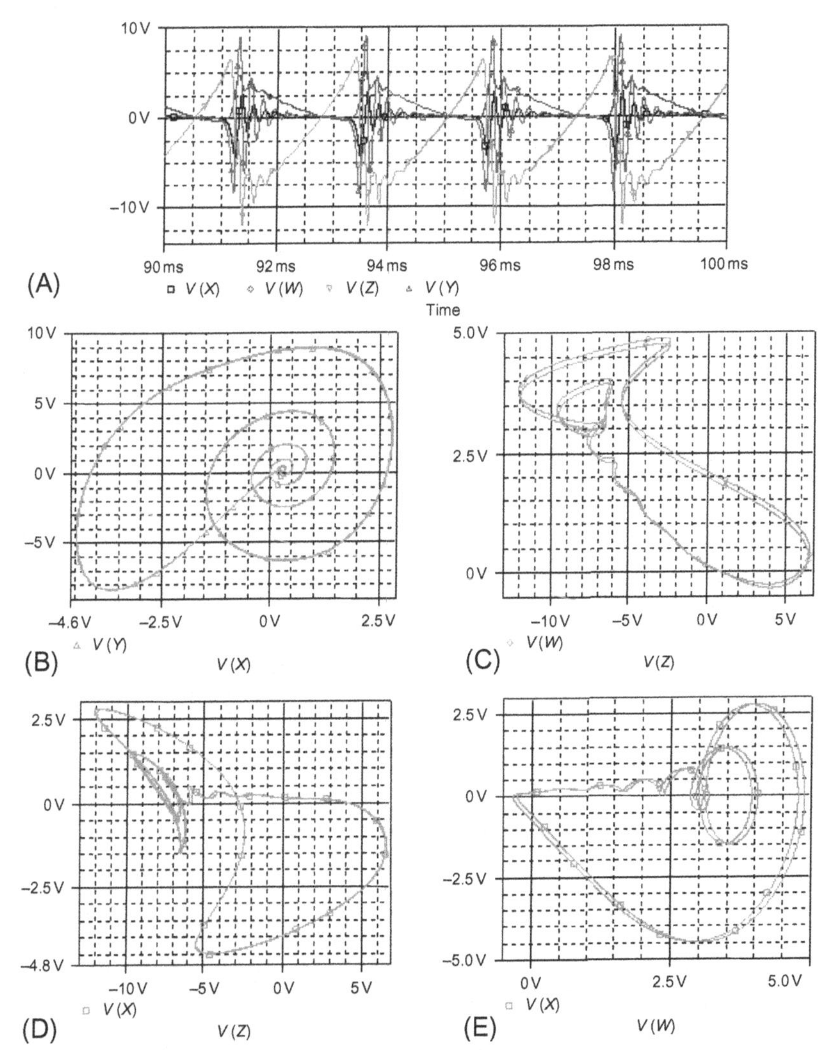

FIG. 14 Pspice simulation results showing the periodic and chaotic bursting oscillations in system (2): (A) time series of the output voltages V_x, V_y, V_z and V_w; phase space projections in the planes (A) (V_x, V_y), (B) (V_z, V_w), (C) (V_z, V_x) and (D) (V_w, V_x) for $R_b = 10\,\text{k}\Omega$. The initial values of output voltages are $(V_x(0), V_y(0), V_y(0), V_w(0)) = (0.1, 0.1, 0.1, 0.1)$.

5 ADAPTIVE FINITE-TIME SYNCHRONIZATION OF PROPOSED AUTONOMOUS SYSTEM WITH HIDDEN ATTRACTOR

According to the interesting features of hidden attractors, in this section an adaptive finite-time synchronization scheme is developed to synchronize two identical structures of system (2) with hidden attractor for text encryption application.

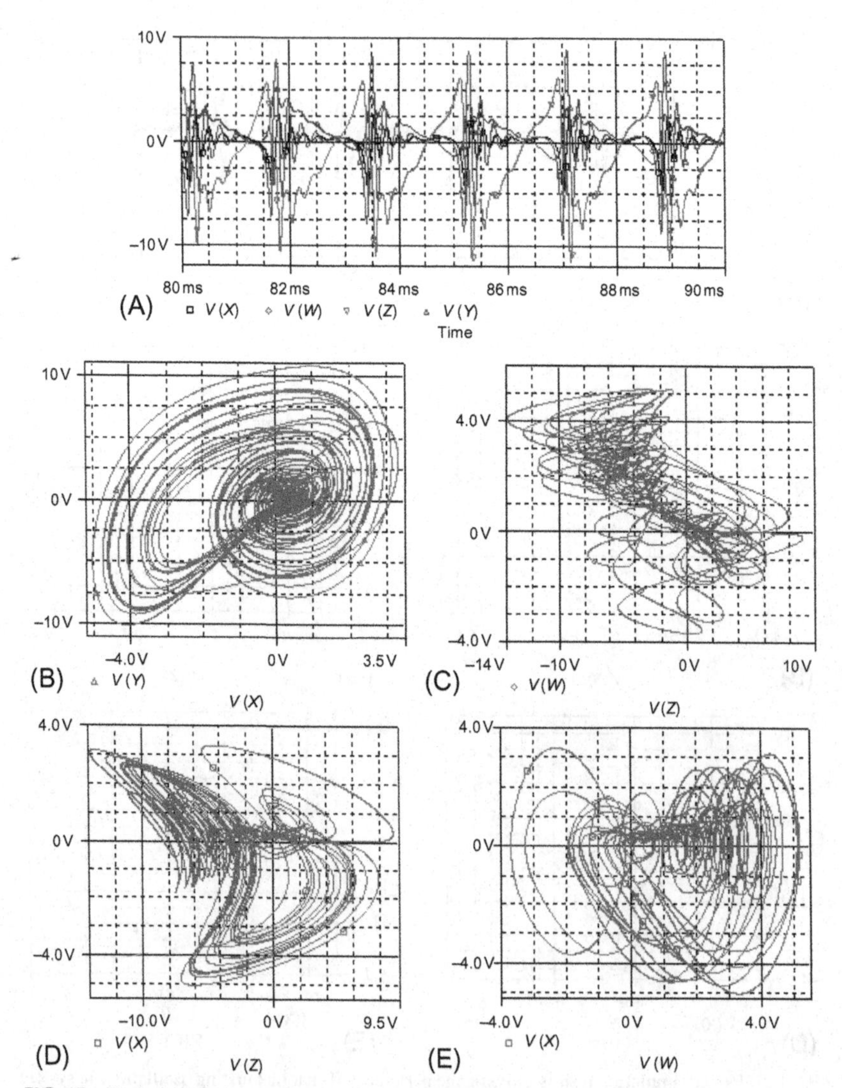

FIG. 15 Pspice simulation results showing the periodic and chaotic bursting oscillations in system (2): (a) time series of the output voltages V_x, V_y, V_z and V_w; phase space projections in the planes (A) (V_x, V_y), (B) (V_z, V_w), (C) (V_z, V_x) and (D) (V_w, V_x) for $R_b = 7.69\,\text{k}\Omega$. The initial values of output voltages are $(V_x(0), V_y(0), V_y(0), V_w(0)) = (0.1, 0.1, 0.1, 0.1)$.

5.1 Preliminaries

The driven system is described by the following set of equations:

$$\begin{cases} \dot{x}_1 = y_1 - x_1 - az_1 \\ \dot{y}_1 = x_1 z_1 - x_1 + w_1 \\ \dot{z}_1 = b - y_1 - x_1 y_1 \\ \dot{w}_1 = -x_1 \end{cases} \tag{8}$$

and the response system is described by:

$$\begin{cases} \dot{x}_2 = y_2 - x_2 - az_2 + u_1(t) \\ \dot{y}_2 = x_2 z_2 - x_2 + w_2 + u_2(t) \\ \dot{z}_2 = b - y_2 - x_2 y_2 + u_3(t) \\ \dot{w}_2 = -x_2 + u_4(t) \end{cases} \tag{9}$$

where $u_1(t)$, $u_2(t)$, $u_3(t)$ and $u_4(t)$ are the controller signals defined as follows:

$$u_1(t) = u_3(t) = u_4(t) = 0 \quad u_2(t) = -k(y_2 - y_1) - ku \tag{10}$$

We set the synchronization error between drive and response systems as:

$$e_1 = x_2 - x_1,\ e_2 = y_2 - y_1,\ e_3 = z_2 - z_1,\ e_4 = w_2 - w_1 \tag{11}$$

The errors system is obtained as follows:

$$\begin{cases} \dot{e}_1 = e_2 - e_1 - ae_3 \\ \dot{e}_2 = f(x,z) - e_1 + e_4 - ke_2 - ku \\ \dot{e}_3 = -e_2 + g(x,y) \\ \dot{e}_4 = -e_1 \end{cases} \tag{12}$$

with the nonlinear functions $f(x,z)$ and $g(x,y)$ defined by:

$$f(x,z) = x_2 z_2 - x_1 z_1 \tag{13}$$

and

$$g(x,z) = x_1 y_1 - x_2 y_2 \tag{14}$$

The response system (9) synchronizes with the drive system (8) in a finite-time if there exists a final time t_{syn} such that $\lim_{t \to t_{syn}} \|e_i(t)\| = 0$.
Assumption The nonlinear functions $f(x,z)$ and $g(x,z)$ respect the condition:

$$\begin{aligned} &\|f(y,z)e_2 + g(x,z)e_3 + e_2 e_4 - e_2 e_3 - e_1 e_4 - ae_1 e_3\| \\ &\le \sum_{i=1}^{4} \xi_i e_i^2 \ \text{ with } \xi_i \ (i = 1, \ldots, 4) \end{aligned} \tag{15}$$

5.2 Main Results

Theorem *The response system* (2) *synchronizes with the drive system* (1) *in the finite-time*

$$t_{syn} \le \frac{1}{2\eta}\left(e_x^2(0) + e_y^2(0) + e_z^2(0) + e_w^2(0)\right) + \frac{|u(0)|}{\eta} \tag{16}$$

if the adaptive feedback controller u is defined as follows:

$$\dot{u} = -\left(\sum_{i=1}^{4} \xi_i e_i^2 - k e_2 u + \eta\right) \operatorname{sign}(u) \tag{17}$$

where η, ξ_i $(i = 1,\ldots,4)$ and k are the positive parameters, and the slave system written as:

$$\begin{cases} \dot{x}_2 = y_2 - x_2 - a z_2 \\ \dot{y}_2 = x_2 z_2 - x_2 + w_2 - k e_2 - k u \\ \dot{z}_2 = b - y_2 - x_2 y_2 \\ \dot{w}_2 = -x_2 \\ \dot{u} = -\left(\sum_{i=1}^{4} \xi_i e_i^2 - k e_2 u + \eta\right) \operatorname{sign}(u) \end{cases} \tag{18}$$

Proof We consider the following Lyapunov candidate function defined by:

$$V(t) = \frac{1}{2}\left(e_1^2 + e_2^2 + e_3^2 + e_4^2\right) + |u(t)| \tag{19}$$

The derivative of (17) is:

$$\dot{V}(t) = e_1 \dot{e}_1 + e_2 \dot{e}_2 + e_3 \dot{e}_3 + e_4 \dot{e}_4 + \dot{u}(t) \operatorname{sign}(u(t)) \tag{20}$$

This derivative along of (10) allows us to obtain:

$$\begin{aligned} \dot{V}(t) = & -e_1^2 + f(x,z) e_2 + g(x,y) e_3 + e_2 e_4 - k e_2^2 - e_2 e_3 - e_1 e_4 - a e_1 e_3 \\ & - k e_2 u + \dot{u} sign(u(t)) \end{aligned} \tag{21}$$

Using assumption given by Eq. (15), Eq. (21) becomes:

$$\dot{V}(t) \leq \sum_{i=1}^{4} \xi_i e_i^2 - k u e_2 + sign(u) \dot{u} \tag{22}$$

If there exist a positive constant η such that

$$\dot{V}(t) \leq -\eta \tag{23}$$

then the update laws for the controller $u(t)$ is designed through the following relations:

$$\dot{u} = -\left(\sum_{i=1}^{4} \xi_i e_i^2 - k e_2 u + \eta\right) sign(u) \tag{24}$$

Therefore, response system (9) synchronizes with the drive system (8) in a finite-time of synchronization given by:

$$t_{syn} \leq \frac{1}{2\eta}\left(e_x^2(0)+e_y^2(0)+e_z^2(0)+e_w^2(0)\right)+\frac{|u(0)|}{\eta} \tag{25}$$

□

5.3 Numerical Verifications

Here, the graphical representation of the synchronization process obtained numerically by carrying out via a numerical integration of drive and response systems. The parameter values of the drive system are selected so that it exhibits chaotic hidden behavior ($a=0$, $b=3$). The initial conditions for the drive and response are chosen as: $(x_1(0), y_1(0), z_1(0), w_1(0)) = (0.1, 0.1, 0.1, 0.1)$ and $(x_2(0), y_2(0), z_2(0), w_2(0)) = (0.1, 0.3, 0.4, 0.5)$, respectively. To guarantee the chaotic synchronization, the other parameters are fixed to be: $k=1.5$, $\eta=0.1$, $\xi_1=1$, $\xi_2=1$, $\xi_3=1$ and $\xi_4=1$. Based on these values, the theoretical settled time of synchronization is $t_{Tsyn}=131.2$. The time dependence of the dynamics of the synchronization errors and the adaptive feedback controller $u(t)$ are displayed in Fig. 16.

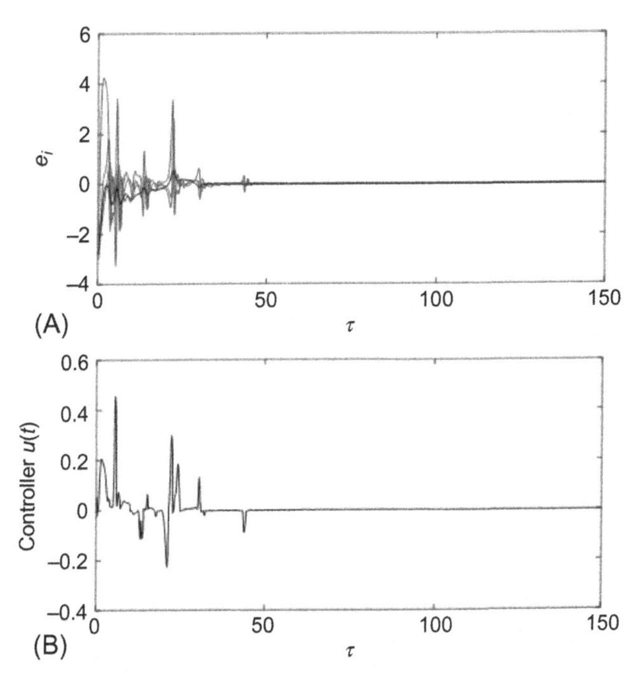

FIG. 16 Time dependence of dynamics errors (A) and adaptive feedback controller $u(t)$ (B) showing synchronization between drive (8) and response (9) systems with adaptive finite time synchronization method.

From Fig. 16, we observe that the numerical time of synchronization is $t_{Nsyn} = 50$. By comparing with theoretical time of synchronization we have $t_{Nsyn} \leq t_{Tsyn}$. This result respects the finite time condition given in [41]. So, we can use these results to realize telecommunication applications.

6 A TEXT ENCRYPTION APPLICATION USING HIDDEN CHAOTIC ATTRACTOR OF PROPOSED AUTONOMOUS SYSTEM WITH ONE OR WITHOUT EQUILIBRIUM POINTS

Based on results from the adaptive finite-time synchronization of two identical systems under scrutiny with hidden attractor, we develop an application on the text encryption. The method of text encryption used in this chapter is reported in [41–43]. In digital cryptography, the letters can be replaced by the numbers which can be assigned by the sender. In this chapter, the message (plaintext) will be replaced by their ASCII codes before the encryption. As the ASCII code table has 128 characters, we define the formula for the ciphertext and decrypted message corresponding to assignment of numbers as follows:

$$\begin{cases} c = p_s + k_s \ (\bmod 128) \\ p_r = c - k_r \ (\bmod 128) \end{cases} \tag{26}$$

where p_s is the plaintext to be encrypted, p_r is the plaintext to be recovered, k_s the secret keys of sender, k_r the secret keys of the receiver and c the ciphertext.

6.1 Proposed Affine Cipher

Let S and R be the sender and the receiver, respectively, in the cryptosystem. We consider the drive system (6) as a sender's system and the response system (7) as a receiver's system. S and R are agreed on a time $t \geq t_{syn}$ and a number of characters of the ASCII codes table which equals 128. The public key $D = hour/Day/Month/year$ is only shared by S and R. These public keys serve to build the two privates keys. The complete procedure between S and R are described in the following subsection.

6.2 Key Generation

To generate the key, the following steps are needed:

(a) The plaintext can be presented into form of integers using the table of the ASCII code. After this assignation, we can group this plaintext in blocks of four terms. If the number of letter constituting the message is a multiple of 4, we will have p_1, p_2, p_3, p_4 for a first bloc and p_5, p_6, p_7, p_8 for a second block, and so on, while, if the number of letter constituting the message is not multiple of 4, then we can add the blank spaces at the end of the sequence to complete the sequence.

(b) For $t \geq t_{syn}$, the sender S has (x_1, y_1, z_1, w_1) as solution of chaotic system (8), and receive R has (x_2, y_2, z_2, w_2) as solution of chaotic system (9). The private keys of the sender and receiver are respectively defined as follows:

$$k_1 = [hour^*(-w_1) + Day^* z_1 + Month^* y_1 + year^* x_1](\bmod 128) \quad (27)$$

and

$$k_2 = [hour^*(-w_2) + Day^* z_1 + Month^* y_2 + year^* x_2](\bmod 128) \quad (28)$$

where $[a]$ is the integer part of a.

(c) The first four sending keys k_{1s}, k_{2s}, k_{3s}, k_{4s} are determined by:

$$Q^{k_1} = \begin{pmatrix} F_{k_1+1} & F_{k_1} \\ F_{k_1} & F_{k_1-1} \end{pmatrix} = \begin{pmatrix} \widetilde{k_1} & \widetilde{k_2} \\ \widetilde{k_3} & \widetilde{k_4} \end{pmatrix} \quad (29)$$

where $k_{1s} = \widetilde{k_1}(\bmod 128)$, $k_{2s} = \widetilde{k_2}(\bmod 128)$, $k_{3s} = \widetilde{k_3}(\bmod 128)$, $k_{4s} = \widetilde{k_4}(\bmod 128)$ and F_n the Fibonacci number, $n \geq 3$ is defined by $F_{n+1} = F_n + F_{n-1}$, taking into account that the initial terms are $F_0 = 0$ and $F_1 = F_2 = 0$.

(d) The first four receiving keys k_{1r}, k_{2r}, k_{3r}, k_{4r} are determined by:

$$Q^{k_2} = \begin{pmatrix} F_{k_2+1} & F_{k_2} \\ F_{k_2} & F_{k_2-1} \end{pmatrix} = \begin{pmatrix} \widetilde{k_1} & \widetilde{k_2} \\ \widetilde{k_3} & \widetilde{k_4} \end{pmatrix} \quad (30)$$

where $k_{1r} = \widetilde{k_1}(\bmod 128)$, $k_{2r} = \widetilde{k_2}(\bmod 128)$, $k_{3r} = \widetilde{k_3}(\bmod 128)$, $k_{4r} = \widetilde{k_4}(\bmod 128)$.

(e) The ciphertext and recovered plaintext are obtained by $c = p_s + k_s$ (mod 128) and $p_r = c - k_r$ (mod 128), respectively. After the treatment of the first block of plaintext, we restart the process for the next block, and so on.

6.3 Numerical Verifications

In this verification, the message to send is "PASSWORD". By following the five steps mentioned above, we obtain the results summarized in Table 1.

From Table 1, it is observed that the sending message is easily received when the synchronization phenomenon is achieved between the drive (8) and response (9) systems. The affine cipher used in this chapter possesses three keys, one public key D, and two private keys, k_1 and k_2. The two public keys have been computed by using the private key while, the four receiver keys and four sending keys have been computed by using the private and sending key as well as the Fibonacci numbers. The receiver cannot recover the original text without the knowledge of the receive private key k_2. Compared to the

TABLE 1 Summary of Sending and Receive Message (PASSWORD)

Times	k_1	Plaintext	Keys k_s	Ciphertext	k_2	Keys k_r	Plaintext Recovered
$t = 100$	67	80 (P)	15	95	67	15	80 (P)
		65 (A)	66	131		66	65 (A)
		83 (S)	66	149		66	83 (S)
		83 (S)	77	160		77	83 (S)
$t = 150$	16	87(W)	85	172	16	85	87(W)
		79 (O)	24	102		24	79 (O)
		82 (R)	24	106		24	82 (R)
		68 (D)	61	129		61	68 (D)

existing works [33,34], our proposed method is fully authenticated and more adapted for telecommunication applications.

7 CONCLUDING REMARKS

Investigating chaotic systems with self-excited and hidden attractors is a great challenge. In this chapter, a four-dimensional chaotic system with self-excited and hidden attractors was introduced. It was found that the structure of the system can change depending on the value of the parameters of the system. The dynamics of such a system have been examined through time series, bifurcation diagrams, Lyapunov exponent plots, and attractor plots. It has been found that the system experienced some interesting behaviors including period-doubling to chaos, reverse period-doubling bifurcation, periodicity, quasi-periodicity, periodic and chaotic bursting oscillations, and controllable amplitude. An analogue electronic circuit capable of mimicking the dynamics of the system under scrutiny with self-excited and hidden attractors was designed and implemented in Pspice software. Both theoretical and Pspice simulation results show a very good agreement. The adaptive finite time synchronization of two identical four-dimensional chaotic systems with hidden attractors was discussed, and some numerical simulations were performed to illustrate the effectiveness of the proposed synchronization method. The results for synchronization were exploited to realize a text encryption application. The simulation verifications were included to illustrate the proposed text encryption technique. It is found that the proposed technique compared to others reported in existing works [44,45] is fully authenticated and more adapted for telecommunication applications. We stress that the approaches followed in this chapter may be exploited rigorously to study any other chaotic system displaying self-excited and hidden attractors.

REFERENCES

[1] Sprott JC. Symmetric time-reversible flows with a strange attractor. Int J Bifurcation Chaos 2015;25(05).
[2] Li C, Hu W, Sprott JC, Wang X. Multistability in symmetric chaotic systems. Eur Phys J Spec Top 2015;224(8):1493–506.
[3] Lü J, Chen G. Generating multiscroll chaotic attractors: theories, methods and applications. Int J Bifurcation Chaos 2006;16(04):775–858.
[4] Brummitt CD, Sprott J. A search for the simplest chaotic partial differential equation. Phys Lett 2009;373(31):2717–21.
[5] Jafari S, Sprott JC. Simple chaotic flows with a line equilibrium. Chaos, Solitons Fractals 2013;57:79–84.
[6] Lorenz EN. Deterministic nonperiodic flow. J Atmos Sci 1963;20(2):130–41.
[7] Rössler OE. An equation for continuous chaos. Phys Lett 1976;57(5):397–8.
[8] Chen G, Ueta T. Yet another chaotic attractor. Int J Bifurcation Chaos 1999;9:14651999.
[9] Lü J, Chen G, Cheng D, Celikovsky S. Bridge the gap between the Lorenz system and the Chen system. Int J Bifurcation Chaos 2002;12(12):2917–26.
[10] Pham VT, Volos C, Jafari S, Kapitaniak T. A simple chaotic circuit with a light-emitting diode. Optoelectron Adv Mater Rapid Commun 2016;10(99):640–6.
[11] Singh JP, Roy BK. Crisis and inverse crisis route to chaos in a new 3-D chaotic system with saddle, saddle foci and stable node foci nature of equilibria. Optik 2016;127(24):11982–2002.
[12] Singh PP, Singh JP, Roy BK. Synchronization and anti-synchronization of Lu and Bhalekar-Gejji chaotic systems using nonlinear active control. Chaos, Solitons Fractals 2014;69:3139.
[13] Lassoued A, Boubaker O. On new chaotic and hyperchaotic systems: a literature survey. Nonlinear Anal Modell Control 2016;21(6):770–89.
[14] Lassoued A, Boubaker O. Dynamic analysis and circuit design of a novel hyperchaotic system with fractional-order terms, Complexity 2017;2017:10 pages. https://doi.org/10.1155/2017/3273408.
[15] Mkaouar H, Boubaker O. Robust control of a class of chaotic and hyperchaotic driven systems. Pramana J Phys 2017;88:9.
[16] Lassoued A, Boubaker O. A new fractional-order jerk system and its hybrid synchronization. In: Fractional order control and synchronization of chaotic systems. Cham: Springer; 2017. p. 699–718.
[17] Lassoued A, Boubaker O. In: Hybrid synchronization of multiple fractional-order chaotic systems with ring connection. Modelling, identification and control (ICMIC). 8th international IEEE conference; 2016. p. 109–14.
[18] Shilnikov LP, Shilnikov AL, Turaev DV. Methods of qualitative theory in nonlinear dynamics part-II. Singapore: World Scientific Publishing Co. Pte. Ltd; 2001.
[19] Wang X, Chen G. A chaotic system with only one stable equilibrium. Commun Nonlinear Sci Numer Simul 2012;17:1264–72.
[20] Molaie M, Jafari S, Sprott JC, Golpayegani SMRH. Simple chaotic flows with one stable equilibrium. Int J Bifurcation Chaos 2013;23.
[21] Kingni ST, Jafari S, Simo H, Woafo P. Three-dimensional chaotic autonomous system with only one stable equilibrium: analysis, circuit design, parameter estimation, control, synchronization and its fractional-order form. Eur Phys J Plus 2014;129:76.
[22] Wei Z. Dynamical behaviors of a chaotic system with no equilibria. Phys Lett A 2011;376:102–8.
[23] Jafari S, Sprott JC, Golpayegani SMRH. Elementary quadratic chaotic flows with no equilibria. Phys Lett A 2013;377:699–702.
[24] Akgul A, Calgan H, Koyuncu I, Pehlivan I, Istanbullu A. Chaos-based engineering applications with a 3D chaotic system without equilibrium points. Nonlinear Dynam 2016;84:481–95.

[25] Gotthans T, Petrzela J. New class of chaotic systems with circular equilibrium. Nonlinear Dynam 2015;73:429–36.

[26] Gotthans T, Sportt JC, Petrzela J. Simple chaotic flow with circle and square equilibrium. Int J Bifurcation Chaos 2016;26(1650):137–8.

[27] Kingni ST, Pham VT, Jafari S, Kol GR, Woafo P. Three-dimensional chaotic autonomous system with a circular equilibrium: analysis, circuit implementation and its fractional-order form. Circ Syst Signal Process 2016;35(19):331–1948.

[28] Pham VT, Jafari S, Wang X, Ma J. A chaotic system with different shapes of equilibria. Int J Bifurcation Chaos 2016;26:1650069.

[29] Jafari S, Sprott JC, Pham VT, Volos C, Li C. Simple chaotic 3D flows with surfaces of equilibria. Nonlinear Dynam 2016;86:1349–58.

[30] Jafari MA, Mliki E, Akgul A, Pham V-T, Kingni ST, Wang X, Jafari S. Chameleon: the most hidden chaotic flow. Nonlinear Dyn 2017;88:2303–17.

[31] Karthikeyan R, Akgul A, Jafari S, Karthikeyan A, Koyuncu I. Chaotic chameleon: dynamic analyses, circuit implementation, FPGA design and fractional-order form with basic analyses. Chaos, Solitons Fractals 2017;103:476–87.

[32] Rajagopal K, Karthikeyan A, Duraisamy P, Chameleon H. Fractional order FPGA implementation. Complexity 2017.

[33] Pehlivan I, Uyaroglu Y. A new 3D chaotic system with golden proportion equilibria: analysis and electronic circuit realization. Comput Electr Eng 2012;38:1777–84.

[34] Kingni ST, Nana B, Ngueuteu SGM, Woafo P, Danckaert J. Bursting oscillations in a 3D system with asymmetrically distributed equilibria: mechanism, electronic implementation and fractional derivation effect. Chaos, Solitons Fractals 2015;71:29–40.

[35] Li C, Sprott JC. Amplitude control approach for chaotic signals. Nonlinear Dyn 2013; 73(3):1335–41.

[36] Li C, Sprott JC. Variable-boostable chaotic flows. Optik Int J Light Electron Opt 2016; 27(22):10389–98.

[37] Kamdoum Tamba V, Rajagopal K, Pham V-T, Hoang DV. Chaos in a system with an absolute nonlinearity and chaos synchronization. Adv Math Phys 2018;2018:12, Article ID 5985489. https://doi.org/10.1155/2018/5985489.

[38] [40] Izhikevich EM. Neural excitability, spiking and bursting. Int J Bifurcation Chaos 2000;10:1171–266.

[39] Izhikevich EM. Dynamical systems in neuroscience: the geometry of excitability and bursting. Cambridge: The MIT Press; 2007.

[40] Vanag VK, Yang L, Dolnik M, Zhabotinsky AM, Epstein IR. Oscillatory cluster patterns in a homogeneous chemical system with global feedback. Nature 2000;406:389–98.

[41] Kengne R, Tchitnga R, Mezatio A, Fomethe A, Litak G. Finite-time synchronization of fractional-order simplest two-component chaotic oscillators. Eur Phys J B 2017;90(5):88.

[42] Muthukumar P, Balasubramaniam P. Feedback synchronization of the fractional order reverse butterfly-shaped chaotic system and its application to digital cryptography. Nonlinear Dyn 2013;74:1169–81.

[43] Muthukumar P, Balasubramaniam P, Ratnavelu K. Fast projective synchronization of fractional order chaotic and reverse chaotic systems with its application to an affine cipher using date of birth (DOB). Nonlinear Dyn 2015;80(4):1883–97.

[44] Muthukumar P, Balasubramaniam P. Feedback synchronization of the fractional order reverse butterfly-shaped chaotic system and its application to digital cryptography. Nonlinear Dyn Int J Nonlinear Dyn Chaos Eng Syst 2013;74(4):69–1181.

[45] Muthukumar P, Balasubramaniam P, Ratnavelu K. Fast projective synchronization of fractional order chaotic and reverse chaotic systems with its application to an aine cipher using date of birth (DOB). Nonlinear Dyn 2015;80(4):1883–97.

Chapter 15

FPGA Implementation of Chaotic Oscillators, Their Synchronization, and Application to Secure Communications

Esteban Tlelo-Cuautle*, Omar Guillén-Fernández*, Jose de Jesus Rangel-Magdaleno*, Ashley Melendez-Cano†, Jose Cruz Nuñez-Perez† and Luis Gerardo de la Fraga‡
*Department of Electronics, INAOE, Puebla, Mexico, †Department of Telecommunications, CITEDI-IPN, Tijuana, Mexico, ‡Department of Computer Sciences, CINVESTAV, Gustavo A. Madero, Mexico

1 INTRODUCTION

Chaotic oscillators have been investigated for many years, and one can find useful references providing a review of theory, methods, and applications, like in [6]. In electronics, for example, Chua's circuit is quite known for the simplicity of construction, by passive circuit elements, of a piecewise-linear (PWL) function generating a double-scroll or multiscroll attractors. Chua's circuit has also been used to implement networks like in [7], where a more deeper analysis is performed to guarantee chaotic behavior. Other kinds of chaotic circuits can have different perturbation functions, that is, Chua's circuit needs a PWL function to generate chaos and other chaotic oscillators need trigonometric, or polynomial functions, and they can be synthesized by using electronic circuits [8]. If the chaotic oscillator is implemented with more than three ordinary differential equations (ODEs), they can have more than one positive Lyapunov exponent, and then, the behavior is said to be hyperchaotic [2, 9]. This is a more complex chaotic behavior. Also, some researchers investigate chaotic oscillators with time-delay in the nonlinear function [10]. If one is interested by increasing the unpredictability or value of the positive Lyapunov exponent, multiscroll

Recent Advances in Chaotic Systems and Synchronization. https://doi.org/10.1016/B978-0-12-815838-8.00015-7

chaotic oscillators are the solution, and more complex behavior can be guaranteed by including time-delay as shown in [11].

A survey on chaotic and hyperchaotic oscillators can be found in [1], and some applications can be found in the literature in [3, 12]. For instance, the synchronization of chaotic oscillators started from the work of Pecora and Carrol [13], in which the authors proposed the general synchronizing scheme by taking a nonlinear system, duplicating a subsystem of this system, and driving the duplicate and the original subsystem with signals from the unduplicated part. The authors demonstrated this by implementing chaotic oscillators with electronic circuits. Currently, new synchronization approaches have been developed for chaotic and hyperchaotic oscillators as shown in [14, 15]. Those synchronization approaches can be implemented with either or both analog and digital circuitry. In this chapter, we show the implementation of the synchronization of two chaotic oscillators by using field programmable gate arrays (FPGAs). Traditional implementations using amplifiers can still be found in recent literature like in [16]. The synchronization of chaotic oscillators is quite useful for the development of applications, like the recent ones given in [5, 17, 18]. However, in this chapter we apply generalized Hamiltonian forms and observer approach, as shown in [19, 20].

Chaotic secure communications can be of great help if they can be implemented with electronic devices like FPGAs. The challenge is how to guarantee high sensitivity to initial conditions using embedded systems or analog circuitry, as shown in [6]. In addition, chaotic behavior has a broad potential not only in digital ciphering and secure communications but also in many engineering areas like in biomedicine, military systems, power systems protection, fluid mixing, optical systems, and so on [12].

From the applications mentioned above, one can wonder what kind of chaotic system can be helpful to implement such engineering applications. In this manner, this chapter details the simulation and FPGA-based implementation of three chaotic oscillators based on PWL functions, namely, saturated nonlinear function series, a function consisting of linear-segments with negative slopes, and sawtooth function [4, 21–23]. The models associated to those PWL-functions-based chaotic oscillators are described by ODEs and solved by three numerical methods. We choose one-step methods like the simple Forward-Euler, Trapezoidal, and fourth-order Runge-Kutta. The discretized equations are then implemented into an FPGA and at the end, two chaotic oscillators are synchronized and connected in a master-slave topology to transmit an image. The rest of this chapter is organized as follows: Section 2 describes the mathematical models of three chaotic oscillators based on PWL functions, and their simulation results by applying three numerical methods, which have been programmed into MATLAB. Section 3 details the hardware description language (HDL) associating the discretized equations of the chaotic oscillator to digital blocks that will synthesize the oscillator into an FPGA. This section also shows cosimulation between Active-HDL and MATLAB-Simulink. Once implemented the chaotic oscillators into the FPGA, the experimental observation of the chaotic attractors and the synchronization of two of them in a master-slave topology are shown in Section 4. The transmission of an image by using the three chaotic oscillators is given in Section 5. Finally, the conclusions are listed in Section 6.

2 SIMULATION OF CHAOTIC OSCILLATORS BASED ON PWL FUNCTIONS

Among all kinds of chaotic oscillators, we choose three examples based on PWL functions. The main advantage is the generation of multiscrolls in more than one direction [6].

The chaotic oscillator based on saturated nonlinear function series system is quite used nowadays due to its simplicity to generate multiscrolls in more than one direction, its model is given by Eq. (1). It consists of three ODEs (x, y, z) having four coefficients (a, b, c, d_1) and one nonlinear function $f(x)$ that can be approached by a PWL function. To generate two scrolls, $f(x)$ can be described as shown in Eq. (2) and sketched in Fig. 1A, where bp is associated to the break-points. If one augments the number of saturation levels in such a function, the number of scrolls beings increased, that is, just one scroll is generated for each saturated level.

$$\begin{aligned} \dot{x} &= y \\ \dot{y} &= z \\ \dot{z} &= -ax - by - cz + d_1 f(x) \end{aligned} \tag{1}$$

$$f(x) = \begin{cases} -k & \text{si } x < -bp_1 \\ mx & \text{si } -bp_1 \leq x \leq bp_1 \\ k & \text{si } x > bp_1 \end{cases} \tag{2}$$

The chaotic oscillator based on a PWL function consisting of negative slopes can be described by Eq. (3). It consists of three state variables (x, y, z), three coefficients (α, β, γ), and the PWL function $f(x)$ that can be increased in its number of linear segments to generate multiscrolls. To generate two scrolls the PWL function $f(x)$ can be described by Eq. (4), and it is sketched in Fig. 1B, where one can see the two negative slopes m_0 and m_1.

$$\begin{aligned} \dot{x} &= \alpha(y - x - f(x)) \\ \dot{y} &= \gamma(x - y + z) \\ \dot{z} &= -\beta y \end{aligned} \tag{3}$$

$$f(x) = m_{2n-1}x + \frac{1}{2}\sum_{i-q}^{2n+1}(m_{i-1} - m_i)(|x + bp_i| - |x - bp_i|) \tag{4}$$

The third chaotic oscillator that is considered herein and can be modeled by a PWL function is the one based on sawtooth function, which can be described

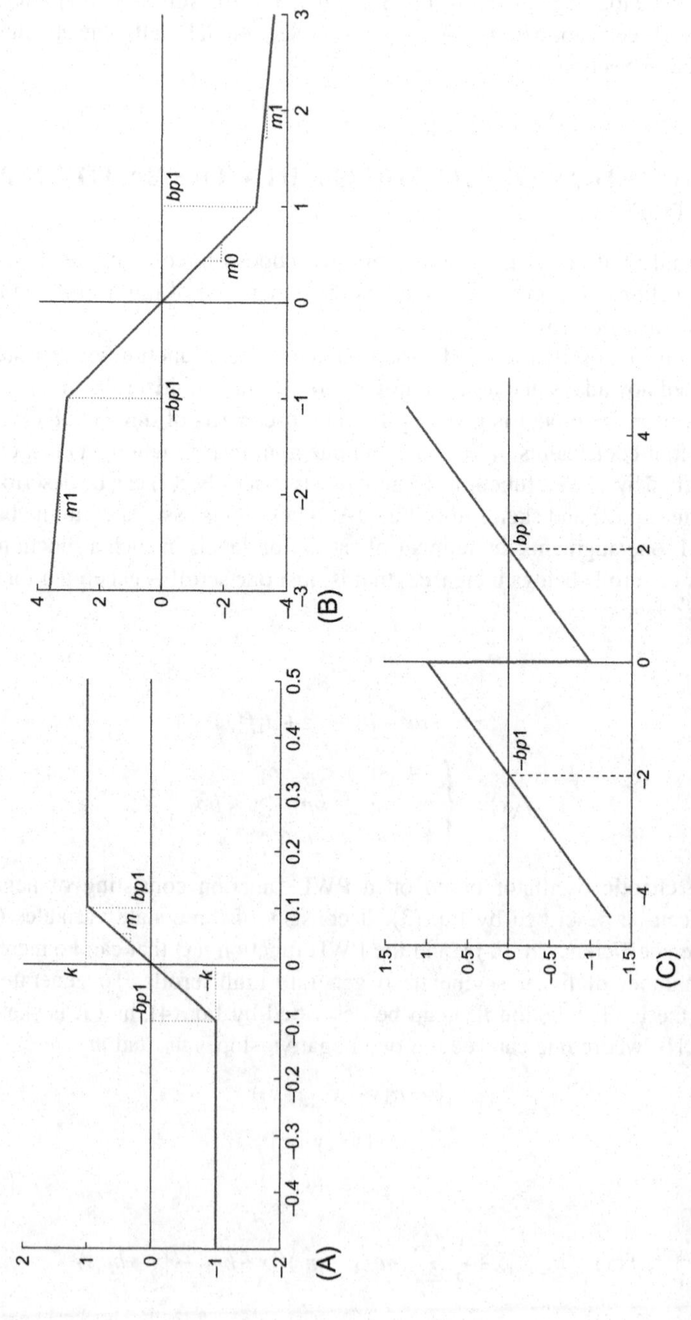

FIG. 1 PWL functions based on: (A) saturated nonlinear series, (B) negative slopes, and (C) sawtooth function.

by Eq. (5), where it consists of three state variables (x, y, z), three coefficients α, β, γ, and the PWL function $f(x)$, which can be increased in the number of linear segments to increase the number of scrolls. To generate two scrolls, $f(x)$ can be described by Eq. (6) and sketched by Fig. 1C.

$$\begin{aligned}\dot{x} &= \alpha y + f(x) \\ \dot{y} &= \alpha x - \gamma y - \alpha z \\ \dot{z} &= \beta y\end{aligned} \tag{5}$$

$$f(x) = \xi(bp \cdot sgn(x) \cdot (2k-1) - x) \quad \text{if } 2kbp \geq |x| > 2(k-1)bp \tag{6}$$

The PWL functions described earlier can also be generated in an irregular form, as shown in [24], where the positive Lyapunov exponent is optimized by applying metaheuristics to generate chaotic attractors with more complex dynamics.

2.1 Numerical Methods to Simulate Chaotic Oscillators

There exists numerical methods that depend of one-step or multisteps to provide a solution, and also some of them can change the order and the step size. In this work, we use three well-known methods, namely, Forward-Euler, Trapezoidal, and fourth-order Runge-Kutta.

The Forward-Euler is the simple numerical method and it has the iterative equation given by Eq. (7).

$$g_{n+1} = g_n + h \cdot f(g_n) \tag{7}$$

where g_n is the vector that represents the state variable values at iteration n, and it requires initial conditions g_0. h is the step size that depends on the eigenvalues to be estimated in order to minimize numerical errors.

It is well known that the Trapezoidal method is more exact than Forward-Euler one because, in this case, it evaluates the average between two values evaluated at the current discrete time $n + 1$ and the previous one n, therefore it has the iterative form given in Eq. (8). In this case, the function $f(g_{n+1})$ can be evaluated by applying Forward Euler.

$$g_{n+1} = g_n + \frac{h}{2} \cdot [f(g_{n+1}) + f(g_n)] \tag{8}$$

The fourth-order Runge-Kutta method has better exactness than the previous ones, and it allows a step-size h value slightly higher than that for the Forward Euler and Trapezoidal methods. Its application requires more resources to perform more operations that are described by Eqs. (9), (10).

$$g_{n+1} = g_n + \frac{h}{6} \cdot (k_1 + 2k_2 + 2k_3 + k_4) \tag{9}$$

$$
\begin{aligned}
k_1 &= h \cdot f(g_n) \\
k_2 &= h \cdot f\left(g_n + \frac{1}{2}h, \frac{1}{2}k_1\right) \\
k_3 &= h \cdot f\left(g_n + \frac{1}{2}h, \frac{1}{2}k_2\right) \\
k_4 &= h \cdot f(g_n + h, k_3)
\end{aligned}
\tag{10}
$$

2.2 Simulation of the PWL-Function-Based Chaotic Oscillators

The chaotic oscillator based on saturated nonlinear functions is simulated by the three numerical methods described earlier where the coefficients were set to [22]: $a = b = c = d_1 = 0.7$. The PWL function has $bp_1 = 0.0165$ and $k = 1$, thus having the slope $m = 60.606$. Fig. 2A shows the attractor by applying the Forward-Euler method with $h = 0.01$ and the initial conditions: $x_0 = 0.1$, $y_0 = 0$, and $z_0 = 0$. By applying the Trapezoidal method with the same conditions as for the application of Forward Euler, the attractor is shown in Fig. 2B. By applying the fourth-order Runge-Kutta again with the same conditions, the attractor is shown in Fig. 2C.

The chaotic oscillator based on a PWL function consisting of negative slopes was simulated by setting $m_0 = -\frac{8}{7}, m_1 = -\frac{5}{7}, \alpha = 10, \beta = 14.286$, $\gamma = 1$, and $bp_1 = 1$. Fig. 3A shows the attractor by applying the Forward-Euler method with $h = 0.01$ and the initial conditions $x_0 = 0.1$, $y_0 = 0$, and $z_0 = 0$. Fig. 3B and C shows the attractors generated by applying the Trapezoidal and fourth-order Runge-Kutta methods, respectively.

To simulate the chaotic oscillator with a PWL function based on sawtooth, the coefficient values were set to [25]: $\alpha = 3$, $\beta = 4$, $\gamma = 1$, $bp_1 = 0.0165$, $\xi = 0.8$, and $k = 1$. Fig. 4A shows the attractor by applying the Forward-Euler method with $h = 0.01$ and the initial conditions $x_0 = 0.1$, $y_0 = 0$, and $z_0 = 0$. Fig. 4B and C shows the attractors generated by applying the Trapezoidal and fourth-order Runge-Kutta methods, respectively.

3 COSIMULATION BETWEEN ACTIVE-HDL AND SIMULINK

Before describing the blocks of the mathematical models of the chaotic oscillators to be synthesized into an FPGA, this section shows the cosimulation between Active-HDL and Simulink. The blocks of the chaotic oscillators were described by the HDL into Active-HDL, and Simulink was used to simulate the oscillators and to verify the exactness of the outputs associated with the state variables.

From the mathematical models, and according to the numerical method being used to simulate chaotic behavior, prior to perform the cosimulation one must describe the digital blocks of the chaotic oscillator. By using VHDL as already highlighted in [22], Fig. 5 shows the block description of the chaotic oscillator based on saturated nonlinear function series and by applying the

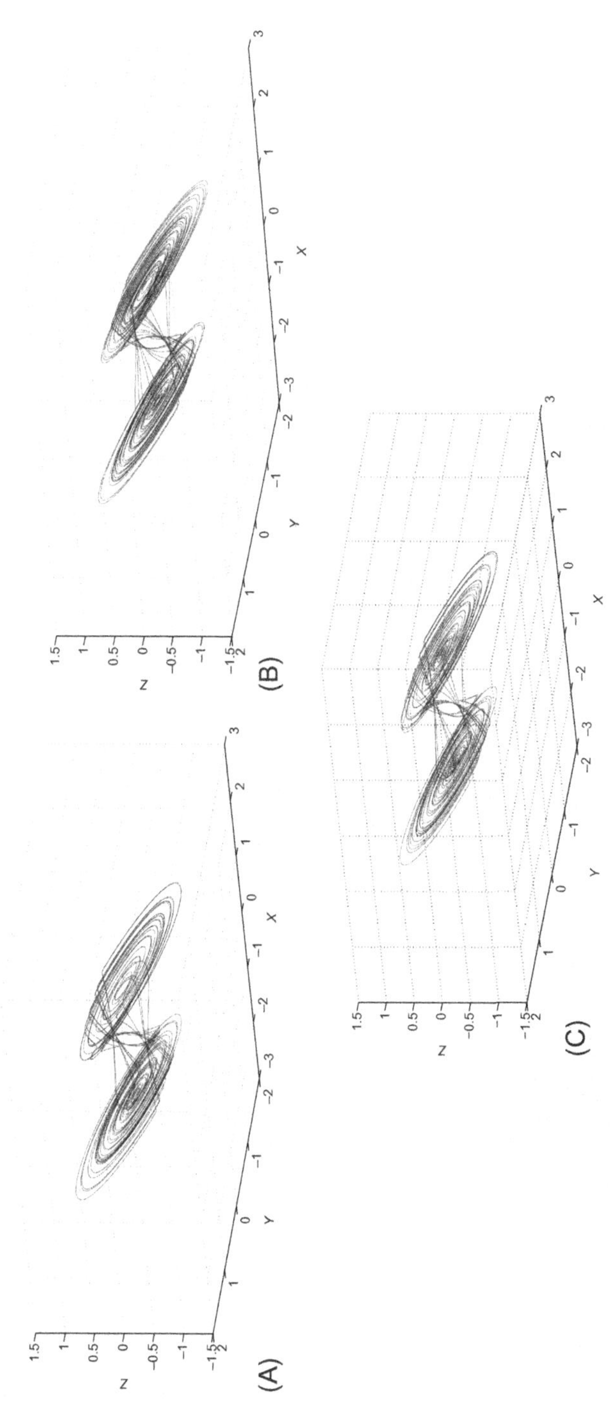

FIG. 2 Two-scroll attractor generated by the oscillator based on saturated nonlinear function series and by applying the numerical method: (A) Forward Euler, (B) Trapezoidal, and (C) fourth-order Runge-Kutta.

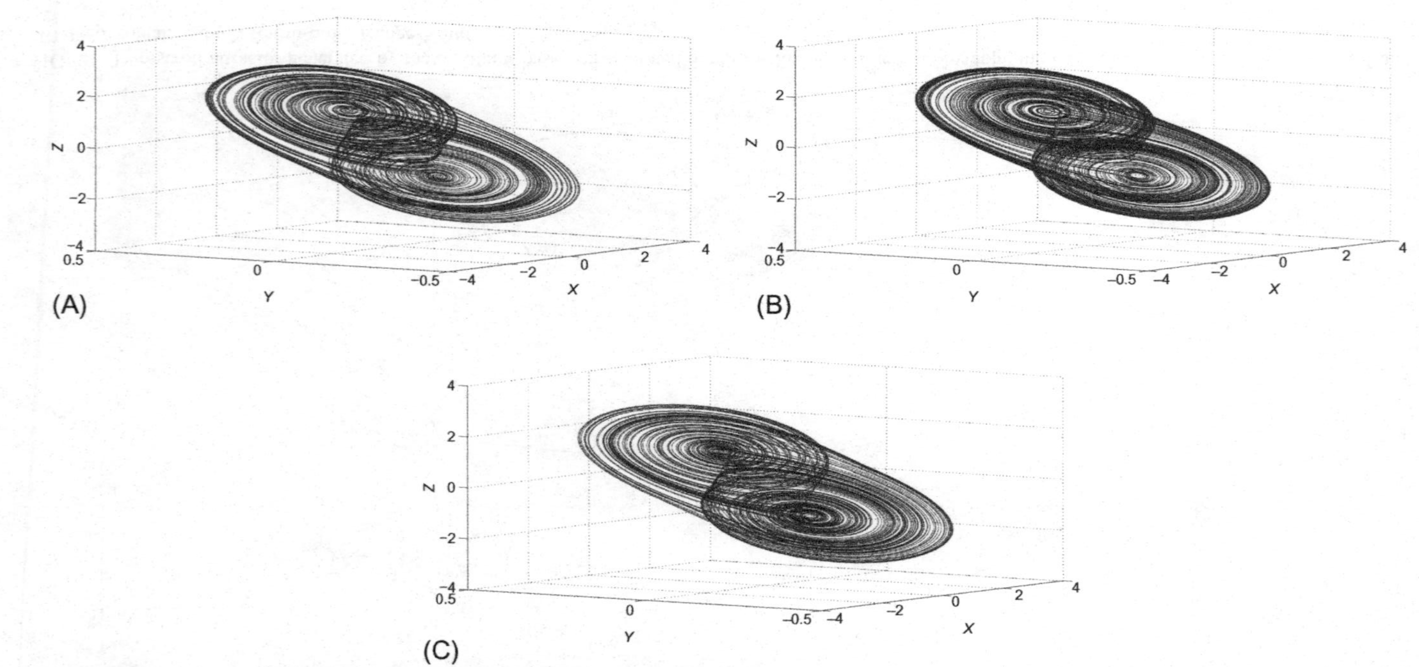

FIG. 3 Two-scroll attractor generated by the oscillator based on negative slopes and by applying the numerical method: (A) Forward Euler, (B) Trapezoidal, and (C) fourth-order Runge-Kutta.

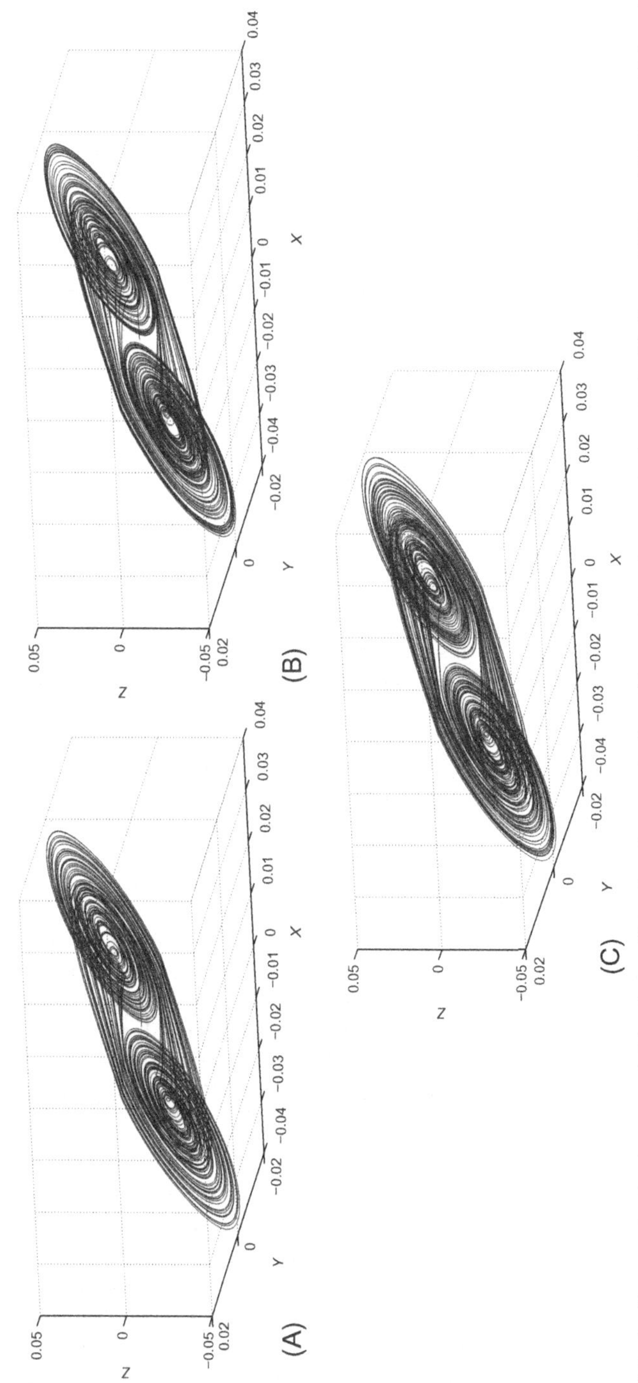

FIG. 4 Two-scroll attractor generated by the oscillator based on sawtooth and by applying the numerical method: (A) Forward Euler, (B) Trapezoidal, and (C) fourth-order Runge-Kutta.

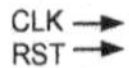

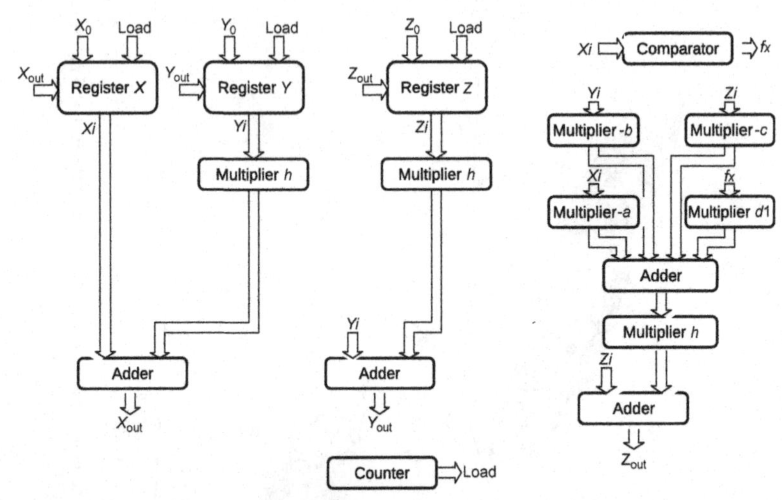

FIG. 5 Block description of the chaotic oscillator based on saturated functions from Eq. (11).

Forward-Euler method. In this manner, from Eq. (1), the state variables can be discretized by applying the Forward-Euler method to give the discrete equations given by

$$\begin{aligned} x_{n+1} &= x_n + hy_n \\ y_{n+1} &= y_n + hz_n \\ z_{n+1} &= x_n + h(-ax_n - by_n - cz_n + d_1 f(x_n)) \end{aligned} \tag{11}$$

In Eq. (11), to generate two scrolls, $f(x_n)$ takes the form shown in Fig. 1A with three linear segments, which values are given in the previous section. The associated blocks are shown in Fig. 5, where it is easy to see that for the first and second equations in Eq. (11), the VHDL descriptions associate the blocks at the left and in the middle, where the output equals to: $X_{out} = X_i + hY_i$ and $Y_{out} = Y_i + hZ_i$, and for the last equation one have $Z_{out} = Z_i + h(-aX_i - bY_i - cZ_i + d_1 f_x)$, and f_x is the PWL function from Fig. 1A, which can be described by using comparator blocks.

The blocks require a clock pulse from the FPGA "CLK," and a reset "RST" pin is assigned to re-start the system. The buses have different number of bits according to the amplitudes of the state variables that can be observed in Figs. 2–4. In this work the computer arithmetic leads us to use a fixed-point format of 28 bits (7.21). The VHDL descriptions for the other chaotic oscillators and by applying the other numerical methods can be done in a similar manner.

Once one have the VHDL descriptions in Active-HDL, then it can be linked to Simulink to perform cosimulation. The resulting chaotic attractor for the oscillator based on saturated nonlinear functions by applying Forward Euler is shown in Fig. 6A. The remaining Fig. 6B and C show the cosimulation results by applying the Trapezoidal and fourth-order Runge-Kutta methods, respectively. The cosimulation results between Active-HDL and Simulink for the oscillators based on negative slopes and sawtooth functions by applying the three numerical methods are shown in Figs. 7 and 8.

4 EXPERIMENTAL OBSERVATION OF CHAOTIC ATTRACTORS AND SYNCHRONIZATION OF TWO CHAOTIC OSCILLATORS

After verifying the correct behavior of the chaotic attractors for the three chaotic oscillators, their FPGA-based implementation was performed by using "Cyclone IV GX EP4CGX150DF31C7" from ALTERA. During the VHDL descriptions of the blocks for the three chaotic oscillators, two kinds of hardware descriptions were performed: Type A and type B. The descriptions classified as Type A include a clock pin CLK to all the blocks, so that all are sequential. The descriptions classified as type B do not include CLK in the blocks performing addition, multiplication and the PWL function, those blocks are combinational.

Table 1 lists the resources of the three chaotic oscillators: based on saturated function series, negative slopes, and sawtooth function. The table lists the three methods for each chaotic oscillator, the Type A or B (combinational), and the resources and maximum frequency response of the blocks that is provided by the synthesizer and depends on the FPGA. The maximum frequency is multiplied by the number of clock cycles that are required to process the data from the input to the output, so that the processing speed or latency is listed in the last column in nanoseconds (ns).

As one can see, the sequential blocks that are mostly used to implement Type A topologies require more resources than the topologies of Type B. In the same manner, as the fourth-order Runge-Kutta numerical method requires more resources than Forward-Euler and the Trapezoidal methods; those implementations have slower time response, as shown in the last column in Table 1. Thus, the maximum processing speed or latency is accomplished by using descriptions of Type B (combinational blocks).

The experimental observation of the chaotic attractors is given in the following. Fig. 9 shows the two-scroll attractor plotting the state variables x-y for the oscillator based on saturated nonlinear function series.

Fig. 10 shows the phase portraits of the chaotic oscillator based on negative slopes, and Fig. 11 shows the chaotic attractor based on sawtooth; in all cases, the FPGA-based implementations were performed by applying the three numerical methods. As one can infer, there are differences due to the numerical methods that were used to discretize the mathematical equations implemented into the FPGA, but in all cases, the two-scroll attractor is appreciated experimentally.

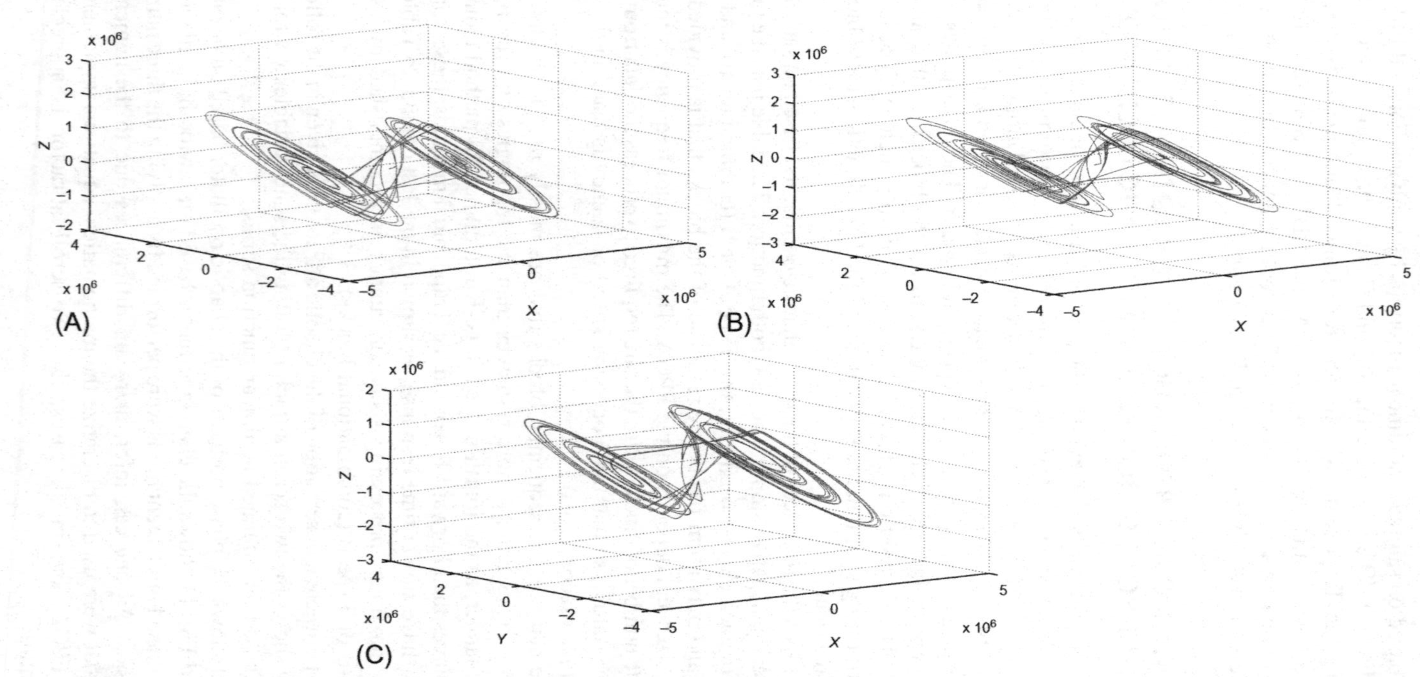

FIG. 6 Two-scroll attractor generated by cosimulation and by applying: (A) Forward Euler, (B) Trapezoidal, and (C) fourth-order Runge-Kutta, for the oscillator based on saturated nonlinear function series.

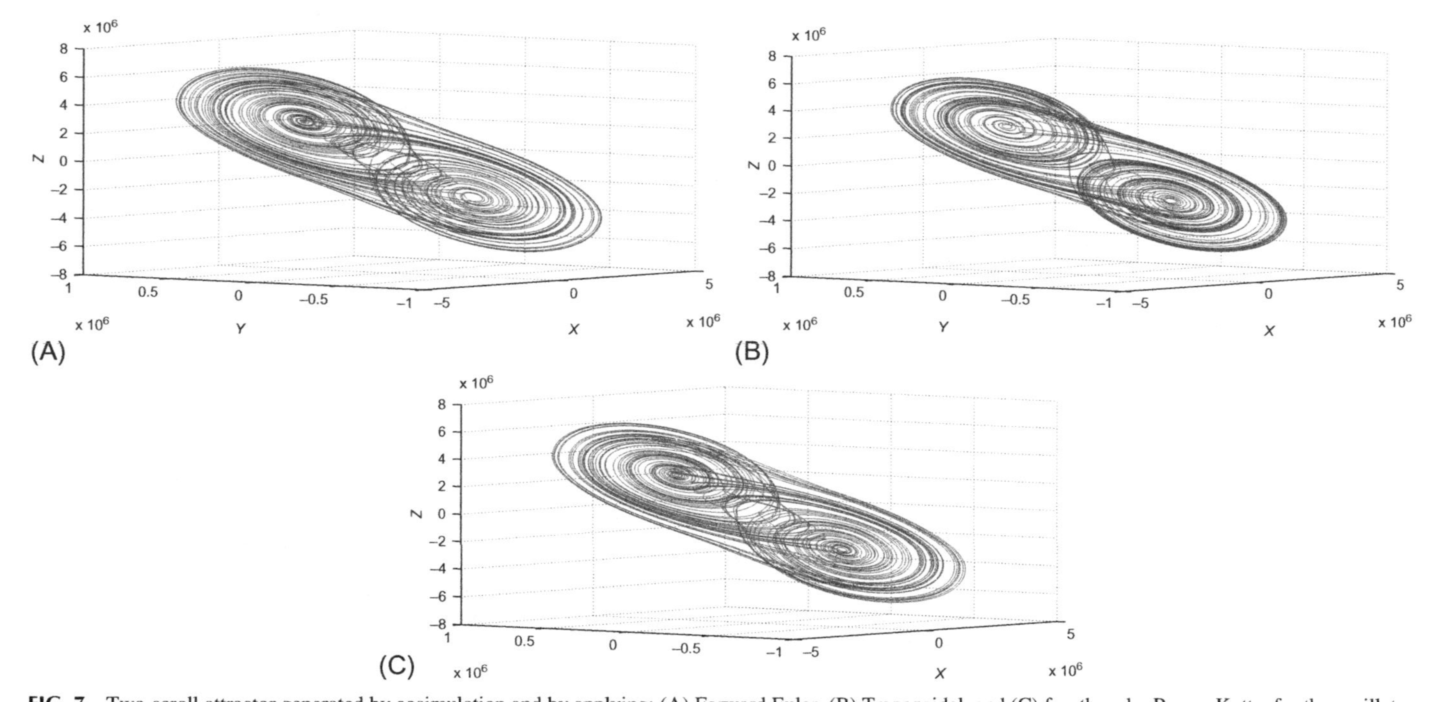

FIG. 7 Two-scroll attractor generated by cosimulation and by applying: (A) Forward Euler, (B) Trapezoidal, and (C) fourth-order Runge-Kutta, for the oscillator based on negative slopes.

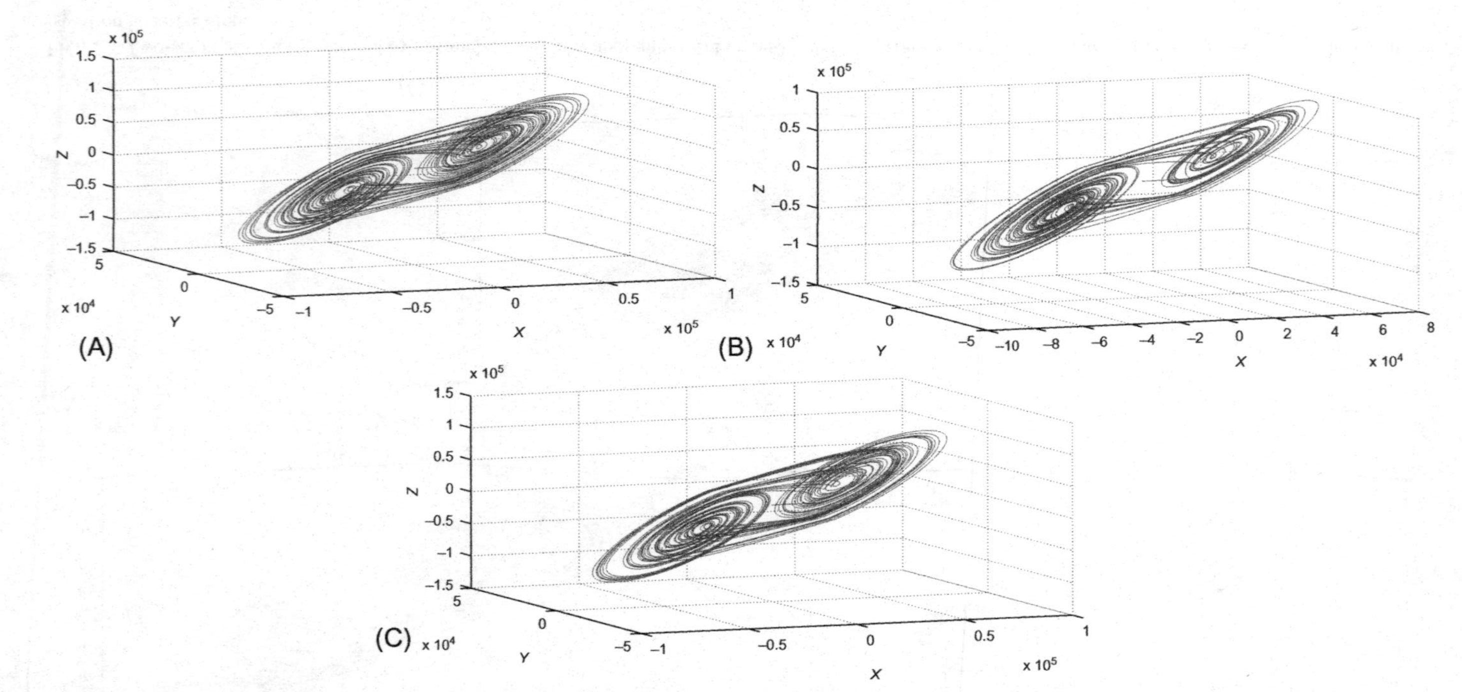

FIG. 8 Two-scroll attractor generated by cosimulation and by applying: (A) Forward Euler, (B) Trapezoidal, and (C) fourth-order Runge-Kutta, for the oscillator based on sawtooth.

TABLE 1 Resources Associated With the Implementations of the Chaotic Oscillators by Using the FPGA Cyclone IV GX EP4CGX150DF31C7

Oscillator	Method	Type	Log Elems	Registers	Max. Freq. (MHz)	Cycles	Latency (ns)
Saturated functions	Forward-Euler	A	911	592	116.04	9	77.58
		B	811	118	33.48	2	60
	Runge-Kutta	A	4745	1106	67.53	17	251
	Fourth order	B	4686	118	14.00	2	143
	Trapezoidal	A	4539	906	66.67	12	179
		B	4507	118	18.00	2	111
Chua neg. slopes	Forward-Euler	A	1238	476	86.72	12	138
		B	1078	118	30.38	2	66
	Runge-Kutta	A	3590	1112	84.08	21	250
	Fourth order	B	3414	118	14.27	2	140
	Trapezoidal	A	2415	918	83.44	20	240
		B	2207	118	14.82	2	134
Sawtooth function	Forward-Euler	A	1272	526	76.76	9	117
		B	1184	118	61.04	2	33
	Runge-Kutta	A	3685	1162	74.72	18	241
	Fourth order	B	3595	118	19.01	2	105
	Trapezoidal	A	2442	1016	75.07	15	199
		B	2333	118	32.01	2	62

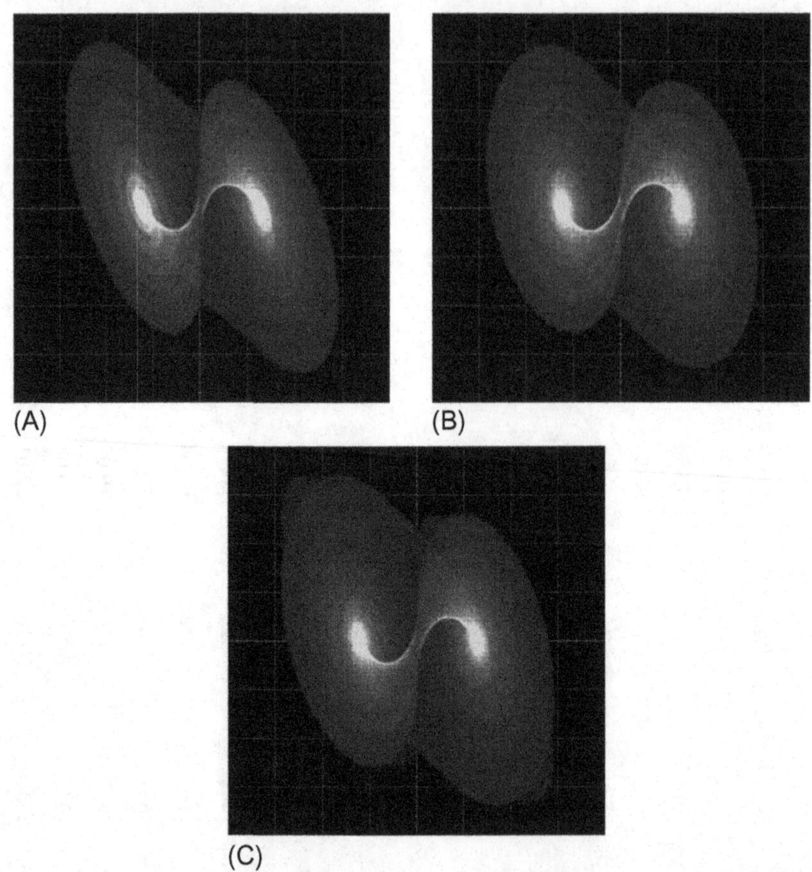

FIG. 9 Phase portraits x-y of the oscillator based on saturated nonlinear function series observed in the oscilloscope with axis $X = 2$ V/div and $Y = 1$ V/div, corresponding to the (A) Forward-Euler, (B) Trapezoidal, and (C) fourth-order Runge-Kutta methods.

4.1 Synchronization in a Master-Slave Topology

Every chaotic oscillator can be described by $\dot{x} = f(x)$; according to the seminal work given in [19], the Hamiltonian approach can be described by

$$\dot{x} = J(x)\frac{\partial H}{\partial x} + S(x)\frac{\partial H}{\partial x}, \quad x \epsilon R^n \tag{12}$$

where ∂H is the gradient vector of the energy function H, positive definite in R^n. H is a quadratic function defined by $H(x) = \frac{1}{2}X^T Mx$, with M as a symmetrical matrix and positive definite. $J(x)$ and $S(x)$ are matrices representing the conservative and nonconservative parts of the system, respectively, and must satisfy: $J(x) + J^T(x) = 0$ and $S(x) = S^T(x)$ [20]. There is the possibility of adding a

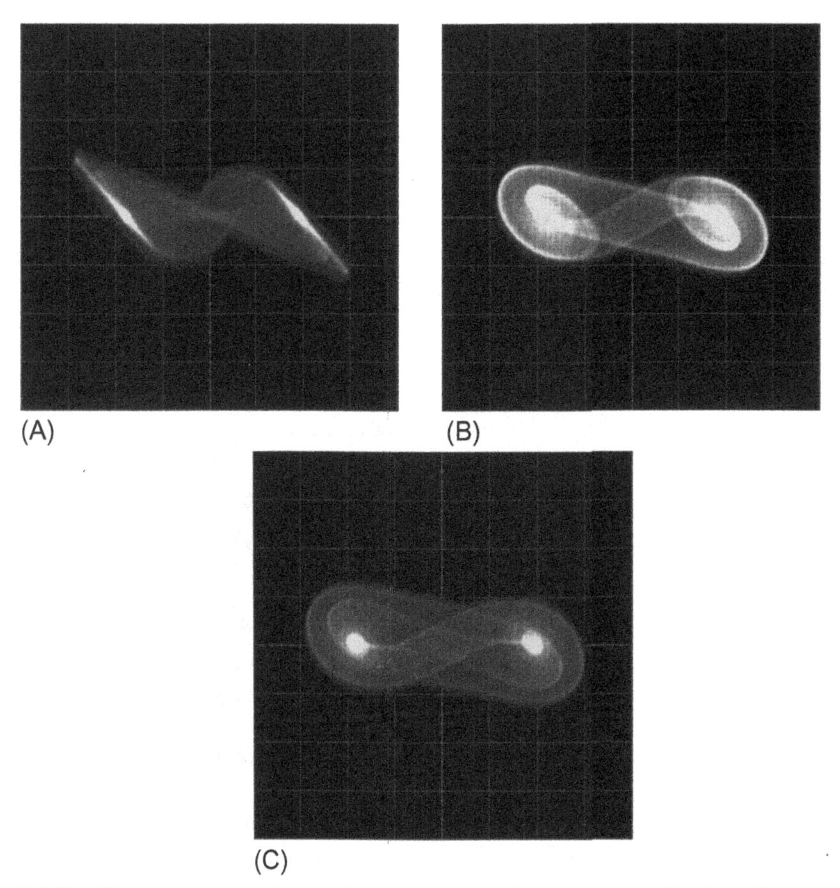

FIG. 10 Phase portraits x-y of the oscillator based on negative slopes observed in the oscilloscope with axis $X=1$ V/div and $Y=0.5$ V/div, corresponding to the (A) Forward-Euler, (B) Trapezoidal, and (C) fourth-order Runge-Kutta methods.

destabilizing vector as $F(x)$, to get the form of a Hamiltonian system, as shown in Eq. (13). This can consider suppositions to get the form given in Eq. (12), without $F(x)$.

$$\dot{x}=J(x)\frac{\partial H}{\partial x}+S(x)\frac{\partial H}{\partial x}+F(x), \quad x \in R^n \tag{13}$$

If one considers the system with destabilizing vector and one linear output, one gets

$$\begin{aligned} \dot{x}&=J(y)\frac{\partial H}{\partial x}+S(y)\frac{\partial H}{\partial x}+F(y), \quad x \in R^n \\ y&=C\frac{\partial H}{\partial x}, \quad y \in R^m \end{aligned} \tag{14}$$

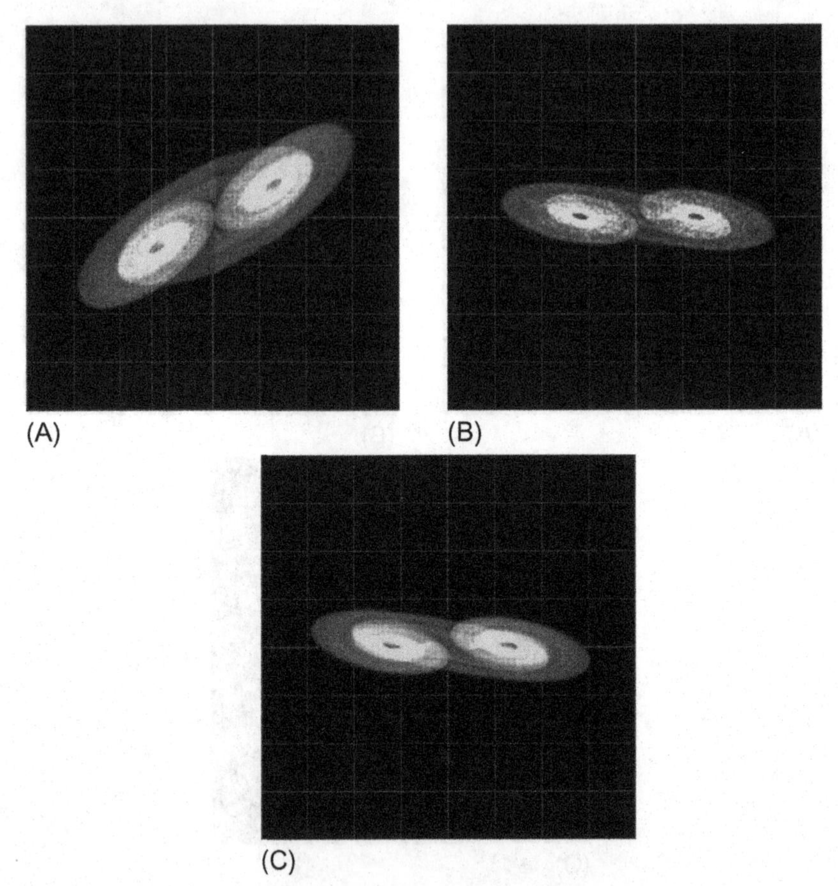

FIG. 11 Phase portraits x-y of the oscillator based on sawtooth observed in the oscilloscope with axis $X = 1$ V/div and $Y = 1$ V/div, corresponding to the (A) Forward-Euler, (B) Trapezoidal, and (C) fourth-order Runge-Kutta methods.

where y is a vector denoting the output of the system. In addition, if ξ is the estimated state vector of x and η the estimated output in terms of ξ, then an observer to Eq. (13) can be given by Eq. (15), where K is a vector of constant gains.

$$\begin{aligned} \dot{\xi} &= J(y)\frac{\partial H}{\partial \xi} + S(y)\frac{\partial H}{\partial \xi} + F(y) + K(y-\eta) \\ \eta &= C\frac{\partial H}{\partial \xi} \end{aligned} \tag{15}$$

The synchronization by Hamiltonian forms is achieved after accomplishing two theorems.

Theorem 1 The state x of the nonlinear system (14) can be global, exponential, and asymptotically estimated by the state ξ of an observer of the form (15), if the pair of matrices (C, S) are observables.

Theorem 2 The state x of the nonlinear system (14) can be global, exponential, and asymptotically estimated by the state ξ of an observer of the form (15), if and only if there exists a constant matrix K such that the symmetric matrix

$$[W-KC]+[W-KC]^T=[S-KC]+[S-KC]^T=2[S-\frac{1}{2}(KC+C^TK^T)] \quad (16)$$

be negative definite [19].

4.2 Synchronization in a Master-Slave Topology for the Three Chaotic Oscillators

In this section the models of the chaotic oscillators are changed in state variables. For example, the oscillator based on saturated nonlinear function series is now described by Eq. (17), for which the proposed energy function is given in Eq. (18).

$$\begin{aligned} \dot{x_1} &= x_2 \\ \dot{x_2} &= x_3 \\ \dot{x_3} &= -ax_1-bx_2-cx_3+d_1f(x) \end{aligned} \quad (17)$$

$$H(x)=\frac{1}{2}[x_1^2+x_2^2+x_3^2] \quad (18)$$

From these equations, the following Hamiltonian system arises:

$$\begin{bmatrix} \dot{x_1} \\ \dot{x_2} \\ \dot{x_3} \end{bmatrix} = \begin{bmatrix} 0 & 1/2 & 7/20 \\ -1/2 & 0 & 17/20 \\ -7/20 & -17/20 & 0 \end{bmatrix} \frac{\partial H}{\partial x} + \begin{bmatrix} 0 & 1/2 & -7/20 \\ 1/2 & 0 & 3/20 \\ -7/20 & 3/20 & -7/10 \end{bmatrix} \frac{\partial H}{\partial x} + \begin{bmatrix} 0 \\ 0 \\ 7f(x)/10 \end{bmatrix} \quad (19)$$

This becomes the master system and its slave system is proposed by adding the gain vector multiplied by the error. The gain vector is obtained verifying that it contains the pair of matrices (C, S) and creating the system given in Eq. (19). In this manner, the gain vector K can be obtained by applying the Sylvester criterion for negative definite matrices [20]. The gains are equal to $k_1=2$, $k_2=4$, $k_3=0$ and the slave system is described by Eq. (20)

$$\begin{bmatrix}\dot{\xi}_1\\ \dot{\xi}_2\\ \dot{\xi}_3\end{bmatrix} = \begin{bmatrix}0 & 1/2 & 7/20\\ -1/2 & 0 & 17/20\\ -7/20 & -17/20 & 0\end{bmatrix}\frac{\partial H}{\partial \xi} + \begin{bmatrix}0 & 1/2 & -7/20\\ 1/2 & 0 & 3/20\\ -7/20 & 3/20 & -7/10\end{bmatrix}\frac{\partial H}{\partial \xi}$$
$$+\begin{bmatrix}0\\ 0\\ 7f(x)/10\end{bmatrix} + \begin{bmatrix}2\\ 4\\ 0\end{bmatrix}(y-\eta) \tag{20}$$

The synchronization error between the master and the slave systems is shown in Fig. 12, it can be seen that the synchronization is done around iteration 400. The synchronization among the state variables of the master and slave systems is shown in Fig. 13.

The chaotic oscillator based on negative slopes is now described by Eq. (21) and its energy function is proposed as Eq. (22).

$$\begin{aligned}\dot{x_1} &= \alpha(x_2 - x_1 - f(x))\\ \dot{x_2} &= \gamma(x_1 - x_2 + x_3)\\ \dot{x_3} &= -\beta x_2\end{aligned} \tag{21}$$

$$H(x) = \frac{1}{2}[x_1^2 + x_2^2 + x_3^2] \tag{22}$$

The master system in Hamiltonian form is given as,

$$\begin{bmatrix}\dot{x_1}\\ \dot{x_2}\\ \dot{x_3}\end{bmatrix} = \begin{bmatrix}0 & 4.5 & 0\\ -4.5 & 0 & 7.64\\ 0 & -7.64 & 0\end{bmatrix}\frac{\partial H}{\partial x} + \begin{bmatrix}-10 & 5.5 & 0\\ 5.5 & -1 & -6.64\\ 0 & -6.64 & 0\end{bmatrix}\frac{\partial H}{\partial x} + \begin{bmatrix}-10f(x)\\ 0\\ 0\end{bmatrix} \tag{23}$$

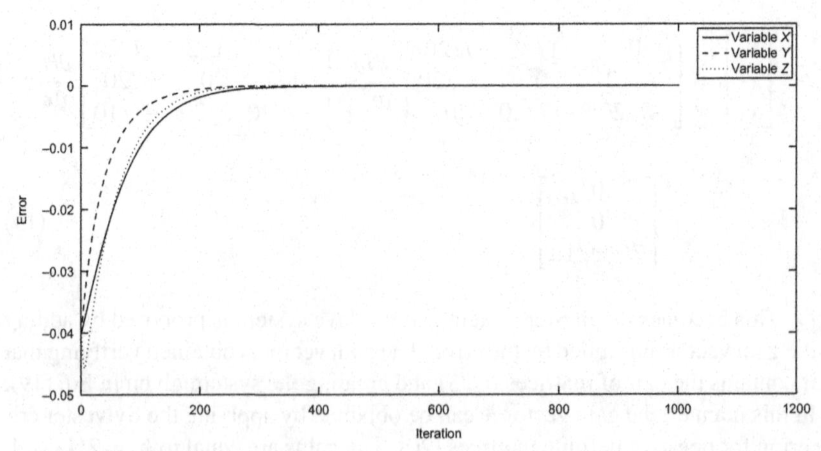

FIG. 12 Synchronization error between the master and slave systems given in Eqs. (19), (20).

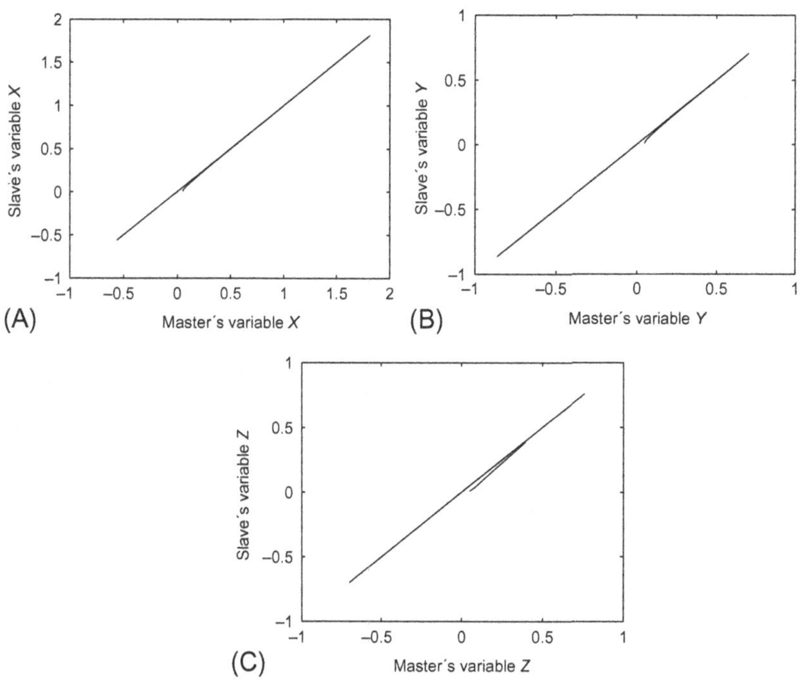

FIG. 13 Synchronization between the state variables of the master and slave systems.

In this case, the gains of the observers ares $k_1 = 3, k_2 = 5, k_3 = 0$ for the slave system that is described by Eq. (24).

$$\begin{bmatrix} \dot{\xi}_1 \\ \dot{\xi}_2 \\ \dot{\xi}_3 \end{bmatrix} = \begin{bmatrix} 0 & 4.5 & 0 \\ -4.5 & 0 & 7.64 \\ 0 & -7.64 & 0 \end{bmatrix} \frac{\partial H}{\partial \xi} + \begin{bmatrix} -10 & 5.5 & 0 \\ 5.5 & -1 & -6.64 \\ 0 & -6.64 & 0 \end{bmatrix} \frac{\partial H}{\partial \xi} + \begin{bmatrix} -10f(x) \\ 0 \\ 0 \end{bmatrix} + \begin{bmatrix} 3 \\ 5 \\ 0 \end{bmatrix} (y - \eta) \tag{24}$$

The synchronization error between the master and the slave systems is shown in Fig. 14, it can be seen that the synchronization is done around iteration 400. The synchronization among the state variables of the master and slave systems is shown in Fig. 15.

The chaotic oscillator based on sawtooth is now described by Eq. (25) and its energy function is proposed as (Eq. 26).

$$\begin{aligned} \dot{x_1} &= \alpha x_2 + f(x) \\ \dot{x_2} &= \alpha x_1 - \gamma x_2 - \alpha x_3 \\ \dot{x_3} &= \beta x_2 \end{aligned} \tag{25}$$

$$H(x) = \frac{1}{2}[x_1^2 + x_2^2 + x_3^2] \tag{26}$$

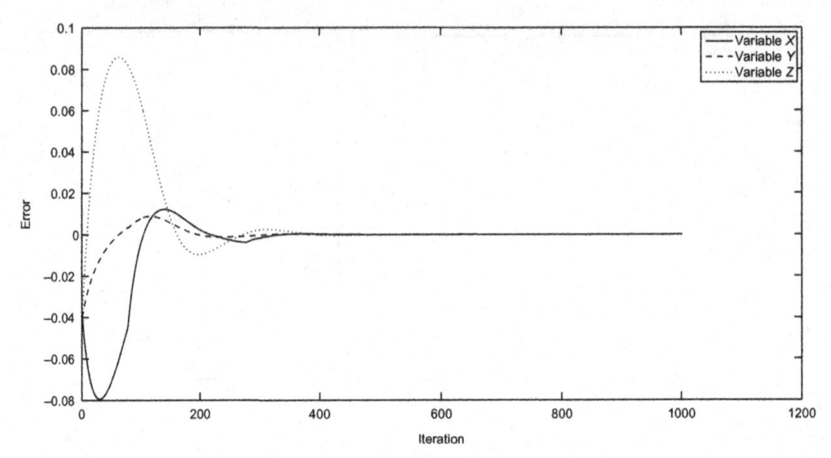

FIG. 14 Synchronization error between the master and slave systems given in Eqs. (23), (24).

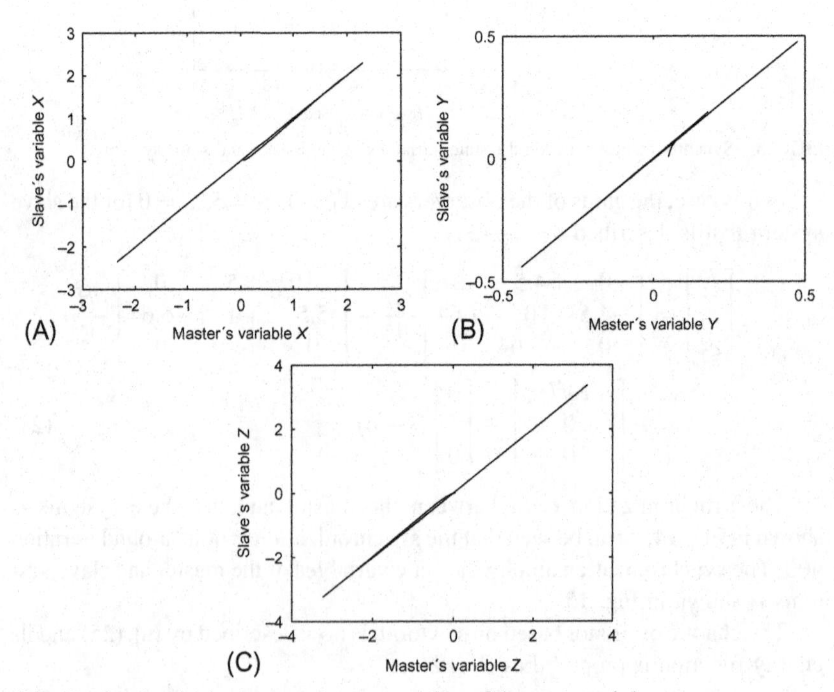

FIG. 15 Synchronization between the state variables of the master and slave systems.

The master system in Hamiltonian form is given as,

$$\begin{bmatrix} \dot{x}_1 \\ \dot{x}_2 \\ \dot{x}_3 \end{bmatrix} = \begin{bmatrix} 0 & 0 & 0 \\ 0 & 0 & -7/2 \\ 0 & 7/2 & 0 \end{bmatrix} \frac{\partial H}{\partial x} + \begin{bmatrix} 0 & 3 & 0 \\ 3 & -1 & 1/2 \\ 0 & 1/2 & 0 \end{bmatrix} \frac{\partial H}{\partial x} + \begin{bmatrix} f(x) \\ 0 \\ 0 \end{bmatrix} \tag{27}$$

In this case, the gains of the observers ares $k_1 = 3, k_2 = 4, k_3 = 0$ for the slave system that is described by Eq. (28).

$$\begin{bmatrix} \dot{\xi}_1 \\ \dot{\xi}_2 \\ \dot{\xi}_3 \end{bmatrix} = \begin{bmatrix} 0 & 0 & 0 \\ 0 & 0 & -7/2 \\ 0 & 7/2 & 0 \end{bmatrix} \frac{\partial H}{\partial \xi} + \begin{bmatrix} 0 & 3 & 0 \\ 3 & -1 & 1/2 \\ 0 & 1/2 & 0 \end{bmatrix} \frac{\partial H}{\partial \xi} + \begin{bmatrix} f(x) \\ 0 \\ 0 \end{bmatrix} + \begin{bmatrix} 3 \\ 4 \\ 0 \end{bmatrix} (y - \eta) \tag{28}$$

The synchronization error between the master and the slave systems is shown in Fig. 16, it can be seen that the synchronization is done around iteration 400. The synchronization among the state variables of the master and slave systems is shown in Fig. 17.

5 APPLICATION TO IMAGE TRANSMISSION

After demonstrating the synchronization of the three chaotic oscillators by Hamiltonian forms and an observer approach, the master-slave topologies can be used to implement a communication system to transmit an image that is encrypted by chaos. The transmission requires modulation and demodulation blocks that are implemented as chaotic masking by addition and subtraction of chaos at the transmitter and reception stages, respectively.

The image to be transmitted is converted to a vector that contains bits representing the information, and in this case, those bits are added to the fixed-point format (7.21) detailed above during the hardware description of the oscillators for FPGA implementation. The chaotic channel will process the mixing chaos

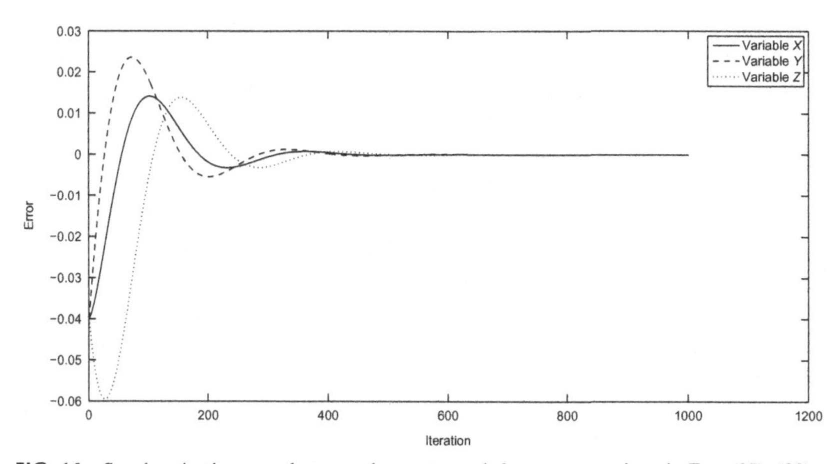

FIG. 16 Synchronization error between the master and slave systems given in Eqs. (27), (28).

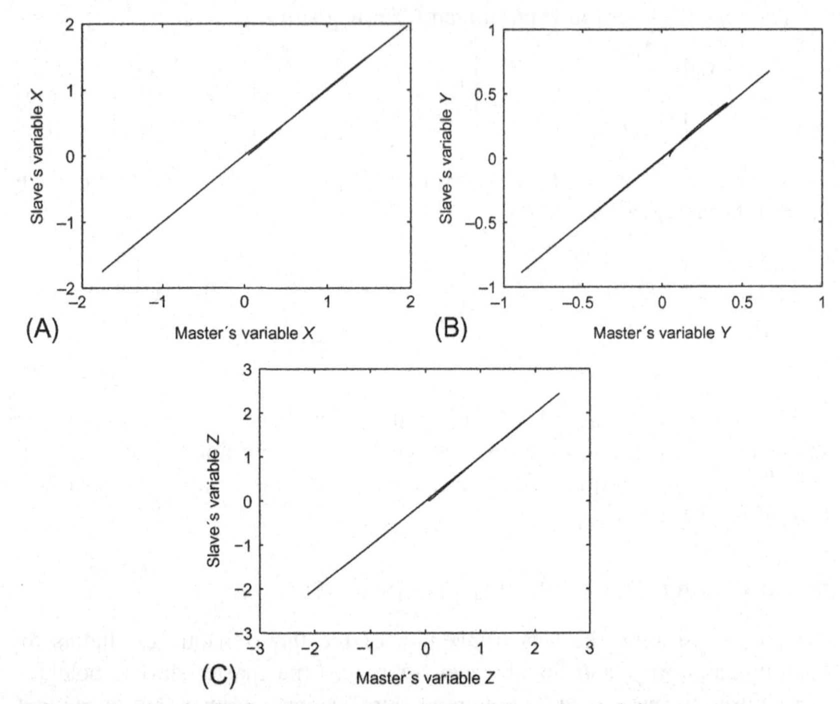

FIG. 17 Synchronization between the state variables of the master and slave systems.

and the image data. Later, at the reception stage, the chaos is subtracted to recover the original data. This process of image transmission is performed herein with the three chaotic oscillators and by using each one of the three state variables (x_1, x_2, x_3) as the communication channel, so that nine experiments are realized for each FPGA-based implementation with different numerical method.

To verify which state variable is the best for image encryption, correlation analysis is performed between the original image data and the chaotic channel when the image is contaminated with chaos. In this manner, the state variable that has the lowest correlation value is the one more suitable for secure transmission. That way, Table 2 lists the computed correlations for the three chaotic oscillators and their respective state variables.

From this table, one can choose the best state variable for each oscillator, and then, the synchronized master-slave topology can be used to implement a chaotic secure communication system. Fig. 18A shows the image transmission by using the state variable x_1 for the oscillator based on saturated nonlinear function series. One can see the original image on the left side, the contaminated image in the chaotic channel in the middle of the figure, and the recovered image after eliminating the chaos on the right side. Fig. 18B shows the image transmission by using the state variable x_3 for the communication system implemented with oscillators based on negative slopes. Finally, again from the results listed in Table 2, the secure system implemented with oscillators based on sawtooth function use the state variable x_3 to transmit and to recover the image, as shown by Fig. 18C.

TABLE 2 Correlation Between the Original Image Data and the Chaotic Channel for Each Chaotic Oscillator and for Each of Their State Variables

Oscillator Based on	State Variable	Correlation
Saturated functions	x_1	0.0124
	x_2	0.0481
	x_3	0.0519
Negative slopes	x_1	0.0350
	x_2	0.1037
	x_3	−0.0042
Sawtooth function	x_1	0.0167
	x_2	0.0497
	x_3	0.0135

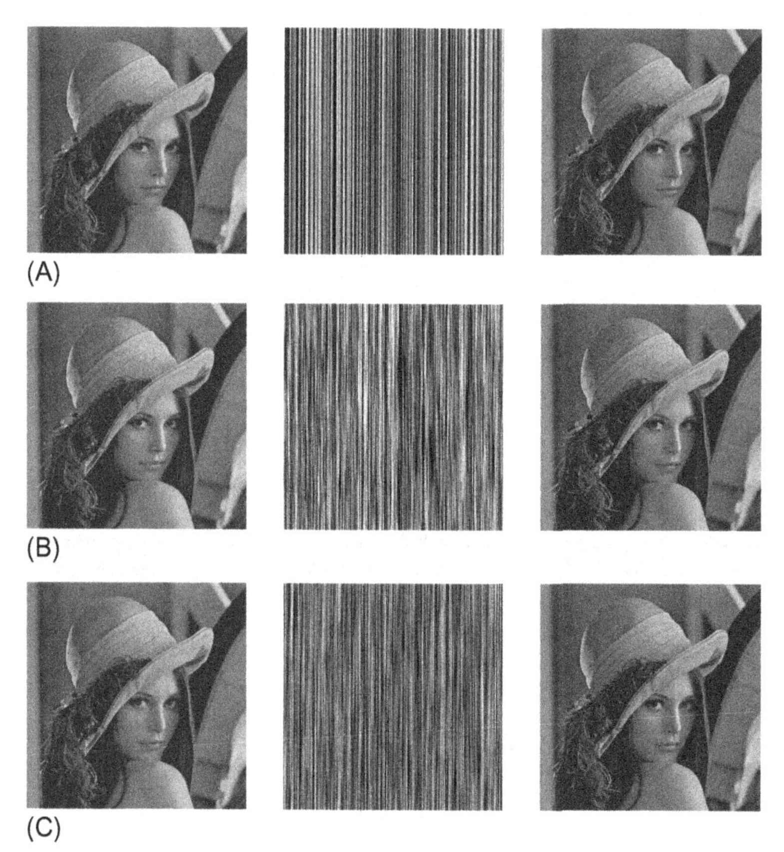

FIG. 18 Images transmitted through the chaotic secure communication system by using chaotic oscillators based on: (A) saturated nonlinear function series, (B) negative slopes, and (C) sawtooth function.

6 CONCLUSION

This chapter has shown the simulation of three chaotic oscillators that are based on PWL functions like saturated nonlinear series, negative slopes, and sawtooth. Three numerical methods were applied to solve the mathematical models, namely, Forward-Euler, Trapezoidal, and fourth-order Runge-Kutta. The simulation results show that the three numerical methods are useful to generate chaotic attractors; however, when the chaotic oscillators are described in the hardware description language (Active-HDL), one can see that the resources depend on the numerical method. Thus, when using FPGAs, it was shown that the fourth-order Runge-Kutta requires more resources than the Forward-Euler method.

The cosimulation between Active-HDL and Simulink provides evidence that the chaotic oscillators will work when they are implemented into the FPGA. That implementation was performed in two ways, by using sequential blocks, where all blocks include a clock signal (CLK), and by combining sequential blocks and combinatorial logic. In both cases, we list the frequency provided by the synthesizer, and it was used to estimate the latency of each chaotic oscillator implemented with sequential combinatorial logic, as shown in Table 1. As one can see, the oscillators implemented from the application of the fourth-order Runge-Kutta method require more resources and time to provide chaotic data.

The synchronization results also show that one can use the three state variables for each chaotic oscillator to transmit an image. However, the correlation analysis listed in Table 2 provides us information on the best state variable to mask the image. Finally, Fig. 18 shows the appropriateness of implementing a secure communication system with FPGAs for image transmission.

ACKNOWLEDGMENTS

This work has been partially supported by CONACyT/México under Grant No. 237991.

REFERENCES

[1] Lassoued A, Boubaker O. On new chaotic and hyperchaotic systems: a literature survey. Nonlinear Anal Modell Control 2016;21(6):770–89.

[2] Lassoued A, Boubaker O. Dynamic analysis and circuit design of a novel hyperchaotic system with fractional-order terms. Complexity 2017;2017: 3273408, 10 pages, https://doi.org/10.1155/2017/3273408.

[3] Mkaouar H, Boubaker O. Robust control of a class of chaotic and hyperchaotic driven systems. Pramana 2017;88(1):9.

[4] Mkaouar H, Boubaker O. Chaos synchronization for master slave piecewise linear systems: application to Chuas circuit. Commun Nonlinear Sci Numer Simul 2012;17(3):1292–302.

[5] Kocamaz UE, Çiçek S, Uyaroğlu Y. Secure communication with chaos and electronic circuit design using passivity-based synchronization. J Circuits Syst Comput 2018;27 (04):1850057.

[6] Lü J, Chen G. Generating multiscroll chaotic attractors: theories, methods and applications. Int J Bifurcation Chaos 2006;16(04):775–858.

[7] Magistris M, Bernardo M, Manfredi S, Petrarca C, Yaghouti S. Modular experimental setup for real-time analysis of emergent behavior in networks of Chua's circuits. Int J Circuit Theory Appl 2016;44(8):1551–71.

[8] Pham VT, Volos C, Kapitaniak T, Jafari S, Wang X. Dynamics and circuit of a chaotic system with a curve of equilibrium points. Int J Electron 2018;105(3):385–97.

[9] Volos C, Maaita JO, Vaidyanathan S, Pham VT, Stouboulos I, Kyprianidis I. A novel four-dimensional hyperchaotic four-wing system with a saddle-focus equilibrium. IEEE Trans Circuits Syst II Express Briefs 2017;64(3):339–43.

[10] Yeniçeri R, Yalçın ME. Asynchronous delay doubler and binary low-pass filter for a time-delay chaotic circuit. Int J Circuit Theory Appl 2016;44(6):1211–21.

[11] Yeniçeri R, Yalçın ME. Multi-scroll chaotic attractors from a generalized time-delay sampled-data system. Int J Circuit Theory Appl 2016;44(6):1263–76.

[12] Chen G, Yu X. Chaos control: theory and applications. vol. 292. New York, NY: Springer Science & Business Media; 2003.

[13] Carroll TL, Pecora LM. Synchronizing chaotic circuits. IEEE Trans Circuits Syst 1991;38 (4):453–6.

[14] Egunjobi AI, Olusola OI, Njah AN, Saha S, Dana SK. Experimental evidence of chaos synchronization via cyclic coupling. Commun Nonlinear Sci Numer Simul 2018;56:588–95.

[15] Ouannas A, Azar AT, Abu-Saris R. A new type of hybrid synchronization between arbitrary hyperchaotic maps. Int J Mach Learn Cybern 2017;8(6):1887–94.

[16] Wang X, Vaidyanathan S, Volos C, Pham VT, Kapitaniak T. Dynamics, circuit realization, control and synchronization of a hyperchaotic hyperjerk system with coexisting attractors. Nonlinear Dyn 2017;89(3):1673–87.

[17] Al-Suhail GA, Tahir FR, Abd MH, Pham VT, Fortuna L. Modelling of long-wave chaotic radar system for anti-stealth applications. Commun Nonlinear Sci Numer Simul 2018;57:80–96.

[18] Yang H, Tang WKS, Chen G, Jiang GP. Multi-carrier chaos shift keying: system design and performance analysis. IEEE Trans Circuits Syst I Reg Papers 2017;64(8):2182–94. https://doi.org/10.1109/TCSI.2017.2685344.

[19] Sira-Ramirez H, Cruz-Hernández C. Synchronization of chaotic systems: a generalized Hamiltonian systems approach. Int J Bifurcation Chaos 2001;11(5):1381–95.

[20] Zambrano-Serrano E, Muñoz-Pacheco JM, Félix-Beltran OG, Trejo-Guerra R, Gómez-Pavón LC, Tlelo-Cuautle E, et al. Synchronization of multi-directional multi-scroll chaos generators: a Hamiltonian approach. In: 2011 Joint 3rd int'l workshop on nonlinear dynamics and synchronization (INDS) and 16th int'l symposium on theoretical electrical engineering (ISTET)IEEE; 2011. p. 1–5.

[21] Munoz-Pacheco JM, Zambrano-Serrano E, Felix-Beltran O, Gomez-Pavon LC, Luis-Ramos A. Synchronization of PWL function-based 2D and 3D multi-scroll chaotic systems. Nonlinear Dyn 2012;70(2):1633–43.

[22] Tlelo-Cuautle E. de Jesus Quintas-Valles A. de la Fraga LG. de Jesus Rangel-Magdaleno J. VHDL descriptions for the FPGA implementation of PWL-function-based multi-scroll chaotic oscillators. PLoS ONE 2016;11(12):e0168300.

[23] Suykens JAK, Huang A. A family of n-scroll attractors from a generalized Chua's circuit. Arch Elektron Ubertrag 1997;51(3):131–7.

[24] Muñoz-Pacheco JM, Guevara-Flores DK, Félix-Beltrán OG, Tlelo-Cuautle E, Barradas-Guevara JE, Volos CK. Experimental verification of optimized multiscroll chaotic oscillators based on irregular saturated functions. Complexity 2018;2018: 3151840, 17 pages, https://doi.org/10.1155/2018/3151840.

[25] Trejo-Guerra R, Tlelo-Cuautle E, Jimenez-Fuentes JM, Sanchez-Lopez C, Munoz-Pacheco JM, Espinosa-Flores-Verdad G, et al. Integrated circuit generating 3- and 5-scroll attractors. Commun Nonlinear Sci Numer Simul 2012;17(11):4328–35.

Chapter 16

On Nonidentical Discrete-Time Hyperchaotic Systems Synchronization: Towards Secure Medical Image Transmission

Narjes Khalifa and Mohamed Benrejeb
University of Tunis El Manar, National Engineering School of Tunis, Automatic Control Research Laboratory, Tunis, Tunisia

1 INTRODUCTION

The main features of chaotic systems are high sensitivity to initial conditions, long-term unpredictability, strong dependence on bifurcation parameters, and random-like behavior [1–6].

Chaotic systems have been widely appreciated in the last decade for a wide variety of applications in various areas such as meteorology, hydrodynamics, electronics, chemical and biophysics engineering, and mainly secure communication systems since the pioneering work of Pecora and Carroll on chaos synchronization [7].

In this context, several approaches have been proposed for different types of synchronization of chaotic systems, namely, complete synchronization [8–12], generalized synchronization [13–16], projective synchronization [17,18], lag synchronization [19], and cluster synchronization [20].

In fact, several works have been done to achieve synchronization between coupled continuous chaotic systems as well as discrete time chaotic systems, using different controller designs, such as feedback synchronization [21–23], synchronization by adaptive controllers [24,25], sliding mode synchronization [26,27], and synchronization-based observer design [28–32].

On the other hand, with the growing use of telemedicine services, a huge number of medical images are transmitted through public channels. Consequently, the protection of patient's information has become an important issue.

Recent Advances in Chaotic Systems and Synchronization. https://doi.org/10.1016/B978-0-12-815838-8.00016-9

Recently, several encryption approaches based on chaos have been proposed, as they have shown their effectiveness compared to conventional encryption algorithms such as Date Encryption Standard (DES), Advanced Encryption Standard (AES), International Data Encryption Algorithm (IDEA), and Rivest Shamir Adleman (RSA), which are not appropriate to protect medical image transmission having some characteristic features such as high correlation among pixels and large sizes.

Based on this consideration, several cryptosystems to secure medical image transmission-based chaos are designed based on specific feature of chaotic systems like mixing property, sensitivity to initial conditions, and to system parameters which are ideal to cryptographic properties [33,34].

These proposed methods are performed based on symmetric chaotic encryption algorithms without synchronization process. However, in real-time application, the decrypted message can't be retrieved correctly because it is not possible to reproduce exactly the same encryption and decryption key generated by chaotic systems in the transmitter and the receiver side which are subject to noise and uncertainties.

From this point of view, the implementation of a secure communication system requires synchronization between the transmitter-receiver pair, which can increase the security of the information transmission. In addition, intensive research works have focused on the complete synchronization study of coupled identical systems, and different applications of this type have been introduced in literature and in our previous published works. In this case, the synchronization process is easy to achieve. Therefore, synchronization between nonidentical chaotic systems is investigated here, in order to fulfill the security issue which is often related to the complexity synchronization scheme.

The main objective of the present work is the elaboration of a general control design which leads to the synchronization of nonidentical chaotic systems, and then, its application to secure medical image transmission.

In this work, the synchronization between nonidentical discrete-time hyperchaotic systems is based on the master-slave unidirectional configuration scheme, such that the considered slave hyperchaotic system is a Luenberger observer. The basic idea of the proposed approach consists of determining the observer gains that lead to the stability property of the coupled master-slave hyperchaotic systems, deduced from Borne and Gentina criterion for stability study [35], associated with Benrejeb arrow form matrix to system description. This arrow-form matrix is such that nonnull elements are located in its diagonal, its first (resp. last) row and first (resp. last) column. Since the seventies, Benrejeb has developed and largely used it for large-scale systems stability study, multimodel control systems, TS fuzzy models stability study, delayed nonlinear systems study [36–41], multimodel chaotic system synchronization [42], and for continuous time and discrete time chaotic systems synchronization [43–46]. The rest of the chapter is organized as follows. In the next section, the proposed synchronization method for nonidentical discrete time hyperchaotic systems is given and tested for the coupled Wang and Hénon Hitzel Zele systems and for the coupled Hénon 3D and Stéfanski systems. Section 3 is dedicated to present a proposed secure medical image cryptosystem, based on the synchronization between Hénon 3D and Stéfanski hyperchaotic systems. Encryption and decryption results and security analysis are, then, given for these medical images in Section 4.

2 PROPOSED METHOD FOR COUPLED NONIDENTICAL CHAOTIC SYSTEMS SYNCHRONIZATION STUDY

In this section, a proposed synchronization method is presented and applied for two cases of coupled master-slave hyperchaotic systems.

2.1 Proposed Synchronization Method: Basic Idea

Let consider the n-dimensional discrete time master system described by:

$$\begin{cases} x_m(k+1) = A(.)x_m(k) + F \\ y_m(k) = Cx_m(k) \end{cases} \tag{1}$$

and a corresponding associate discrete time state slave observer given by:

$$\begin{cases} x_s(k+1) = A'(.)x_s(k) + G + u(k) \\ y_s(k) = Cx_s(k) \end{cases} \tag{2}$$

$x_m(k) = [x_{m1}(k)x_{m2}(k)\cdots x_{mn}(k)]^T \in R^n$ and $x_s(k) = [x_{s1}(k)x_{s2}(k)\cdots x_{sn}(k)]^T \in R^n$ are, respectively, the state vectors of the master and the slave systems, $y_m(k) = [y_{m1}(k)y_{m2}(k)\cdots y_{mp}(k)]^T \in R^p$ and $y_s(k) = [y_{s1}(k)y_{s2}(k)\ldots y_{sp}(k)]^T \in R^p$, respectively, the output vectors of the master and the slave systems, $A(.) = \{a_{ij}(.)\} \in R^{n\times n}$, $A' = \{a_{ij}'(.)\} \in R^{n\times n}$ two nonlinear matrices of appropriate dimensions, $F \in R^n$, $G \in R^n$ and $C \in R^{p\times n}$ constant matrices.

$u(k) = [u_1(k)u_2(k)\cdots u_n(k)]^T \in R^n$ is the control vector of the slave observer system, such that:

$$u(k) = L(y_m(k) - y_s(k)) - (A'(.) - A(.))x_s(k) + F - G \tag{3}$$

and L the matrix observer gain to be determined, $L = \{L_{ij}\} \in R^{n\times p}$.

For the error $e(k)$ defined by:

$$e(k) = x_m(k) - x_s(k) \tag{4}$$

it comes the following dynamic error system, such as:

$$e(k+1) = A_\alpha(.)e(k) \tag{5}$$

with:

$$A_\alpha(.) = A(.) - LC \tag{6}$$

The synchronization between the drive system (1) and the observer system (2) controlled by (3) is achieved if the error system (5) is stable, that is:

$$\lim_{k\to\infty} \|x_m(k) - x_s(k)\| = 0 \tag{7}$$

Instantaneous characteristic matrix $A_\alpha(.)$, appropriated to Borne and Gentina stability criterion application [35], is called Benrejeb arrow form matrix of type I, if $A_\alpha(.)$ is such that:

$$\begin{pmatrix} a_{11}(\cdot) & & & & a_{1n}(\cdot) \\ & \ddots & & & \vdots \\ & & & & \\ & & & a_{n-1,n-1}(\cdot) & a_{n-1,n}(\cdot) \\ a_{n1}(\cdot) & \cdots & & a_{n,n-1}(\cdot) & a_{nn}(\cdot) \end{pmatrix} \tag{8}$$

and called Benrejeb arrow form matrix of type II, if $A_\alpha(.)$ is such that:

$$\begin{pmatrix} a_{11}(\cdot) & a_{12}(\cdot) & \cdots & & a_{1n}(\cdot) \\ a_{21}(\cdot) & a_{22}(\cdot) & & & \\ \vdots & & \ddots & & \\ & & & a_{n-1,n-1}(\cdot) & \\ a_{n1}(\cdot) & & & & a_{nn}(\cdot) \end{pmatrix} \tag{9}$$

Theorem 1. *The error dynamic system* (5) *is asymptotically stable, that is, the system* (1) and (2) *are synchronized, if the following conditions are satisfied* [35]:

(i) *the nonconstant elements the matrix $A_\alpha(.)$ are isolated in only one row or one column*;
(ii) *there exists $\varepsilon > 0$ such that:*

$$(I_n - M(A_\alpha(.)))\begin{pmatrix} 1 & 2 & \cdots & i \\ 1 & 2 & \cdots & i \end{pmatrix} > \varepsilon > 0, \ \forall i = 1,2,\ldots,n \tag{10}$$

with: $A_\alpha(.) = \{a_{ij}(.)\}$, $M(A_\alpha(.)) = \{m_{ij}(.)\}$, $m_{ij} = |a_{ij}(.)|$, $\forall\, i, j = 1, 2, \ldots, n$.
Proof. The choice of a comparison system characterized by the overvaluing matrix $M(A_\alpha(.))$, relatively to the vector norm $p(z(k)) = [|z_1(k)|\cdots|z_n(k)|]$ leads, when the nonlinearities are isolated in one row or one column of $A_\alpha(.)$, to the sufficient conditions (10),

by the use of Borne and Gentina practical stability criterion [35], which generalizes the Kotelyanski lemma for nonlinear systems and defines large scale classes of systems for which the linear conjecture can be verified for the comparison system defined by:

$$p(z(k+1)) = M(A_\alpha(.))p(z(k)) \tag{11}$$

This theorem is useful and easy to use for the synchronization studies when the instantaneous characteristic matrix is in Benrejeb arrow form matrices of Types I or II.

2.2 Case of the Synchronization of Wang System Coupled to Hénon Hitzl Zele System

In this subsection, let consider the master hyperchaotic Wang system [47] given by:

$$\begin{cases} x_{m1}(k+1) = a_3 x_{m2}(k) + (a_4+1)x_{m1}(k) \\ x_{m2}(k+1) = a_1 x_{m1}(k) + x_{m2}(k) + a_2 x_{m3}(k) \\ x_{m3}(k+1) = (a_7+1)x_{m3}(k) + a_6 x_{m2}(k)x_{m3}(k) + a_5 \end{cases} \tag{12}$$

with the following bifurcation parameters: $(a_1, a_2, a_3, a_4, a_5, a_6, a_7) = (-1.9, 0.2, 0.5, -2.3, 2, -0.6, -1.9)$,
which can be described, in the state space, in the form (1), by:

$$\begin{aligned} A(.) &= \begin{bmatrix} -1.3 & 0.5 & 0 \\ -1.9 & 1 & 0.2 \\ 0 & -0.6x_{m3}(k) & -0.9 \end{bmatrix} \\ F &= \begin{bmatrix} 0 \\ 0 \\ 2 \end{bmatrix} \\ C &= \begin{bmatrix} 0 & 1 & 0 \\ 1 & 0 & 0 \end{bmatrix} \end{aligned} \tag{13}$$

The slave Hénon Hitzl Zele system [48] is described by:

$$\begin{cases} x_{s1}(k+1) = -a x_{s2}(k) + u_1(k) \\ x_{s2}(k+1) = 1 + x_{s3}(k) - b x_{s2}^2(k) + u_2(k) \\ x_{s3}(k+1) = a x_{s2}(k) + x_{s1}(k) + u_3(k) \end{cases} \tag{14}$$

with the following bifurcation parameters a and b given by: $(a, b) = (0.3, 1.07)$, and, in the state space, in the form (2), by:

$$\begin{aligned} A'(.) &= \begin{bmatrix} 0 & -0.3 & 0 \\ 0 & -1.07x_{S2}(k) & 1 \\ 1 & 0.3 & 0 \end{bmatrix} \\ G &= \begin{bmatrix} 0 \\ 1 \\ 0 \end{bmatrix} \\ C &= \begin{bmatrix} 0 & 1 & 0 \\ 1 & 0 & 0 \end{bmatrix} \end{aligned} \tag{15}$$

For the nonlinear matrix observer gain L such that:

$$L = \begin{bmatrix} L_{11}(.) & L_{12}(.) \\ L_{21}(.) & L_{22}(.) \\ L_{31}(.) & L_{32}(.) \end{bmatrix} \tag{16}$$

the instantaneous characteristic matrix of dynamic error, defined in the form (5), becomes:

$$\begin{aligned} A_\alpha(.) &= A(.)-L(.)C \\ &= \begin{bmatrix} -1.3-L_{12}(.) & 0.5-L_{11}(.) & 0 \\ -1.9-L_{22}(.) & 1-L_{21}(.) & 0.2 \\ -L_{32}(.) & -0.6x_{m3}(k)-L_{31}(.) & -0.9 \end{bmatrix} \end{aligned} \tag{17}$$

For the gains L_{11} and L_{22} chosen constant such as:

$$\begin{cases} L_{11}=0.5 \\ L_{22}=-1.9 \end{cases} \tag{18}$$

the matrix $(I - M(A_\alpha(.)))$ is in Benrejeb arrow form matrix of Type I, such that:

$$I-M(A_\alpha(.))=\begin{bmatrix} 1-|-1.3-L_{12}(.)| & 0 & 0 \\ 0 & 1-|1-L_{21}(.)| & -|0.2| \\ -|-L_{32}(.)| & -|-0.6x_{m3}(k)-L_{31}(.)| & 1-|-0.9| \end{bmatrix} \tag{19}$$

This matrix presents a nonlinearity in the last row, then the application of the Theorem 1 leads to the chosen observer gains:

$$\begin{cases} L_{12}=-1.3 \\ L_{21}=1 \\ L_{31}(.)=-0.6x_{m3}(k) \end{cases} \tag{20}$$

satisfying the stability property of the error system characterized by the matrix, defined in (5).

Then, the gain matrix $L(.)$, synchronizing the two coupled hyperchaotic master-slave systems, is as follows:

$$L(.)=\begin{bmatrix} 0.5 & -1.3 \\ 1 & -1.9 \\ -0.6x_{m3}(k) & 0 \end{bmatrix} \tag{21}$$

and the corresponding control law such that:

$$\begin{cases} u_1(k) = -1.3e_1(k)+0.5e_2(k)-1.3x_{s1}(k)+0.8x_{s2}(k) \\ u_2(k) = \begin{cases} -1.9e_1(k)+e_2(k)-1.9x_{s1}(k)+ \\ (1.07x_{s2}(k)+1)x_{s2}(k)-0.8x_{s3}(k)-1 \end{cases} \\ u_3(k) = \begin{cases} -0.6x_{m3}(k)e_2(k)-x_{s1}(k)- \\ (0.3+0.6x_{m3}(k))x_{s1}(k)-0.9x_{s3}(k)+2 \end{cases} \end{cases} \tag{22}$$

As a result, Figs. 1 and 2 show the finite time synchronization of the state variables between the two studied hyperchaotic systems of Wang and Hénon Hitzl Zele for initial conditions $(x_{m1}(0),x_{m2}(0),x_{m3}(0))=(0.5,0.7,0.01)$ and $(x_{s1}(0),x_{s2}(0),x_{s3}(0))=(0.2,0.5,0.1)$ and for a sampling time equal to 0.01 s.

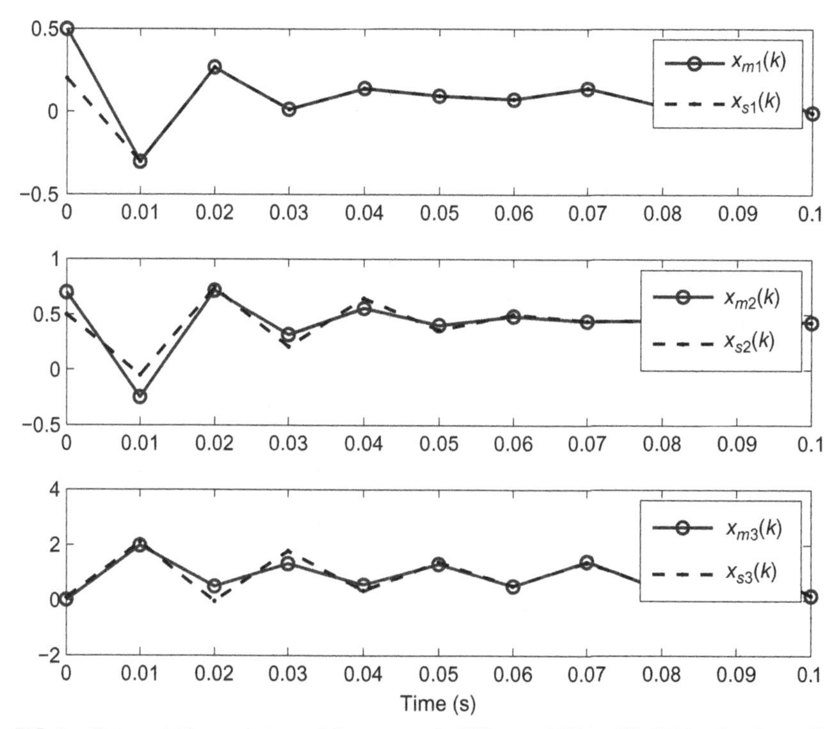

FIG. 1 State variables evolutions of the two coupled Wang and Hénon Hitzl Zele after the application of the proposed control law (22).

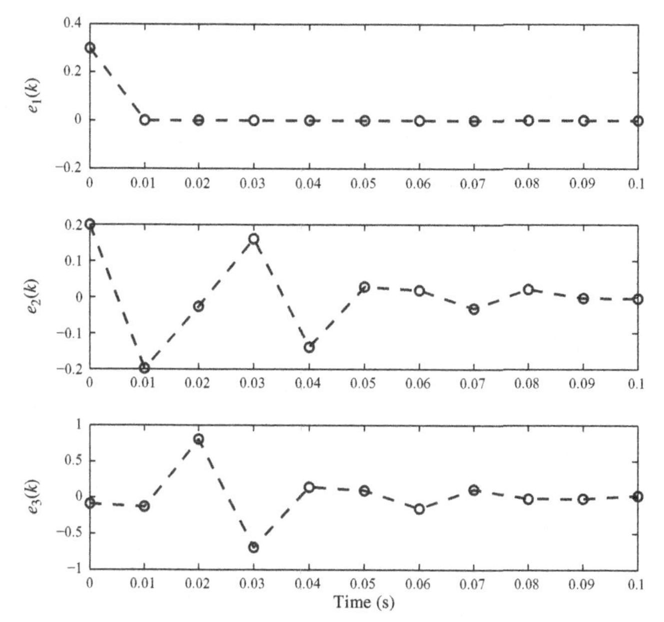

FIG. 2 Evolutions of dynamic errors of the two coupled Wang and Hénon Hitzl Zele after the application of the control law (22).

2.3 Case of the Synchronization of Generalized Hénon 3D System Coupled to Stéfanski System

In this subsection, the master hyperchaotic Hénon 3D system [49] given by:

$$\begin{cases} x_{m1}(k+1) = a - x_{m2}(k)^2 - bx_{m3} \\ x_{m2}(k+1) = x_{m1}(k) \\ x_{m3}(k+1) = x_{m2}(k) \end{cases} \tag{23}$$

with the following bifurcation parameters a and b given by: $(a,b) = (1.76, 0.1)$, is described, in the state space, in the form (1) by:

$$\begin{aligned} A(.) &= \begin{bmatrix} 0 & -x_{m2}(k) & -0.1 \\ 1 & 0 & 0 \\ 0 & 1 & 0 \end{bmatrix} \\ F &= \begin{bmatrix} 1.76 \\ 0 \\ 0 \end{bmatrix} \\ C &= \begin{bmatrix} 0 & 1 & 0 \end{bmatrix} \end{aligned} \tag{24}$$

The slave hyperchaotic Stéfanski system [50] given by:

$$\begin{cases} x_{s1}(k+1) = 1 + x_{s3}(k) - \alpha x_{s2}^2(k) + u_1(k) \\ x_{s2}(k+1) = 1 + \beta x_{s2}(k) - \alpha x_{s1}^2(k) + u_2(k) \\ x_{s3}(k+1) = \beta x_{s1}(k) + u_3(k) \end{cases} \tag{25}$$

with the following bifurcation parameters α and β such as: $(\alpha, \beta) = (1.4, 0.2)$, is described in the state space, in the form (2), by:

$$\begin{aligned} A'(.) &= \begin{bmatrix} 0 & -1.4x_{s2}(k) & 1 \\ -1.4x_{s1}(k) & 0.2 & 0 \\ 0.2 & 0 & 0 \end{bmatrix} \\ G &= [1\ 1\ 0]' \\ C &= \begin{bmatrix} 0 & 1 & 0 \end{bmatrix} \\ L(.) &= [L_1(.)\ L_2(.)\ L_3(.)]' \end{aligned} \tag{26}$$

Then, the dynamic error system is characterized by the matrix $A_\alpha(.)$ such that:

$$\begin{aligned} A_\alpha(.) &= A(.) - L(.)C \\ &= \begin{bmatrix} 0 & -x_{m2}(k) - L_1(.) & -0.1 \\ 1 & -L_2(.) & 0 \\ 0 & 1 - L_3(.) & 0 \end{bmatrix} \end{aligned} \tag{27}$$

For the gain: $L_3 = 1$, the matrix $(I - M(A_\alpha(.)))$ can be in Benrejeb arrow form matrix of type II such that:

$$I - M(A_\alpha(.)) = \begin{bmatrix} 1 & -|x_{m2}(k) + L_1(.)| & -0.1 \\ -1 & 1 - |L_2(.)| & 0 \\ 0 & 0 & 1 \end{bmatrix} \tag{28}$$

For $L_2(.)$ chosen constant such that $L_2 = 0$, this matrix presents a nonlinearity in the first row, and for the choice of the nonlinear gain $L_1(.)$ such that $L_1(.) = -x_{m2}(k)$, the stability conditions of the Theorem 1 are satisfied.

Then, the error system characterized by the matrix (27) is stable for the following vector observer gain:

$$L = \begin{bmatrix} -x_{m2}(k) \\ 0 \\ 1 \end{bmatrix} \tag{29}$$

and for the corresponding control law:

$$\begin{cases} u_1(k) & = & \begin{cases} -x_{m2}(k)(x_{m2}(k) - x_{s2}(k)) + \\ (1.04x_{s2}(k) - x_{m2}(k))x_{s2}(k) - 1.1x_{s3}(k) + 0.76 \end{cases} \\ u_2(k) & = & (1.4x_{s1}(k) + 1)x_{s1}(k) - 0.2x_{s2}(k) - 1 \\ u_3(k) & = & -0.2x_{s1}(k) + x_{s2}(k) + e_2(k) \end{cases} \tag{30}$$

As a result, Figs. 3 and 4 show the finite time synchronization property between the hyperchaotic Hénon 3D and the Stéfanski systems for initial conditions: $(x_{m1}(0), x_{m2}(0), x_{m3}(0)) = (0.02, 0.5, 0.01)$ and $(x_{s1}(0), x_{s2}(0), x_{s3}(0)) = (0.09, 0.68, 0.001)$ and for a sampling time equal to 0.01 s.

The proposed synchronization approach is conducted, with success, for nonidentical hyperchaotic systems having identical dimension. The obtained results are acceptable and can be used to increase the security level of the communication systems, based on chaotic systems synchronization.

Table 1 shows the synchronization times of the studied coupled hyperchaotic Wang and Hénon Hitzl Zele systems on one hand and Hénon 3D and Stéfanski hyperchaotic systems used to secure medical image transmission as shown in the next section. These times can be minimized, if necessary, by the best choice of observer gain parameters.

3 PROPOSED CRYPTOSYSTEM TO SECURE MEDICAL IMAGES BASED ON COUPLED HYPERCHAOTIC HÉNON 3D AND STEFANSKI SYSTEMS SYNCHRONIZATION

3.1 Cryptosystem Design: Problem Statement

Currently, the implementation of remote medical data visualization interfaces provides access to patient records containing textual data and medical images

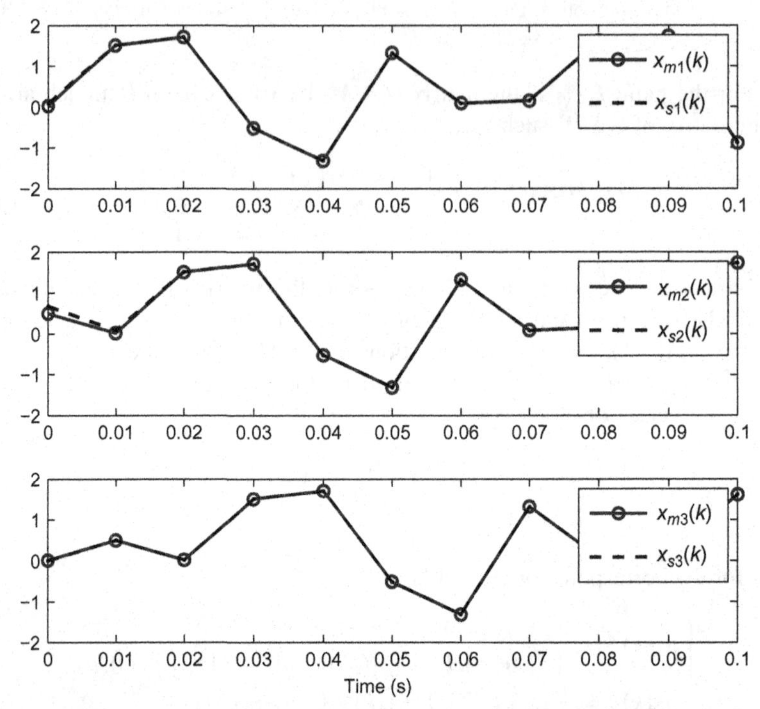

FIG. 3 State variables evolutions of the two coupled systems Hénon 3D and Stéfanski after the application of the control law (30).

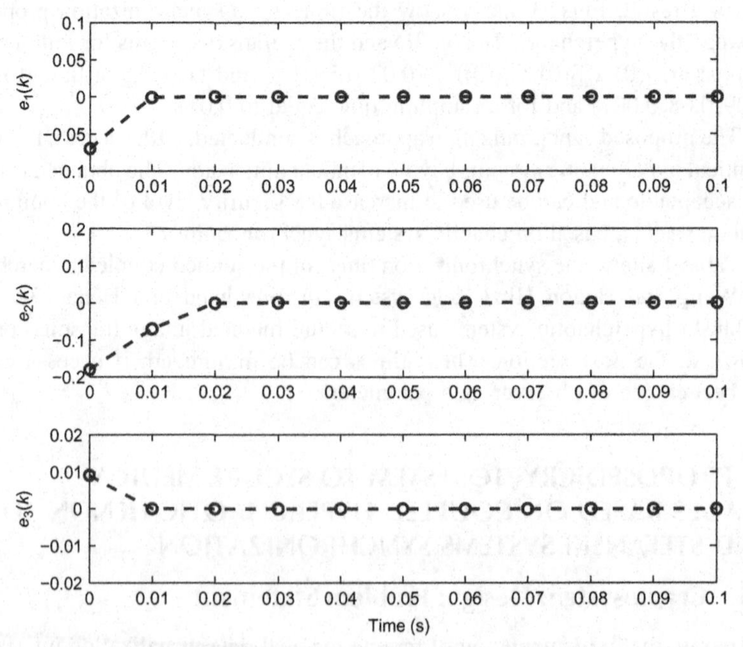

FIG. 4 Evolutions of dynamic errors of the two coupled Hénon 3D and Stéfanski after the application of the control law (30).

TABLE 1 Finite Time Synchronization for Coupled Hyperchaotic Studied Systems

Coupled Hyperchaotic Systems	Synchronization Time
Wang and Hénon Hitzl Zele	0.08 s
Hénon 3D and Stefanski	0.02 s

(MRI, X-ray scanner, etc.) which are considered confidential personal health data. Then, robust encryption scheme has to be implemented to secure digital medical images in order to allow great remote diagnostics from several distant experts.

In this work, a new approach to provide the security for digital medical images using the synchronization of nonidentical coupled 3D Hénon and Stéfanski hyperchaotic systems is proposed.

The implementation of the proposed synchronization schema, the medical image encryption algorithm and the different analysis are done using Matlah/Simulink software.

The encryption process is achieved using XOR encryption function taking as inputs the plain image and the encryption key generated by the master Hénon 3D system; then, when the synchronization process is occurring between both master and slave systems, the decrypted image is retrieved by using XOR decryption function.

The proposed approach for medical image encryption and decryption and the obtained results are detailed in the following subsection.

3.2 Encryption and Decryption Process and Results

The encryption and decryption process is given in the following seven steps.

Step 1. Initialization of the master and the slave hyperchaotic systems: select the initial conditions of the master hyperchaotic 3D Hénon system and the slave Stéfanski system.
Step 2. Encryption key generation: generate a sequence of size $M \times N$ from Hénon 3D system, which should have the same size of the plain image to be encrypted. M is the number of columns and N is the number of rows.
Step 3. Encryption key reshaping: reshape the obtained sequence to a matrix of size $M \times N$ and multiply it by a gain.
Step 4. Encryption process: perform an XOR operation between the plain image with the encryption key.
Step 5. Synchronization process: implement the synchronization of the nonidentical master and slave Hénon 3D and Stéfanski hyperchaotic systems.

The details of the proposed synchronization between these two systems are given in Section 2.3.

Step 6. Decryption key generation: generate a sequence of size $M \times N$ from the slave Stéfanski 3D system; reshape it on 2-D matrix and multiply it by a gain. The decryption key should have the same size of the encrypted image to be decrypted.

Step 7. Decryption process: perform an XOR operation between the decryption key and the obtained encrypted image in order to obtain the decrypted image.

Standard gray medical images, having the different sizes: 256 × 256, 512*512 and 232*232, are considered as the original images as shown, respectively, in Fig. 5A, D, and G. Their corresponding encryption images are shown, respectively, in Fig. 5B, E, and H. Fig. 5C, F, and I show the corresponding decrypted images.

The obtained medical image encryption and decryption results show the effectiveness of our proposed cryptosystem based on the synchronization of discrete time nonidentical hyperchaotic Hénon 3D and Stéfanski systems.

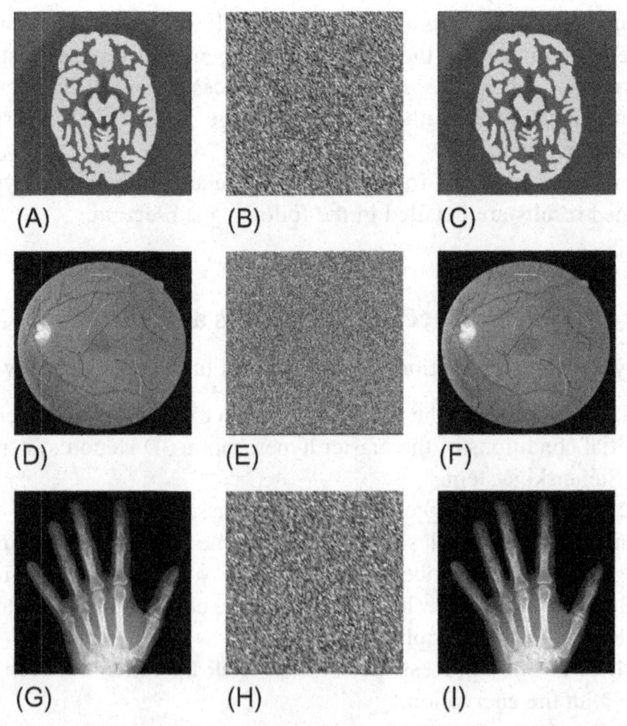

FIG. 5 Medical images encryption and decryption results; (A), (D) and (G) plain images; (B), (E) and (H) encrypted images; (C), (F) and (I) decrypted images.

3.3 Measurement of Encryption and Decryption Quality

Mean Square Error (MSE), Peak Signal-to-Noise Ratio (PSNR), and Structural Similarity Index Measure (SSIM) (see Appendix A) are used to measure the proposed encryption and decryption algorithms quality [51].

In this context, MSE, PSNR, and SSIM are performed between plain and encrypted images, then between plain and decrypted images. The obtained values of SSIM are very close to zero, those of PSNR low, and the MSE ones >30 dB, which denotes the difference between original and encrypted images, as shown in Table 2.

Furthermore, for plain and decrypted images, the obtained values of SSIM are very close to 1, those of PSNR high, and MSE values close to 0, as shown in Table 3. As a result, the different resultant values of MSE, PSNR, and SSIM show the effectiveness of the proposed encryption algorithm applied to secure medical image transmission.

3.4 Security Analysis

To show the security and efficiency of the proposed image cryptosystem, different security analyses are performed including key space analysis, correlation coefficient between plain images and their corresponding encrypted ones, correlation coefficient between adjacent pixels, and information entropy analysis, which are presented in details in the following subsections.

TABLE 2 MSE, PSNR, and SSIM Values Between Plain and Encrypted Images

Images	MSE	PSNR	SSIM
Fig. 5A and B	93.2738	8.1372	0.0023
Fig. 5D and E	75.8129	7.7288	0.0041
Fig. 5G and H	42.2462	5.9944	0.0034

TABLE 3 MSE, PSNR, and SSIM Values Between Plain and Decrypted Images

Images	MSE	PSNR	SSIM
Fig. 5A and C	6.1×10^{-5}	49.9718	0.9998
Fig. 5D and F	0	55.8007	0.9999
Fig. 5G and I	0	48.9251	0.9995

TABLE 4 Key Space Sizes

Image	Key Space Size
Fig. 5A	$2^{256\times256\times8}$
Fig. 5D	$2^{512\times512\times8}$
Fig. 5G	$2^{232\times232\times8}$

3.4.1 *Key Space Analysis*

The encryption key is a matrix of size $M \times N$ generated by the hyperchaotic Hénon 3D system, such that every value of the matrix is coded on 8 bytes. The size of the encryption key is equal to $M \times N \times 8$ (bytes); indeed, the space of key is of the order $2^{M\times N\times 8}$. In our case, Table 4 shows the key lengths of our proposed method for different images, which are $>2^{100}$, widely sufficient and robust against brute force attack [52].

3.4.2 *Statistical Analysis*

- Histograms analysis

  In image processing, the histogram of an image denotes to a histogram of the pixel intensity values and shows the distribution of pixels grayscale values. In security context, the histogram of encrypted images has to be uniform. As shown in Fig. 6B, E, and H, histograms of encrypted medical images are uniform and different from those of plain images.
- Correlation analysis

  Correlation coefficient (cc) measures the relation intensity between two images [53]. It is calculated using the following relation:

$$\text{cc} = \frac{\sum_m \sum_n \left(A_{mn} - \overline{A}\right)\left(B_{mn} - \overline{B}\right)}{\sqrt{\sum_m \sum_n \left(A_{mn} - \overline{A}\right)^2 \left(B_{mn} - \overline{B}\right)^2}} \tag{31}$$

  $\overline{A}$ and $\overline{B}$ denote, respectively, the mean of A and B. If the coefficient of correlation (cc) is equal to zero, the two images are different. If it is equal to 1, it means that the two images are identical. In this case, the correlation coefficient is computed between plain and encrypted images, then between plain and decrypted images. As a result, correlation coefficients, between plain and encrypted images, shown in Table 5, are close to 0 which means that they are different. Whereas correlation coefficient between plain and decrypted images are close to 1 as shown in Table 6.
- Correlation analysis of adjacent pixels

  Correlation analysis is an important tool to test the relationship between adjacent pixels in an image (see Appendix B). For plain images, adjacent

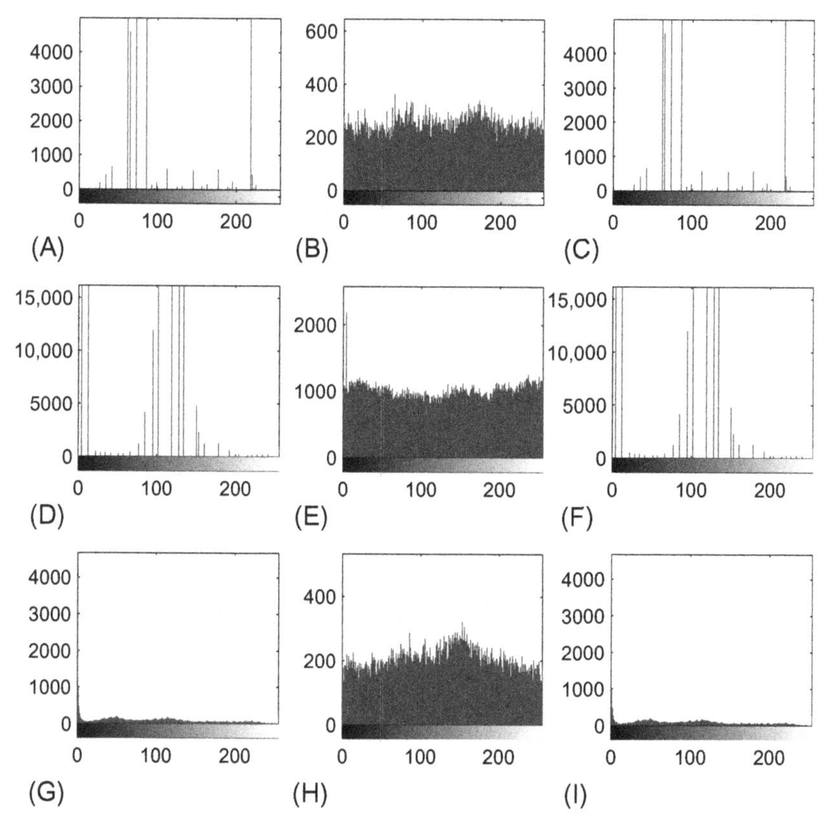

FIG. 6 Histograms: (A), (D) and (G) plain images; (B), (E) and (H) encrypted images; (C), (F) and (I) decrypted images.

TABLE 5 Correlation Coefficient Between Plain and Encrypted Images

Images	Correlation Coefficient
Fig. 5A and B	−0.0144
Fig. 5D and E	0.0028
Fig. 5G and H	0.0087

TABLE 6 Correlation Coefficient Between Plain and Decrypted Images

Images	Correlation Coefficient
Fig. 5A and C	0.9999
Fig. 5D and F	1.0000
Fig. 5G and I	0.9999

pixels have high correlation. However, the correlation coefficient of adjacent pixels in encrypted images should be low.

The correlation coefficients of the adjacent pixels of plain images in horizontal, vertical, and in diagonal directions are significantly decreased after applying the proposed encryption algorithm, as shown in Table 7.

- Information entropy analysis

The entropy of an information $E(m)$ is calculated as follows:

$$E(m) = \sum_{i=0}^{2^{N-1}} p(m_i) \log \left(\frac{1}{p(m_i)} \right) \tag{32}$$

N is the number of bytes representing a symbol m_i and $p(m_i)$ the probability of the symbol m_i, therefore, the entropy is expressed in bytes. For a random source emitting 2^N, the entropy is $E(m) = N$.

For encrypted message, entropy should have the value N. With an entropy less than N, there exists a certain degree of predictability which can menace the transmission security.

For $N = 8$, Table 8 gives the entropy values of encrypted studied images which are close to 8.

TABLE 7 Correlation Analysis of Adjacent Pixels

	Correlation Coefficients		
Image	*Horizontal*	*Vertical*	*Diagonal*
Fig. 5A	0.9521	0.9611	0.9277
Fig. 5B	0.0238	0.0445	0.5624
Fig. 5D	0.9436	0.9466	0.9399
Fig. 5E	0.0216	0.0251	0.3630
Fig. 5G	0.9564	0.9957	0.9559
Fig. 5H	−0.0034	−0.0126	0.5991

TABLE 8 Entropy Test for the Encrypted Images

Image	Entropy
Fig. 5B	7.9891
Fig. 5E	7.9915
Fig. 5H	7.9811

4 CONCLUSION

In this chapter, encryption and decryption algorithms have been proposed and applied to secure medical image transmission. They are based on a new synchronization method of nonidentical Hénon 3D and Stéfanski hyperchaotic systems. The synchronization study is based on the use of Borne and Gentina criterion for nonlinear systems stability, associated with Benrejeb arrow form matrix for system description and achieved in finite time. The corresponding built asymmetric cryptosystem has shown its effectiveness against brute force and statistical attacks for different sizes of input medical images.

APPENDIX A [53]

- Mean Square Error (MSE) analysis

  The Mean Square Error (MSE) is used in this paper to measure the distance between plain and encrypted or decrypted images.

  The high value of (MSE) corresponds to a high difference between plain and encrypted images. It is given by:

$$\mathrm{MSE}=\frac{1}{M\times N}\sum_{i=1}^{M}\sum_{j=1}^{N}(P(i,j)-C(i,j))^2 \tag{33}$$

  $P(i,j)$ and $C(i,j)$ represent respectively the plain and the encrypted or decrypted images of size $M \times N$.

  For a MSE $\geq$30 dB, there is a difference between the plain and encrypted images.

- Peak Signal-to-Noise Ratio (PSNR) analysis

  The Peak Signal-to-Noise Ratio (PSNR) defines the ratio between the maximum intensity and the MSE of an image. It can be used to measure the conformity between plain and encrypted images. For an image of size $M \times N$, it is given by:

$$\mathrm{PSNR}=20\log_{10}\left[\frac{I_{\max}}{\sqrt{\mathrm{MSE}}}\right] \tag{34}$$

  For a good encryption algorithm, the PSNR should be as low as possible between the plain and encrypted images.

- Structural Similarity Index Measure (SSIM)

  Structural Similarity Index Measure (SSIM) is a tool to measure the similarity between images. It has been designed to improve PSNR and MSE metrics, proved to be inconsistent with human eye perception. SSMI is calculated in various windows of an image for which the measure between two windows x and y of size $M \times N$ is given by:

$$\mathrm{SSIM}(x,y)=\frac{(2\mu_x\mu_y+c_1)(2\sigma_{xy}+c_2)}{\left(\mu_x^2+\mu_y^2+c_1\right)\left(\sigma_x^2+\sigma_y^2+c_2\right)} \tag{35}$$

μ_x is the average of x, μ_y the average of y, ${\sigma_x}^2$ the variance of x, ${\sigma_y}^2$ the variance of y, σ_{xy} the covariance of x and y, $c_1 = (k_1L)^2$ and $c_2 = (k_2L)^2$ two variables to stabilize the division with weak denominator, L the range of the pixel value, $k_1 = 0.01$ and $k_1 = 0.03$, chosen by default. The SSIM value in the range $[-1\ 1]$ reaches 1, in the case of two identical images.

APPENDIX B [54]

The correlation coefficient of adjacent pixels $r_{p,\ c}$ of each pair is calculated according to the following formulas. $P(i,j)$ and $C(i,j)$ are gray values of the plain pixel and the encrypted one. $E(P)$, $E(C)$, $D(P)$ and $D(C)$ are mathematical expectation of plain pixels $P(i,j)$, mathematical expectation of cipher pixels $C(i,j)$, variance of plain pixels $P(i,j)$ and variance of cipher pixels $C(i,j)$, respectively.

$$E(P) = \frac{\sum_{i=1}^{M}\sum_{j=1}^{N} P(i,j)}{M \times N},\ E(C) = \frac{\sum_{i=1}^{M}\sum_{j=1}^{N} C(i,j)}{M \times N} \tag{36}$$

$$D(P) = \frac{\sum_{i=1}^{M}\sum_{j=1}^{N} [P(i,j) - E(P)]^2}{M \times N} \tag{37}$$

$$\text{cov}(P, C) = \frac{\sum_{i=1}^{M}\sum_{j=1}^{N} [P(i,j) - E(P)][C(i,j) - E(C)]}{M \times N} \tag{38}$$

$$r_{p,c} = \frac{\text{cov}(PC)}{\sqrt{D(P)}\sqrt{D(C)}} \tag{39}$$

REFERENCES

[1] Hunt GMK. Determinism, predictability and chaos. Analysis 1987;47(3):129–33.

[2] Glasnert E, Weiss B. Sensitive dependence on initial conditions. Nonlinearity 1993;6: 1067–75.

[3] Layek GC. An introduction to dynamical systems and chaos. India: Springer; 2015.

[4] Allali S, Khalifa N, Benrejeb M. On mixing chaotic systems via TS fuzzy modeling. In: The International Conference on Control. Malte: Decision and Information Technologies; 2016.

[5] Allali S, Khalifa N, Benrejeb M. Generating a new strange attractor by mixing discrete and discretized continuous TS fuzzy model. In: The International Conference on Control. Barcelona: Decision and Information Technologies; 2017.

[6] Trabelsi H, Benrejeb M. Control of chaos in permanent magnet synchronous motor with parameter uncertainties: a Lyapunov approach. Int J Innov Sci Res 2015;13(1):279–85.

[7] Pecora LM, Carroll TL. Synchronization in chaotic systems. Phys Rev Lett 1990;64(8):821–4.

[8] Yu-Zhu X, Wei X, Xiu-Chun L, Su-Fang T. Complete synchronization of uncertain chaotic dynamical network via a simple adaptive control. Chinese Phys B 2008;17(1):80.

[9] Zhang Q, Lu J, Chen S. Coexistence of anti-phase and complete synchronization in the generalized Lorenz system. Commun Nonlinear Sci Numer Simul 2010;15(10):3067–72.

[10] Al-Sawalha MM, Noorani MSM. Antisynchronization of two hyperchaotic systems via nonlinear control. Commun Nonlinear Sci Numer Simul 2009;14(8):3402–11.

[11] Yao C, Zhao Q, Yu J. Complete synchronization induced by disorder in coupled chaotic lattice. Phys Lett A 2013;377(5):370–7.

[12] Shooshtari BK, Forouzanfar AM, Molaei MR. Identical synchronization of a non-autonomous unified chaotic system with continuous periodic switch. Springer Plus 2016;5(1):1667.

[13] Abarbanel HDI, Rulkov NF, Sushchik MM. Generalized synchronization of chaos: the auxiliary system approach. Phys Rev E 1996;53:4528–35.

[14] Yang SS, Duan CK. Generalized synchronization in chaotic systems. Chaos, Solitons Fractals 1998;9:1703–7.

[15] Inoue M, Kawazoe T, Nishi Y, Nagadome M. Generalized synchronization and partial synchronization in coupled maps. Phys Lett A 1998;249:69–73.

[16] Wang YW, Guan ZH. Generalized synchronization of continuous chaotic system. Chaos, Solitons Fractals 2006;27:97–101.

[17] Chang CM, Chen HK. Chaos and hybrid projective synchronization of commensurate and incommensurate fractional-order Chen-Lee systems. Nonlinear Dyn 2010;62(4):851–8.

[18] Xin B, Wu Z. Projective synchronization of chaotic discrete dynamical systems via linear state error feedback control. Entropy 2015;17:2677–87.

[19] Diao M, Yu YG, Wang S, Yu JZ. Hybrid lag function projective synchronization of discrete chaotic systems with different orders. Int J Nonlinear Sci Numer Simul 2010;11(7):503–8.

[20] Ma Z, Liub Z, Zhang CG. A new method to realize cluster synchronization in connected chaotic networks. Chaos 2016;16(2):023103.

[21] Zhang J, Tang W. Control and synchronization for a class of new chaotic systems via linear feedback. Nonlinear Dyn 2009;58(4):675.

[22] Muthukumar P, Balasubramaniam P. Controlling chaotic discrete system via the improved closed-loop control. Nonlinear Dyn 2013;74(4):1169–81.

[23] Wang J, Shen H, Park JH, Wu Z-G. Finite-time $l_2 - l_\infty$ synchronization for discrete-time nonlinear chaotic systems via information-constrained delayed feedback. Complexity 2014;21:138–46.

[24] Wang X, Wang Y. Adaptive control for synchronization of a four-dimensional chaotic system via a single variable. Nonlinear Dyn 2011;65(3):311–6.

[25] Zhang J, Zhang L, An X, Luo H, Yao KE. Adaptive coupled synchronization among three coupled chaos systems and its application to secure communications. EURASIP J Wire Commun Netw 2016;2016(1):134.

[26] Yassen MT. Chaos control of chaotic dynamical system using backstepping design. Chaos, Solitons Fractals 2006;27:537–48.

[27] Salarieh H, Alasty A. Control of stochastic chaos using sliding mode method. J Comput Appl Math 2009;225(1):135–45.

[28] Grassi G, Mascolo S. Nonlinear observer design to synchronize hyperchaotic systems via a scalar signal. IEEE Trans Circ Syst I Fundam Theory Appl 1997;44(10):1011–4.

[29] Chen MJ, Li DP, Zhang AJ. Chaotic synchronization based on nonlinear state-observer and its application in secure communication. J Mar Sci Appl 2004;3:64–70.

[30] Chen M, Wu Q, Jiang C. Disturbance-observer-based robust synchronization control of uncertain chaotic systems. Nonlinear Dyn 2012;70(4):2421–32.

[31] Tran XT, Kang HJ. A novel observer-based finite time control method for modified function projective synchronization of uncertain chaotic (hyperchaotic) systems. Nonlinear Dyn 2015;80(1):905–16.
[32] Filali RL, Benrejeb M, Borne P. On observer-based secure communication design using discrete-time hyperchaotic systems. Commun Nonlinear Sci Numer Simul 2014;19 (5):1424–32.
[33] Fu C, Meng W-h, Zhan Y-F, Zhu Z-g, Lau FCM, Tse CK, Ma H-F. An efficient and secure medical image protection scheme based on chaotic maps. Comput Biol Med 2013;3(8): 1000–10.
[34] Zhou W, Liu W-Q, Wang D-L, Zhu G-X, Hu Y-J, Zhan Y-F. An efficient medical image protection scheme with parallel chaotic key stream generation. Inf Technol J 2014;13:1602–11.
[35] Borne P. Nonlinear system stability. Vector norm approach. Syst Control Encyclopedia 1987;5:3402–6.
[36] Benrejeb M. Sur la synchronisation des systèmes continus non linéaires en régime forcé. Doctor Engineer ThesisUniversité des Sciences et Technique de Lille; 1976.
[37] Benrejeb M, Borne P. On an algebraic stability criterion for nonlinear process. Interpretation in the frequency domain, In: Proceedings of MECO'78 Congress, Athens; 1978. p. 678–82.
[38] Benrejeb M, Borne P, Laurent F. Sur une application de la représentation en flèche à l'analyse des processus. RAIRO Automat 1982;16(2):133–46.
[39] Benrejeb M. Stability study of two level hierarchical nonlinear systems. IFAC Proc Vol 2010;43(8):30–41.
[40] Benrejeb M. On the use of arrow form matrices for processes stability and stabilizability studies, Plenary lecture, In: 2nd International Conference on Systems and Computer Science, Villeneuve d'Ascq; 2013.
[41] Benrejeb M, Gasmi M. On the use of an arrow form matrix for modelling and stability analysis of singularly perturbed non-linear systems. Syst Anal Model Simul 2001;40(4):509.
[42] Dridi A, Filali RL, Benrejeb M. On discrete-time chaotic multimodels synchronization using aggregation techniques for encryption. Int J Comput Sci Netw Secur 2018;18(2):81–90.
[43] Filali RL, Benrejeb M, Borne P. On observer synchronization of non-identical discrete-time hyperchaotic maps based on aggregation techniques and arrow form matrix. Int J Comput Commun Control 2015;10(3):307–18.
[44] Khalifa N, Filali RL, Benrejeb M. On secure image transmission combining chaotic encryption and watermarking using dead beat synchronization of 4D Henon maps, In: 3rd International Conference on Control, Engineering and Information Technology (CEIT); 2015.
[45] Khalifa N, Filali RL, Benrejeb M. A fast selective image encryption using discrete wavelet transform and chaotic systems synchronization. Inf Technol Control 2016;45(3):235–42.
[46] Wang L. 3-scroll and 4-scroll chaotic attractors generated from a new 3-d quadratic autonomous system. Nonlinear Dyn 2009;56:453–62.
[47] Hitzl DL, Zele F. An exploration of the Hénon quadratic map. Physica D 1985;14(3):305–26.
[48] Yan Z. Q-S synchronization in 3D Hénon-like map and generalized Hénon map via a scalar controller. Phys Lett A 2005;342(4):309–17.
[49] Stéfanski K. Modelling chaos and hyperchaos with 3-D maps. Chaos, Solitons Fractals 1998;9 (1):83–93.
[50] Norouzi B, Mirzakuchaki S, Seyedzadeh SM, Mohammad RM. A simple, sensitive and secure image encryption algorithm based on hyper-chaotic system with only one round diffusion process. Multimed Tools Appl 2014;71:1469.
[51] Zhai Y, Lin S, Zhang Q. Improving image encryption using multi-chaotic map. Proc Power Electron Intell Transp Syst 2008;143–8.

[52] Zhang X, Chen W. A new chaotic algorithm for image encryption. Proc Int Audio Lang Image Process 2008;889.
[53] Wang Z, Bovik AC, Sheikh HR, Simoncelli EP. Image quality assessment: from error visibility to structural similarity. IEEE Trans Image Process 2004;13(4).
[54] Pareek NK, Patidar V, Sud KK. Image encryption using chaotic logistic map. Image Vis Comput 2006;24(9):926–34.

FURTHER READING

[55] Fu C, Zhang G-Y, Bian O, Lei W-M, Hong-feng M. A novel medical image protection scheme using a 3-dimensional chaotic system. PLoS One 2015;9(1).

Chapter 17

Fractional-Order Hybrid Synchronization for Multiple Hyperchaotic Systems

Abir Lassoued and Olfa Boubaker
National Institute of Applied Sciences and Technology, Tunis, Tunisia

1 INTRODUCTION

Chaos synchronization has been widely studied since the pioneer work of Pecora and Carroll [2] and many synchronization approaches have been proposed including phase synchronization [3], lag synchronization [4], and generalized synchronization [5]. In this framework, hybrid synchronization (HS) [6] for which the complete synchronization (CS) [7] and the complete antisynchronization (AS) [8] coexist under the same designed controllers is considered one of the most interesting approaches. In fact, HS schema for chaotic systems can eliminate the influence of uncertainties with a small error bound [9].

On the other hand, fractional calculus is an old concept [10–12]. Recently, new research works have proven that fractional-order (FO) models can describe more effectively complex real problems than classical integer-order models. Furthermore, many basic chaotic and hyperchaotic systems [13] have been generalized into their corresponding FO models generating stranger attractors such as the FO Duffing system [14], the FO Chua's system [15], and the FO Lorenz system [16].

Furthermore, until now, chaos synchronization of FO systems has been considered a challenging problem. Various synchronization schemes have been successfully applied for FO systems. Indeed, HS was applied between two identical [17] or nonidentical [18] FO chaotic systems. However, HS for multiple coupled FO systems has still not yet been fully investigated. Actually, the synchronization of several FO systems is considered more useful in engineering applications as it is more effective to enhance the security in digital communication and leads to a brighter future in multilateral communications.

Otherwise, several HS behaviors for multiple integer-order chaotic systems are proposed in the literature such as HS with chain connection structure [19]

Recent Advances in Chaotic Systems and Synchronization. https://doi.org/10.1016/B978-0-12-815838-8.00017-0

and HS with ring connection structure [20]. In the later schema, all systems are coupled on a chain and are correlative [20]. Indeed, the first FO hyperchaotic system antisynchronizes the second one and also the first FO hyperchaotic system synchronizes the third one. In additional, the $(N-2)$th FO hyperchaotic system synchronizes the Nth one and the $(N-1)$th FO hyperchaotic system antisynchronizes the Nth one.

The purpose of this chapter is to design an HS via ring connection for multiple FO hyperchaotic systems. This work extends the approach reported in [20] for integer-order chaotic systems and in [1, 21] for FO chaotic systems.

The chapter is organized in five sections. In Section 2, first, the basic definition of fractional calculus is described. Then, the HS problem for multiple FO hyperchaotic systems is proposed. In Section 3, the main results of the proposed synchronization schema are given. In Section 4, HS between multiple identical FO hyperchaotic systems is achieved. Finally, in Section 5, HS between multiple nonidentical FO hyperchaotic systems is also presented and realized.

2 PRELIMINARIES AND PROBLEM POSITION

In this section, the fundamental mathematical definitions of FO derivatives are presented. The problem position of HS of multiple FO hyperchaotic systems coupled with ring connection is also exposed.

2.1 Preliminaries

The generalized concept of FO integral differential operator could be defined as follows

$$
{}_aD_t^q = \begin{cases} \dfrac{d^q}{dt^q}, & R(q) > 0 \\ 1, & R(q) = 0 \\ \displaystyle\int_a^t d\tau, & R(q) < 0 \end{cases} \tag{1}
$$

where q is the FO and $R(q)$ describes the real part of q. The parameters t and a are the limit of the operation with $a < t$.

Note that there are different definitions for FO calculus such as the Riemann-Liouville (R-L) definition and the Caputo definition. In the rest of this chapter, we adopt the Caputo definition because it considers the initial conditions which can interpret physical signals in engineering applications.

Definition 1. The Caputo FO derivative is described by the following equation [22]

$$
D^q x(t) = J^{m-q} x^{(m)}(t), \quad q > 0 \tag{2}
$$

where $x^{(m)}$ is the m-order derivative of x with m the integer part of q such as $m = [q]$. J is the integral operator described by the following expression

$$J^{\beta}x(t) = \frac{1}{\Gamma(\beta)} \int_{0}^{\infty} (t-\sigma)^{\beta-1} y(\sigma)\, \mathrm{d}\sigma \tag{3}$$

where β is a rational number and $\Gamma(\beta)$ is the Gamma function defined as

$$\Gamma(\beta) = \mathrm{e}^{-t} t^{\beta-1} \ \ \mathrm{d}t \tag{4}$$

In the rest of this chapter, we consider FO model written as [23]

$$\frac{d^q x(t)}{\mathrm{d}t^q} = D^q x(t) = f(x(t)) \tag{5}$$

where D^q is the Caputo fractional derivative operator and q is the corresponding FO such as $q \in]0, 1[$ and $x \in \mathbb{R}^n$.

The stability analysis of FO systems is more complex than the integer ones which needs particular definitions, introduced below.

Definition 2. For system (5), the equilibrium points of $f(x(t))$ are asymptotically stable if all eigenvalues λ_i of the Jacobian matrix $J = \partial f(x(t))/\partial x(t)$ evaluated at the equilibrium points satisfy $|\arg(\lambda_i)| > (q\pi)/2$.

Definition 3. (Chen et al. [24]). If there exists a positive definite Lyapunov function $V(x)$ such that $D^q V(x) < 0$ for all $t \geq t_0$, then the solutions of system (5) are asymptotically stable.

Lemma 1 (Chen et al. [24]). *If the FO system satisfies Definition 3, then system (5) satisfies Definition 2.*

2.2 Problem Position

Let consider multiple FO hyperchaotic systems with ring connection described by [21]

$$\begin{cases} D^q x_1 &= A_1 x_1 + g_1(x_1) + D_1(x_N - x_1) \\ D^q x_2 &= A_2 x_2 + g_2(x_2) + D_2(x_1 - x_2) \\ &\vdots \\ D^q x_N &= A_N x_N + g_N(x_N) + D_N(x_{N-1} - x_N) \end{cases} \tag{6}$$

where the state vector x_i is defined as $x_i = (x_{i1}, x_{i2}, x_{i3}, \ldots, x_{in})$ with $i \in [1, \ldots, N]$ and n the system dimension. The parameter N designs the number of coupled FO chaotic systems. The g_i functions and the matrices A_i express the nonlinear part and the linear one of each coupled system, respectively. $D_i = \mathrm{diag}(d_{1i}, \ldots, d_{N1})$ are N-dimensional diagonal matrices to be designed such that the d_{ij}, $i \in [1\ldots N]$ and $j \in [1\ldots n]$, represent the coupled positive parameters of the diagonal matrices.

To achieve the HS, nonlinear control laws are designed as

$$\begin{cases} D^q x_1 = A_1 x_1 + g_1(x_1) + D_1(x_N - x_1) \\ D^q x_2 = A_2 x_2 + g_2(x_2) + D_2(x_1 - x_2) + u_1 \\ \vdots \\ D^q x_N = A_N x_N + g_N(x_N) + D_N(x_{N-1} - x_N) + u_{N-1} \end{cases} \tag{7}$$

For system (7), the AS error vector $e_A = [e_{A1} \ldots e_{A(N-1)}]$ and the CS error vector $e_C = [e_{C1} \ldots e_{C(N-1)}]$ are depicted, respectively, as

$$\begin{aligned} e_{Ai} &= x_i(t) + x_{i+1}(t) \\ e_{Cj} &= x_{j+2}(t) - x_j(t) \end{aligned} \tag{8}$$

where both indices i and j are included in $[1 \ldots N - 1]$.

Definition 4. The HS is achieved if the error vectors e_A and e_C satisfy the following conditions

$$\begin{aligned} \lim_{t \to +\infty} \|e_{Ai}\| &= \lim_{t \to +\infty} \|x_i(t) + x_{i+1}(t)\| = 0, \quad i \in [1 \ldots N-1] \\ \lim_{t \to +\infty} \|e_{Cj}\| &= \lim_{t \to +\infty} \|x_{j+2}(t) - x_j(t)\| = 0, \quad j \in [1 \ldots N-1] \end{aligned} \tag{9}$$

Elsewhere, we note from Definition 4 that the dynamical errors e_A and e_C, related to system (7), are asymptotically stable if the conditions (9) are satisfied.

The main objective of this chapter is to design the control laws u_j ($j \in [1 \ldots N - 1]$) satisfying the conditions (9) and also designing the matrices D_i ($i \in [1 \ldots N]$).

3 MAIN RESULTS

Theorem 1. *For the integer-order control laws $u_i(t)$, $i \in [1 \ldots N - 1]$,*

$$\begin{cases} u_1 = v_1 - [(-1-(-1)^N)D_1 + 2D_2 - (A_2 - A_1)]x_1 - g_2(x_2) + g_1(x_1) \\ u_2 = v_2 - [-2(D_2 - D_3) - (A_3 - A_2)]x_2 - g_3(x_3) + g_2(x_2) + u_1 \\ \vdots \\ u_{N-1} = v_{N-1} - [-2(D_{N-1} - D_N) - (A_N - A_{N-1})]x_{N-1} \\ \qquad + g_N(x_N) + g_{N-1}(x_{N-1}) + u_{N-2}. \end{cases} \tag{10}$$

the AS errors dynamics given by

$$D^q e_A = D^q x_i + D^q x_{i+1} = (M + H)e_A = L e_A \tag{11}$$

are asymptotically stable if the matrices L, L_1, L_2, and H satisfy the following conditions

$$\begin{aligned} & L = L_1 + L_2 \\ & L_1^T = -L_1 \\ & L_2 = diag(l_1, \ldots, l_N) \quad and \quad l_i < 0, i \in [1 \ldots N] \end{aligned} \tag{12}$$

where H is a constant matrix such as $[v_1 \ldots v_{N-1}]^T = H[e_{A1} \ldots e_{A(N-1)}]^T$ *and*

$$M = \begin{bmatrix} K_1 & K_2 & K_3 & & D_1 \\ D_2 & A_3 - D_3 & 0 & \cdots & 0 \\ 0 & D_3 & A_4 - D_4 & & 0 \\ & \vdots & & \ddots & \vdots \\ 0 & 0 & D_{N-1} & & A_N - D_N \end{bmatrix}$$

where

$$\begin{aligned} K_1 &= A_2 - (-1)^{N-1} D_1 - D_2 \\ K_2 &= -(-1)^{N-2} D_1 \\ K_3 &= -(-1)^{N-3} D_1 \\ K_4 &= A_N - D_N \end{aligned}$$

Proof. For the AS error described by the first equation in system (8), the proposed control laws (10) are introduced in the expression $x_i(t)$ and $x_{i+1}(t)$. The related FO derivative is described by

$$D^q e_A = (M + H) e_A = L e_A \tag{13}$$

Thus, the relation (11) of Theorem 1 is obtained. □

Then, the Lyapunov function V is defined as $V = (1/2) e_A^T e_A$. The related FO derivative is written as

$$\begin{aligned} D^q V &= (1/2) D^q (e_A^T e_A) \\ &= (1/2)(e_A^T L^T e_A + e_A^T L e_A) \\ &= (1/2) e_A^T (L^T + L) e_A \end{aligned} \tag{14}$$

Let $L = L_1 + L_2$ and then $L^T = L_1^T + L_2^T$. Based on Definition 3, the FO function $D^q V$ is negative definite if

$$L^T + L < 0$$

However, if $L_1 = -L_1^T$ and $L2$ = diag(li) with $li < 0$ then we have

$$D^q V = e_A^T L_2 e_A < 0 \tag{15}$$

Thus, the equilibrium $e_A = 0$ of system (13) is asymptotically stable. Then, the AS is correctly applied under the designed controllers.

Theorem 2. *If the AS error is asymptotically stable under the control laws (10), the CS error is asymptotically stable. Therefore, the HS is realized.*

Proof. According to the second equation of system (8), the CS error vector e_C can be defined as, with $j \in [1...N-1]$

$$e_{Cj}(t) = x_{j+2}(t) - x_{j+1}(t) + x_{j+1}(t) - x_j(t) \tag{16}$$

Using Definition 4, if the AS error dynamic e_A is asymptotically stable, we obtain

$$\lim_{t \to +\infty} ||x_{j+2} + x_{j+1}|| - \lim_{t \to -\infty} ||x_{j+1} + x_j|| = 0 \tag{17}$$

Therefore, the CS under the same controllers, Eq. (10) is achieved. As a conclusion, the HS of multiple coupled FO systems is also realized. □

4 HYBRID SYNCHRONIZATION BETWEEN IDENTICAL FO HYPERCHAOTIC SYSTEMS

4.1 FO Hyperchaotic Systems

Consider the following four FO hyperchaotic systems with FO terms described by [25]

$$\begin{cases} D^q x_{11} &= x_{12} + d_{11}(x_{41} - x_{11}) \\ D^q x_{12} &= x_{13} + d_{12}(x_{42} - x_{12}) \\ D^q x_{13} &= -a x_{13} - b x_{12} + G(x_{11}) + d_{13}(x_{43} - x_{13}) \\ D^q x_{14} &= k x_{12} - h x_{14} + G(x_{11}) + d_{14}(x_{44} - x_{14}) \end{cases} \tag{18}$$

$$\begin{cases} D^q x_{21} &= x_{22} + d_{21}(x_{11} - x_{21}) + u_{11} \\ D^q x_{22} &= x_{23} + d_{22}(x_{12} - x_{22}) + u_{12} \\ D^q x_{23} &= -a x_{23} - b x_{22} + G(x_{21}) + d_{23}(x_{13} - x_{23}) + u_{13} \\ D^q x_{24} &= k x_{22} - h x_{24} + G(x_{21}) + d_{24}(x_{14} - x_{24}) + u_{14} \end{cases} \tag{19}$$

$$\begin{cases} D^q x_{31} &= x_{32} + d_{31}(x_{21} - x_{31}) + u_{21} \\ D^q x_{32} &= x_{33} + d_{32}(x_{22} - x_{32}) + u_{22} \\ D^q x_{33} &= -a x_{33} - b x_{32} + G(x_{31}) + d_{33}(x_{23} - x_{33}) + u_{23} \\ D^q x_{34} &= k x_{32} - h x_{34} + G(x_{31}) + d_{34}(x_{24} - x_{34}) + u_{24} \end{cases} \tag{20}$$

$$\begin{cases} D^q x_{41} &= x_{42} + d_{41}(x_{31} - x_{41}) + u_{31} \\ D^q x_{42} &= x_{43} + d_{42}(x_{32} - x_{42}) + u_{32} \\ D^q x_{43} &= -a x_{43} - b x_{42} + G(x_{41}) + d_{43}(x_{33} - x_{43}) + u_{33} \\ D^q x_{44} &= k x_{42} - h x_{44} + G(x_{41}) + d_{44}(x_{34} - x_{44}) + u_{34} \end{cases} \tag{21}$$

where q is a rational number as $0 < q < 1$. u_1, u_2, and u_3 are the control inputs with $u_i = [u_{i1}, u_{i2}, u_{i3}, u_{i4}]$, $i = 1...3$. $G(x_{i1})$ are nonlinear functions defined as

$G(x_{i1}) = -cx_{i1}^2 + d|x_{i1}|x_{i1} + m|x_{i1}|^r x_{i1}^{-1}$ with $i = 1...3$. $(a, b, c, d, k, h, m, r) =$ (0.93, 1.11, − 0.11, − 0.21, 14, 0.001, 6.26, 1.32) are the system parameters.

For each system, we have $A_i = \begin{bmatrix} 0 & 1 & 0 & 0 \\ 0 & 0 & 1 & 0 \\ 0 & -b & -a & 0 \\ 0 & k & 0 & -h \end{bmatrix}$ and $g_i(x_{i1}) = \begin{bmatrix} 0 \\ 0 \\ G(x_{i1}) \\ G(x_{i1}) \end{bmatrix}$

with $i = 1...4$.

The AS error vector is defined as $e_A = [e_{A1}e_{A2}e_{A3}]$ such that $e_{Ai} = x_i(t) + x_{i+1}(t)$, $i \in [1...3]$.

4.2 Control Design

Using Eq. (10), the control laws are written as

$$\begin{cases} u_1 &= H_1 e_A + 2(D_1 - D_2)x_1 - g_2(x_2) - g_1(x_1) \\ u_2 &= H_2 e_A + 2(D_2 - D_3)x_2 - g_3(x_3) - g_2(x_2) - u_1 \\ u_3 &= H_3 e_A + 2(D_3 - D_4)x_3 - g_4(x_4) - g_3(x_3) - u_2 \end{cases} \tag{22}$$

According to the conditions (11), H_1, H_2, and H_3 are given by

$$H_1 = \begin{bmatrix} 0 & 0 & 0 & 0 & 0 & 0 & 0 & 0 & 0 & 0 & 0 & 0 \\ -1 & 0 & b & -k & 0 & 0 & 0 & 0 & 0 & 0 & 0 & 0 \\ 0 & -1 & 0 & 0 & 0 & 0 & 0 & 0 & 0 & 0 & 0 & 0 \\ 0 & 0 & 0 & 0 & 0 & 0 & 0 & 0 & 0 & 0 & 0 & 0 \end{bmatrix}$$

$$H_2 = \begin{bmatrix} N_1 & 0 & 0 & 0 & 0 & 0 & 0 & 0 & -d_{31} & 0 & 0 & 0 \\ 0 & N_2 & 0 & 0 & -1 & 0 & b & -k & 0 & -d_{32} & 0 & 0 \\ 0 & 0 & N_3 & 0 & 0 & -1 & 0 & 0 & 0 & 0 & -d_{33} & 0 \\ 0 & 0 & 0 & N_4 & 0 & 0 & 0 & 0 & 0 & 0 & 0 & -d_{43} \end{bmatrix}$$

with $N_1 = d_{11} - d_{21}$, $N_2 = d_{12} - d_{22}$, and $N_3 = d_{13} - d_{23}$,

$$H_3 = \begin{bmatrix} -d_{11} & 0 & 0 & 0 & 0 & 0 & 0 & 0 & 0 & 0 & 0 & 0 \\ 0 & -d_{12} & 0 & 0 & 0 & 0 & 0 & 0 & -1 & 0 & b & -k \\ 0 & 0 & -d_{13} & 0 & 0 & 0 & 0 & 0 & 0 & -1 & 0 & 0 \\ 0 & 0 & 0 & -d_{14} & 0 & 0 & 0 & 0 & 0 & 0 & 0 & 0 \end{bmatrix}$$

As a conclusion, the dynamical errors are expressed by the following equation

$$D^q e_A = L(e_A)e_A = (L_1 + L_2)e_A \quad (i = 1...3 \text{ and } j = 1...3)$$

where

$$L_1 = \left(\begin{array}{cccc|cccc|cccc}
0 & 1 & 0 & 0 & -d_{11} & 0 & 0 & 0 & d_{11} & 0 & 0 & 0 \\
-1 & 0 & 1+b & -k & 0 & -d_{12} & 0 & 0 & 0 & d_{12} & 0 & 0 \\
0 & -1-b & 0 & 0 & 0 & 0 & -d_{13} & 0 & 0 & 0 & d_{13} & 0 \\
0 & k & 0 & 0 & 0 & 0 & 0 & -d_{14} & 0 & 0 & 0 & d_{14} \\
\hline
d_{11} & 0 & 0 & 0 & 0 & 1 & 0 & 0 & -d_{31} & 0 & 0 & 0 \\
0 & d_{12} & 0 & 0 & -1 & 0 & 1+b & -k & 0 & -d_{32} & 0 & 0 \\
0 & 0 & d_{13} & 0 & 0 & -1-b & 0 & 0 & 0 & 0 & -d_{33} & 0 \\
0 & 0 & 0 & d_{14} & 0 & k & 0 & 0 & 0 & 0 & 0 & -d_{34} \\
\hline
-d_{11} & 0 & 0 & 0 & d_{31} & 0 & 0 & 0 & 0 & 1 & 0 & 0 \\
0 & -d_{12} & 0 & 0 & 0 & d_{32} & 0 & 0 & -1 & 0 & 1+b & -k \\
0 & 0 & -d_{13} & 0 & 0 & 0 & d_{33} & 0 & 0 & -1-b & 0 & 0 \\
0 & 0 & 0 & -d_{14} & 0 & 0 & 0 & d_{34} & 0 & k & 0 & 0
\end{array}\right)$$

and

$$L_2 = \left(\begin{array}{cccc|cccc|cccc}
d_{11}-d_{21} & 0 & 0 & 0 & 0 & 0 & 0 & 0 & 0 & 0 & 0 & 0 \\
0 & d_{12}-d_{22} & 0 & 0 & 0 & 0 & 0 & 0 & 0 & 0 & 0 & 0 \\
0 & 0 & -a+d_{13}-d_{23} & 0 & 0 & 0 & 0 & 0 & 0 & 0 & 0 & 0 \\
0 & 0 & 0 & -h+d_{14}-d_{24} & 0 & 0 & 0 & 0 & 0 & 0 & 0 & 0 \\
\hline
0 & 0 & 0 & 0 & -d_{31} & 0 & 0 & 0 & 0 & 0 & 0 & 0 \\
0 & 0 & 0 & 0 & 0 & -d_{32} & 0 & 0 & 0 & 0 & 0 & 0 \\
0 & 0 & 0 & 0 & 0 & 0 & -a-d_{33} & 0 & 0 & 0 & 0 & 0 \\
0 & 0 & 0 & 0 & 0 & 0 & 0 & -h-d_{34} & 0 & 0 & 0 & 0 \\
\hline
0 & 0 & 0 & 0 & 0 & 0 & 0 & 0 & -d_{41} & 0 & 0 & 0 \\
0 & 0 & 0 & 0 & 0 & 0 & 0 & 0 & 0 & -d_{42} & 0 & 0 \\
0 & 0 & 0 & 0 & 0 & 0 & 0 & 0 & 0 & 0 & -a-d_{43} & 0 \\
0 & 0 & 0 & 0 & 0 & 0 & 0 & 0 & 0 & 0 & 0 & -h-d_{44}
\end{array}\right)$$

In order to satisfy the stability conditions (12) of the AS errors, the diagonal parameters must be defined as

$$\begin{array}{l}
d_{11}-d_{21}<0 \\
d_{12}-d_{22}<0 \\
-a+d_{13}-d_{23}<0 \\
-h+d_{14}-d_{24}<0 \\
-d_{31}<0 \\
-d_{32}<0 \\
-a-d_{33}<0 \\
-h-d_{34}<0 \\
-d_{41}<0 \\
-d_{42}<0 \\
-a-d_{43}<0 \\
-h-d_{44}<0
\end{array}$$

4.3 Simulation Results

The parameters (d_{11}, d_{12}, d_{13}, d_{14}, d_{21}, d_{22}, d_{23}, d_{24}, d_{31}, d_{32}, d_{33}, d_{34}, d_{41}, d_{42}, d_{43}, d_{44}) are equal to (0, 0, 0, 0, 1, 1, 1, 1, 10, 10, 18, 18, 10, 10, 10, 10). For numerical simulations, the initial conditions are fixed as $x_i(0) = (1, 1, 1, 1)$ with $i = 1 \ldots 4$ and $j = 1 \ldots 3$. Finally, the parameter q is equal to 0.98.

The dynamical evolution of the AS error vector is illustrated in Fig. 1. In fact, the errors e_{Ai} converge to zero as $t \to +\infty$. Figs. 2 and 3 describe the state trajectories of the coupled FO systems. In Fig. 2, we note that the couples ($x_1(t)$, $x_2(t)$), ($x_2(t)$, $x_3(t)$) and ($x_3(t)$, $x_4(t)$) express the AS where as the couples ($x_1(t)$, $x_3(t)$) and ($x_2(t)$, $x_4(t)$) express the CS. In Fig. 3, displaying a zoom on the

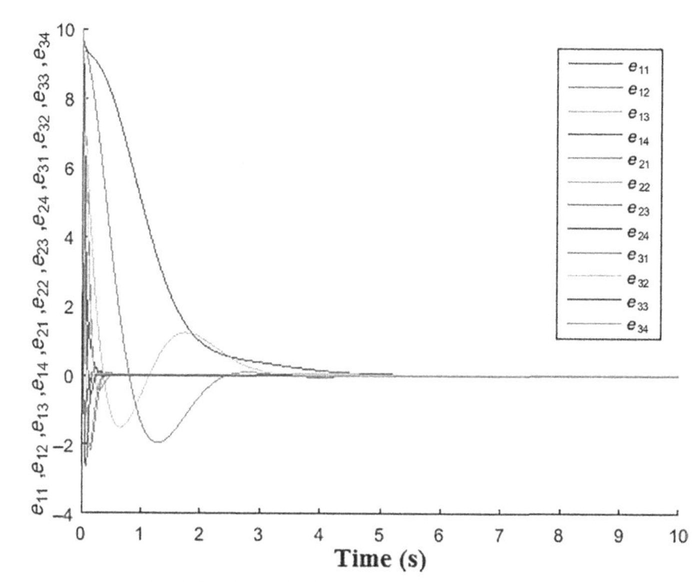

FIG. 1 State trajectories of the dynamical errors $e_A = [e_{ij}]$, $i = 1\ldots3$ and $j = 1\ldots4$.

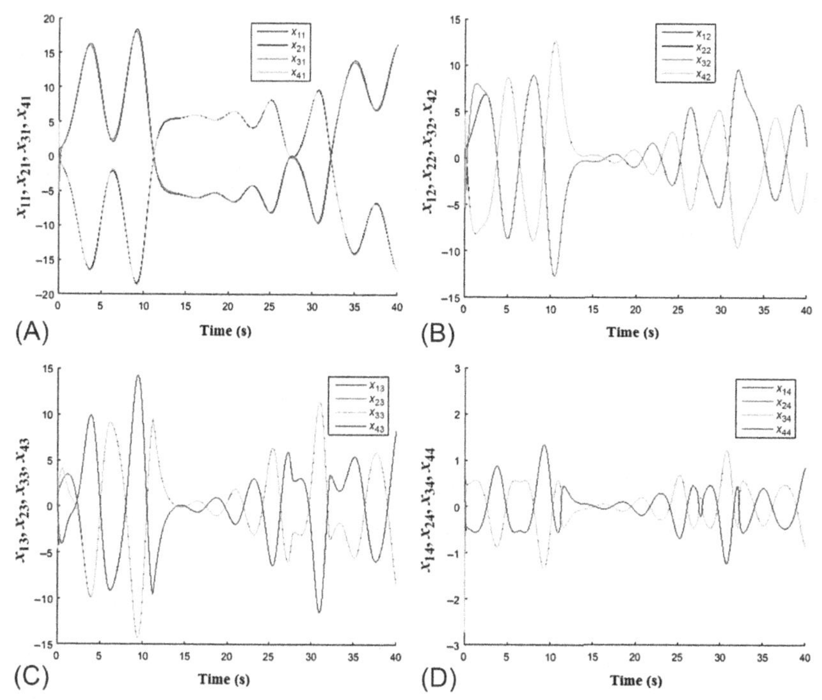

FIG. 2 State trajectories of each coupled hyperchaotic systems: (A) x_{11}, x_{21}, x_{31}, x_{41}; (B) x_{12}, x_{22}, x_{32}, x_{42}; (C) x_{13}, x_{23}, x_{33}, x_{43}; (D) x_{14}, x_{24}, x_{34}, x_{44}.

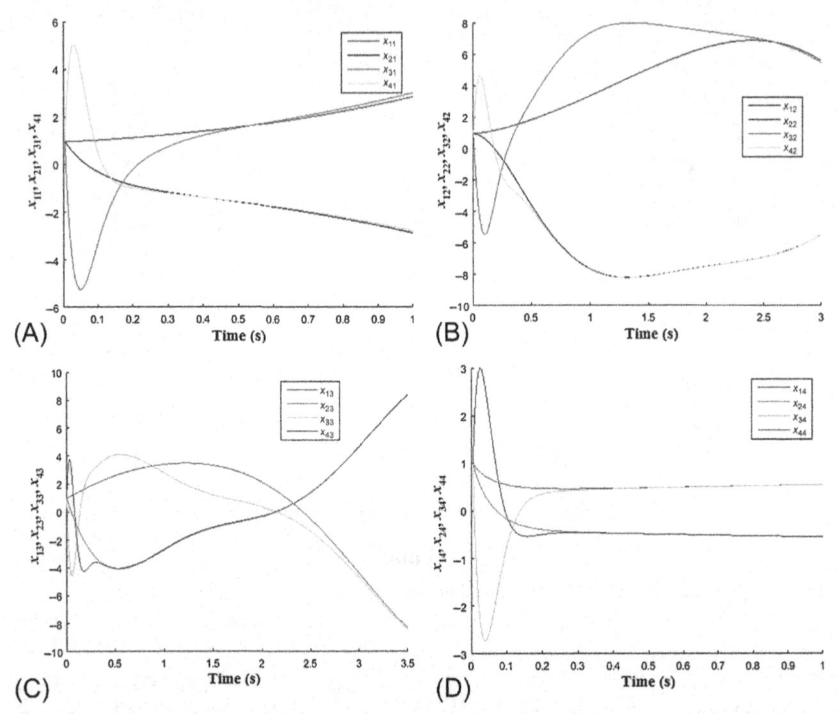

FIG. 3 Zoom on the state trajectories of each coupled hyperchaotic systems: (A) $x_{11}, x_{21}, x_{31}, x_{41}$; (B) $x_{12}, x_{22}, x_{32}, x_{42}$; (C) $x_{13}, x_{23}, x_{33}, x_{43}$.

dynamical modes, we note that the four state trajectories are different in the beginning, but after a short period they converge in pairs. Thus, the HS is achieved under the proposed control laws u_1, u_2, and u_3.

5 HYBRID SYNCHRONIZATION BETWEEN NONIDENTICAL FO HYPERCHAOTIC SYSTEMS

In this section, HS is achieved between three different FO hyperchaotic systems.

5.1 Different FO Hyperchaotic Systems

Let consider here three FO coupled hyperchaotic systems including a hyperjerk system [26] and two FO hyperchaotic systems with FO terms [25]. These systems are described, respectively, by

$$\begin{cases} D^q x_{11} = x_{12} + d_{11}(x_{31} - x_{11}) \\ D^q x_{12} = x_{13} + d_{12}(x_{32} - x_{12}) \\ D^q x_{13} = x_{14} + d_{13}(x_{33} - x_{13}) \\ D^q x_{14} = -x_{14} - 5.2x_{13} - 2.7x_{12} + 4.5(x_{11}^2 - 1) + d_{14}(x_{34} - x_{14}) \end{cases} \tag{23}$$

$$\begin{cases} D^q x_{21} &= x_{22} + d_{21}(x_{11} - x_{21}) + u_{11} \\ D^q x_{22} &= x_{23} + d_{22}(x_{12} - x_{22}) + u_{12} \\ D^q x_{23} &= -ax_{23} - bx_{22} + G(x_{21}) + d_{23}(x_{13} - x_{23}) + u_{13} \\ D^q x_{24} &= kx_{22} - hx_{24} + G(x_{21}) + d_{24}(x_{14} - x_{24}) + u_{14} \end{cases} \tag{24}$$

$$\begin{cases} D^q x_{31} &= x_{32} + d_{31}(x_{21} - x_{31}) + u_{21} \\ D^q x_{32} &= x_{33} + d_{32}(x_{22} - x_{32}) + u_{22} \\ D^q x_{33} &= -ax_{33} - bx_{32} + G(x_{31}) + d_{33}(x_{23} - x_{33}) + u_{23} \\ D^q x_{34} &= kx_{32} - hx_{34} + G(x_{31}) + d_{34}(x_{24} - x_{34}) + u_{24} \end{cases} \tag{25}$$

where q is the order of the FO derivative, such as $0 < q < 1$. u_1 and u_2 are the control inputs such as $u_i = [u_{i1}, u_{i2}, u_{i3}, u_{i4}]$, $i = 1\ldots2$. $D_i = \text{diag}(d_{i1}, \ldots, d_{i4})$, $i = 1\ldots4$, are the coupled matrices.

For the linear part of each coupled system, the following matrices are considered

$$A'_1 = \begin{bmatrix} 0 & 1 & 0 & 0 \\ 0 & 0 & 1 & 0 \\ 0 & 0 & 0 & 1 \\ 0 & -2.7 & -5.2 & -1 \end{bmatrix}$$

$$A'_2 = A'_{23} = \begin{bmatrix} 0 & 1 & 0 & 0 \\ 0 & 0 & 1 & 0 \\ 0 & -b & -a & 0 \\ 0 & k & 0 & -h \end{bmatrix}$$

Similarly, for the nonlinear part of each coupled system, the following nonlinear vectors are considered

$$g_1(x_{11}) = \begin{bmatrix} 0 \\ 0 \\ 0 \\ 4.5(x_{11}^2 - 1) \end{bmatrix}, \; g_i(x_{i1}) = \begin{bmatrix} 0 \\ 0 \\ G(x_{i1}) \\ G(x_{i1}) \end{bmatrix} \text{ with } i = 2\ldots3$$

5.2 Control Design

The AS error vector is written as $e_A = [e_{A1} e_{A2}]$, and the control laws are defined as

$$\begin{cases} u_1 &= H_1 e_A - (2D_2 - (A_2 - A_1))x_1 - g_2(x_2) - g_1(x_1) \\ u_2 &= H_2 e_A + 2(D_2 - D_3)x_2 - g_3(x_3) - g_2(x_2) - u_1 \end{cases} \tag{26}$$

According to the conditions (11), H_1, H_2 are given by

$$H_1 = \begin{bmatrix} 0 & 0 & 0 & 0 & 0 & 0 & 0 & 0 \\ -1 & 0 & 0 & 2.7 & 0 & 0 & 0 & 0 \\ 0 & -1 & 0 & 5.2 & 0 & 0 & 0 & 0 \\ 0 & 0 & -1 & 0 & 0 & 0 & 0 & 0 \end{bmatrix}$$

$$H_2 = \begin{bmatrix} d_{11}-d_{21} & 0 & 0 & 0 & 0 & 0 & 0 & 0 \\ 0 & d_{12}-d_{22} & 0 & 0 & -1 & 0 & b & -k \\ 0 & 0 & d_{13}-d_{23} & 0 & 0 & -1 & 0 & 0 \\ 0 & 0 & 0 & d_{14}-d_{24} & 0 & 0 & 0 & 0 \end{bmatrix}$$

As a result, we can rewrite the dynamical error such as $D^q e_A = L e_A = (L_1 + L_2) e_A$ where

$$L_1 = \left(\begin{array}{cccc|cccc} 0 & 1 & 0 & 0 & d_{11} & 0 & 0 & 0 \\ -1 & 0 & 1 & 2.7 & 0 & d_{12} & 0 & 0 \\ 0 & -1 & 0 & 6.2 & 0 & 0 & d_{13} & 0 \\ 0 & -2.7 & -6.2 & 0 & 0 & 0 & 0 & d_{14} \\ \hline d_{11} & 0 & 0 & 0 & 0 & 1 & 0 & 0 \\ 0 & d_{12} & 0 & 0 & -1 & 0 & 1+1.11 & -h \\ 0 & 0 & d_{13} & 0 & 0 & -1-1.11 & 0 & 0 \\ 0 & 0 & 0 & d_{14} & 0 & h & 0 & 0 \end{array}\right)$$

and

$$L_2 = \left(\begin{array}{cccc|cccc} -d_{11}-d_{21} & 0 & 0 & 0 & 0 & 0 & 0 & 0 \\ 0 & -d_{12}-d_{22} & 0 & 0 & 0 & 0 & 0 & 0 \\ 0 & 0 & -d_{13}-d_{23} & 0 & 0 & 0 & 0 & 0 \\ 0 & 0 & 0 & -d_{14}-d_{24}-1 & 0 & 0 & 0 & 0 \\ \hline 0 & 0 & 0 & 0 & -d_{31} & 0 & 0 & 0 \\ 0 & 0 & 0 & 0 & 0 & -d_{32} & 0 & 0 \\ 0 & 0 & 0 & 0 & 0 & 0 & -a-d_{33} & 0 \\ 0 & 0 & 0 & 0 & 0 & 0 & 0 & -h-d_{34} \end{array}\right)$$

According to the conditions (12), if the following stability conditions

$$\begin{aligned} &-d_{11}-d_{21} < 0 \\ &-d_{12}-d_{22} < 0 \\ &-d_{13}-d_{23} < 0 \\ &-d_{14}-d_{24}-1 < 0 \\ &-d_{31} < 0 \\ &-d_{32} < 0 \\ &-a-d_{33} < 0 \\ &-h-d_{34} < 0 \end{aligned}$$

are satisfied, then the AS and the CS for three different FO hyperchaotic systems are correctly achieved.

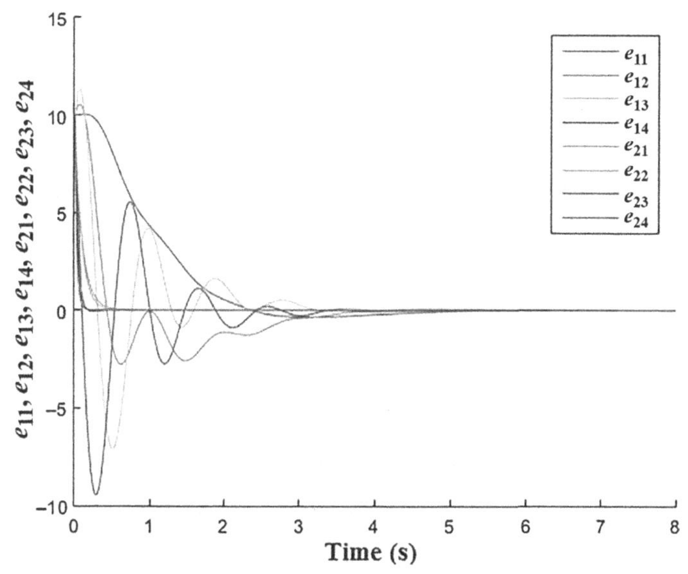

FIG. 4 State trajectories of the dynamical errors $e_A = [e_{ij}]$, $i = 1...3$ and $j = 1...3$.

5.3 Simulation Results

For simulation computations, the parameters are chosen as $(d_{11}, d_{12}, d_{13}, d_{14}, d_{21}, d_{22}, d_{23}, d_{24}, d_{31}, d_{32}, d_{33}, d_{34}) = (1, 1, 1, 1, 5, 5, 5, 5, 10, 10, 20, 15)$ and the initial conditions are fixed as $x_1(0) = (0.001, 0.001, 0.001, 4)$, $x_2(0) = x_3(0) = (1, 1, 1, 1)$. The parameter q is equal to 0.98.

The dynamical evolution of the AS error vector is illustrated in Fig. 4. In fact, the errors e_{Ai} with $i = [1...3]$ converge to zero as $t \to +\infty$. In Fig. 5, note that the couples $(x_1(t), x_2(t))$ and $(x_2(t), x_3(t))$ express the AS whereas the couple $(x_1(t), x_3(t))$ expresses the CS. In Fig. 6, displaying a zoom on the dynamical modes, note that the three state trajectories are different in the beginning but after a short period they converge, in pairs. Thus, the HS is achieved under the proposed control laws u_1 and u_2.

6 CONCLUSION

In this chapter, the HS problem is solved for multiple FO hyperchaotic systems coupled with ring connection. The CS and the AS approaches coexist at the same time for such synchronization schema. Two case studies between four and three FO hyprerchaotic systems are conducted. Finally, numerical simulations have been carried out in order to prove the well achievement of the HS problem. The implementation of the proposed control laws is envisaged thereafter to realize a real application in engineering fields.

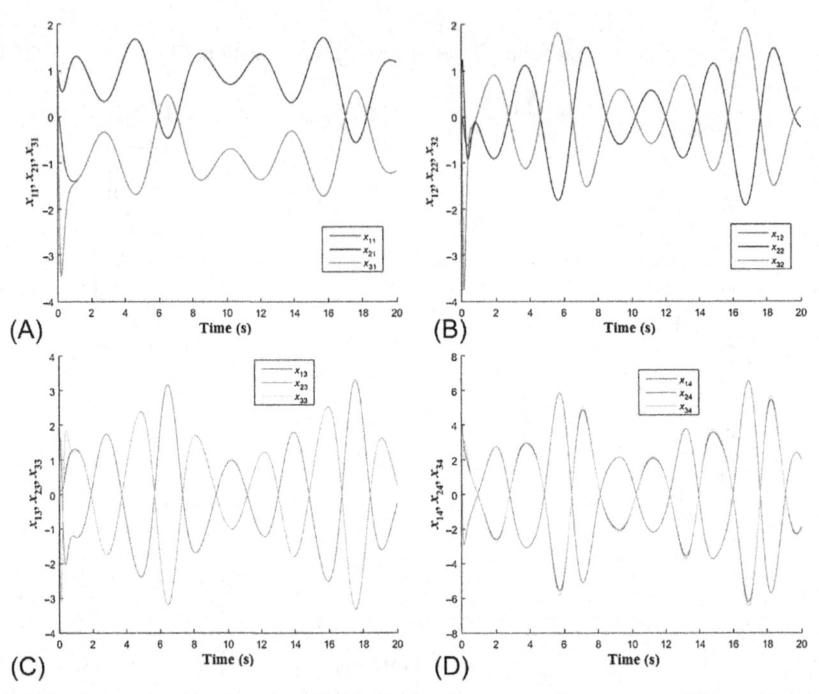

FIG. 5 State trajectories of each coupled hyperchaotic systems: (A) x_{11}, x_{21}, x_{31}; (B) x_{12}, x_{22}, x_{32}; (C) x_{13}, x_{23}, x_{33}; (D) x_{14}, x_{24}, x_{34}.

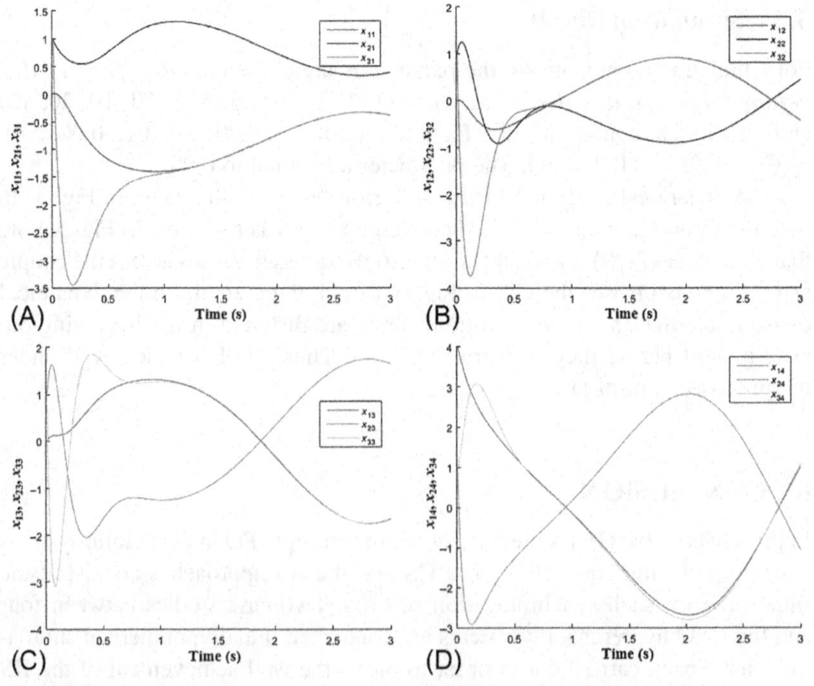

FIG. 6 Zoom on the state trajectories of each coupled hyperchaotic systems: (A) x_{11}, x_{21}, x_{31}; (B) x_{12}, x_{22}, x_{32}; (C) x_{13}, x_{23}, x_{33}.

REFERENCES

[1] Lassoued A, Boubaker O. Hybrid chaotic synchronization between identical and non-identical fractional-order systems. Int J Model Identif Control 2017(to appear).

[2] Pecora LM, Carroll TL. Synchronization in chaotic systems. Phys Rev Lett 1990;64(8):821–4.

[3] Skardal PS, Sevilla-Escoboza R, Vera-Vila VP, Buld JM. Optimal phase synchronization in networks of phase-coherent chaotic oscillators. Chaos 2017;27(1):013111.

[4] Yuan X, Li C, Huang T. Projective lag synchronization of delayed chaotic systems with parameter mismatch via intermittent control. Int J Nonlinear Sci 2017;23(1):3–10.

[5] Ali-Tahir S, Sari M, Bouhamidi A. Generalized synchronization of identical and nonidentical chaotic dynamical systems via master approaches. Int J Optimization Control Theories Appl 2017;7(3):248–54.

[6] Liu H, Li Y, Yu H, Zhu L. Hybrid synchronization in unified chaotic systems with unknown parameters. J Northeastern Univ 2011;32(10):1394–7.

[7] Chen X, Qiu J, Song Q, Zhang A. Synchronization of N coupled chaotic systems with ring connection based on special anti-symmetric structure. Abstr Appl Anal 2016;2013:10.

[8] Chen X, Wang C, Qiu J. Synchronization and anti-synchronization of N different coupled chaotic systems with ring connection. Int J Mod Phys C 2014;25(5):12.

[9] Liu P, Liu S. Robust adaptive full state hybrid synchronization of chaotic complex systems with unknown parameters and external disturbances. Nonlinear Dyn 2012;70(1):585–99.

[10] Kilbas AA, Srivastava M, Trujillo J. Theory and applications of fractional differential equations. New York: Elsevier Science Limited; 2006.

[11] Soukkou A, Leulmi S. A fractional computational algorithm for designing advanced feedback controllers of dynamical nonlinear systems. Int J Model Identif Control 2016;26(2):90–109.

[12] Dalir M, Bashour M, Liu X, Ma X. Applications of fractional calculus. Appl Math Sci 2010;4 (21):1021–32.

[13] Lassoued A, Boubaker O. On new chaotic and hyperchaotic systems: a literature survey. Nonlinear Anal Modell Control 2016;21(6):770–89.

[14] Gao X, Yu J. Chaos in the fractional order periodically forced complex Duffing's oscillators. Chaos Solitons Fractals 2005;24(4):1097–104.

[15] Hartley T, Lorenzo C, Qammer H. Chaos in a fractional order Chua's system. IEEE Trans Circuits Syst I 1995;42(8):485–90.

[16] Yu Y, Li H, Wang S, Yu J. Dynamic analysis of a fractional-order Lorenz chaotic system. Chaos Solitons Fractals 2009;42(2):1181–9.

[17] Liu X, Jiang J. Hybrid projective synchronization of fractional-order chaotic systems with time delay. Discret Dyn Nat Soc 2013;8:459801.

[18] Agrawal SK, Srivastava M, Das S. Hybrid synchronization between different fractional order hyperchaotic systems using active control method. J Nonlinear Syst Appl 2013;4(3):70–67.

[19] Liu Y, Lu L. Synchronization of N different coupled chaotic systems with ring and chain connections. Appl Math Mech 2008;10:1181–90.

[20] Chen X, Qiu J, Cao J, He H. Hybrid synchronization behavior in an array of coupled chaotic systems with ring connection. Neurocomputing 2016;173:1299–309.

[21] Lassoued A, Boubaker O. Hybrid synchronization of multiple fractional-order chaotic systems with ring connection. In: ICMIC 2016: 8th international conference on modeling, identification and control; 2016. p. 109–14.

[22] Podlubny I. Fractional differential equations. New York, NY: Academic Press; 1999.

[23] Matignon D. Stability results for fractional differential equations with applications to control processing. Proc Comput Eng Syst Appl 1996;2:963–8.

[24] Chen D, Zhang R, Liu X, Ma X. Fractional order Lyapunov stability theorem and its applications in synchronization of complex dynamical networks. Commun Nonlinear Sci Numer Simul 2014;19(12):4105–21.

[25] Lassoued A, Boubaker O. Dynamic analysis and circuit design of a novel hyperchaotic system with fractional-order terms. Complexity 2017;2017:11.

[26] Konstantinos E, Sprott JC. Chaotic hyperjerk systems. Chaos Solitons Fractals 2006;28(3):739–46.

FURTHER READING

[27] Sun J, Shen Y, Zhang GD. Transmission projective synchronization of multi-systems with non-delayed and delayed coupling via impulsive control. Chaos 2012;22:043107.

Index

Note: Page numbers followed by *f* indicate figures, and *t* indicate tables.

A

Active-HDL, 306–311, 309*f*, 312–314*f*
Adaptive finite synchronization, with hidden attractor
  adaptive feedback controller, 293
  Lyapunov candidate function, 293
  numerical verifications, 295–296
  preliminaries, 292–293
Adomian polynomials, 67–68, 69*t*
Akaike information criterion (AIC), 111, 115
Allee effects
  branch point bifurcation, 212
  Hopf bifurcation, 212
  noninteger polynomials, 212
  strong Allee effect
    bifurcation diagram, 217, 220*f*
    dynamical analysis, 213–217
    eigenvalues of system, 215, 216*t*
    Lyapunov exponents, 228*f*
    numerical continuation, 217–221, 221*f*
    predator-prey system, 217–220
    projection of chaotic attractor, 228*f*
    three-dimensional strange attractor, 227*f*
  weak Allee effect
    bifurcation diagram, 223, 223*f*
    dynamical analysis, 213–217
    eigenvalues of system, 215, 216*t*
    Lyapunov exponent, 230*f*
    numerical continuation, 223, 224–226*f*
    predator-prey system, 221–226
    projection of strange attractor, 229*f*
    three-dimensional strange attractor, 229*f*
AMD 486 Series Processor, 154
Antimonotonicity, 32–33
Approximate model-based control, 265, 270
Assumed mode method (AMM), 234, 237
Attention deficit disorder (ADD) model, 166, 175–181
Autocorrelation coefficients, 173
Avatar, 130

B

Backstepping control, 236, 242–244
Bayesian information criterion (BIC), 111, 115
Bistability, 26, 34*f*
Borland, 154
Boundary surface (BS), 128–130
  autostereoscopic display, 133*f*
  3D printing, 134*f*
  visualization, 145–150, 148–150*f*

C

Cave automatic virtual environment (CAVE), 135–137
Chameleon chaotic system, 278
Chaos, 63–64, 323
Chaos synchronization, 261–262
Chaos theory, 187
Chaotic attractor (CHA), 128–129, 129*f*
Chaotic/hyperchaotic systems
  flexible manipulator
    control techniques, 234–236
    desired trajectory, 234, 234*t*
    modeling methods, 234
    soft computing techniques, 234–236
    suppression of link deflection, 234
  hidden attractors, 233–234
  self-excited attractors, 233–234
Chaotic maps
  attention deficit disorder model, 175–181
  quasi-Gaussian distribution, 166–174
Chaotic oscillators, 23–25, 261
  Active-HDL and Simulink, 306–311, 309*f*, 312–314*f*
  block description, 310, 310*f*
  chaotic attractors
    on negative slopes, 317*f*
    resources, 311, 315*t*
    on sawtooth, 318*f*
  image transmission, 323–325, 325*f*
  master-slave topology, 319–323
    Hamiltonian approach, 316–318
    negative slopes, 320
    sawtooth function, 321
    synchronization error, 320, 323
  numerical methods
    Forward-Euler method, 305

Chaotic oscillators *(Continued)*
fourth-order Runge-Kutta method, 305–306
Trapezoidal method, 305
two-scroll attractor, 307–309*f*, 312–314*f*
piecewise-linear functions, 303–306
negative slopes, 303, 304*f*, 306
saturated nonlinear series, 303, 304*f*, 307*f*
sawtooth function, 303–306, 309*f*
state variables, 303–305
synchronization, 311–323
VHDL descriptions, 310–311
Chaotic trajectory tracking control
composite control technique, 244, 244*f*
with 0.3 kg payload, 248–251
with nominal payload (0.145 kg), 245–248
Chua's circuit, 4, 23–25, 127, 301–302
attractors, 128–129, 129*f*
boundary surface, 129–130
C-SBS calculation
CPUs, 154–158, 155–156*f*
GPGPU technology, 159
GRID technology, 159
discovery, 128
virtual reality systems, 130
input data, 131
3D displays, 133–134, 133*f*
3D modeling, 131–132
3D printing, 134
3D virtualization sequence, 130, 131*f*
visualization, 132
visualization
of boundary surface, 145–150, 148–150*f*
CAVE system, 137
of chaotic attractor, 138–145, 139–144*f*, 146–148*f*
mixed reality, 136, 136*f*
of state space, 137–138
virtual CAVE environment, 151, 152–154*f*
Chua's diode, 23–25, 128*f*
Classic display, 133–134
Continuous dynamical systems, 165
Continuous singular terminal sliding-mode (CSTSM) controller, 262
Controllable amplitude, 287–288
CPU architecture, 154–158, 155–156*f*
CPU Core2 Quad, 157
CPU i7 2670QM, 157
Cross-section of BS (C-SBS) calculation
CPUs, 154–158, 155–156*f*
GPGPU technology, 159
GRID technology, 159
Cryptosystem design, 337–339
Cybernaut, 130

D

Determinism, 187
Digital cryptography, 296
Dimple formation, 33
Dirac delta function, 166
Discrete (map) systems, 165
Djgpp, 154
Double-scroll chaotic attractor, 26, 30, 35
dSPACE board, 270
Dynamical system, 187–188
Dynamic surface control, 236, 241–242

E

Eigenvalues
Allee effect, 215, 216*t*
hyperjerk memristive system, 93, 94*f*
jerk system, 9, 9*t*
Electronic circuit implementation
hidden chaotic attractors, 289, 290*f*
Kirchhoff's laws, 288–289, 288*f*
Pspice simulation results, 289, 289–292*f*
self-excited chaotic attractors, 288–289, 289*f*
theoretical model, 288–289, 288*f*
Euler's discretization scheme, 270
Excitable tissue, 188, 194–198
Extreme multistability, 90, 96–100

F

Field programmable gate arrays (FPGAs)
Active-HDL and Simulink, 306–311
Cyclone IV GX EP4CGX150DF31C7, 311–323
hardware description, of oscillators, 311–323
piecewise-linear functions, 303–306
resources utilization, 82*t*
RTL block, 81, 81*f*
strange attractor, 81, 82*f*
Finite element method (FEM), 234
5D multistable chaotic system
circuit design
operational-amplifier approach, 82, 83*f*
PSpice projections, 83, 84*f*
field programmable gate arrays
resources utilization, 82*t*
RTL block, 81, 81*f*
strange attractor, 81, 82*f*

hidden attractors, 78, 79*f*
Lyapunov exponents, 78, 80*f*
Follower system, 52
Forward-Euler method, 305
4D chaotic system with one and without equilibrium points. *See also* No equilibrium chaotic system
adaptive finite synchronization
adaptive feedback controller, 293
Lyapunov candidate function, 293
numerical verifications, 295–296
preliminaries, 292–293
electronic circuit implementation
hidden chaotic attractors, 289, 290*f*
Kirchhoff's laws, 288–289, 288*f*
Pspice simulation results, 289, 289–292*f*
self-excited chaotic attractors, 288–289, 289*f*
theoretical model, 288–289, 288*f*
hidden attractor, 277–278, 281–286
self-excited attractor, 277–278, 280–281
text encryption application
affine cipher, 296
key generation, 296–297
numerical verifications, 297–298
3D autonomous system, 279
Fractional order hyperjerk system (FOHJS), 65
Adomian polynomials, 67–68, 69*t*
bifurcation, 68–70, 70*f*
2D phase portraits, 68, 69*f*
Fuzzy Reality, 130

G

Gaussian distribution, 166, 170. *See also* White Gaussian noise
Gaussian mixture model (GMM), 105–106
chaotic system
evaluation data, length of, 116–120, 119–121*f*
1D, 115–116, 118*f*
2D, 120–122, 121–122*f*
components, number of, 111
cost functions, 112–113, 114*f*
evaluation phase, 112–113
learning phase, 112
expectation, 110
initial setting, 109–110
likelihood score evaluation, 110
maximization, 110
parameters, 109
strange attractor, 107–109
Global uniform relative degree, 263
GPGPU technology, 159
GRID technology, 159

H

Hamiltonian approach, 316–318
Hardware description language (HDL), 306–311, 309*f*, 312–314*f*
Head-mounted displays (HMDs), 135, 135*f*
Hénon 3D System, 336–337
Hénon Hitzl Zele System, 333–335
Hidden attractor, 41–42, 63–64, 78, 165–166, 277–278
5D multistable system, 79*f*, 82*f*, 84*f*
hyperjerk memristive system, 89–90
text encryption application
affine cipher, 296
key generation, 296–297
numerical verifications, 297–298
High-performance computer clusters, 159
Hindmarsh-Rose (H-R) neuronal model
bifurcation analysis, 191–193, 196–203*f*
differential equations, 190
Lyapunov exponent, 191, 192*f*, 193, 196*f*
2D network system, 190–191
wave propagation, 194–205, 206*f*
Hodgkin-Huxley (H-H) model, 190
Hopf bifurcation, 25–26, 28
Hybrid synchronization
multiple FO hyperchaotic systems, 356–360
nonidentical FO hyperchaotic systems, 360–363
Hyperchaotic Wang system, 333
Hyperjerk systems, 90. *See also* Memristive hyperjerk system

I

IBM Cyrix processors, 154
Image transmission, 323–325, 325*f*
Intel Pentium, 154

J

Jacobian matrix
hyperjerk memristive system, 93
RC circuits, 28
3D chaotic system, 44–45
Jerk, 90
Jerk systems, 3–4
chaotic behaviors, 10, 10–11*f*
circuit design, 13–17, 15–16*f*
comparative analysis, 12–13, 13*t*
equilibrium analysis, 5–9

Jerk systems *(Continued)*
experimental analysis
electrical assembly, 17, 17*f*
periodic attractors, 17, 19–20*f*
strange attractor, 17, 18*f*, 20*f*
time series of state variables, 17, 18*f*
Lyapunov exponent analysis, 11–12, 12*f*
nonlinear function, 4–5
oscillator circuit, 4
phase portraits, 5, 6*f*
stability analysis, 9, 9*t*
state variables, 5, 7–8*f*

K

Kaplan-Yorke dimension
hyperjerk memristive system, 92–93
jerk systems, 12
3D systems, 57*f*, 58*t*
Kirchhoff's circuit laws, 45–48

L

Leader system, 52
Likelihood score, 109–110
Log-likelihood, 111, 115
Lorenz system, 41, 77–78
Lumped parameter method (LPM), 234
Lyapunov exponent (LE)
5D multistable system, 78, 80*f*
Hindmarsh-Rose neuronal model, 191, 192*f*, 193, 196*f*
hyperjerk memristive system, 92–93, 96, 96*f*
jerk systems, 11–12, 12*f*
strong Allee effect, 228*f*
3D chaotic system, 45, 48*f*, 58*t*
weak Allee effect, 230*f*

M

Master-slave synchronization, 262
control approaches, 262
definition, 264
nonlinear system, 263
output-feedback-based CSTSM controller design
finite-time sliding-mode observer, 266–267
numerical simulation, 267–270
practical relative degree, 265
tracking errors, 265
problem statement, 263–264
Master-slave topology, 319–323
Hamiltonian approach, 316–318
negative slopes, 320
sawtooth, 321
synchronization error, 320, 323
MATLAB-Simulink, 306–311, 309*f*, 312–314*f*
Maximum likelihood estimation (MLE), 109
Mean Square Error (MSE), 341, 345
Memristive hyperjerk system, 90–91
dynamical behavior, 92–93
bifurcation diagram, 94–96, 95*f*
Lyapunov exponents, 96, 96*f*
phase portraits, 94–96, 95*f*, 97*f*
eigenvalues, 93, 94*f*
equilibrium points, 93
extreme multistability
bifurcation and Lyapunov exponents, 96–98, 97–100*f*
phase portraits, 96–98, 98–101*f*
hysteresis loops, 91, 91–92*f*
Jacobian matrix, 93
Memristor, 89
Memristor-based fractional order system, 65
Microsoft Hololens, 135, 135*f*
Mixed reality system, 136, 136*f*
Multiple FO hyperchaotic systems
control design, 357–358
dynamical errors, 357–358
hybrid synchronization, 356–357
integral differential operator, 352
Lyapunov function, 355
N-dimensional diagonal matrices, 353
simulation results, 358–360
MultiSIM software, 4, 17

N

Neuron, 188
Neuronal system, 188. *See also* Hindmarsh-Rose (H-R) neuronal model
Newton-Raphson method, 27
No equilibrium chaotic system
dynamical behavior, 65, 66–67*f*
fractional order hyperjerk system, 65
Adomian polynomials, 67–68, 69*t*
bifurcation, 68–70, 70*f*
circuit implementation, 70–71, 71*f*
2D phase portraits, 68, 69*f*
model parameters, 64, 64*t*
strange attractor, 64, 65*f*
Nonidentical discrete-time hyperchaotic systems

cryptosystem design, 337–339
encryption and decryption process, 339–340
Hénon 3D System, 336–337
Hénon Hitzl Zele System, 333–335
Hyperchaotic Wang system, 333
measurement of encryption and decryption, 341
MSE analysis, 341, 345
proposed synchronization method, 331–332
PSNR analysis, 341, 345
security analysis
correlation analysis, 342
correlation analysis by adjacent pixels, 342, 344*t*, 346
histograms analysis, 342
key space analysis, 342
statistical analysis, 342–344
Nonidentical FO hyperchaotic systems
control design, 361–362
AS error vector, 361
hybrid synchronization, 360
simulation results, 363
Nonlinear dynamical systems, 89, 165

O

Offline data visualization, 132
1D parameter estimation, 115–116, 118*f*
One-scroll chaotic attractor, 26, 30
Operational-amplifiers, 82, 83*f*
Optitrack system, 137
Ordinary differential equation (ODE), 90, 106–107, 301–302

P

Parametric density estimation
bifurcation, 107, 108*f*
chaotic system
evaluation data, length of, 116–120, 119–121*f*
GMM modeling, 113–115, 115–117*f*
1D, 115–116, 118*f*
2D, 120–122, 121–122*f*
differential equations, 106–107
Gaussian mixture model
components, number of, 111
cost functions, 112–113
expectation, 110
initial setting, 109–110
likelihood score evaluation, 110
maximization, 110
parameters, 109
strange attractor, 107–109
state variables, 106–107, 107*f*
Peak Signal-to-Noise Ratio (PSNR) analysis, 341, 345
Phone classification, 106–109
Piecewise-linear (PWL) functions, 303–306
negative slopes, 303, 304*f*, 306
saturated nonlinear series, 303, 304*f*, 307*f*
sawtooth function, 303–306, 309*f*
Poincare-Andronov-Hopf bifurcation, 213
Poincaré map, 45, 47*f*
Polygonization, 131
Practical relative degree (PRD), 265
Predator-prey BB model, bifurcation analysis
Allee effects
branch point bifurcation, 212
Hopf bifurcation, 212
noninteger polynomials, 212
impulsive effect, 211
mathematical modeling, 212–213
preliminaries, 213
seasonal effect, 211
seasonally perturbed system, 227–230
strong Allee effect
bifurcation diagram, 217, 220*f*
dynamical analysis, 213–217
eigenvalues of system, 215, 216*t*
Lyapunov exponents, 228*f*
numerical continuation, 217–221, 221*f*
predator-prey system, 217–220
projection of chaotic attractor, 228*f*
three-dimensional strange attractor, 227*f*
time delay effect, 211
weak Allee effect
bifurcation diagram, 223, 223*f*
dynamical analysis, 213–217
eigenvalues of system, 215, 216*t*
Lyapunov exponent, 230*f*
numerical continuation, 223, 224–226*f*
predator-prey system, 221–226
projection of strange attractor, 229*f*
three-dimensional strange attractor, 229*f*
Primary bubble, 33
Proposed synchronization method, 331–332
Pseudo-Gaussian distribution, 167
PSpice
3D chaotic system, 50*f*
4D chaotic system, 71*f*
5D multistable system, 83, 84*f*

R

Realtime data visualization, 132
Reconstructed phase space (RPS), 106, 109
Regions of attraction (RAs), 129–130
Resistance and capacitance (RC) circuit
  differential equations, 25–26
  dynamical behaviors
    bifurcation diagram, 31–33, 31*f*, 33–34*f*
    bistability, 33, 34*f*
    oscilloscope observations, 35, 35–36*f*
    phase portraits, 30, 30*f*
    regions of, 29–30, 29*f*
  equilibrium points, 26
    Jacobian matrix, 28
    Newton-Raphson method, 27
    phase portraits, 29*f*
    stability analysis, 27, 27*f*
  experimental set-up, 25–26, 25*f*
Rössler system, 41
Routh-Hurwitz criterion, 27, 279
Runge-Kutta method, 9, 113–114, 305–306

S

Secure communication, 302, 323–325, 325*f*
Security analysis
  correlation analysis, 342
  correlation analysis by adjacent pixels, 342, 344*t*, 346
  histograms analysis, 342
  key space analysis, 342
  statistical analysis, 342–344
Self-excited attractors, 165–166, 277–278
Shilnikov criteria, 277–278
Signum function, 43
Simulink, 270, 306–311, 309*f*, 312–314*f*
Singular perturbation (SP) technique
  two link flexible robot manipulator (TLFM), 234–236, 239–241
    backstepping control, 236
    composite controllers, 234–236, 235*t*
    dynamic surface control, 236
    fast subsystem, 239–241
    slow subsystem, 239–240
  two-time scale separation principle, 234
Spatiotemporal patterns, 188–189, 207
Stable limit cycle (SLC), 128–129, 129*f*
Static patterns, 188–189
Step function, 167, 169–170*f*
Strange attractors, 105–106
Structural Similarity Index Measure (SSIM), 341, 345

T

Tent map, 166
Text encryption application, hidden chaotic attractor
  affine cipher, 296
  key generation, 296–297
  numerical verifications, 297–298
3D chaotic system
  equilibrium points, 41–42, 42*t*
  parabolic equilibrium
    bifurcation diagram, 45, 47*f*
    chaotic behavior, 45, 46*f*
    circuit design, 45–48, 49*f*
    controlled states, 48–52, 52–53*f*
    equilibrium points and stability, 44–45
    Lyapunov exponents, 45, 48*f*
    nonlinear functions, 43, 44*f*, 55, 57*f*, 58*t*
    Poincaré map, 45, 47*f*
    PSpice phase portraits, 50*f*
    synchronization, 52–55, 56–57*f*
3D printing, 134
3D virtualization sequence, 130, 131*f*
Trapezoidal method, 305–306, 311
2D cross-section of BS (2D C-SBS), 148, 151–154
2D parametric estimation, 120–122, 121–122*f*
Two link flexible robot manipulator (TLFM)
  composite control
    fast subsystem, backstepping control technique, 242–244
    results and discussion, 244–251
    slow subsystem, dynamics surface control, 241–242
  control techniques, 234–236
  desired trajectory, 234, 234*t*, 244
  dynamic model, 237–239
  modeling methods, 234
  schematic representation, 236*f*
  soft computing techniques, 234–236
  SP modeling method, 234–236, 239–241
    composite controllers, 234–236, 235*t*
    dynamic surface control, 236
    fast subsystem, 239–241
    slow subsystem, 239–240
  suppression of link deflection, 234
  variables, 236, 237*t*

V

Van der Pol oscillator systems
  circuit diagram, 271*f*
  experimental study, 270–272
  master-slave synchronization, 272*f*
Virtual reality (VR) systems, 130
  input data, 131
  3D displays, 133–134, 133*f*
  3D modeling, 131–132
  3D printing, 134
  3D virtualization sequence, 130, 131*f*
  visualization, 132
Visual C compiler, 154

W

Wave propagation, 189, 194–205, 206*f*
White Gaussian noise, 166
  bifurcation diagram, 169*f*
  exponential function, 167–170
  map plots, 166, 167*f*
  step function, 167, 169–170*f*
  2-D number generation, by Tent map, 166–167, 168*f*
White noise, 166
Wien bridge chaotic oscillator, 23–25
WinChip, 154

Printed in the United States
By Bookmasters